INTERNATIONAL CENTRE FOR MECHANICAL SCIENCES

COURSES AND LECTURES - No. 310

CRACK DYNAMICS IN METALLIC MATERIALS

EDITED BY

J.R. KLEPACZKO
UNIVERSITY OF METZ

Springer-Verlag Wien GmbH

Le spese di stampa di questo volume sono in parte coperte da
contributi del Consiglio Nazionale delle Ricerche.

This volume contains 286 illustrations.

ISBN 978-3-211-82226-5 ISBN 978-3-7091-2824-4 (eBook)
DOI 10.1007/978-3-7091-2824-4

Originally published by CISM, Udine in 1990.

Springer-Verlag Wien

In order to make this volume available as economically and as rapidly as possible the authors' typescripts have been reproduced in their original forms. This method unfortunately has its typographical limitations but it is hoped that they in no way distract the reader.

PREFACE

This book, with its four chapters, is a pubblication of the lecture notes prepared by the lectures for an advanced course: "Crack Dynamic in Metallic Materials". The course has been organized by the International Centre for Mechanical Sciences in Udine, Italy.

The objective of this pubblication is to provide a fundamental knowledge on current status of a growing branch of fracture mechanics called generally "Crack Dynamics". The word "Dynamic" is not necessarily associated with inertia forces but also with rate sensitivity of a material under consideration. Those effects, when superimposed, result in a highly non-linear responses of an isolated cracks to different loading rates.

This volume has been prepared by four lecturers: D. R. CURRAN, J. F. KALTHOFF, J. R. KLEPACZKO, and F. NILSSON, and containts four following chapters:

1. *"Dynamic fracture theory" by F. Nilsson;*
2. *"Experimental fracture dynamics" by J. F. Kalthoff;*
3. *"Dynamic crack initiation, some experimental methods and modelling" by J. R. Klepaczko;*
4. *"Microstatistical fracture mechanics in dynamic fracture" by D. R. Curran*

While the book in its present format may not be appropriate as a textbook, it provides some knowledge, useful as a text supplement in lecturing on fracture mechanics. On the other hand it may serve as an instructional aid in engineering of fracture prevention under impact. In addition, some references have been provided in each chapter to enhance informations on the state-of-the-art in fracture dynamics.

Finally, the editor would like to acknowledge, on behalf of all lecturers and participants, an excellent organization of the course by the CISM staff.

Janusz R. Klepaczko

CONTENTS

DYNAMIC FRACTURE THEORY

F. Nilsson
University of Uppsala, Uppsala, Sweden

ABSTRACT

Dynamic (inertia and strain rate) effects influence crack growth processess if the applied loading rate is sufficiently high or if the crack tip moves with a speed that is a significant fraction of the wave velocities. In the paper basic results as well as solutions to boundary value problems incorporating dynamic effects are discussed. The integral expression for the energy flow to a moving crack tip is derived and related path-area integrals are discussed. It is shown that these quantities are non-trivial only if the energy density behaves as $O(r^{-1/2})$. The J-integral emerges as a special case of the general formulations. For linear dynamic problems a line integral defined in the Laplace transform space is introduced. The asymptotic field of a crack growing dynamically in a linear elastic material is derived and the stress-intensity factors are defined. K_I -solutions are discussed for a number of problems involving both stationary cracks under dynamic loading as well as moving tips. Asymptotic solutions are given for both stationary and moving tips in different non-linear materials with either rate-independent or rate-dependent elasto-plastic behaviour. These solutions provide the basis for a discussion of dynamic crack growth criteria. Finally, some aspects on numerical modelling of dynamic crack problems are briefly discussed.

1. INTRODUCTION

With the term dynamic fracture it is here meant cases when inertia or strain rate effects are of importance for the fracture behaviour. It is well known from common experience that materials may fracture more easily during rapid loading than when the load is applied slowly. A rapid load application may cause stress intensification due to inertia effects and the fracture toughness properties may be affected by the loading rate. The conditions under which crack growth initiation takes place during transient loading are of obvious interest for many technical applications. One example is provided by material testing where impact loading is used such as the Charpy test. If a quantitative understanding of such tests is desired then a dynamic fracture theory is needed.

A crack propagating in a body often reaches very high velocities. These may be significant in comparison with the wave speeds of the material even if the applied loading is of quasi-static nature. This is obviously a highly dynamic event where inertia as well as strain-rate effects are of importance for the process. The engineering interest is mainly focussed on the conditions under which a rapidly growing crack may be arrested. Large experimental programs such as the Heavy Section Steel Technology project [1] are underway with the purpose of investigating the arrest capabilities of materials for the nuclear industry. Applications of dynamic fracture theory are also of primary interest in seismology, since an earth-quake is essentially a dynamic crack growth process on a large scale.

In these notes the general problem is formulated for two-dimensional cases. General relations concerning the energy relation are derived. These can easily be generalized to three-dimensional cases, but for notational simplicity only the two-dimensional forms are discussed here. The reader is referred to the work by Strifors [2] - [3] for discussion of the more general formulations. Small deformations and rotations will be assumed. Obviously large deformations occur in the vicinity of the crack tip, but it will be assumed that the large strains are confined to a very small region wherein the behaviour is controlled by the outer field where small deformations prevail. This assumption is almost always made in fracture mechanics and has been proven successful. Extensive discussions of the general relations for the large deformation case can be found in [2] - [3] and in the work by Lidström [4]. The boundary value problems discussed will also be of two-dimensional character, since three-dimensional solutions are very scarce. Two-dimensional problems can be of four types: plane strain, plane stress, antiplane strain or axially symmetric. We will mostly consider the plane strain problem but occasionally the other types will be discussed.

2. PROBLEM DEFINITION

A two-dimensional body (fig. 1) under plane strain contains a crack defined by the crack surfaces $S_+(t)$ (the upper one) and $S_-(t)$ (the lower one). The crack tip position is denoted by $x_i^c(t)$.The problem is to find the crack tip motion when the body is subjected to given loadings. The equations of motion of a regular point of the body is given by (2.1) provided that the body forces are absent,

$$\sigma_{ij,j} = \rho \ddot{u}_i , \quad i = 1, 2 \quad . \tag{2.1}$$

σ_{ij} denotes the stress tensor, u_i the displacement vector and ρ the mass density. A dot over the letter denotes time differentiation. The strain tensor ε_{ij} is defined by (2.2)

$$\varepsilon_{ij} = \frac{1}{2} (u_{i,j} + u_{j,i}) \tag{2.2}$$

All constitutive relations that appear in this article can be written in the form

$$\dot{\varepsilon}_{ij} = F_{ij} (\dot{\sigma}_{kl} , \sigma_{mn} , \varepsilon_{pq} , \dot{\varepsilon}_{rs}) \quad . \tag{2.3}$$

Normally, constitutive relations of the form (2.3) can not describe the decohesive processes that occur at the crack tip. We somewhat loosely define the <u>process zone</u> as the region where significant deviations from the assumed constitutive relation occur. It shall here be assumed that the process zone is so small that the behaviour within it is completely controlled by quantities that are obtained from a solution of the problem without modelling this zone. This is the assumption of <u>autonomy</u> (Barenblatt [5], Broberg [6]), which is the basic postulate in virtually all fracture theories. The question whether autonomy prevails in a certain situation or not is complicated and so far not very extensively studied. A discussion of these matters is outside the scope of the present article. In addition to the field equations (2.1) - (2.3), boundary conditions are needed to complete the problem statement. The outer boundary of the body S is divided into S_t and S_u. On these surfaces the traction and the displacement history respectively are assumed to be given.

$$\sigma_{ij}\, n_j = T_i(t) \quad \text{on} \quad S_T \tag{2.4}$$

$$u_i = u_i^* \,(t) \quad \text{on} \quad S_u \quad . \tag{2.5}$$

Here n_j is the outward pointing normal vector, T_i the time-dependent traction vector and u_i^* the time-dependent displacements. On the crack surfaces $S_\pm$ the following boundary con-ditions are assumed, ($n_i^\pm$ are normal vectors of the crack surfaces),

$$\sigma_{ij}\, n_j = 0 \quad \text{if} \quad u_i^+ n_i^+ + u_i^- n_i^- < 0, \text{ on } S_\pm \tag{2.6}$$

or

$$u_i^+ n_i^+ + u_i^- n_i^- = 0, \qquad \text{on } S_\pm \qquad (2.7)$$

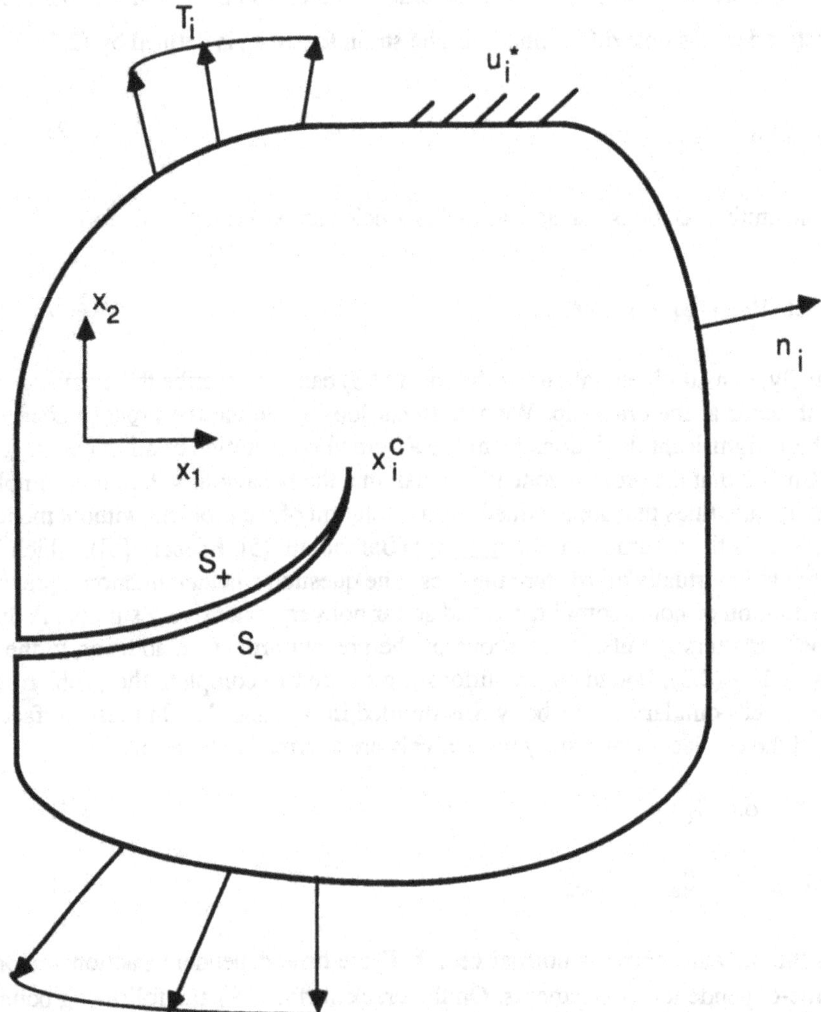

Fig. 1, The crack growth problem.

Eq. (2.7) is the case when crack closure occurs and the boundary conditions must then be supplemented by relations for the tangential forces that may act on S_+ and S_- because of friction. The case of crack closure will not be considered further.

If the motion x_i^c of the tip is given, then the equations (2.1) - (2.6) completely define the problem which can be solved by the usual methods of solid mechanics. However, in most cases the motion of the tip is not known a priori, but is part of the solution to the problem. The above set of equations must then be complemented by relations that govern the tip motion, so called fracture criteria. The formulation of these is the most important part of fracture research.

3. KINEMATICAL RELATIONS

It is convenient for the further discussion to introduce a local coordinate system attached to the tip (fig. 2). Let, thus, the origin of the Cartesian system x_i' coincide with the tip. The x_1'- axis is chosen as the tangent of the crack surface. By letting β_{ij} denote the direction cosines between the x_i' - and the x_i - system, the transformation between the systems read

$$x_i = \beta_{ij}\, x_j' + x_i^c \,, \tag{3.1}$$

or

$$x_i' = \alpha_{ij}\,(x_j - x_j^c) \,, \tag{3.2}$$

where α_{ij} is the inverse of β_{ij} and thus

$$\beta_{ij}\,\alpha_{jm} = \delta_{im} \,. \tag{3.3}$$

δ_{im} is the Kronecker delta.

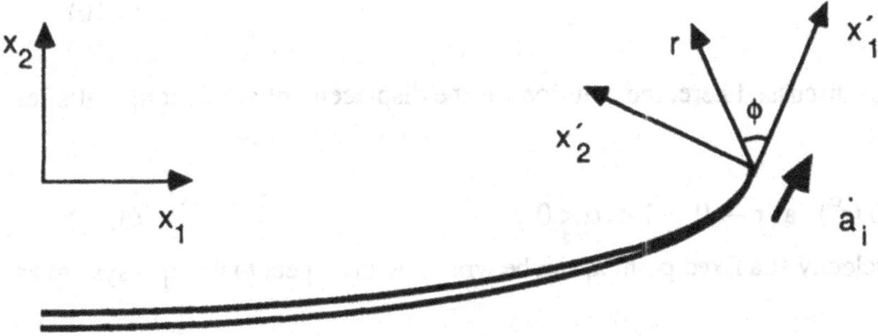

Fig. 2, Definition of coordinate systems.

Vectors transform covariantly so that

$$u_i = \beta_{ij} u_j' \ , \tag{3.4}$$

$$u_i' = \alpha_{ij} u_j \ , \tag{3.5}$$

where u_i' are the vector components with respect to the x_i' - system.

It is also convenient to introduce a polar coordinate system connected to the x_i'-system as

$$x_1' = r \cos \varphi \ , \tag{3.6}$$

$$x_2' = r \sin \varphi \ . \tag{3.7}$$

The velocity of the tip is obtained by time differentiation of x_i^c.

$$\dot{a}_i = \frac{\partial}{\partial t} x_i^c = \dot{a}\, l_i^v \quad , \tag{3.8}$$

$$\dot{a} = (\dot{a}_i \dot{a}_i)^{1/2} \quad . \tag{3.9}$$

$\dot{a}$ is the measure of crack-tip velocity and l_i^v the unit vector giving the direction of propagation. For most of the discussion we will assume smooth crack propagation, i.e the velocity vector coinciding with the x_1' - axis

$$l_i^v n_i^{\pm} = 0 \ . \tag{3.10}$$

For most cases discussed here, the solution for the displacement gradient $u_{i,j}$ satisfies the condition

$$(u_{i,j} u_{i,j})^{1/2} = 0\,(r^{\alpha}) \quad \text{as } r \to 0, \ -1 < \alpha < 0 \ . \tag{3.11}$$

The particle velocity at a fixed point x_j can be written with respect to the x_i' - system as

$$\dot{u}_i = \frac{\partial u_i}{\partial t}\,|_{x_j} = \frac{\partial u_i}{\partial x_m'} \frac{\partial x_m'}{\partial t} + \frac{\partial u_i}{\partial t}\,|_{x_k'} \quad . \tag{3.12}$$

Utilizing the coordinate transformation (3.2), we obtain

$$\dot{u}_i = \frac{\partial u_i}{\partial x_m{}'}\left[\frac{\partial \alpha_{mq}}{\partial t}(x_q - x_q^c) - \alpha_{mq}\dot{x}_q^c\right] + \frac{\partial u_i}{\partial t}|x_k{}' =$$

$$= \frac{\partial u_i}{\partial x_m{}'}\frac{\partial \alpha_{mq}}{\partial t}\beta_{qk}x_k{}' - \frac{\partial u_i}{\partial x_m{}'}\alpha_{mq}\dot{x}_q^c + \frac{\partial u_i}{\partial t}|x_k{}' \ . \tag{3.13}$$

Because of the condition (3.11), the first and the last term of (3.13) are non-singular as $r \to 0$, while the second term behaves as $0(r^\alpha)$. Thus, close to the tip we obtain

$$\dot{u}_i \to -\frac{\partial u_i}{\partial x_m{}'}\alpha_{mq}\dot{x}_q^c \qquad \text{as } r \to 0 \ . \tag{3.14}$$

Utilizing the coordinate transformations (3.1) - (3.5) we obtain with respect to the fixed x_i-system

$$\dot{u}_i \to \frac{\partial u_i}{\partial x_k}\beta_{km}\alpha_{mq}\dot{x}_q^c = -\frac{\partial u_i}{\partial x_k}\dot{x}_k^c \ , \quad \text{as } r \to 0 \tag{3.15}$$

or with respect to the $x_i{}'$ - system, for smooth propagation,

$$\dot{u}_i{}' \to -\frac{\partial u_i{}'}{\partial x_1{}'}\dot{a} \ , \qquad \text{as } r \to 0 \ . \tag{3.16}$$

Using a similar reasoning for the particle accelerations, we obtain

$$\ddot{u}_i \to u_{i,mk}\,\dot{a}_m\,\dot{a}_k \qquad \text{as } r \to 0, \tag{3.17}$$

$$\ddot{u}_i{}' \to u'_{i,11}\,\dot{a}^2 \qquad \text{as } r \to 0 \ . \tag{3.18}$$

In the resulting expressions (3.15) - (3.18), time enters only through the crack velocity $\dot{a}_i$ and no other time derivatives remain. The crack tip is then said to be in a state of <u>asymptotically steady motion</u>.

4. ENERGY RELATIONS

Ever since the pioneering work by Griffith [7], the energy relations at a moving tip have been considered important. The number of contributions to this subject is quite large.It seems that the correct expression for the energy flow to a moving tip valid for arbitrary material behaviour is due to Kostrov and Nikitin [8].

Consider a fixed area (fig. 3) with the perimeter C′ enclosing the tip and in addition a control curve C which is fixed with respect to and moving with the tip. The normal unit vector n_j' is chosen as positive when pointing out from the area A between C and C′. The region A is free from singularities and we shall consider the power balance of it.

Some care has to be taken when performing time differentiation of integrals over areas with a changing boundary. Consider two neighbouring time instants t and t+Δt. We want the time-derivative of a quantity Φ integrated over the region A. Let A′ and A′′ denote the current areas at times t and t+Δt respectively. The material derivative is then obtained from

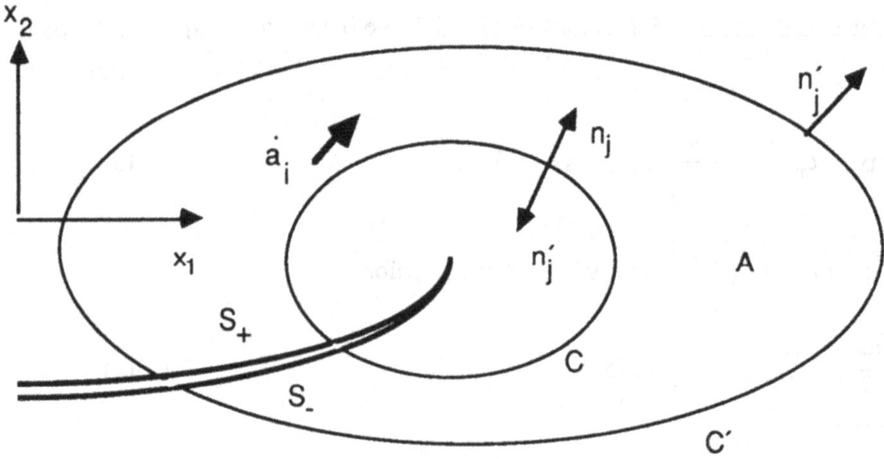

Fig. 3, Control curves for energy flow calculation.

$$\frac{D}{Dt}\int_A \Phi\, dA = \lim_{\Delta t\to 0}\frac{1}{\Delta t}\left[\int_{A''}\left(\Phi+\frac{\partial\Phi}{\partial t}\Delta t\right)dA - \int_{A'}\Phi\, dA\right]$$

$$= \lim_{\Delta t\to 0}\frac{1}{\Delta t}\left[\int_{A''\cap A'}\frac{\partial\Phi}{\partial t}\Delta t\, dA + \int_{A''-A'}\Phi\, dA\right] \quad . \tag{4.1}$$

∩ denotes the common part of A′′ and A′. The part which is an integral over A′′- A′ is obtained by considering a curve element ds of the boundary where s is a line coordinate.

The normal velocity of a point on the boundary C is its propagation velocity $\dot{a}_i$ projected on the surface normal n_j. The area element generated is $\dot{a}_i n_i ds\Delta t$. Introducing it into (4.1) and performing the limit process yields

$$\frac{D}{Dt}\int_A \Phi\, dA = \int_A \dot{\Phi}\, dA + \int_C \Phi\, \dot{a}_i\, n_i' ds \ . \tag{4.2}$$

The line integral here and those appearing below are to be taken counter-clock wise.

Let e be the internal energy per unit mass. Furthermore, let h_i denote the heat flux vector. For a regular point without energy sources or sinks we have the local power balance

$$\rho\dot{e} = \sigma_{ij}\, \dot{\varepsilon}_{ij} - h_{i,i} \ . \tag{4.3}$$

Defining $\dot{\Gamma}$ as the net power per unit thickness transferred to the material currently occupying the region within C, we have the power balance for the area A,

$$\frac{D}{Dt}\left[\int_A \rho\left(e+\frac{1}{2}\dot{u}_i\dot{u}_i\right) dA\right] + \dot{\Gamma} = \int_{C'} (\sigma_{ij}\, \dot{u}_i - h_j)\, n_j'\, ds \ . \tag{4.4}$$

Since C moves with the velocity $\dot{a}_j$ and $S_\pm$ are stationary, use of (4.2) gives

$$\dot{\Gamma} = -\int_A \rho\, (\dot{e}+\dot{u}_i\ddot{u}_i)\, dA + \int_C \rho\left(e+\frac{1}{2}\dot{u}_i\dot{u}_i\right) \dot{a}_j\, n_j\, ds + \int_{C'} (\sigma_{ij}\, \dot{u}_i - h_j)\, n_j'\, ds \tag{4.5}$$

Here $n_j = -n_j'$ on the curve C (see fig. 3).

Assuming that no heat transfer occurs across the crack surfaces and since these are traction-free we obtain ($C_{tot} = C' + S_\pm + C$)

$$\dot{\Gamma} = -\int_A \rho\, (\dot{e}+\dot{u}_i\ddot{u}_i)\, dA + \int_C \rho\left(e+\frac{1}{2}\dot{u}_i\dot{u}_i\right)\dot{a}_j n_j\, ds$$

$$+ \int_{C_{tot}} (\sigma_{ij}\dot{u}_i - h_j)\, n_j'\, ds + \int_C (\sigma_{ij}\dot{u}_i - h_j)\, n_j\, ds \ . \tag{4.6}$$

The divergence theorem can be used to convert the line integral over C_{tot} into an area integral

$$\dot{\Gamma} = -\int_A [\rho\,(\dot{e}+\dot{u}_i\ddot{u}_i)-(\sigma_{ij}\,\dot{u}_i - h_j\,)_{,j}]\,dA+\int_C \left[\rho\left(e+\frac{1}{2}\dot{u}_i\dot{u}_i\right)\dot{a}_j+\sigma_{ij}\dot{u}_i-h_j\right] n_j\,ds \qquad (4.7)$$

The area integral in (4.7) is further simplified into

$$\int_A [\,]\,dA = \int_A [-\rho\dot{e}+\sigma_{ij}\,\dot{\varepsilon}_{ij}-h_{j,j} - \rho\dot{u}_i\ddot{u}_i + \sigma_{ij,j}\dot{u}_i]\,dA\ . \qquad (4.8)$$

On account of the equations of motion (2.1) and the local energy balance (4.3), the area integral vanishes. The power $\dot{\gamma}$ supplied to the crack tip is obtained by letting the curve C shrink to zero

$$\dot{\gamma} = \lim_{C\to 0}\int_C \left[\rho\left(e+\frac{1}{2}\dot{u}_i\dot{u}_i\right)\dot{a}_j+\sigma_{ij}\dot{u}_i-h_j\right] n_j\,ds\ . \qquad (4.9)$$

The energy flow per unit crack advance γ is obtained by dividing $\dot{\gamma}$ by the absolute value of the tip velocity $\dot{a}$. Introducing the local steady state results (3.15) we get

$$\gamma = \lim_{C\to 0}\int_C \left[\rho\left(e+\frac{1}{2}\dot{a}_m\dot{a}_k u_{i,m}u_{i,k}\right)\overset{v}{l}_j-\sigma_{ij}\overset{v}{l}_k u_{i,k}-h_j/\dot{a}\right] n_j\,ds\ . \qquad (4.10)$$

Note that this form is only possible if the displacement field satisfies condition (3.11). However, as will be seen below, if (3.11) is not satisfied there is no non-zero energy flow to the tip and so (4.10) is valid for all cases of interest. For the particular case of smooth crack propagation (4.10) is simplified if the x_1-axis is chosen to coincide with the propagation direction

$$\gamma = \lim_{C\to 0}\int_C \left[\rho\left(e+\frac{1}{2}\dot{a}^2 u_{i,1}u_{i,1}\right)\delta_{1j} - \sigma_{ij}u_{i,1} - h_j/\dot{a}\right] n_j\,ds\ . \qquad (4.11)$$

γ, as defined by (4.11), expresses the total amount of mechanical and thermal energy supplied to the tip. Of interest is to separate these contributions. In order to do this we assume that the mechanical power can be divided into one elastic, purely reversible part and one inelastic, purely dissipative part.

$$\dot{w} = \sigma_{ij}\dot{\varepsilon}_{ij} = \sigma_{ij}\dot{\varepsilon}_{ij}^{(e)} + \sigma_{ij}\dot{\varepsilon}_{ij}^{(ie)} = \dot{w}^{(e)} + \dot{w}^{(ie)} \quad . \tag{4.12}$$

The local entropy balance can be found in textbooks on thermodynamics

$$\rho\dot{s}_e = -\left(\frac{h_j}{\theta}\right)_{,j} + \rho\dot{s}_e^{(i)} \quad . \tag{4.13}$$

s_e is the entropy per unit mass, θ the temperature and $\dot{s}_e^{(i)}$ the internal entropy production rate per unit mass. Because of the assumption (4.12), the internal entropy production consists of two parts.

$$\rho\dot{s}_e^{(i)} = \frac{\dot{w}^{(ie)}}{\theta} - h_j\frac{\theta_{,j}}{\theta^2} \geq 0 \quad . \tag{4.14}$$

The above inequality states the Clausius-Duhem´s principle of non-negative entropy production. Use of (4.12)-(4.14) together with the local energy balance (4.3) gives the following result for the internal energy.

$$\rho\dot{e} = \sigma_{ij}\dot{\varepsilon}_{ij} + \rho\theta\dot{s}_e - \dot{w}^{(ie)} = \dot{w}^{(e)} + \rho\theta\dot{s}_e \quad . \tag{4.15}$$

The energy flow to the tip γ can now be written as

$$\gamma = \lim_{C\to 0}\left\{\int_C\left[\left(w^{(e)} + \frac{1}{2}\rho\dot{a}^2 u_{i,1}u_{i,1}\right)\delta_{1j} - \sigma_{ij}u_{i,1}\right]n_j\, ds\right.$$

$$\left. + \int_C[w^{(ie)}\delta_{1j} - h_j/\dot{a}]n_j\, ds\right\} = \gamma_m + \gamma_t \quad . \tag{4.16}$$

The first part γ_m represents the flow of mechanical energy while the second γ_t represents the thermal energy flow.

We shall now investigate under which conditions the above given expressions are independent of the path C. Consider two different paths C′ and C″ both enclosing the tip. Together with the crack surfaces these paths define a closed area. Using the divergence theorem, the line integral around this area is transformed into an area integral. Thus, we obtain for γ

$$\int_{C'+C''+S_\pm} [\ ]n_j = \int_A \left[\rho\left(e+\frac{1}{2}\dot{a}^2 u_{i,1}u_{i,1}\right)\delta_{1j} - \sigma_{ij}u_{i,1} - h_j/\dot{a}\right]_{,j} dA \ . \tag{4.17}$$

The curve integral will not depend on C if the integrand of the area integral vanishes. The condition becomes

$$\rho e_{,1} + \rho\dot{a}^2 u_{i,1}u_{i,11} - \sigma_{ij,j}u_{i,1} - \sigma_{ij}\varepsilon_{ij,1} - h_{j,j}/\dot{a} = 0 \ . \tag{4.18}$$

<u>If steady state conditions</u> prevail, i.e. the state with respect to the moving x_i' system is time-independent, then time derivatives can be replaced by spatial derivatives.

$$\rho e_{,1} - \sigma_{ij}\varepsilon_{ij,1} - h_{j,j}/\dot{a} = \frac{1}{\dot{a}}(-\rho\dot{e} + \sigma_{ij}\dot{\varepsilon}_{ij} - h_{j,j}) \ . \tag{4.19}$$

The right member vanishes on account of the local energy balance (4.3). Similarly, on account of the equations of motion (2.1) we have

$$\rho\dot{a}^2 u_{i,1}u_{i,11} - \sigma_{ij,j}u_{i,1} = \rho u_{i,1}\ddot{u}_i - \rho\ddot{u}_i u_{i,1} = 0 \ . \tag{4.20}$$

Thus, the integral (4.11) will be path-independent for steady state conditions. In a general case we can not expect path-independence for arbitrary paths, but the results (3.16) and (3.18) for locally steady state motion show that (4.11) is <u>asymptotically</u> path-independent. This is important since otherwise the value of γ would depend on how the limiting process was performed.

Performing similar calculations on γ_m, the mechanical energy flow, we end with the following condition

$$w_{,1}^{(e)} + \rho\dot{a}^2 u_{i,1}u_{i,11} - \sigma_{ij,j}u_{i,1} - \sigma_{ij}\varepsilon_{ij,1} = 0 \ . \tag{4.21}$$

We note that unless

$$w^{(e)}_{,1} = \sigma_{ij}\varepsilon_{ij,1}\ , \quad \text{as } r \to 0\ , \tag{4.22}$$

the integral for γ_m will not be asymptotically path-independent. (4.22) implies that the material in the vicinity of the tip must possess a strain energy function that is a unique function of the strains.

It is instructive to evaluate (4.11) for circular paths with radius r

$$\gamma = \lim_{r\to 0} \int_{\varphi=-\pi}^{\pi} \left[\rho\left(e + \frac{1}{2}\dot{u}_{i,1}\dot{u}_{i,1} \right)\delta_{1j} - \sigma_{ij}u_{i,1} - h_j / \dot{a} \right] n_j r\, d\varphi\ . \tag{4.23}$$

It is immediately seen that in order for γ to be finite and non-zero the integrand must behave as $0\,(r^{-1})$ as r goes to zero. A stronger singularity leads to an infinite value of γ and a weaker one leads to zero energy flow. The solutions for singularities at moving crack tips that so far have been obtained show the r^{-1} -singularity in energy only for cases where the crack tip state is elastic. Thus energy criteria for crack growth can only be used for these cases.

For stationary cracks the integral (4.11) loses its interpretation of energy flow to the tip. The integral will in general give non-trivial results for stationary crack tips, which are a measure of the strength of the singularity.

5. INTEGRALS INDEPENDENT OF THE INTEGRATION REGION

Integrals that are not dependent on the choice of integration region have been the subject of much attention in the theory of fracture. The above discussed integral for the energy flow was found to be dependent on the integration path in the general case. We shall in this section consider some integrals that are not dependent on the integration regions. These integrals may in some cases give the same result as the energy flow integral. This is, however, not the case in general so we will consider the integrals discussed below separately from the energy flow integrals.

Guided by many previous works (c.f. Strifors [9], Aoki and Kishimoto [10], Atluri [11], Moran and Shih [12]) we consider the following form.

$$I = \int_C \left[\left(w + \frac{1}{2}\rho\dot{u}_i\dot{u}_i \right) n_k' l_k - \sigma_{ij}u_{i,k}n_j' l_k \right] ds + \int_A Z\, dA\ . \tag{5.1}$$

Here, C is a closed curve around an area A which contains no singularities. w is the deformation work

$$w = \int_0^t \dot{w}\, dt \ , \tag{5.2}$$

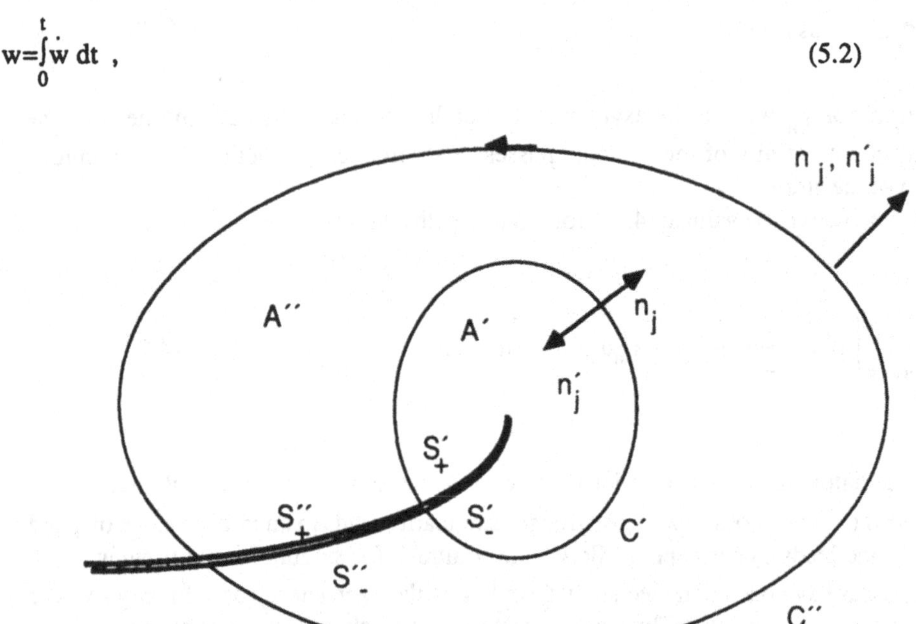

Fig. 4, Different integration regions.

n_j' the outward pointing unit normal vector of C. l_k is a constant unit vector on which we, for the moment, place no restrictions. Z is to be chosen so that if I is taken over a regular area, it becomes zero. By the divergence theorem it is readily seen that I becomes zero if Z is chosen as

$$Z = -l_k\left[\left(w+\frac{1}{2}\rho\dot{u}_i\dot{u}_i\right)_{,k} - (\sigma_{ij}u_{i,k})_{,j}\right] = -l_k[w_{,k} - \sigma_{ij}\varepsilon_{ij,k} + \rho\dot{u}_i\dot{u}_{i,k} - \rho\ddot{u}_i u_{i,k}] \ . \tag{5.3}$$

Here the equations of motion (2.1) and the definition of strain (2.2) have been used. By the choice of (5.3) for Z the integral I is zero around a closed area. Let now I(C) denote the value of I taken along the curve C enclosing the crack tip including the contribution from the crack surfaces between the intersection of C and the crack surfaces and the crack tip. Consider now two curves C″and C′ (fig. 4) with their areas A″and A′ respectively. $S''_\pm$ and $S'_\pm$ are the corresponding parts of the crack surfaces. Let $n_j = n_j'$ on $C''$and $S_\pm$ and $n_j = -n_j'$ on C′. We have

$$I(C'')-I(C')=\int_{C''}[\ \ ]n_j'ds+\int_{S_\pm''}[\ \ ]n_j'\,ds+\int_{A''}ZdA$$

$$+\int_{C'}[\ \ ]n_j'\,ds-\int_{S_\pm'}[\ \ ]n_j'\,ds-\int_{A'}Z\,dA\ . \tag{5.4}$$

C'', C' and $S_\pm'' - S_\pm'$ form a curve enclosing the area A''-A'. Thus $I(C'')$ - $I(C')$ becomes zero and we can conclude that

$$I=\int_C\left[\left(w+\frac{1}{2}\rho\dot{u}_i\dot{u}_i\right)n_k-\sigma_{ij}u_{i,k}n_j\right]l_k\,ds+\int_{S_\pm}\left(w+\frac{1}{2}\rho\dot{u}_i\dot{u}_i\right)n_kl_kds+\int_A ZdA \tag{5.5}$$

is not dependent of C provided that the crack surfaces are traction free.

In order for (5.5) to be meaningful the different integrals must not be divergent. For the area integral this requires that

$$Z=0(r^{-\kappa_1})\text{ as } r\to 0\ ,\ \kappa_1<2\ , \tag{5.6}$$

and for the integral along the crack surface

$$n_kl_k\to 0\ \text{ as } r\to 0\ , \tag{5.7}$$

or

$$w+\frac{1}{2}\rho\dot{u}_i\dot{u}_i=0(r^{-\kappa_2})\ \text{ as } r\to 0,\ \kappa_2<1\ , \tag{5.8}$$

or

$$\lim_{r\to 0}\left(w+\frac{1}{2}\rho\dot{u}_i\dot{u}_i\right)_{|S_+}=\lim_{r\to 0}\left(w+\frac{1}{2}\rho\dot{u}_i\dot{u}_i\right)_{|S_-}\ . \tag{5.9}$$

The condition (5.8) can not be satisfied in cases when the integrand is $0(r^{-1})$ and the last condition (5.9) is only satisfied under special conditions. Thus in the general case it must be required that (5.7) is satisfied i.e. l_k must be chosen as the unit tangent of the crack surfaces.

For moving crack tips the results for asymptotically steady motion can be employed.

$$\dot{w} = -\dot{a}_k w_{,k} = -\dot{a} l_k^v w_{,k} \quad \text{as } r \to 0 \ , \tag{5.10}$$

$$\sigma_{ij}\dot{\varepsilon}_{ij} = -\dot{a}_k \sigma_{ij}\varepsilon_{ij,k} = -\dot{a} l_k^v \sigma_{ij}\varepsilon_{ij,k} \quad \text{as } r \to 0 \ . \tag{5.11}$$

(5.10) and (5.11) show that the first two terms of the integrand of the area integral vanish as $r \to 0$ if l_k is equal to l_k^v thus implying smooth crack growth. The same arguments can be used for the remaining terms and thus the area integral is convergent. It is noted that if the entire problem is <u>steady-state</u> then the area integral vanishes identically. The line integral part of I is thus path-independent for steady-state conditions.

For <u>stationary</u> crack tips all terms in (5.3) with possible exception of the first two of the area integral are $0(r^{-1})$ or of weaker singularity. The first parenthesis vanishes provided

$$w_{,k} \to \sigma_{ij}\, \varepsilon_{ij,k} \quad \text{as } r \to 0 \ . \tag{5.12}$$

This condition is satisfied for elastic materials. To which extent we can expect (5.12) to be valid for other materials is discussed below.

We note, as for the energy integral, that a finite, non-trivial value of I is possible only if the integrand of the curve integral is $0(r^{-1})$ as $r \to 0$.

For elastic materials (5.12) is identically satisfied. Thus,

$$I = l_k \left\{ \int_C \left[\left(w + \frac{1}{2}\rho \dot{u}_i \dot{u}_i \right) n_k - \sigma_{ij} u_{i,k} n_j \right] ds \right.$$

$$\left. + \int_{S_\pm} \left(w + \frac{1}{2}\rho \dot{u}_i \dot{u}_i \right) n_k ds - \int_A \rho (\dot{u}_i \dot{u}_{i,k} - \ddot{u}_i u_{i,k}) ds \right\} \ . \tag{5.13}$$

In the <u>linearly</u> elastic case it is easily verified that the symmetry condition (5.9) is satisfied so that I is convergent for arbitrary l_k. The contribution from $S_\pm$, however, does not vanish in general for finite integration paths.

For <u>linear</u>, elastic or visco-elastic materials, a surface integral that is independent of the integration surface can be derived that is expressed in terms of the Laplace transforms of the quantities (Nilsson [13]). Assume a constitutive law of the relaxation form

$$\sigma_{ij}(t) = \int_{\tau=0}^{t} C_{ijkl}(t-\tau)\dot{\varepsilon}_{kl}(\tau) d\tau \ . \tag{5.14}$$

Here, the relaxation functions C_{ijkl} are assumed to possess the symmetries with respect to the indices that are necessary for defining a potential. Let now a bar over a letter define the Laplace transform of the corresponding quantity. Transforming (5.14) yields

$$\bar{\sigma}_{ij}=\bar{C}_{ijkl}\,\bar{\varepsilon}_{kl}\ , \tag{5.15}$$

and the following functional of $\bar{\varepsilon}_{kl}$ is defined

$$\bar{U}=\frac{1}{2}\bar{\sigma}_{ij}\,\bar{\varepsilon}_{ij}=\frac{1}{2}\bar{C}_{ijkl}\,\bar{\varepsilon}_{ij}\,\bar{\varepsilon}_{kl}\ . \tag{5.16}$$

We note immediately from (5.16) that

$$\bar{\sigma}_{ij}=\frac{\partial\bar{U}}{\partial\bar{\varepsilon}_{ij}}\ . \tag{5.17}$$

Assuming quiescent conditions and transforming the equations of motion and the definition of strains, we obtain (p is the transform parameter)

$$\bar{\sigma}_{ij,j}=\rho p^2\bar{u}_i\ , \tag{5.18}$$

$$\bar{\varepsilon}_{ij}=\frac{1}{2}(\bar{u}_{i,j}+\bar{u}_{j,i})\ . \tag{5.19}$$

Consider now the following integral over the surface S with l_k now being a constant vector field

$$\bar{I}_L=l_k\left\{\int_C\left[\left(\bar{U}+\frac{1}{2}\rho p^2\bar{u}_i\bar{u}_i\right)n_k-\bar{\sigma}_{ij}\bar{u}_{i,k}\,n_j\right]ds+\int_{S_\pm}\left(\bar{U}+\frac{1}{2}\rho p^2\,\bar{u}_i\bar{u}_i\right)n_k\,ds\right\}\ . \tag{5.20}$$

Let, for the moment, C be a closed curve enclosing a regular volume A. Application of the divergence theorem yields

$$\int_C l_k[\quad]ds=\int_A l_k[\bar{U}_{,k}+\rho p^2\bar{u}_i\bar{u}_{i,k}-\bar{\sigma}_{ij,j}\bar{u}_{i,k}-\bar{\sigma}_{ij}\bar{u}_{i,kj}]\,dA=$$

$$\int_A l_k\left[\bar{U}_{,k}+\rho p^2\bar{u}_i\bar{u}_{i,k}-\rho p^2\bar{u}_i\bar{u}_{i,k}-\frac{\partial\bar{U}}{\partial\bar{\varepsilon}_{ij}}\frac{\partial\bar{\varepsilon}_{ij}}{\partial x_k}\right]dA=0\ . \tag{5.21}$$

By the previously given arguments, $\bar{I}_L$ is thus independent of C if the crack surfaces are traction-free. The possible use of $\bar{I}_L$ shall be discussed below. It should be pointed out that a straight-forward physical interpretation of $\bar{I}_L$ is not possible. The definition (5.20) may also be expressed in the real time domain, as pointed out by Gurtin [14]. The products in (5.20) between two transformed quantities are then replaced by convolution integrals.

The main usefulness of integrals of the kind presented here is that they can be used to determine the strength of a singular field provided the form of this is known. By using a surface integral the region in the border vicinity can be avoided in the integration process. This is obviously not possible when volume integrals have to be calculated. Therefore, in many applications of fracture mechanics one uses only the line integral even for cases when this is not strictly path-independent. For quasi-static conditions, the line integral part of (5.5) with l_2=0 is termed the J-integral, i.e.

$$J = \int_C (w\, \delta_{1j} - \sigma_{ij}\, u_{i,1}) n_j \, ds \qquad (5.22)$$

The J-integral is obviously strictly path-independent only if (5.12) holds identically. However, in practice, it is found that in many applications an approximate path-independence is obtained and for this reason the J-integral has been used extensively outside its theoretical region of validity.

6. GENERAL ASPECTS ON CRACK TIP FIELDS AND FRACTURE CRITERIA

For simplicity, we will in this section, first consider only the symmetric (mode I) growth of cracks in two-dimensional bodies and focus the attention to the stress-, strain- and displacement-fields in the vicinity of the tip. All material models considered here admit a solution for the stress of the form

$$\sigma_{ij}=Q_\sigma(t)f_\sigma(r)\Sigma_{ij}(\varphi,\dot{a})+\hat{\sigma}_{ij}(r,\varphi,t) \qquad (6.1)$$

where f_σ and Σ_{ij} are independent of the geometry and loading conditions of the body. The first term dominates for small values of r so that

$$\lim_{r\to 0} (\sigma_{ij}\hat{\sigma}_{ij}/\sigma_{kl}\sigma_{kl})=0 \qquad (6.2)$$

Q_σ is a scalar quantity which depends on the entire problem. Similarly, the strains can be written

$$\varepsilon_{ij}=Q_\varepsilon(t)f_\varepsilon(r)\,E_{ij}(\varphi,\dot{a})+\hat{\varepsilon}_{ij}(r,\varphi,t), \tag{6.3}$$

$$\lim_{r\to 0}(\varepsilon_{ij}\hat{\varepsilon}_{ij}/\varepsilon_{kl}\varepsilon_{kl})=0 \ , \tag{6.4}$$

with the exception of the case when the material is rate independent perfectly plastic, in which case the function E_{ij} is not unique. The magnitude parameter Q_ε can be related to Q_σ by the constitutive relation. Finally, a similar form for the displacements is assumed

$$u_i=Q_u(t)f_u(r)U_i(\varphi,\dot{a})+\hat{u}_i(r,\varphi,t) \ , \tag{6.5}$$

compatible with the crack tip strain field. In all cases considered here, with the exception of the perfectly plastic case, the radial dependence is of the form

$$f_\sigma(r) = r^{-\alpha} \quad 0<\alpha<1 \ , \tag{6.6}$$

$$f_\varepsilon(r) = r^{-\beta} \quad 0<\beta\leq 1 \ , \tag{6.7}$$

$$f_u(r) = r^{1-\beta} \ . \tag{6.8}$$

We conclude immediately that when (6.6) -(6.8) prevail, the asymptotically steady state result is valid.

As mentioned earlier the idealized material models used for describing the bulk deformation can not be expected to be valid very near to the crack tip. Thus, there will be a zone, the process zone, where the above description fails. Suppose that the region with different material behaviour and possibly also large deformation effects can be enclosed within a curve C_p. We will now define the concept of autonomy (Barenblatt [5]) as when the state within C_p is entirely controlled by the singularity parameters Q_σ, Q_ε and Q_u obtained from a solution to the problem without consideration of the process zone. In the following, it will be assumed that autonomy prevails and this assumption is the basis for virtually all fracture theories. The question of whether autonomy really prevails or not in a certain situation is difficult and theoretical analyses of these problems are scarce and will not be discussed here.

Since we can not a priori exclude history effects on the processes within C_p, the instantaneous values of singularity parameters do not necessarily control the state of the process region. In general, the time histories of these parameters need to be considered. For a propagating crack tip it can be assumed, that what is occurring in the process region at a time instant t will not affect the state at a time instant t_0 when the distance travelled by the crack during the time interval t_0 - t is sufficiently large. Denote this necessary distance that the tip has to move l_{min}. Obviously l_{min} will be of the same order as the linear extent of the process zone. Assume that Q_σ and Q_ε are interchangeable so that the choice between these is immaterial. In time-dependent materials the relation between Q_σ and Q_ε may be more complicated, but in principle one of them can be expressed in terms of the other through the constitutive equation. Expand the time history of Q_σ about t_0 in a Taylor series

$$Q_\sigma(t)=Q_\sigma(t_0)+\sum_{k=1}^{\infty}\frac{1}{k!}\frac{\partial^k Q_\sigma}{\partial t^k}\Big|_{t_0}(t-t_0)^k \quad . \tag{6.9}$$

The state of the process zone can then be formulated as a function of a sufficient number of the time derivates of Q_σ and of $\dot{a}$. The state parameter that is of interest is the velocity $\dot{a}$ of the tip. Thus, the crack growth equation can be expressed as

$$\dot{a}=h\left(\frac{\partial^{k_1}Q_\sigma}{\partial t^{k_1}},\frac{\partial^{k_2}a}{\partial t^{k_2}}\right) \quad , \quad k_i=0,\ 1,\ 2,\ 3.....; \quad i=1,2. \tag{6.10}$$

A derivative can be neglected in (6.10) if its contribution to (6.9) during the time interval $l_{min}/\dot{a}$ preceding t_0 is negligible, i.e.

$$\left|\frac{1}{Q_\sigma(t_0)}\frac{1}{k!}\frac{\partial^k Q_\sigma}{\partial t^k}\left(\frac{l_{min}}{\dot{a}}\right)^k\right|<<1 \quad , \quad k=0,1,2,3... \tag{6.11}$$

and a similar expression for holds for $\dot{a}$ and its time derivatives..Very little information is known about the practical implications of these aspects. Virtually all research is based on the assumption that l_{min} is negligible.The crack tip velocity, $\dot{a}$, then depends only on the instantaneous value of Q_σ and the growth equation can be written in the alternative forms

$$\dot{a} = h_c(Q_\sigma) \tag{6.12}$$

$$Q_\sigma = Q_\sigma^c(\dot{a}) \quad . \tag{6.13}$$

It is easily realized that the reduction of (6.10) into (6.12) and (6.13) may not necessarily result in unique functions h_c and Q_σ^c, since the function h in (6.10) may not be mono tonous and $\dot{a}$ appears in both members of (6.10). Thus, there may exist different branches of h_c and Q_σ^c. If this is the case, it is likely that derivatives of Q_σ and $\dot{a}$ must be accounted for deciding which branch to follow.

For the case of initiation from a stationary crack, the above reasoning does not apply and in principle the entire time history of Q_σ needs to be considered. In conclusion, the assumption of autonomy leads to that the crack growth criteria should be formulated in terms of the singularity parameters appropriate for the chosen constitutive model. Other criteria are occasionally used in fracture theory. One popular choice is to consider some field quantity such as the normal stress or the effective plastic strain at some fixed distance in front of the tip. The criterion is taken in the form that the tip grows so that the chosen quantity is maintained constant and equal to a critical value during growth. The distance is chosen with respect to the physical properties of the material such as the inclusion spacing. The motive for using criteria of this type is usually to avoid to base a fracture criterion on the singular field, since it is felt that process zone effects may render this solution meaningless. However, it can be argued that if this is the case, the field quantities at small finite distances from the tip are also unreliable since the presence of the process zone will as well disturb the state as obtained from solutions where this zone is not present.. Similar objections can be raised against criteria based on the critical crack opening displacement or angle.

Criteria based on the energy flow to the tip have a long standing in fracture mechanics. As we have seen, γ is completely determined by the singular solution so that the introduction of an energy criterion does not differ from a criterion based on the singularity parameter. In fact, except for the case when an elastic singularity prevails, γ seems to be equal to zero and can thus be of no help in formulating growth equations.

In quasi-static fracture mechanics the J-integral which is the first term of (5.1) is often used as fracture parameter. This is however only a variant of the singularity parameter criterion since it is assumed that the singular field is controlled by a field which is measurable by the J-integral. As pointed out earlier, the full form of the I-integral is of no value for crack growth in some non-elastic materials since the singular fields in those cases render the integral identically zero. For stationary cracks a non-zero value usually results, so that for <u>initiation</u> the I-integral may be of value.

Turning now to the question of the crack growth direction it is usually assumed that the tip grows so that the tip is locally in mode I motion. This is termed the principle of <u>local symmetry</u> (Gol´dstein and Salganik [15]) and for general materials no theoretical foundation for this assumption is available. For <u>linearly elastic</u> materials it can be shown (Strifors [3]) that such a motion is the only one that is stable, if an energy criterion is applied and the material is assumed to be isotropic with respect to its fracture properties. Experimental observations tend to support the criterion of local symmetry, even though it is difficult to distinguish between different hypotheses.

The crack tip may in some cases, especially for high velocities, experience abrupt changes in the propagation direction either by <u>kinking</u> or <u>branching</u>. Several theoretical analyses of these phenomena mostly assuming elastic conditions have been performed, but a definite explanation does not seem to have been advanced. One hypothesis is that the crack branches when it is energetically favourable. This concept does however violate an important physical principle, in that the event of crack branching is determined by something that occurs after the event. Thus, it has been proposed that crack branching or kinking must depend on the structure of the near tip field before the event occurs. We will not, in the following, discuss either non-symmetric smooth crack growth or discontinuous direction changes because of the few conclusive analyses that exist for these cases.

7. SOME BASIC ELASTODYNAMIC RESULTS

The motion of an isotropic linearly elastic body is governed by the equations of motion (2.1), the definition of strain (2.2) and the constitutive relation (7.1). For isothermal conditions we have

$$\sigma_{ij} = 2G\left(\varepsilon_{ij} + \frac{\nu}{1-2\nu}\varepsilon_{kk}\delta_{ij}\right) . \tag{7.1}$$

G is the shear modulus and ν Poisson´s ratio. Young´s modulus E is given by

$$E = 2G(1+\nu) . \tag{7.2}$$

For the case of plane strain it is convenient to represent the displacements in terms of the potentials ϕ and ψ

$$u_1 = \phi_{,1} + \psi_{,2} , \tag{7.3}$$

$$u_2 = \phi_{,2} - \psi_{,1} , \tag{7.4}$$

$$u_3 \equiv 0 . \tag{7.5}$$

It is easily verified (c.f. [16]) that the necessary equations are satisfied if ϕ and ψ are solutions to the wave equations (7.6)-(7.7)

$$\phi_{,ii} = \frac{1-2\nu}{2(1-\nu)}\frac{\rho}{G}\ddot{\phi} = \frac{1}{c_1^2}\ddot{\phi} , \tag{7.6}$$

$$\psi_{,ii} = \frac{\rho}{G}\ddot{\psi} = \frac{1}{c_2^2}\ddot{\psi} . \tag{7.7}$$

c_1 is the velocity of irrotational waves and c_2 that of equivoluminal waves. The case of plane stress is simply obtained by changing $\nu \rightarrow \nu/(1-\nu)$ in the above expressions.

If u_1 and u_2 vanish identically and u_3 only dependent of x_1 and x_2, it is easily found that u_3 must satisfy (7.8)

$$u_{3,jj} = \frac{1}{c_2^2}\ddot{u}_3 \quad . \tag{7.8}$$

This condition is known as antiplane strain and we note that it is uncoupled from the plain strain condition and can thus be considered separately.

8. CRACK TIP FIELDS IN ELASTIC MATERIALS

Guided by the previous discussion we seek solutions of the wave equations that satisfy the conditions (3.11). It is readily found that the singular solution must be a combination of plane and anti-plane strain. Thus, we first consider the plane strain case. It is convenient to use a coordinate system moving with the tip. Assuming smooth propagation, the tip velocity will coincide with the x_1'-axis. Utilizing the results (3.18) the wave equations (7.9)-(7.10) assume the following forms near the moving tip

$$\phi_{,ii} = \frac{\dot{a}^2}{c_1^2}\phi_{,11} \quad , \tag{8.1}$$

$$\psi_{,ii} = \frac{\dot{a}^2}{c_2^2}\psi_{,11} \quad . \tag{8.2}$$

In unabridged form, (8.1) and (8.2) become

$$\beta_1^2\,\phi_{,11} + \phi_{,22} = 0 \quad , \tag{8.3}$$

$$\beta_2^2\,\psi_{,11} + \psi_{,22} = 0 \quad , \tag{8.4}$$

where

$$\beta_k^2 = 1-(\dot{a}/c_k)^2, \quad k = 1, 2 \quad , \tag{8.5}$$

and it is assumed that $\dot{a}$ is smaller than c_2.

Introduce the polar coordinates

$$x_1' = r_k \cos \varphi_k \quad , \tag{8.6}$$

$$\beta_k x_2' = r_k \sin \varphi_k \quad , \tag{8.7}$$

and (8.3) - (8.4) transform to

$$\frac{\partial^2 \phi}{\partial r_1^2} + \frac{1}{r_1}\frac{\partial \phi}{\partial r_1} + \frac{1}{r_1^2}\frac{\partial^2 \phi}{\partial \varphi_1^2} = 0 \quad , \tag{8.8}$$

$$\frac{\partial^2 \psi}{\partial r_2^2} + \frac{1}{r_2}\frac{\partial \psi}{\partial r_2} + \frac{1}{r_2^2}\frac{\partial^2 \psi}{\partial \varphi_2} = 0 \quad . \tag{8.9}$$

We assume solutions of the form $\phi = r_1^{\alpha} f_1(\varphi_1)$ and $\psi = r_2^{\alpha} f_2(\varphi_2)$. (8.8) then reduces to

$$f_1''(\varphi_1) + \alpha^2 f_1(\varphi_1) = 0 \quad , \tag{8.10}$$

with the solution

$$f_1 = A_1 \sin \alpha\varphi_1 + B_1 \cos \alpha\varphi_1 \quad . \tag{8.11}$$

Similarly, we have for (8.7)

$$f_2 = A_2 \sin \alpha\varphi_2 + B_2 \cos \alpha\varphi_2 \quad . \tag{8.12}$$

The boundary conditions to be satisfied on the crack surface are those of vanishing tractions. The stresses can be expressed by using (7.1), (7.3), (7.4) and the reduced wave equations (8.3)-(8.4) as

$$\sigma_{11} = G\,[(1+2\beta_1^2-\beta_2^2)\phi_{,11} + 2\psi_{,12}] \quad , \tag{8.13}$$

$$\sigma_{22} = G\,[-(1+\beta_2^2)\,\phi_{,11} - 2\psi_{,12}] \quad , \tag{8.14}$$

$$\sigma_{12} = G\,[2\phi_{,12} + (1+\beta_2^2)\psi_{,11}] \quad . \tag{8.15}$$

The boundary conditions take the forms

$$\sigma_{22} = -(1+\beta_2^2)\,\phi_{,11} - 2\psi_{,12} = 0 \qquad \text{on } \varphi_1 = \varphi_2 = \pm\pi \ , \tag{8.16}$$

$$\sigma_{12} = 2\phi_{,12} + (1+\beta_2^2)\,\psi_{,11} = 0 \qquad \text{on } \varphi_1 = \varphi_2 = \pm\pi \ . \tag{8.17}$$

(8.16) and (8.17) can be integrated with respect to x_1 to yield

$$-(1+\beta_2^2)\,\phi_{,1} - 2\psi_{,2} = 0 \qquad \text{on } \varphi_1 = \varphi_2 = \pm\pi \ , \tag{8.18}$$

$$2\phi_{,2} + (1+\beta_2^2)\,\psi_{,1} = 0 \qquad \text{on } \varphi_1 = \varphi_2 = \pm\pi \tag{8.19}$$

In polar coordinates, (8.18)-(8.19) read on the crack surfaces

$$\mp (1+\beta_2^2)\,\phi_{,r_1} \pm \frac{2\beta_2}{r_2}\,\psi_{,\varphi_2} = 0 \ , \tag{8.20}$$

$$\mp \frac{2\beta_1}{r_1}\,\phi_{,\varphi_1} \pm (1+\beta_2^2)\,\psi_{,r_2} = 0 \ , \tag{8.21}$$

where the upper sign refers to the upper crack surface and the lower sign to the lower crack surface.

It is advantageous to make some symmetry considerations. Define mode I as

$$u_1(x_1, x_2) = u_1(x_1, -x_2) \tag{8.22}$$

$$u_2(x_1, x_2) = -u_2(x_1, -x_2) \tag{8.23}$$

and mode II as

$$u_1(x_1, x_2) = -u_1(x_1, -x_2) \tag{8.24}$$

$$u_2(x_1, x_2) = u_2(x_1, -x_2) \ . \tag{8.25}$$

We see that mode I case is obtained if ϕ is symmetric and ψ antisymmetric with respect to $x_2' = 0$. The reverse is true for the mode II case. Considering first the symmetric case, (8.20) and (8.21) yield

$$-(1+\beta_2^2)\,A_1 \sin \alpha\pi - 2\beta_2\,B_2 \sin \alpha\pi = 0 \tag{8.26}$$

$$-2\beta_1 A_1 \cos \alpha\pi - (1+\beta_2^2)\,B_2 \cos \alpha\pi = 0 \tag{8.27}$$

The eigenvalue problem yields the following secular equation

$$\sin 2\alpha\pi = 0 \quad , \tag{8.28}$$

which is satisfied for

$$\alpha_n = {}^{n}/2 \quad , \quad n = 1,2,3,........ \quad . \tag{8.29}$$

The first term $\alpha_1 = {}^1/2$ gives displacements that behave as $r^{-1/2}$ and thus become unbounded as $r \rightarrow 0$. This term is ruled out on physical grounds and the next term with $\alpha_2 = {}^3/2$ becomes dominant. Solving (8.26) for A_1, the solutions for ϕ and ψ are obtained.

$$\phi = B_2 \frac{2\beta_2}{(1+\beta_2^2)} r_1^{3/2} \sin \varphi_1 /2 \quad , \tag{8.30}$$

$$\psi = B_2\, r_2^{1/2} \cos \varphi_2 /2 \quad . \tag{8.31}$$

The constant B_2 can not be determined from the asymptotic analysis. Instead, the entire problem has to be solved to obtain B_2. Performing the necessary differentiations, we obtain for the displacements and the stresses

$$u_1 \rightarrow \frac{K_I}{GN}\left(\frac{2}{\pi}\right)^{1/2} [(\beta_2^2+1)\, r_1^{1/2} \cos \varphi_1 /2 - 2\beta_1\beta_2 r_2^{1/2} \cos \varphi_2 /2] \quad \text{as } r \rightarrow 0 \quad , \tag{8.32}$$

$$u_2 \rightarrow \frac{K_I}{GN}\left(\frac{2}{\pi}\right)^{1/2} [-(\beta_2^2+1) r_1^{1/2} \sin \varphi_1/2 + 2\beta_1\beta_2\, r_2^{1/2} \sin \varphi_2/2] \quad \text{as } r \rightarrow 0 \quad , \tag{8.33}$$

$$\sigma_{11} \rightarrow \frac{K_I}{N} (2\pi)^{-1/2} [(\beta_2^2+1)(2\beta_1^2-\beta_2^2+1) r_1^{-1/2} \cos \varphi_1/2$$
$$- 4\beta_1\beta_2 r_2^{-1/2} \cos \varphi_2/2] \quad \text{as } r \rightarrow 0 \quad , \tag{8.34}$$

$$\sigma_{22} \to \frac{K_I}{N}(2\pi)^{-1/2}\,[-(\beta_2^2+1)^2 r_1^{-1/2}\cos\varphi_1/2 + 4\beta_1\beta_2 r_2^{-1/2}\cos\varphi_2/2] \quad \text{as } r\to 0\,, \tag{8.35}$$

$$\sigma_{12} \to \frac{K_I}{N}(2\pi)^{-1/2}\,[2\beta_1(\beta_2^2+1)r_1^{-1/2}\sin\varphi_1/2 - 2\beta_1(\beta_2^2+1)r_2^{-1/2}\sin\varphi_2/2] \quad \text{as } r\to 0, \tag{8.36}$$

$$N = 4\beta_1\beta_2 - (1+\beta_2^2)^2 \ . \tag{8.37}$$

Instead of the constant B_2 in (8.30), we have, analogously to the static case, introduced the stress intensity factor K_I as

$$K_I = \lim_{x_1'\to +0} \sigma_{22}\,(2\pi\, x_1')^{1/2} \ , \quad x_2' = 0 \ . \tag{8.38}$$

for the mode I case.

Performing the same type of calculations for the antisymmetric case (mode II) we obtain

$$u_1 \to \frac{K_{II}}{N}\left(\frac{2}{\pi}\right)^{1/2}\,[2\beta_2 r_1^{1/2}\sin\varphi_1/2 - (\beta_2^2+1)\beta_2 r_2^{1/2}\sin\varphi_2/2] \quad \text{as } r\to 0 \ , \tag{8.39}$$

$$u_2 \to \frac{K_{II}}{GN}\left(\frac{2}{\pi}\right)^{1/2}\,[2\beta_1\beta_2 r_1^{1/2}\cos\varphi_1/2 - (\beta^2+1)\beta_2 r_2^{1/2}\cos\varphi_2/2] \quad \text{as } r\to 0 \ , \tag{8.40}$$

$$\sigma_{11} \to \frac{K_{II}}{N}(2\pi)^{-1/2}\,[2\beta_2(\beta_2^2-1-2\beta_1^2)r_1^{-1/2}\sin\varphi_1/2 + 2\beta_2(\beta_2^2+1)r_2^{-1/2}\sin\varphi_2/2] \quad \text{as } r\to 0 \ , \tag{8.41}$$

$$\sigma_{22} \to \frac{K_{II}}{N}(2\pi)^{-1/2}\,[2\beta_2(\beta_2^2+1)r_1^{-1/2}\sin\varphi_1/2 - 2\beta_2(\beta_2^2+1)r_2^{-1/2}\sin\varphi_2/2] \quad \text{as } r\to 0 \ , \tag{8.42}$$

$$\sigma_{12} = \frac{K_{II}}{N}(2\pi)^{-1/2}\,[4\,\beta_1\beta_2\, r_1^{-1/2}\cos\varphi_1/2 - (\beta_2^2+1)^2\, r_2^{-1/2}\cos\varphi_2/2] \quad \text{as } r\to 0 \ , \tag{8.43}$$

$$K_{II} = \lim_{x_1'\to 0} \sigma_{12}\,(2\pi x_1')^{1/2} \ , \quad x_2' = 0 \ . \tag{8.44}$$

The anti-plane strain (mode III) case can be solved by similar but considerably simpler calculations

$$u_3 \rightarrow \frac{K_{III}}{G\beta_2}\left(\frac{2r_2}{\pi}\right)^{1/2} \sin \varphi_2/2 \qquad \text{as } r \rightarrow 0 \ , \tag{8.45}$$

$$\sigma_{13} \rightarrow K_{III}(2\pi\, r_2)^{-1/2} \sin \varphi_2/2 \qquad \text{as } r \rightarrow 0 \ , \tag{8.46}$$

$$\sigma_{23} \rightarrow K_{III}(2\pi\, r_2)^{-1/2} \cos \varphi_2/2 \qquad \text{as } r \rightarrow 0 \ , \tag{8.47}$$

$$K_{III} = \lim_{x_1' \rightarrow 0} \sigma_{23}\,(2\pi\, x_1')^{-1/2} \ , \quad x_2' = 0 \ . \tag{8.48}$$

The equations (8.32)-(8.48) completely determine the state at a moving tip in terms of the scalar quantities K_I, K_{II}, K_{III} and the velocity factors β_1 and β_2. They were first given by Sih [17] for the steady-state case. Nilsson [18], Clifton and Freund [19], Achenbach and Bazant [20] later pointed out that these relations are also valid for arbitrary crack motion. It can also be verified that the equations reduce to the appropriate quasi-static forms as $\dot{a} \rightarrow 0$.

By inserting (8.32) - (8.48) into the integral for the energy-flow to the tip (4.17) we obtain, for smooth crack motion

$$\gamma = \frac{1}{2G}\left[\frac{1-\beta_2^2}{N}(\beta_1 K_I^2 + \beta_2 K_{II}^2) + \frac{1}{\beta_2} K_{III}^2\right] = \sum_{m=1}^{3} g_m(\dot{a}/c_2, \nu)\, K_m^2 \ , \tag{8.49}$$

where

$$g_1(0) = g_2(0) = \frac{1-\nu}{2G} \ , \tag{8.50}$$

$$g_3(0) = \frac{1}{2G} \ . \tag{8.51}$$

Thus, γ is uniquely related to the stress-intensity factors and the propagation velocity. We note that γ becomes unbounded when $N \rightarrow 0$, if K_I or K_{II} is non-zero. The equation $N=0$ is the secular equation for the velocity of Rayleigh surface waves. The velocity of these waves is denoted by c_R. Similarly, when $\dot{a} \rightarrow c_2$ the last term of (8.49) becomes unbounded if K_{III} is non-zero. The different velocity functions g_i in (8.49) are shown in fig. 5. For moderate velocities of the order of a tenth of the wave velocities they do not differ much from unity.

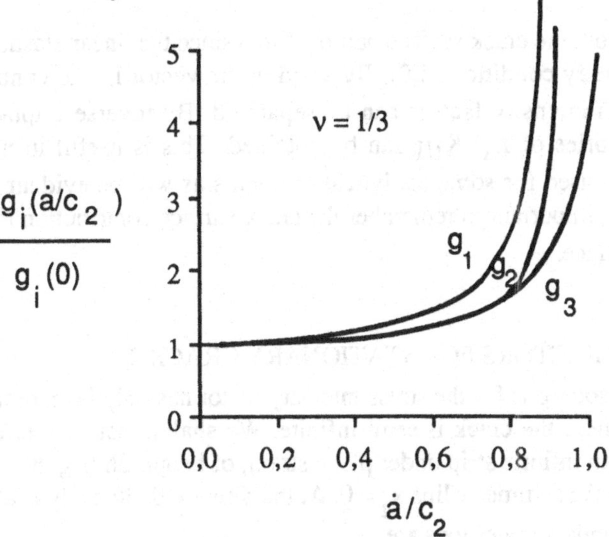

Fig. 5, The velocity functions for γ.

We shall now investigate the relation between the Laplace transformed integral $\bar{I}_L$ and the stress intensity factors. Unfortunately no useful results can be obtained for moving cracks. The equation (5.20) is, as mentioned before, equivalent to a convolution integral in time. Thus, the state at a certain time-instant has to be connected to the states at the particular point for all preceding times. For a moving crack, these states are outside the singular region of the crack position considered and thus no simple results are obtainable. For a stationary crack, we can write the fields in the form

$$\sigma_{ij} \rightarrow (2\pi r)^{-1/2} \left[K_I(t)\, \Sigma^e_{ijI}(\varphi) + K_{II}(t)\, \Sigma^e_{ijII}(\varphi) + K_{III} \Sigma^e_{ijIII}(\varphi)\right] \quad , \tag{8.52}$$

$$u_i \rightarrow \frac{1}{G}(2\pi r)^{1/2} \left[K_I(t)\, U^e_{iI}(\varphi) + K_{II}(t)\, U^e_{iII}(\varphi) + K_{III}\, U^e_{iIII}(\varphi)\right] \quad . \tag{8.53}$$

Here, the angle functions can be obtained from (8.32) - (8.48) by letting $\dot{a}$ tend to zero and performing the limiting process. By inserting into (5.20) the Laplace transformed versions of (8.52)- (8.53) and integrating around a circle with the radius r, we obtain, if $l_3 = 0$,

$$\bar{I}_L = \frac{1-\nu}{2G}\left[\left(\bar{K}_I^{\,2} + \bar{K}_{II}^{\,2} + \frac{1}{1-\nu}\bar{K}_{III}^{\,2}\right)l_1 - 2\bar{K}_I\bar{K}_{II}l_2\right] \quad . \tag{8.54}$$

No contribution arises from the crack surface part of (5.20) since the linear elastic singular field satisfies the symmetry condition (5.9). By varying the vector l_k the contributions from the different stress intensity factors can be separated. By inverse Laplace transformation the time histories of K_I- K_{III} can be obtained. This is useful in numerical analysis and can also be used for some analytical problems as will be evident from the following. It is, however, important to remember the crack surface contribution if l_k is not a tangent to the crack surface.

9 STRESS-INTENSITY FACTORS FOR STATIONARY CRACK TIPS

Analytical dynamic solutions for the stress intensity factor has only been obtained for a few idealized cases where the crack is semi-infinite. We shall consider the following simple example ([13]). An infinite strip under plane strain, of height 2h (fig. 6) contains a semi-infinite crack along the symmetry line $x_2 = 0$. At the time t = 0, the body is at rest and for positive times the boundary conditions are

$$u_2 = \pm v_0\, q(t) \quad \text{on } x_2 = \pm h\,, \tag{9.1}$$

$$u_1 = 0 \quad \text{on } x_2 = \pm h\ , \tag{9.2}$$

$$\sigma_{22} = \sigma_{12} = 0 \quad \text{on } x_2 = 0 \ \ , \ \ x_1 < 0\ . \tag{9.3}$$

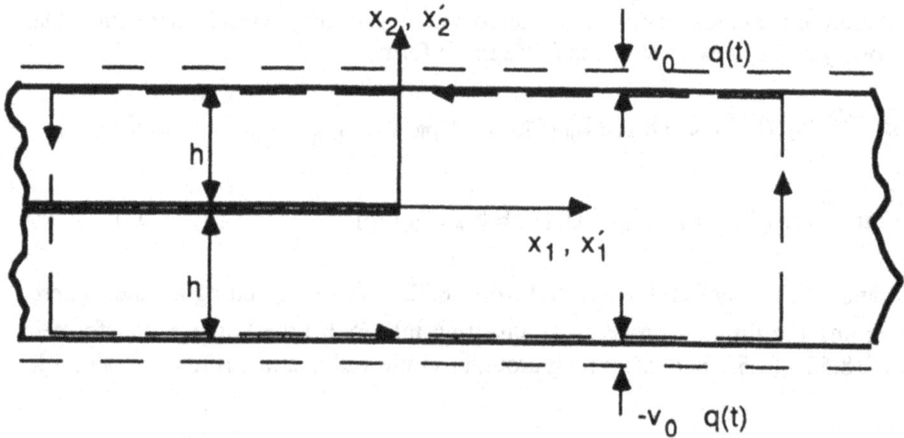

Fig. 6, The infinite strip with displacement boundary conditions.

v_0 is a constant and q(t) a dimensionless function. For every time instant the state is independent of the existence of the tip at distances from the tip greater than c_1t. We now consider the Laplace transformed version of the problem. At $|x_1|>c_1t$ the state is independent of x_1 and the wave equations (7.6) - (7.7) take the forms

$$\bar{\phi}_{,22}=\frac{p^2}{c_1^2}\bar{\phi} \ , \qquad (9.4)$$

$$\bar{\psi}_{,22}=\frac{p^2}{c_2^2}\bar{\psi} \ . \qquad (9.5)$$

The solutions of (9.4) - (9.5) is readily obtained, viz

$$\bar{\phi}=A_3 \sinh\frac{p}{c_1}x_2+B_3\cosh\frac{p}{c_1}x_2 \ , \qquad (9.6)$$

$$\bar{\psi}=A_4 \sinh\frac{p}{c_2}x_2+B_4\cosh\frac{p}{c_2}x_2 \ . \qquad (9.7)$$

The transformed boundary conditions can be written in terms of $\bar{\phi}$ and $\bar{\psi}$

$$\bar{u}_1=\bar{\phi}_{,1}+\bar{\psi}_{,2}=0 \ , \quad x_2=\pm h \ , \qquad (9.8)$$

$$\bar{u}_2=\bar{\phi}_{,2}-\bar{\psi}_{,1}=\pm v_0\,\bar{q}\,(p) \ , \quad x_2=\pm h \ , \qquad (9.9)$$

$$\bar{u}_2=\bar{\phi}_{,2}-\bar{\psi}_{,2}=0 \ , \quad x_2=0, \ \ x_1>c_1t \ , \qquad (9.10)$$

$$\bar{\sigma}_{22}=2G\left[\bar{\phi}_{,22}-\bar{\psi}_{,12}+\frac{\nu}{1-2\nu}(\bar{\phi}_{,11}+\bar{\phi}_{,22})\right]=0, x_2=0, \ \ x_1<0 \ , \qquad (9.11)$$

$$\bar{\sigma}_{12}=G\,(2\bar{\phi}_{,12}+\bar{\psi}_{,22}-\bar{\psi}_{,11})=0 \ , \quad x_2=0 \ . \qquad (9.12)$$

Insertion of (9.6) - (9.7) into the boundary conditions yields the solution

$$\bar{u}_2 = v_0\,\bar{q}(p)\,\frac{\sinh(px_2/c_1)}{\sinh(ph/c_1)} \quad , \quad x_1 > c_1 t \ , \tag{9.13}$$

$$\bar{\varepsilon}_{22} = \frac{1-2\nu}{2(1-\nu)G}\,\bar{\sigma}_{22} = v_0\,\bar{q}\,(p)\,\frac{p}{c_1}\,\frac{\cosh(px_2/c_1)}{\sinh(ph/c_1)} \quad , x_1 > 0 \ , \tag{9.14}$$

$$\bar{u}_2 = v_0\,\bar{q}(p)\,\frac{\cosh(px_2/c_1)}{\cosh(ph/c_1)} \quad , \quad x_1 < -c_1 t \ , \tag{9.15}$$

$$\bar{\varepsilon}_{22} = \frac{1-2\nu}{2(1-\nu)G}\,\bar{\sigma}_{22} = v_0\,\bar{q}(p)\,\frac{p}{c_1}\,\frac{\sinh(px_2/c_1)}{\cosh(ph/c_1)} \quad , \quad x_1 < 0 \ . \tag{9.16}$$

The remaining components are either zero or give no contribution to the $\bar{I}_L$-integral for the path chosen (see fig. 6). Choosing l_2=0 it is immediately realized that no contributions come from the integration along $x_2 = \pm h$ since here $n_1 = 0$ and $u_{i,1}$= 0. The non-trivial contributions from the paths at $x_1 = \pm d > \pm c_1 t$ are easily calculated and the result is

$$\bar{I}_L = v_0^2\,\bar{q}^2(p)p\rho\,\frac{2c_1}{\sinh(2ph/c_1)} \quad . \tag{9.17}$$

No contributions emerge from the crack surfaces since $l_2 = 0$ and thus the crack surface term in (5.20) zero.

After some trivial manipulations we obtain the transformed stress intensity factor

$$\bar{K}_I(p) = \frac{2v_0 G}{(h\,(1-2\nu))^{1/2}}\,\bar{q}(p)\left[\frac{2ph/c_1}{\sinh(2ph/c_1)}\right]^{1/2} \quad . \tag{9.18}$$

Eqn. (9.18) can then be inverted for each specific form of $\bar{q}(p)$. It is seen that the time-dependence of K_I is simply given by the convolution between q(t) and the inverse of the factor within brackets. Consider for example a linearly increasing q, i.e.

$$q(t) = t/t_0 \tag{9.19}$$

$$\bar{q}(p) = \frac{1}{p^2\,t_0} \tag{9.20}$$

The inverse of (9.18) is readily obtained by use of a Laplace transform table as

$$K_I(t) = \frac{2v_0 G}{(h(1-2\nu))^{1/2}} \frac{h}{t_0 c_1} \frac{4}{\sqrt{\pi}} \sum_{0 \le n \le (c_1 t/4h)} (-1) \binom{-1/2}{n} \left(\frac{c_1 t}{h} - 1 - 4n \right)^{1/2} \quad (9.21)$$

Here n is an integer and $\binom{-1/2}{n}$ denotes generalized binomial coefficients. Since (9.21) for any given t is a finite series, it gives the exact result for K_I. In fig. 7 $K_I(t) / K_{Iqs}$ is shown as function of dimensionless time where

$$K_{Iqs} = \frac{2v_0 G}{(h(1-2\nu))^{1/2}} \quad . \quad (9.22)$$

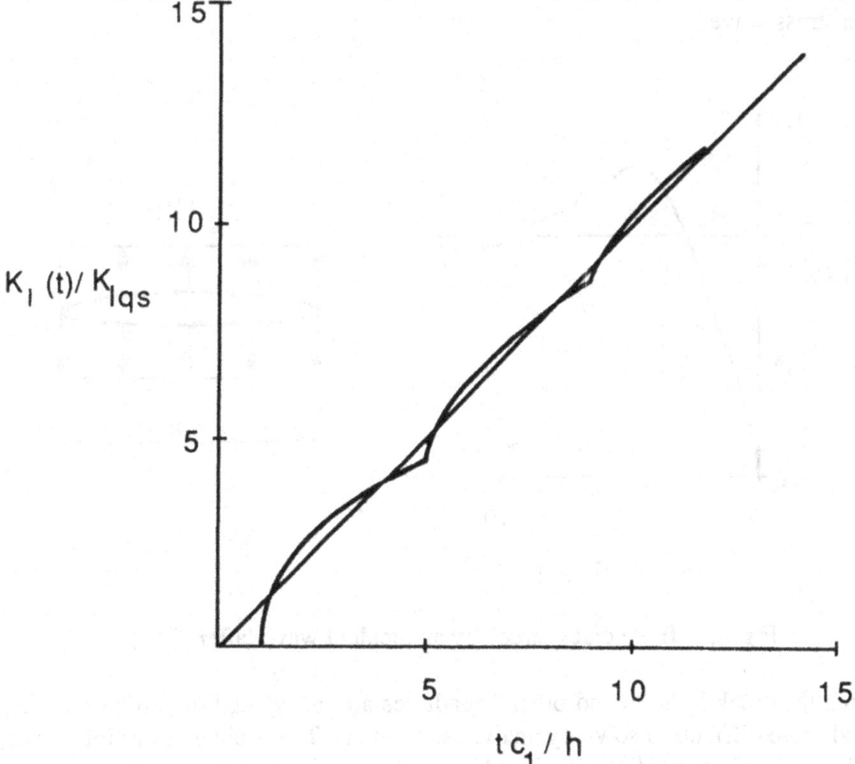

Fig. 7, The stress intensity factor for the strip problem.

$K_I(t)$ oscillates around the quasi-static result $(t/t_0) K_{Iqs}$ with a diminishing amplitude. The behaviour is fairly typical for dynamic K_I - solutions and we note that for $c_1 t/h > 15$ the dynamic effects on K_I are negligible.

For $c_1 t'/h < 5$, the situation is equivalent to that of two plane irrotational wave fronts moving in an unbounded medium towards each other and meeting at the crack plane. The waves have a stress discontinuity σ_0 at the front and we find from (9.21) that $K_I(t)$ due to the action of one of these waves

$$K_I = \frac{2\sigma_0}{\sqrt{\pi}} \frac{(1-2\nu)^{1/2}}{(1-\nu)} (c_1 t')^{1/2} \quad . \tag{9.23}$$

t' is taken as zero at the instant the wave hits the crack plane. This result was first obtained by Maue [21] using a much more complicated method. The square-root dependence on t' is typical for K_I solutions for semi-infinite cracks in unbounded media struck by a step function stress-wave.

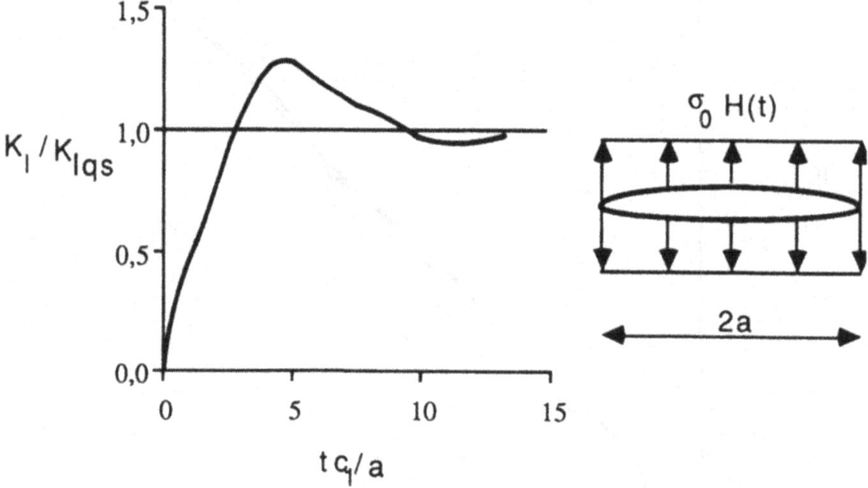

Fig. 8, A finite crack struck by an incident wave (after [24]).

When the crack is finite and other boundaries are introduced the problems become more and more difficult to solve by analytical methods. The problem of a finite crack of length 2a in an infinite medium that is subjected to an incoming wave with a propagation direction perpendicular to the crack plane (fig. 8) has been considered by several authors (Sih et al. [22], Thau and Lu [23], Kim [24]). Kim [24] solved this and similar problems by an analytical technique based on integral transform methods. The analysis leads to a linear equation system in which the coefficients are complicated integrals. In fig. 8 the result for the normalized stress intensity factor is shown for the case $\nu = 0,25$ and we note the same general features as for the previous example.

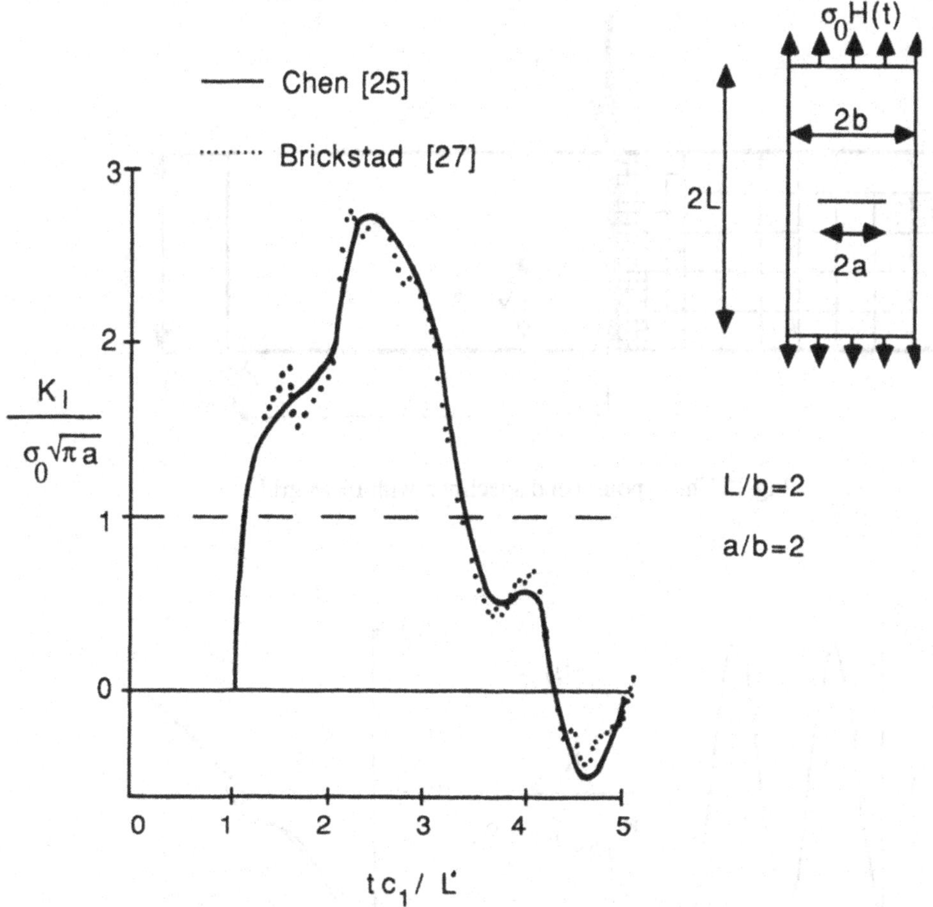

Fig.9, The problem considered in [25]-[27].

When the body is finite in all directions, direct numerical methods such as the finite element method have to be used. We shall, in a later section of this article, discuss how such analyses are to be performed and this section is ended by showing some examples of practical interest.

The geometry and loading situation depicted in fig. 9 have served the purpose of being a bench-mark problem. It was originally considered by Chen [25] and is a rectangular plate subjected to uniform tension σ_0 with a step function time dependence. Numerical results for K_I have been obtained by different authors ([25], Aberson et al [26], Brickstad [27]) and some are shown in fig. 9. We note a much larger dynamic overshoot than in the preceding examples which shows that inferences about the dynamic behaviour drawn from solutions for infinite geometries can be misleading.

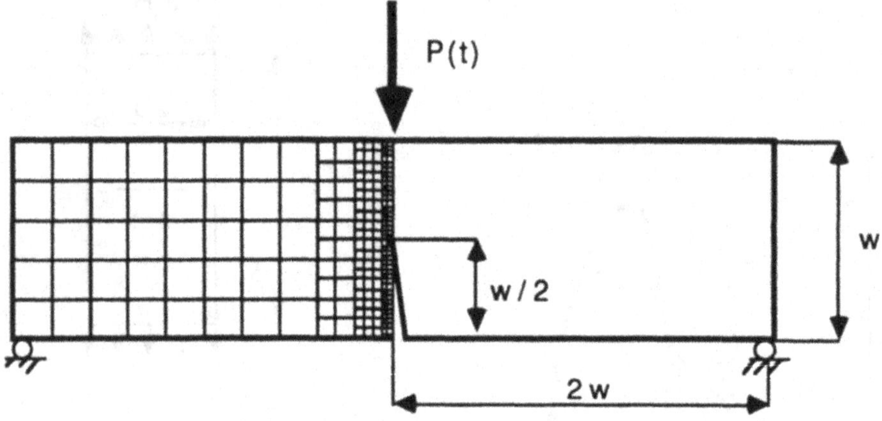

Fig 10 Three-point bend specimen with FEM grid.

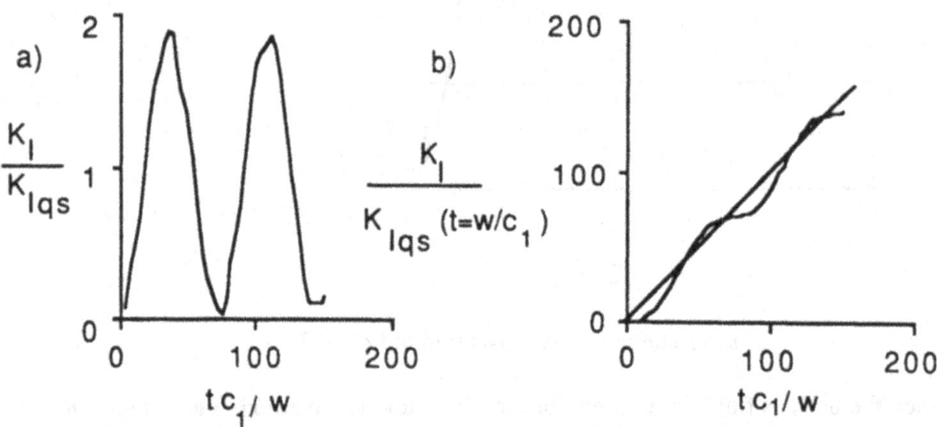

Fig. 11, K_I -results for three-point bend specimen
a) Step time loading b) Linear ramp loading.

A geometry that has a wide-spread use in the experimental work is the three-point bend specimen (fig. 10). In fig. 11 the stress-intensity factor for two different time histories is shown. Fig. 11 a) shows K_I for a suddenly applied force that is afterwards maintained constant and fig. 11b) the corresponding result for a force increasing linearly with time.The problem was analysed by the present author by use of the program system ABAQUS [28] and thefinite element model is sketched in fig. 10. We note that the response to the step time load is very similar to that of a single degree harmonic oscillator.

10. STRESS-INTENSITY FACTORS FOR MOVING CRACK TIPS

The problems that have been considered for K_I-solutions can be divided into steady-state, self-similar and general transient problems.

Steady-state problems are characterized by that the state as observed in a moving coordinate system, is time independent. Thus, the state is dependent on x_i' only, which is defined by

$$x_i' = x_i - \dot{a}\, t\, \delta_{1i} \;, \tag{10.1}$$

where the tip velocity $\dot{a}$ is constant. Steady-state problems admit a simpler analysis than transient problems, but unfortunately few of the steady-state solutions have any practical interest. We will consider the infinite strip with a semi-infinite crack (fig. 6) (Nilsson[29]) moving with a constant velocity and steady state conditions are assumed to prevail. The boundaries at $x_2=\pm h$ are clamped and displaced by an amount $\pm v_0$ respectively. Analogously to the previously considered problem for a stationary tip, this problem admits a simple analysis. At large distances in front of the tip, the state is undisturbed by the crack. For the plane strain case we have

$$\varepsilon_{22} = v_0 / h = \sigma_{22} \frac{(1-2\nu)}{2G\,(1-\nu)} \quad . \tag{10.2}$$

At large distances behind the tip the stresses vanish. By choosing the integration path for the I-integral as sketched in fig. 6 and using (5.13) where the volume integral is identically zero due to the steady-state assumption, it is easily obtained that

$$I = \gamma = \frac{2 v_0^2\, G\,(1-\nu)}{h\,(1-2\nu)} \quad . \tag{10.3}$$

The energy flow is thus independent of the velocity $\dot{a}$. By use of (8.47) the stress-intensity factor is obtained

$$K_I = \frac{2 v_0\, G}{(h\,(1-2\nu))^{1/2}} \left(\frac{N\,(1-\nu)}{\beta_1\,(1-\beta_2^2)} \right)^{1/2} = K_{Iqs}\, f_N\,(\dot{a} / c_2, \nu) \;\; . \tag{10.4}$$

The velocity function f_N is shown in fig. 12, together with some other velocity functions discussed below. f_N decreases from unity at $\dot{a} = 0$ to zero at the Rayleigh velocity c_R. The strip geometry can serve as a useful approximation for some geometries that have been used in experimental investigations (c.f. Dahlberg et al. [30]).

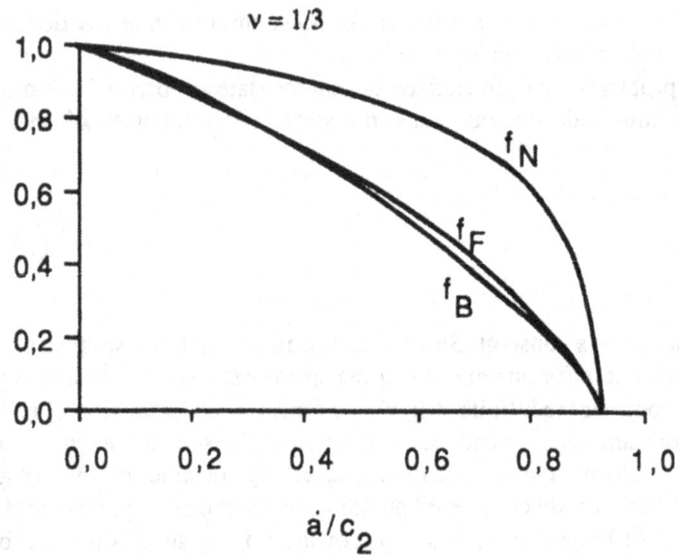

Fig. 12, The velocity dependence of K_I for some problems.

Self-similar problems are such that they possess no characteristic length and that the time dependence of the loading has a certain form. Systematical treatments of this type of problems can be found in Willis [31] and Cherepanov and Afas´anev [32]. Assuming the following form for the potentials for a plane problem,

$$\phi = t^n \bar{\phi}(\xi_i) \ , \tag{10.5}$$

$$\psi = t^n \bar{\psi}(\xi_i) \ , \tag{10.6}$$

$$\xi_i = x_i / t \ , \tag{10.7}$$

it is evident that the displacements and the stresses are of the forms

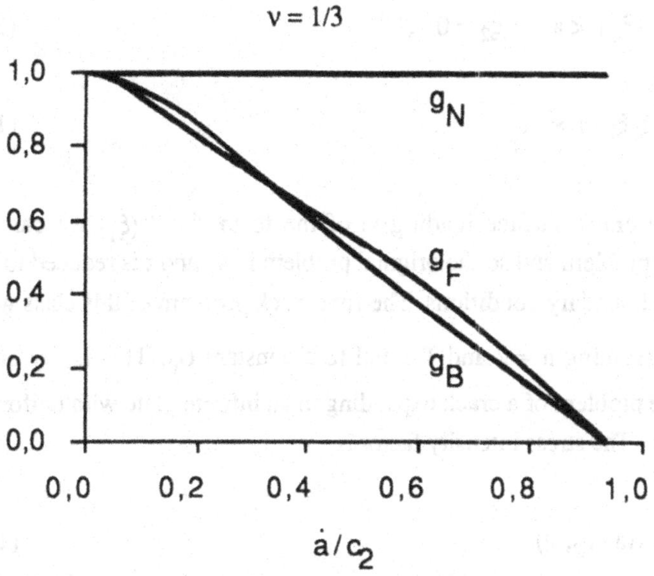

Fig. 13, The energy flow for the problems in fig. 12.

$$u_i = t^{n-1}\,\bar{u}_i\,(\xi_k) \quad , \tag{10.8}$$

$$\sigma_{ij} = t^{n-2}\,\bar{\Sigma}_{ij}\,(\xi_k) \quad . \tag{10.9}$$

Performing the differentiations the wave equations are reduced to

$$\bar{\phi}_{,ii} = \frac{1}{c_1^2}\,(\xi_k\,\xi_k\,\delta_{ik}\,\bar{\phi}_{,ii} + 2(n-1)\,\xi_i\,\bar{\phi}_{,i} + n(n-1)\bar{\phi}) \quad , \tag{10.10}$$

$$\bar{\psi}_{,ii} = \frac{1}{c_2^2}\,(\xi_k\,\xi_k\,\delta_{ik}\,\bar{\psi}_{,ii} + 2(n-1)\,\xi_i\,\bar{\psi}_{,i} + n(n-1)\,\bar{\psi}) \quad . \tag{10.11}$$

Consider now a crack expanding from zero length with a constant velocity a and subjected to normal tractions on the crack surfaces, we can state the boundary conditions as

$$\bar{u}_i = 0 \quad , \quad |\xi_1| \geq \dot{a} \quad , \quad \xi_2 = 0 \quad , \tag{10.12}$$

$$\bar{\Sigma}_{22} = \bar{T}(\xi_1) \quad , \quad |\xi_1| < \dot{a} \quad , \quad \xi_2 = 0 \quad , \tag{10.13}$$

$$\bar{u}_i \, , \bar{\Sigma}_{ij} \rightarrow 0 \quad , \quad \xi_i \xi_i \rightarrow \infty \quad . \tag{10.14}$$

Thus, if the crack surface loading is of the form $t^{n-2}\bar{T}(\xi_1)$ we can expect self-similarity of the problem and so the original problem in x_i and t is reduced to a problem in ξ_i with constant boundary conditions. The first crack problem of this class was solved by Broberg [33], assuming n = 2 and $\bar{T}$ equal to a constant σ_0. This is, by superposition, equivalent to the problem of a crack expanding in an infinite plate with uniform tension σ_0 acting at infinity . The stress intensity factor is

$$K_I = \sigma_0 \sqrt{\pi \dot{a} t} \, f_B(\dot{a} / c_2, \nu) \quad . \tag{10.15}$$

K_I is thus the product of the static stress-intensity factor corresponding to the instantaneous crack length and a velocity dependent function shown in fig. 12. This function also approaches zero at c_R. Other self-similar problems have the same structure for the solution, although the velocity function differs between the problems. Employing eqn. (8.49) the energy flow γ can be obtained. As can be seen from fig. 13, γ differs considerably from the constant value obtained for the strip problem.

Closed form solutions can also be obtained for some transient problems with non-uniform crack velocity. Consider e.g. the following problem solved by Freund [34]. An unbounded body (fig. 14) contains a semi-infinite crack and the body is subjected to static loading remotely from the tip. Let the crack tip be at rest and placed at $x_1 = 0$ for $t = 0$. The remote loading gives the stresses $\sigma_{22} = p(x_1)$ for $x_2 = 0$ in front of the tip. If the crack tip now starts to move with an arbitrary velocity, the solution for the stress-intensity factor is

$$K_I = f_F(\dot{a} / c_2, \nu) \int_0^{a(t)} \frac{p(x_1)}{(a(t) - x_1)^{1/2}} \, dx_1 \quad . \tag{10.16}$$

The function f_F is depicted in fig. 12 and as can be seen it is fairly close to the function f_B.

The structure of (10.16) is remarkably simple. It consists of static part which in no way depends on the crack growth history and a function which depends only on the instantaneous velocity. The solution (10.16) can also be used for finite bodies during a limited time interval. The validity ceases when a wave emitted at the start of crack growth returns to the tip after having been reflected at the nearest boundary or the other crack tip.

It is important to realize that the static part of (10.16) is not in general equal to the quasi-static stress-intensity factor for the current crack length. This is due to the fact that in a finite body the stresses are redistributed during quasi-static crack growth. Eshelby [35], who previously to Freund obtained a mode III solution to the same problem, termed the integral in (10.16) as the reflection-less stress-intensity factor.

The energy flow γ for the present problem is shown in fig. 13. A reasonably good approximation is given by the simple relation

$$\gamma / \gamma_{stat} = 1 - \dot{a} / c_R \quad . \tag{10.17}$$

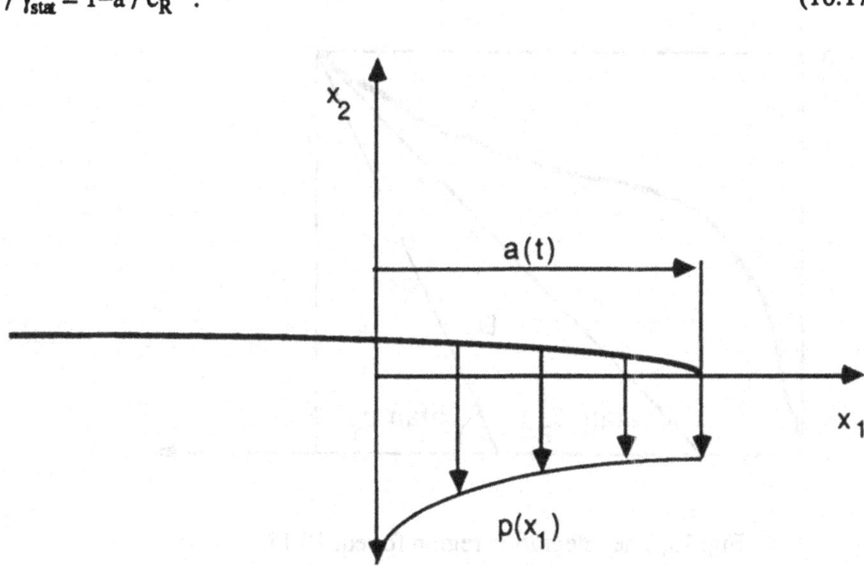

Fig. 14, Freund's problem.

Various extensions to this problem have been given by Freund [36] - [38] among others. These can be summarized in the solution given by Burridge [39] and Kostrov [40]. For arbitrary dynamic loading conditions in an infinite body with a semi-infinite crack in the same way as for the previous problem, the stress-intensity factor is given by

$$K_I(t) = f_F(\dot{a}/c_2, \nu) \iint_D p(x_1, t')\, g_B(a(t) - x_1, t - t')\, dx_1 dt' \quad , \tag{10.18}$$

$p(x_1, t)$ are the time dependent σ_{22}-stresses that would have resulted if the crack tip had remained stationary at $x_1 = 0$. g_B is a function given by Burridge [39] and D the region in the x_1-t space defined by

$$D: \quad \begin{array}{l} t' < t \\ a(t) - c_1 (t-t') < x_1 < a(t) - c_R (t-t') \, . \end{array} \tag{10.19}$$

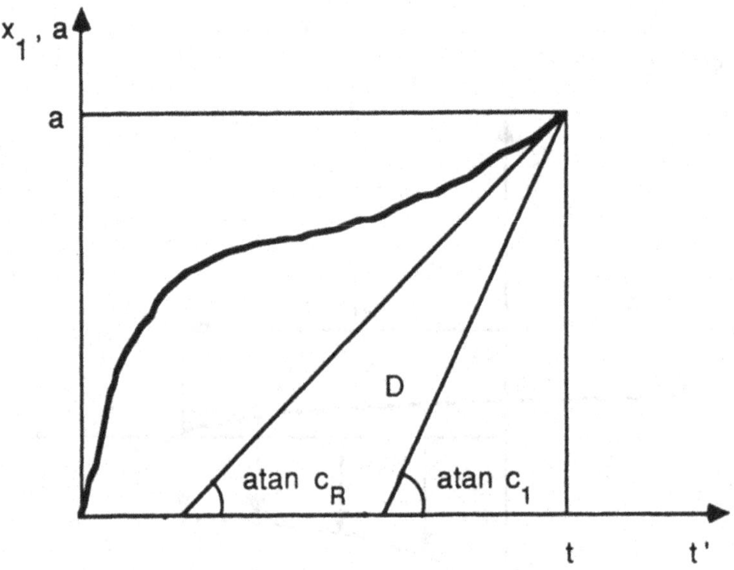

Fig. 15, The integration region for eq. 10.18.

The integration region D is sketched in fig. 15. The expression for g_B is rather complicated and is given in the form of a non-elementary integral in [39]. Its functional dependence together with the integration space show, however, that the integral in (10.18) does not depend on the crack growth history. As stated by Freund [37]: "The stress intensity factor for mode I extension of a half-plane crack is given by the universal function of crack-tip velocity, f_F, times the stress intensity factor appropriate for a crack of fixed length, equal to the instantaneous length, subjected to the applied loading, whether this loading is time-independent or time-dependent."

In applications to finite bodies, (10.18) is subject to the same restrictions as was described above. We can, however, obtain some interesting results. Suppose that a crack tip has been propagating in a finite body. Imagine now that the tip is stopped at some time instant t_0. Then under a time interval limited by the first reflection of wave emitted from the tip at t_0, the stresses in front of the tip can in principle be obtained from the instantaneous state of the body at t_0. Denote the end-point of this interval t_d. During $t_0 < t < t_d$ we can then use (10.18) since $p(x_1, t)$ is in principle determined. Assume now that instead of stopping at t_0 the crack tip continues to propagate. The crack tip motion during this time interval will not affect the integral in (10.18). The stress-intensity factor will thus not depend on events happened after t_0. This result enabled Nilsson [41] to obtain solutions for arbitrary velocity variations occurring after a state of steady or self-similar crack tip motion. A special result can be immediately deduced. Suppose that a crack tip abruptly changes its velocity from $\dot{a}^-$ to $\dot{a}^+$. Then the following relation holds for the stress intensity factor immediately before (K_I^-) and immediately after (K_I^+) the velocity change

$$K_I^+ / K_I^- = f_F(\dot{a}^+) / f_F(\dot{a}^-) \tag{10.20}$$

The relation (10.20) can e.g. be useful in investigating the effects that a discontinuity in fracture properties may induce.

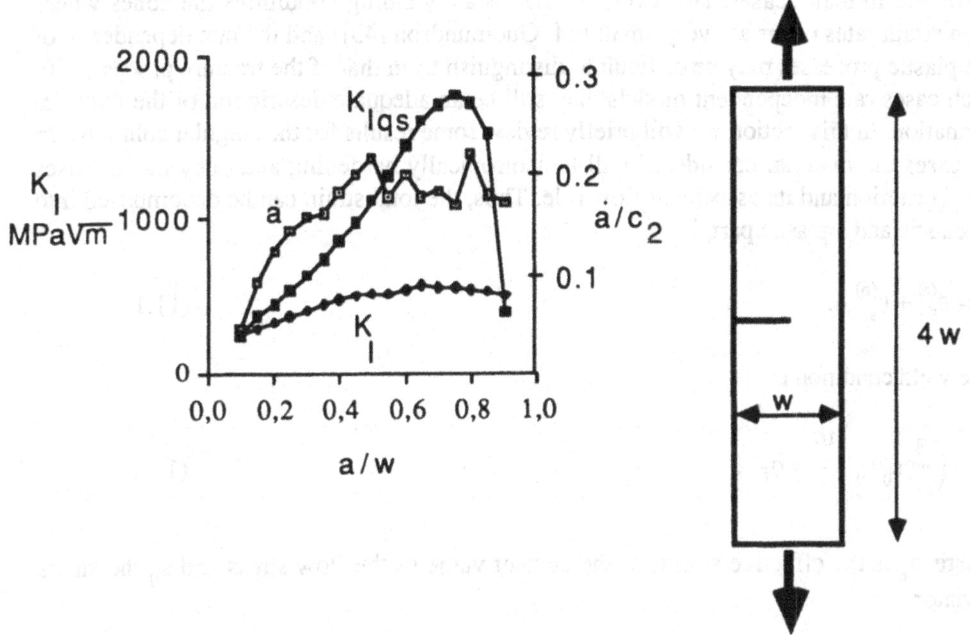

Fig. 16, Results from a FEM calculation [42]

The relation (10.18) is the most general analytic result that has been obtained for dynamic crack problems. For finite bodies it is in general not possible to use analytical methods to obtain the complete solution, instead numerical methods have to be employed.

In fig. 16 an example from a FEM calculation of a particular event is shown (Brickstad and Nilsson [42]). The difference between the static K_I and the dynamic one is much larger than one would expect from the nature of the previously discussed solutions for infinite bodies. It is typical that dynamic effects manifest themselves for much lower crack velocities in a finite body than in an infinite. Otherwise it is difficult to say anything in general for crack growth in finite bodies since the behaviour is complicated and very problem dependent on the particular boundary value problem. Some aspects on numerical methods are discussed below.

11.SINGULAR SOLUTIONS FOR RATE INDEPENDENT MATERIALS

In most metallic materials plastic deformation occurs around the tip at loads where fracture may happen. This plastic deformation may obviously affect the fracture behaviour considerably and consequently a great deal of attention has been devoted to these questions. In dynamic fracture high strain rates occur and rate independent plasticity models may not be sufficient in many cases. However, at small scale yielding conditions the zones where high strain rates occur are very small (c.f. Gudmundson [43]) and the rate dependence of the plastic processes may be difficult to distinguish from that of the fracture processes. In such cases rate independent models may still be an adequate description of the bulk deformation. In this section we will briefly review some results for the singular solutions. In all cases the material considered will be isotropically hardening and obey the v. Mises yield criterion and its associated flow rule. Thus, the total strain can be decomposed into an elastic and a plastic part, i.e.

$$\varepsilon_{ij} = \varepsilon_{ij}^{(e)} + \varepsilon_{ij}^{(p)} \quad . \tag{11.1}$$

The yield condition is

$$\sigma_e = \left(\frac{3}{2} s_{ij} s_{ij} \right)^{1/2} = \sigma_f \quad , \tag{11.2}$$

where σ_e is the effective stress, σ_f the current value of the flow stress and s_{ij} the stress deviator

$$s_{ij} = \sigma_{ij} - \frac{1}{3} \delta_{ij} \sigma_{kk} \quad . \tag{11.3}$$

In a perfectly plastic material, the flow stress is constant and equal to the initial stress σ_Y, while the flow rule can be written as

$$\dot{\varepsilon}_{ij}^{(p)} = \begin{cases} \dot{\lambda} s_{ij} & \sigma_e = \sigma_Y \ , \ \dot{\sigma}_e = 0 \\ 0 & \sigma_e < \sigma_Y \text{ or } \sigma_e = \sigma_Y \text{ and } \dot{\sigma}_e < 0 \end{cases} \tag{11.4}$$

Here $\dot{\lambda}$ is non-negative but otherwise undetermined by the constitutive assumption.

For a hardening material, the flow rule is

$$\dot{\varepsilon}_{ij}^{(p)} = \begin{cases} \frac{3}{2} \frac{\dot{\sigma}_e}{\sigma_e \sigma_f' (\varepsilon_e^p)} s_{ij} & \sigma_e = \sigma_f \text{ and } \dot{\sigma}_e > 0 \\ 0 & \sigma_e < \sigma_f \text{ or } \sigma_e = \sigma_f \text{ and } \dot{\sigma}_e \leq 0 \end{cases} \tag{11.5}$$

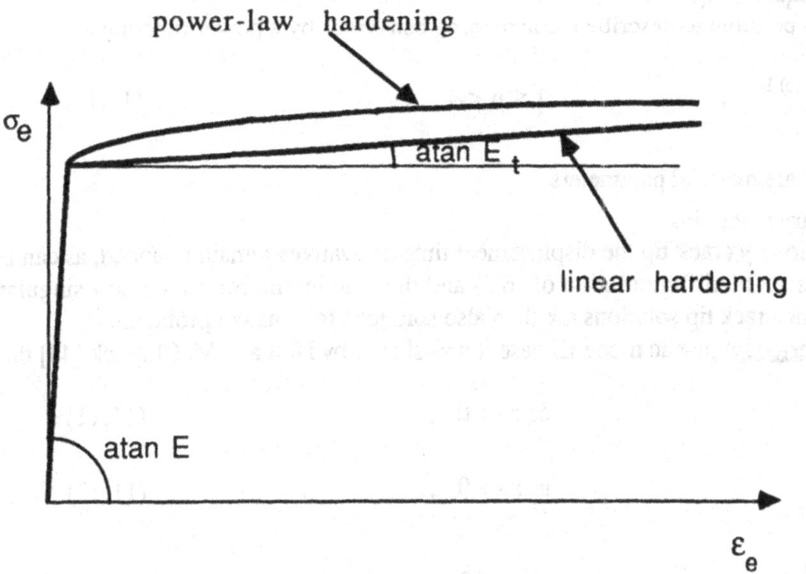

Fig. 17, Different strain hardening assumptions.

Here σ_f is assumed to be a function of the effective plastic strain ε_e^p only, where

$$\varepsilon_e^{(p)} = \int_0^t \left(\frac{2}{3} \dot{\varepsilon}_{ij}^{(p)} \dot{\varepsilon}_{ij}^{(p)} \right)^{1/2} dt \ . \tag{11.6}$$

For a linearly strain-hardening material with the tangent modulus E_t defined as in fig. 17 we have

$$\sigma_f = \sigma_Y + E_t (\varepsilon_e - \sigma_Y / E) \tag{11.7}$$

or

$$\sigma_f \left(1 - \frac{E_t}{E} \right) = \sigma_Y + E_t (\varepsilon_e^{(p)} - \sigma_Y / E) \tag{11.8}$$

and thus

$$\sigma_f' (\varepsilon_e^{(p)}) = \frac{E_t}{1 - E_t / E} \ . \tag{11.9}$$

For typical metallic materials the ratio E_t / E is almost always smaller than 0.01 so that σ_f' is practically equal to E_t.

It is often possible to describe the hardening behaviour by a power function e.g.

$$\sigma_f = \sigma_Y + \sigma_0 \varepsilon_e^{(p)\,1/n} \ , \qquad 1 \le n < \infty \ . \tag{11.10}$$

Here σ_0 and n are material parameters.

11.1. Stationary crack tips

For a stationary crack tip the displacement time derivatives remain bounded, as can be immediately seen by differentiation of (6.5) and thus the inertia forces are non-singular. The quasi-static crack tip solutions are then also solutions to dynamic problems.

For the perfectly plastic mode III case it was shown by Hult and McClintock [44] that

$$\sigma_{3\varphi} = \tau_Y \ , \qquad \text{as } r \to 0 \ , \tag{11.11}$$

$$\sigma_{3r} = 0 \ , \qquad \text{as } r \to 0 \ , \tag{11.12}$$

$$\varepsilon_{3\varphi}^{(p)} = Q_\varepsilon \frac{g(\varphi)}{r} \qquad \text{as } r \to 0 \ , \tag{11.13}$$

$$\varepsilon_{3r}^{(p)} = 0 \ , \qquad \text{as } r \to 0 \ , \qquad (11.14)$$

where τ_Y is the yield stress in pure shear and $g(\varphi)$ an function undetermined by the asymptotic solution . The particular form of g will be determined by the solution to the entire problem. Thus, in this case a unique problem independent singularity description is not possible. In the quasi-static case, $g(\varphi) = R(\varphi)$, where $R(\varphi)$ is the distance to the elastic plastic boundary. This may not be the case in a transient problem.

The mode I solution to the corresponding problem has a similar structure (Rice [45])

$$\sigma_{ij} = \sigma_Y \Sigma_{ij}^{pp}(\varphi) \qquad \text{as } r \to 0 \ , \qquad (11.15)$$

$$\varepsilon_{r\varphi}^{(p)} = Q_\varepsilon \frac{g(\varphi)}{r} \qquad \text{as } r \to 0 \ . \qquad (11.16)$$

The remaining plastic strain components are zero. Again $g(\varphi)$ is undetermined by the asymptotic solution while Σ_{ij}^{pp} is given by the Prandtl indentation plastic slip line field.

For a linearly hardening material the crack tip solution is given by the elastic solutions (8.32-8.37) where the Young's modulus E shall be replaced by $(EE_t)^{1/2}$, ν put to 1/2 and $\dot{a} = 0$.

For a power-law hardening material, the asymptotic solutions were given by Hutchinson [46], Rice and Rosengren [47]. These solutions, for mode I, are of the form

$$\sigma_{ij} \to Q_\sigma r^{-\frac{1}{1+n}} \Sigma_{ij}^{HRR}(\varphi, n) \ , \qquad (11.17)$$

$$\varepsilon_{ij} \to Q_\varepsilon r^{-\frac{n}{1+n}} E_{ij}^{HRR}(\varphi, n) \ . \qquad (11.18)$$

The functions Σ_{ij}^{HRR} and E_{ij}^{HRR} are problem-independent and can be found in [46]-[47]. From the constitutive relation (11.10) it is easily found that if Σ_{ij} and E_{ij} are suitably scaled,

$$Q_\sigma = \sigma_0 Q_\varepsilon^{1/n} \ . \qquad (11.19)$$

11.2. Moving crack tips

The solutions for the moving crack tip are considerably more complicated than for the stationary tip. Furthermore the time derivatives of the displacements are now singular as discussed earlier (eq. (3.17) - (3.18)) and different solutions will be obtained for the case when inertia effects are considered in comparison with the quasistatic case. Only steady-state solutions will be considered here although there is a possibility that the result of asymptotically steady-state motion may not be applicable.

For the mode III perfectly plastic quasi-static case, Rice [45] showed that the near tip stress distribution is the same as for the stationary tip (eq. (11.15)). The strain component ε_{31} is given by

$$\varepsilon_{31} = -\frac{\tau_Y}{2G} \sin\varphi \left[1 + \ln \frac{R(\varphi)}{r} \right] \quad . \tag{11.20}$$

The strain component ε_{32} has a complicated distribution, but along the crack line we have

$$\varepsilon_{32}(x_1, 0) = \frac{\tau_Y}{2G} \left\{ 1 + \ln\left(\frac{R(0)}{x_1}\right) + \frac{1}{2}\left[\left(\ln\left(\frac{R(0)}{x_1}\right)\right]^2\right\} \quad . \tag{11.21}$$

If inertia effects are considered the solution is much more complicated, but the asymptotic result for the strain ε_{32} in front of the crack tip was given by Slepyan 48] as

$$\varepsilon_{32}(x_1, 0) \to Q_\varepsilon \left[(1-\beta_2^2)^{-1/2} - 1\right] \ln \frac{R_0}{x_1} \quad \text{as } r \to 0 \quad . \tag{11.22}$$

Here R_0 is a length parameter. We note immediately that (11.22) is different from (11.21) and does not reduce to the quasi-static form as $\dot{\alpha} \to 0$ ($\beta_2 \to 1$). An exact solution for $\varepsilon_{32}(x_1, 0)$ was derived by Freund and Douglas [49]. They showed for any given x_1, the exact solution tends to (11.21) as $\dot{a} \to 0$, while the behaviour (11.22) is approached for arbitrarily small values of x_1 for any $\dot{a} \neq 0$. Thus the region of validity of the dynamic solution (11.22) vanishes as $\dot{a} \to 0$. Freund and Douglas [49] estimate that for $\dot{a}/c_2 < 0.3$, the range of validity is typically smaller than $R_p/100$, where R_p is a typical dimension of the plastic region.

The corresponding problems for mode I are more involved. Drugan, Rice and Sham [50] have given the solution for the quasi-static problem. The stresses are bounded and strain varies at most as ln (r). As an example of these results, we cite the expression for the crack surface displacement.

$$u_2 \to \xi \frac{\sigma_Y}{E} r \ln\left(\frac{r_t}{r}\right) \quad , \quad \varphi = \pm\pi, r \to 0 \quad , \tag{11.23}$$

where

$$r_t = R_0 \exp\left(1+\frac{\alpha E}{\xi\sigma_Y^2}\frac{dL}{da}\right) \quad . \tag{11.24}$$

R_0 is a length parameter which is undetermined by the asymptotic analysis, as is the dimensionless number α. ξ is a problem-dependent constant depending on Poissons's ratio ($\xi \approx 5{,}46$ for $\nu = 0{,}3$). L is some suitable measure of the outer load. The parameters R_0 and ξ will be dependent on the particular choice of L. In the small-scale yielding case L can be chosen as the value of J taken around a path in the elastic region.

For the <u>dynamic</u> case progress has been made for the incompressible problem ($\nu = 0{,}5$) . Leighton, Champion and Freund [51] show that stresses as well as strains are bounded in the open interval $0 < \dot{a} < c_2$. They conclude a similar behaviour as for the mode III case i.e. the region of validity of the dynamic solution vanishes as $\dot{a} \to 0$.

For <u>linearly hardening</u> materials asymptotic solutions of the following type can be derived for the case when inertia effects are considered (Achenbach et al [52], Östlund and Gudmundson [53]).

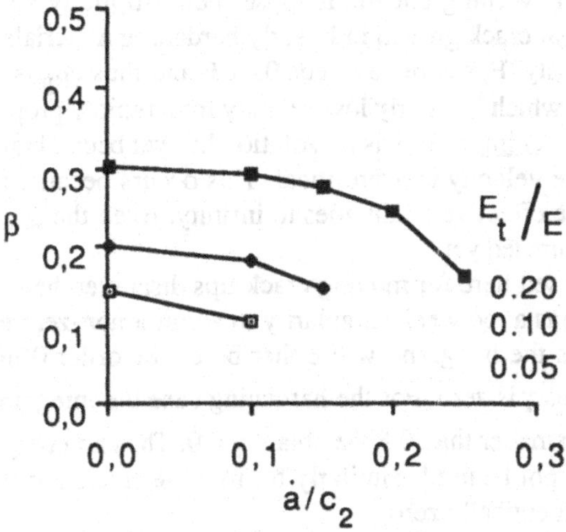

Fig. 18, Exponent for the singularity in a linearly hardening material (after [54]).

$$\sigma_{ij} \to Q_\sigma r^{-\beta}\Sigma_{ij}^{lh}(\varphi, \dot{a}/c_2) \qquad \text{as } r \to 0 \quad , \tag{11.25}$$

$$\varepsilon_{ij} \to Q_\sigma r^{-\beta} E_{ij}^{lh}(\varphi, \dot{a}/c_2) \qquad \text{as } r \to 0 \ , \tag{11.26}$$

$$\dot{u}_i \to Q_\sigma \dot{a} r^{-\beta} u_i^{lh}(\varphi, \dot{a}/c_2) \qquad \text{as } r \to 0 \ . \tag{11.27}$$

In [52] the problem was considered without regard to the possibility of reversed plastic deformation, while this was considered in [53]. The results are, however, not much affected by reversed plasticity. The exponent β depends on the ratio E_t/E, ν and $\dot{a}/c_2$ as shown for the mode I plane strain case in fig. 18. It is noted that β is always smaller than 0.5. The functions Σ_{ij}^{lh} etc. are problem independent and are given in the cited references.

The results reduce to the quasi-static solutions for $\dot{a} \to 0$ so the peculiarities that occurred in the perfectly plastic case do not enter here. It should however be pointed out that it was not possible to find solutions of the presented type for crack speeds exceeding c_{Rp}, where c_{Rp} is the pseudo Rayleigh velocity i.e.

$$c_{Rp} = 0.96\left(\frac{E_t}{3\sqrt{3}\,\rho}\right)^{1/2} \tag{11.28}$$

Above this limit the governing equations loose their ellipticity and no solution to the problem of supercritical crack growth in linearly hardening materials has been found.yet. As mentioned previously, E_t seldom exceeds 0.01 E and thus c_{Rp} is usually smaller than about 220 m/s in steel which is a fairly low velocity for a rapidly propagating crack.

For power law hardening materials no solution has yet been obtained for the dynamic case, since every finite velocity is overcritical. This occurs because the tangent modulus approaches zero as the effective strain goes to infinity. Even the quasi-static case seems not to have been fully treated yet.

In all cases discussed here for moving crack tips discussed here the integrand of the energy flow integral has a too weak singularity to admit a non-zero energy flow. For the perfectly plastic cases the integrand will either be of the order 0(ln r) or 0(1) (mode I dynamic case) and thus γ is zero. For the hardening case the integrand is of order $0(r^{-2\beta})$ and since β always is smaller than 0.5 we obtain $\gamma = 0$. Thus an energy criterion for crack growth can obviously not be used. Similarly for moving crack tips the integral I defined by eqn. (5.5), will be identically zero.

To what extent we can use the integral J, which is clearly not path-independent for moving tips, will be discussed in later section.

For stationary tips the singularity of the integrand of I is $0(r^{-1})$ as can be seen both from (11.15) - (11.16) and from (11.17) - (11.18). Thus a finite value for I will result. We shall investigate if asymptotic path-independence of the J-integral holds i.e. whether the condition (5.12) is satisfied. For all the considered cases, stresses can asymptotically be written as $Q_\sigma \sigma_{ij}^0$ where σ_{ij}^0 does not depend on time. Likewise for the strains we have $Q_\varepsilon \varepsilon_{ij}^0$. For large strains, if the elastic ones are neglected, we have

$$\sigma_f = \sigma_e = \sigma_0 \, \varepsilon_e^{1/n} \quad , \tag{11.29}$$

where, due to the condition of proportional loading, the effective strain is

$$\varepsilon_e = \left(\frac{2}{3} \varepsilon_{ij} \varepsilon_{ij} \right)^{1/2} \quad . \tag{11.30}$$

The constitutive relation (11.5) after insertion of (11.29), time integration and the assumptions of proportional loading can be written

$$\varepsilon_{ij}^{(p)} = \frac{3}{2} \frac{s_{ij}}{\sigma_0 \, \varepsilon_e^{1/n-1}} = \frac{3}{2} \frac{s_{ij}}{\sigma_0^n \, \sigma_e^{1-n}} \quad . \tag{11.31}$$

The plastic deformation work is given by

$$w = \int \sigma_e \dot{\varepsilon}_e \, dt = \frac{1}{1/n+1} \sigma_0 \, \varepsilon_e^{1/n+1} \quad . \tag{11.32}$$

Use of the chain rule for differentiation gives

$$w_{,k} = \frac{\partial w}{\partial \varepsilon_{ij}} \varepsilon_{ij,k} = \frac{\partial w}{\partial \varepsilon_e} \frac{\partial \varepsilon_e}{\partial \varepsilon_{ij}} \varepsilon_{ij,k} \quad . \tag{11.33}$$

Differentiation of (11.30) and use of (11.31) leads then to

$$w_{,k} = s_{ij} \varepsilon_{ij,k} = \left(\sigma_{ij} + \frac{1}{3} \delta_{ij} \sigma_{mm} \right) \varepsilon_{ij,k} \quad . \tag{11.34}$$

Since the plastic deformation is incompressible, (11.34) reduces to the condition (5.12) and thus the J-integral is asymptotically path-independent. The area integral of eq. (5.5) is then not divergent and we can conclude that the I-integral is well-behaved.

A common problem with most of the singular solutions given here, is that their range of validity is often extremely limited as discussed by Nilsson and Ståhle [54] and Gudmundson [43]. Thus their relevance to fracture criteria can be limited since at least for small yielding situations the extent of the singular region is much smaller than the fracture process zone.

12.SINGULAR SOLUTIONS FOR STRAIN-RATE DEPENDENT MATERIALS

High strain rates occur in dynamically loaded cracked bodies at least in the tip vicinity. It is therefore important to investigate the effects of strain rates on the crack tip fields.

The effect of high strain rates on plastic deformation is complicated and far from fully understood, especially at the extremely high rates that occur in the crack tip vicinity. Many different types of constitutive relations have been suggested, some of which are very complicated. Here, only one particular constitutive relation is to be discussed. It is a special case of a more general theory suggested by Perzyna [55]. As in the rate independent case we assume elastic and viscoplastic strains to be additive. The total strain rate can then be written

$$\dot{\varepsilon}_{ij} = \dot{\varepsilon}_{ij}^{(e)} + \dot{\varepsilon}_{ij}^{(vp)} \quad , \tag{12.1}$$

$$\dot{\varepsilon}_{ij}^{(vp)} = \lambda \left[H \left(\frac{\sigma_e}{\sigma_f(\varepsilon_e^p)} - 1 \right) \right]^m \frac{s_{ij}}{\sigma_e} \quad . \tag{12.2}$$

Here, H denotes the Heaviside step function, λ is a fluidity material parameter and m also a material parameter. Plastic deformation is, according to this model, possible only if the effective stress σ_e exceeds the current flow stress σ_f which assumed to be a function of effective plastic strain as in the rate-independent case. Indeed as $\lambda \to \infty$ this model approaches the previously discussed strain-rate independent model.

12.1 Stationary crack tip

We shall here, as before, assume a hardening behaviour at large strains of the power-law type, thus including linear hardening (n = 1) and perfectly plastic ($n = \infty$) behaviour as special cases. Assume that the asymptotic solutions can be written as

$$\sigma_{ij} \to Q_\sigma \Sigma_{ij}(\varphi) r^{-\alpha} \qquad \text{as } r \to 0 \ , \ 0 < \alpha < 1 \ , \tag{12.3}$$

$$\varepsilon_{ij} \to Q_\varepsilon E_{ij}(\varphi) r^{-\beta} \qquad \text{as } r \to 0 \ , \ 0 < \beta < 1 \ . \tag{12.4}$$

From (12.3) it is readily seen that the displacements are of order $O(r^{1-\beta})$. Time differentiation then yields that accelerations are $O(r^{1-\beta})$ and thus bounded.

Insertion of (12.3) and (12.4) into (12.1) and (12.2) gives the strain rates i.e.

$$\dot{\varepsilon}_{ij}^{(e)} = \frac{1}{2G} \dot{Q}_\sigma \left(\Sigma_{ij} - \frac{\nu}{1+\nu} \Sigma_{kk} \delta_{ij} \right) r^{-\alpha} \quad , \tag{12.5}$$

$$\dot{\varepsilon}_{ij}{}^{(vp)} = \lambda \sigma_0^{-m} Q_\sigma^m Q_\varepsilon^{-m/n} r^{-m(\alpha-\beta/n)} \Sigma_e^{m-1} S_{ij} E_{ij}^{-m/n} \quad . \tag{12.6}$$

Σ_e and S_{ij} denote the angle distributions of σ_e and s_{ij} respectively. From (12.5) and (12.6) it is directly obtained that the elastic strain rates will dominate for any finite $\dot{Q}_\sigma$, provided that

$$m < \frac{1}{1-\beta/\alpha n} \quad . \tag{12.7}$$

We thus conclude that an elastic singularity will prevail and $\alpha = \beta = 1/2$, if

$$m < \frac{1}{1-1/n} \quad . \tag{12.8}$$

Since the accelerations are bounded, this singularity will have the same angle distribution as the quasi-static one. The result (12.8) was first obtained for the perfectly plastic case by Hui and Riedel [56].

For m values larger than what is given by (12.8), it is in general fairly complicated to obtain a solution. For the perfectly plastic case, (12.6) is simplified into

$$\dot{\varepsilon}_{ij}{}^{(vp)} = \lambda \sigma_Y^{-m} Q_\sigma^m r^{-m\alpha} \Sigma_e^{m-1} S_{ij} \quad . \tag{12.9}$$

Obviously the visco-plastic strain rates will dominate over the elastic ones if $m > 1$. Comparing with eq. (11.31), it is noted that the right members are completely analogous apart from a multiplicative constant. Solution of the problem would thus yield a strain-rate field of the same form as the strain field resulting from (11.31) with n substituted by m. However, the only time-dependent part of the strain rate field is the amplitude parameter $\dot{Q}_\varepsilon$. A time integration then yields that the stress and strain-fields are identical to that of a power-law hardening material. (eq. (11.17) - (11.18))

$$\sigma_{ij} \to Q_\sigma r^{-\frac{1}{1+m}} \Sigma_{ij}^{HRR}(\varphi, m) \quad , \tag{12.10}$$

$$\varepsilon_{ij} \to Q_\varepsilon r^{-\frac{m}{1+m}} E_{ij}^{HRR}(\varphi, m) \quad . \tag{12.11}$$

From the constitutive equation (12.2), we have that

$$\dot{Q}_\varepsilon = (\lambda / \sigma_Y)\, Q_\sigma^m \quad . \tag{12.12}$$

Insertion of the fields (12.10) - (12.11) into the J-integral (5.22) yields an expression of the form

$$J = (\lambda / \sigma_Y)\, Q_\sigma\, Q_\varepsilon\, \eta\, (m) \quad . \tag{12.13}$$

Here, the dimensionless constant η depends on m. The absolute numerical value will of course depend on how the functions Σ_{ij}, E_{ij} are normalized.

Insertion of (12.12) into (12.13) gives

$$J = \left(\frac{\lambda}{\sigma_Y}\right)^{1-1/m} \dot{Q}_\varepsilon^{\ 1/m}\, Q_\varepsilon\, \eta \quad . \tag{12.14}$$

This differential equation is easily solved to yield

$$Q_\varepsilon = \left[(m+1)\eta^{-m} \left(\frac{\sigma_Y}{\lambda}\right)^{m-1} \int_0^t J^m\, dt \right]^{1/(m+1)} \quad . \tag{12.15}$$

Thus knowing J(t), Q_ε can be obtained by numerical integration and also can Q_σ from (12.13). In author's experience, it is however difficult to evaluate J from a finite element calculation, since the region where the singularity prevails is very small and this results in path-dependence of J. A possible remedy is to use the I-integral which is strictly independent of the region of integration. No results of such calculations for this material model have however appeared in the literature, to the knowledge of the author.

The remaining cases i.e. $n < \infty$ together with $m > n/(n-1)$ and $m = n/(n-1)$ are so far not resolved.It should also be pointed out here that other constitutive assumptions, such as e.g. the Bodner-Partoum model, will give different results for the singularity. For the Bodner-Partoum model an elastic singularity will always prevail.

12.2 Moving crack tips

We shall perform a similar analysis for the moving tip as it was done for the stationary tip. The main difference is that time differentiation now has to be performed according to (3.16) and (3.18). Thus, accelerations will not be bounded but of the order $0(r^{\beta-1})$. The elastic strain rate now becomes

$$\dot{\varepsilon}_{ij}^{(e)} = -\dot{a}\, \varepsilon_{ij,1}^{(e)} = -\frac{1}{2G}\, \dot{a}\, Q_\sigma \left(\Sigma_{ij}' - \frac{\nu}{1+\nu}\, \Sigma_{kk}'\, \delta_{ij}\right) r^{-\alpha-1} \quad , \tag{12.16}$$

where Σ_{ij}' collect the differentiated forms of Σ_{ij}.

Since the viscoplastic strain rate will retain the form (12.6), it is clear from the previous reasoning that an elastic singularity will prevail, if

$$m < \frac{3}{1-1/n} \quad . \tag{12.17}$$

Since accelerations now have to be considered this singularity will be one given for a moving tip ((8.32)-(8.37)).

For m-values not satisfying the condition (12.17), the solution is more complicated. Lo [57] showed that in this case the visco-plastic and the elastic strain-rates will be of the same order and solved the singular problem. A particular feature of this solution is that it does not contain any arbitrary constant and thus is completely autonomous. Consequently, this singular solution is of no use for a fracture criterion.

Recently Freund and Yang [58] have reconsidered this problem with a modified constitutive titutive model and for their particular behaviour the singular solution permits an arbitrary constant. It is, however, felt by the present author that the uncertainties of the effects of very high strains are so large that any more far-reaching conclusions can not be made at this stage.

13. CRACK GROWTH CRITERIA

13.1. Initiation of crack growth

In materials like metals where the response to small strain deformation is primarily elastic, it is reasonable as a first approximation to assume the body to be entirely linearly elastic. As has been seen above, the field at the crack tip is then characterized by stress-intensity factor K_I and followingly the criterion for crack growth initiation is based on this parameter. Thus, in accordance with the discussion in section 6, crack growth initiation is assumed to occur if

$$K_I(t) \geq K_{Id}\left(\frac{\partial^{k_1} K_I}{\partial t^{k_2}}\right)\Big|_t \quad , \quad k_1 = 1,2,3..... \quad . \tag{13.1}$$

According to the discussion in section 6, the crack tip state is assumed to be expressible in terms of the function K_{Id} of the time-derivatives of K_I. When this state reaches a critical level the tip starts to move. The functional character of K_{Id} can not be determined without considering the non-elastically deforming zone behaviour. Heuristically one can however argue that the lowest time derivatives are the most important for characterizing the momentary state. In practice only the first derivative is retained so that the growth condition reduces to

$$K_I(t) \geq K_{Id}(\dot{K}_I) \quad \text{for growth} \quad . \tag{13.2}$$

The function K_{Id} is termed the <u>dynamic fracture toughness</u> and is assumed to be a material function. By careful experimentation it should be possible to determine K_{Id} and by studying different geometries it should be possible to verify or falsify (13.2). Unfortunately, well-controlled dynamic fracture experiments are difficult to perform and there is in fact very little solid experimental verification of (13.2), as was discussed by Nilsson [59]. The most difficult part of a dynamic fracture test is to determine when crack growth initiation occurs. To the author's knowledge, no really accurate method has yet been developed.

As the load to fracture increases, the criterion (13.2) will sooner or later anyway loose its validity because the non-linear zone becomes too large. When the applicability of (13.2) is lost is not well-known, but it can be expected that a relation of the following type may give the applicability region of (13.2)

$$l_{min} > \omega \left(\frac{K_I}{\sigma_Y} \right)^2 , \tag{13.3}$$

l_{min} is here the smallest characteristic dimension of the structure and ω a dimensionless number. For quasi-static conditions this is the well-known ASTM-condition and in this case $\omega = 2.5$. Its value in dynamic situations is not known.

Outside the applicability region of the linear theory fairly little is known regarding dynamic fracture. Elasto-plastic, preferably strain rate dependent constitutive, laws have to be used in modelling the body. According to the previous discussion we found that the I-integral was not dependent on the integration region irrespective of the material behaviour. It was furthermore noted that for a stationary tip, the I-integral will be finite and non-zero for the material models discussed here. It thus appears reasonable to use the following criterion, in analogy with (13.2),

$$I(t) \geq I_{cd}(\dot{I}) \ . \tag{13.4}$$

If, as may often happen, the state of stress is near that of proportional loading, the I-integral reduces to the J-integral. To be on the safe side, it appears to be advantageous to evaluate I in order to meet situations when the fields do not meet the condition of proportional loading, which is likely under transient loading. The experimental verification of the I-(or J-) criterion for dynamic situations is at the same level as for the linear criterion i.e. no systematic studies have been performed. If one wants to predict the functional dependence of say K_{Id} on $\dot{K}_I$, modelling of the non-linearly deforming region is necessary. Assume e.g. a small-scale yielding situation (fig. 19), where the outer boundary conditions are given by

$$\sigma_{ij} \rightarrow K_I(t)\,(2\pi r)^{-1/2}\,\Sigma^e_{ij}(\varphi) \quad \text{as } r \rightarrow \infty \tag{13.5}$$

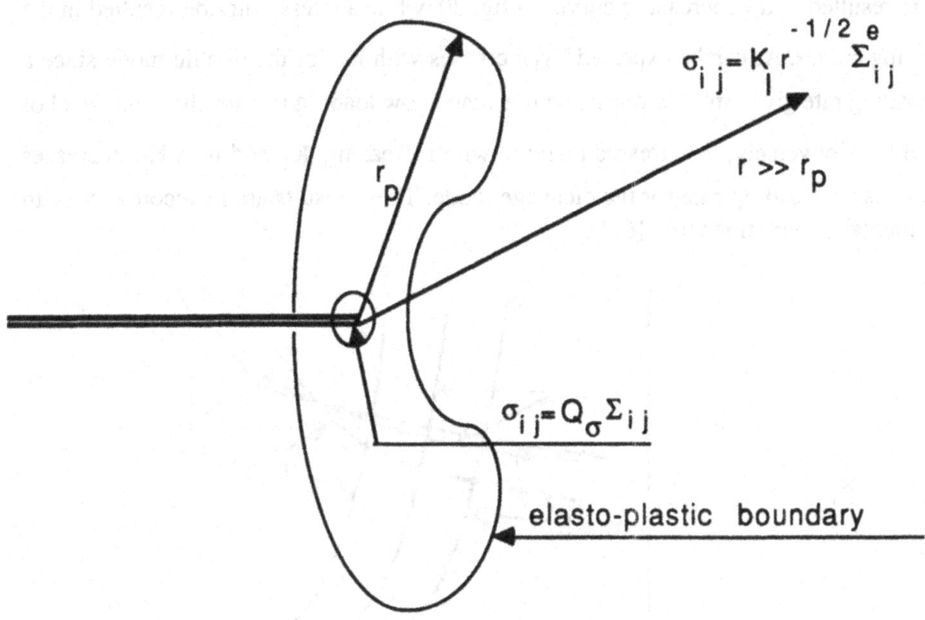

Fig. 19, Illustration of a small-scale yielding situation.

Let us adopt a strain rate dependent plasticity theory of the form discussed in section 12 with an exponent n > 1. The singular stress field will then be characterized by Q_σ and the singular strain field by Q_ε. From material models of fracture it appears that the cleavage mode of failure is chiefly governed by the stress state. It is thus suggested that the following criterion may be suitable

$$Q_\sigma (t) \geq Q_{\sigma c} \ , \quad \text{for cleavage fracture.} \qquad (13.6)$$

Here, $Q_{\sigma c}$ is a critical material property. Similarly, it is often assumed that ductile fracture is governed by the strain state and analogously we assume the following criterion,

$$Q_\varepsilon(t) \geq Q_{\varepsilon c} \ , \quad \text{for ductile fracture,} \qquad (13.7)$$

where $Q_{\varepsilon c}$ is a material property.

Fracture in a certain situation is governed by the criterion which is first satisfied by (13.6) or (13.7). The stated problem has not yet been solved, but Nilsson et al. [60] have analysed a strip yield approximation of it. The results for K_{Id} are summarized in fig. 20. K_{Ic} here denotes the quasi-static fracture toughness and $\dot{K}_{ref}$ is a reference value of the stress-intensity factor rate containing material parameters. A displacement criterion for fracture resulted in the increasing curves in fig. 20, while a stress criterion resulted in the decreasing curves. As can be expected K_{Id} increases with $\dot{K}_I$ for the ductile mode since a high loading rate gives smaller plastic strains than a low loading rate for the same level of applied K_I. Conversely, the stresses increase with increasing $\dot{K}_I$ and thus K_{Id} decreases with increasing loading rate for the cleavage mode. These results are in accordance with experimental observations (c.f. [61]).

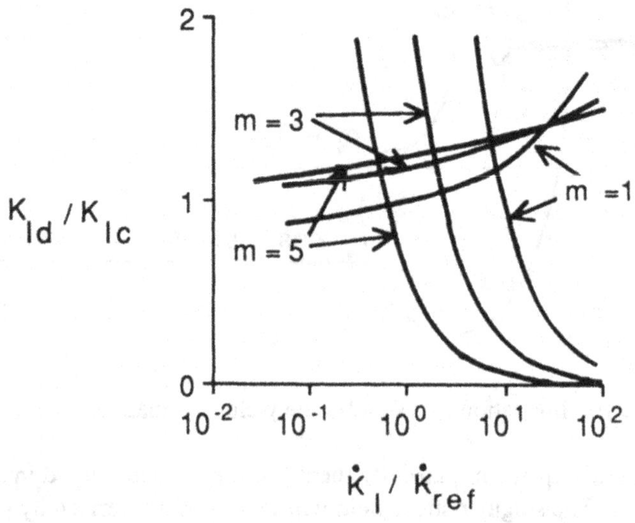

Fig. 20, The rate dependence of K_{Id} as predicted from a strip yield model.

13.2 Crack growth and arrest

The crack growth succeeding initiation in a transient situation is often relatively fast. In some cases it may happen that the fast crack propagation is preceded by a stage of relatively slow growth. In a quasi-static process this phase is often termed stable crack growth. In a fully dynamic treatment the distinction between stable and unstable crack growth is diffuse. Guided by the discussion in section 6, we may assume a relation governing the crack growth of the form

$$Q(t) = Q_c(a-a_0, \dot{a}, \dot{Q}) \quad . \tag{13.8}$$

Here, Q is the appropriate singularity parameter. As was discussed above for initiation there may well be competing mechanisms for growth so that several equations of the form (13.8) have to be considered that contain different singularity parameters e.g. Q_σ and Q_ε. The mechanism, which is currently the most easy to satisfy, is then assumed to occur. The appearance of $a-a_0$ in the argument of the Q_c function makes it possible to describe slow (stable in a quasi-static situation) crack growth. As $a-a_0$ gets larger the influence of this argument ought to diminish.

As far as the present author knows, no one has tried to investigate a growth criterion of the form (13.8) in its full generality. Special cases have, however, been studied. If $\dot{a}$ is negligible, the above equation represents a resistance curve with a possible dependence on loading rate. If, on the other hand, the dependence of $a-a_0$ and $\dot{Q}$ is negligible, (13.8) is of the form frequently used in analyses of dynamic crack propagation.

For situations when the non-linearly deforming zone is small, the stress-intensity factor K_I can be used as the characterizing parameter and this is in fact done in most experimental investigations. As for the crack growth initiation the validity of the K_I-approach is limited. For those cases when the linear theory is applicable, the crack growth criterion reduces to

$$K_I(t) = K_{pc}(\dot{a}) \quad . \tag{13.9}$$

In (13.9) any influence of $a-a_0$ or $\dot{Q}$ on K_{pc} has been dropped. Experience from quasi-static fracture mechanics suggests that it is unlikely that the stress-intensity factor approach is applicable for the stage when the fracture resistance depends on the amount of crack growth. Furthermore, experimental results suggest that the effect of crack velocity is much more pronounced than the effect of $\dot{K}_I$.

The region of validity of the K_I-criterion for crack growth is not known. Few systematic studies have been devoted to this problem. Dahlberg et al. [30] have found indications that the conditions for K_I-validity may be more restrictive than in the static case.

In the non-linear range obvious candidates for Q are the singularity parameters for the chosen constitutive model. As mentioned previously it may however be that the zone where the singular fields dominate is much smaller than the process zone and thus no autonomy with respect to these parameters exists. How to resolve this problem has not been very well investigated. One alternative may be to model the fracture process region in some simple way e.g. by a cohesive zone. Another alternative may be to use higher order terms of a series expansion of the crack tip fields. A more practical problem with the singular parameters for a non-linear material is the difficulty involved in extracting these from a numerical analysis. The I-integral cannot be used for many materials since the fields have a singularity strength that results in I being identically zero, as discussed previously.

The J-integral is thus <u>not</u> a measure of the singularity strength in these cases, but rather a measure of the outer field outside that plastically deforming region. Whether J is an appropriate measure or not for dynamic cases has not been investigated.

The non-linear material model that has recently been most popular for dynamic crack growth description is the visco-plastic model either of the Perzyna type with $n < 3$ or the Bodner-Partoum type. Both these models have the advantage that the singularity is of the elastic type, as was shown in section 12. Thus, K_I for the inner singularity can be used in the crack growth law. Brickstad [62] analysed a series of experiments performed on different geometries of the same material using the Perzyna model. The values of γ (or K_I) calculated from the experimental results fell nicely on a common line, while a purely linear analysis resulted in a wide scatter.

Analytical studies of the functional form of K_{pc} in (13.9) have been performed for perfectly plastic rate independent and rate dependent materials by Freund and Douglas [49], [63], Freund et al. [64] and Mataga et al. [65]. These authors assume that ductile fracture is governed by a criterion based on a fixed critical plastic strain at a fixed distance in front of the tip. For the rate-independent and for the rate-dependent materials, this model results in a K_{pc} that increases monotonously with increasing velocity. For the cleavage mode of crack growth, it is assumed that the growth is governed by a critical value of the inner K_I. This results in a $K_{pc}(\dot{a})$ which initially decreases and after passing through a minimum increases for high velocities. Obviously, steady crack growth with K_{pc} decreasing with $\dot{a}$ can never occur in a experiment since it is an unstable situation.

14. SOME ASPECTS ON NUMERICAL METHODS

Most problems in dynamic fracture mechanics require numerical modelling. By far the most widely used is the finite element method (FEM). These are today available for a wide variety of material models and with a range of element types.

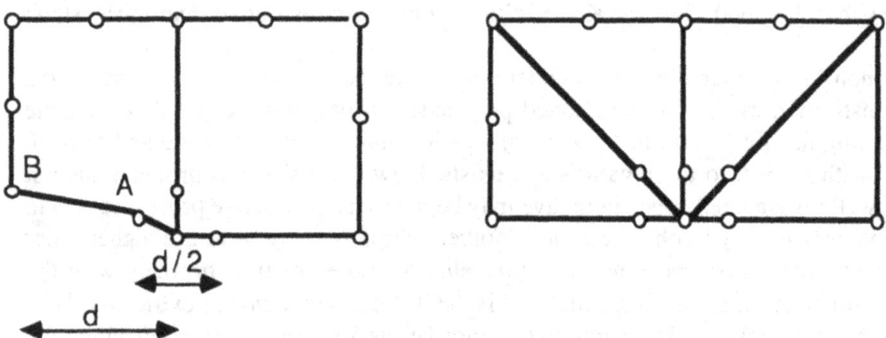

Fig. 21, Simulation of a $r^{-1/2}$ singularity with isoparametric elements.

The particular problem in performing FEM-analysis is the representation of the crack tip. For stationary crack tips this is a fairly straight-forward problem. With the quadratic isoparametric elements that are now common in most commercial FEM-programs singularities of the type $r^{-1/2}$ and r^{-1} can easily be introduced. Consider (fig. 21) either an eight-noded quadrilateral or a six-noded triangular element. By moving the midside nodes on the edges of the elements surrounding the tip, to the quarter point position indicated in fig. 21, it has been shown ([66], [67]) that a strain singularity of the form $r^{-1/2}$ is created. Some initial doubts of the applicability of this transformation for the quadrilateral elements seem now to have been resolved. The stress intensity factor can be evaluated either by the I-integral or from the displacements of the nodes along the crack surface in the element adjoining the tip. If the vertical displacements are v_A and v_B at points A and B respectively, we get

$$K_I = \frac{E}{4(1-\nu^2)}\left(\frac{2\pi}{d}\right)^{1/2}(4v_A - v_B)\ , \quad \text{for plane strain,} \tag{14.1}$$

and a corresponding relation for K_{II} where instead the horizontal displacements are to be used. In the author's experience, the accuracy using (14.1) or a region independent integral are comparable. It is also possible to obtain a strain singularity of the form r^{-1} by collapsing one of the sides of a isoparametric quadrilateral (fig. 22), but still retaining the all the degrees of freedom of the three nodes now occupying the same global position. By collapsing one side and moving the midside nodes of the adjacent edges to the quarter-point positions, both the $r^{-1/2}$ and r^{-1} behaviours of strains are retained, which may well approximate singularities with intermediate exponents.

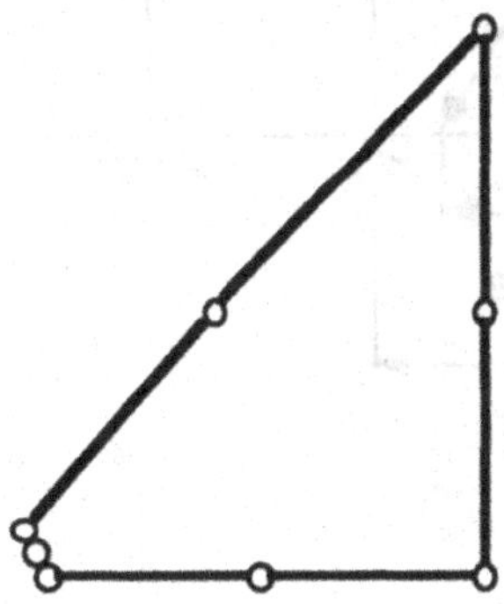

Fig. 22. Simulation of a r^{-1} singularity by an isoparametric element.

Another method suggested by Akin [68] is to modify the shape-functions directly. By Akin's method it is possible to obtain elements that contain a $r^{-\alpha}$-singularity which are compatible with the surrounding elements. This scheme was used by Thesken and

Gudmundson [69] for analysis of dynamic crack problems.

The problem of modelling moving crack tips is much more complicated. The probably most widely used method is nodal relaxation (fig. 23) . Regular elements are used and the crack motion is simulated by gradually releasing the nodal forces in the crack plane. Let F_i denote the current value of the force on the node i that is under relaxation. During the time interval for the crack to travel from this node to the next

$$F_i\,(t) = F_{i0}\, r\,(\Delta a/d)\ . \tag{14.2}$$

Here F_{i0} is the force on the node i, just before the relaxation starts, Δa the distance travelled by the tip within the element, d the length of the element edge and r a suitably chosen function such that r (0) = 1 and r (1) = 0. Rydholm et al. [70] suggested the form

$$r\,(\Delta a/d) = (\ 1 - \Delta a/d\ )^{1/2} \tag{14.3}$$

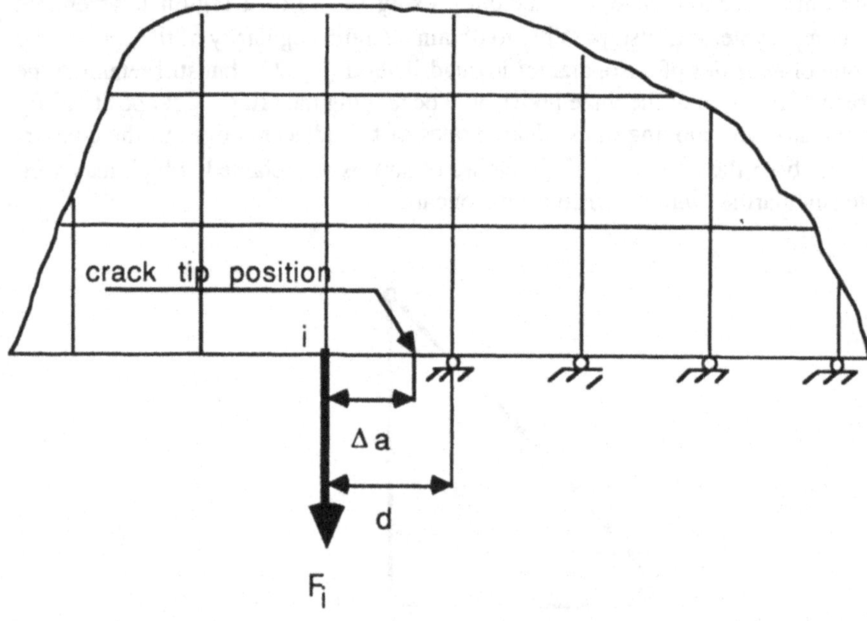

Fig. 23, Nodal relaxation

This particular choice has proven successful and has been used in several investigations (c.f. [27], [63]). Malluck and King [71] investigated different forms of r(Δa/d) and found that the particular choice did not affect the final results much. Some commercial programs (e.g. ABAQUS [28]) have a linear form implemented as standard.

The energy flow is evaluated by an integration of the force-displacement relation of the relaxing node. Thus an approximation of the energy flow to the tip is

$$\gamma = \frac{1}{d}\int_{t_i}^{t_{i+1}} F_i \, dv \ , \tag{14.4}$$

where v is the normal displacement of the node i.The scheme of nodal relaxation can obviously also be used for three-dimensional problems although this seems not to have been done so far.

A drawback of the nodal relaxation method is that the crack tip is at most time not situated at the tip, thus precluding singularity elements or other careful description of the tip region. One possible way of achieving of this is to use so called convective meshes. The entire problem is transformed by the coordinate transformation (3.1). Thus the element mesh remains stationary with respect to the moving x_i'-system. The spatial derivatives occurring in the expressions for velocities (eq. 3.12) and accelerations cause extra contributions in the finite element equations. Thus the stiffness matrix must be modified and an additional damping matrix appears. These matrices are non-symmetric and the special methods for effective solution of the finite element equations cannot be used. Furthermore, for non-linear problems the integration of the plastic strain-rates becomes more complicated since the integration points in the numerical scheme do not correspond to constant material points. The method has been used by Bazant et al. [72] and Östlund and Gudmundsson [73] for elastic problems with success. Freund and Douglas [64] and Mataga et al. [65] have employed convective meshes for analysis of elasto-plastic problems. A drawback with these methods is that boundaries that are not parallel with the crack plane can not easily be modelled. As far as the author knows no one has yet analysed problems for bodies with boundaries inclined to the crack plane by the moving mesh technique.

Another way of having the tip always placed at a node is to use elements that deform with time. As for the convective methods, additional contributions to the finite element equations arise since the shape-functions will be time-dependent. No results for this type of elements have yet been published.

Further developments are needed especially if non-linear problems must be solved for three-dimensional conditions to the accuracy needed for calculating singularity strengths.

ACKNOWLEDGEMENT

The author wants to thank Dr. A. Giannakopoulos and Mr. S. Östlund for careful reading of the manuscript and valuable suggestions for improvements.

REFERENCES

1. Pugh, C.E., Bass, B.R., Naus, D.J., Nanstad, R.K., deWit, R., Fields, R.J. and S.R. Low III: Crack run-arrest behaviour in wide SEN plates of a LWR pressure vessel Material, in: Trans. 9th Int. Conf. on Structural Mechanics in Reactor Technology, (Ed. F.H. Wittman), Vol. G, 1987, 21-26.
2. Strifors, H.: Thermomechanical theory of fracture, kinematics and physical principles, Department of Strength of Materials, Royal Institute of Technology, Stockholm, Sweden, Report 27, 1980.
3. Strifors, H.: Thermomechanical theory of fracture based upon the linear strain tensor, Department of Strength of Materials, Royal Institute of Technology, Stockholm, Sweden, Report 28, 1980.
4. Lidström, P.: Equations of balance and related problems in continuum fracture mechanics, Lund University, Lund, Sweden, LUTFD2/(TFME-1001)/1-11/,(1985), 1985.
5. Barenblatt, G.I.: The formation of equilibrium cracks during brittle fracture. General ideas and hypotheses. Axially-symmetric cracks, J. Appl. Math. Mech. PMM, 23 (1959), 622-635.
6. Broberg, K.B.: Mathematical methods in fracture mechanics, in: Trends in Applications of Pure Mathematics to Mechanics,Vol III, (Ed.H. Zorski), Pitman Publ. Ltd, 1979, 57-78.
7. Griffith, A.A.; The phenomena of rupture and flow in solids, Phil. Trans. Roy. Soc., A221 (1921), 163-173.
8. Kostrov, B.V. and L.V.Nikitin: Some general problems of mechanics of brittle fracture, Arch. Mech. Stos., 22 (1970), 749-776.
9. Strifors, H. C.: A generalized force measure of conditions at crack tips, Int. J. Solids and Structures, 10 (1974), 1389-1404.
10. Kishimoto, K., Aoki, S. and M.Sakata: Dynamic stress intensity factors using $\hat{J}$ -integral and finite element method, Eng. Fract. Mech., 13 (1980), 387-394.
11. Atluri, S.N.: Path-independent integrals in finite elasticity and inelasticity, with body force, inertia and arbitrary crack-face conditions, Eng. Fract. Mech., 16 (1982), 341-364.
12. Moran, B. and C.F.Shih: Crack tip and associated domain integrals from momentum and energy balance, Eng. Fract. Mech., 27 (1987), 615-642.
13. Nilsson, F.: A path-independent integral for transient crack problems, Int. J. of Solids and Structures, 9 (1973), 1107-1115.
14. Gurtin, M.E.: On a path-independent integral for elastodynamics, Int. J. of Fracture, 12 (1976), 643-644.
15. Gol´dstein, R.V. and R.L.Salganik: Brittle fracture of solids with arbitrary cracks, Int. J. of Fract., 10 (1974), 507-523.
16. Achenbach, J.D: Wave Propagation in Elastic Solids, North-Holland, Amsterdam, 1973.

17. Sih, G.C.: Dynamic aspects of crack propagation, in: Inelastic Behaviour of Solids (Ed. Jaffee, R.I. and M.F.Kanninen), McGraw-Hill, New York, 1970,
18. Nilsson, F.: A note on the stress-singularity at a non-uniformly moving crack tip, J.of Elasticity, 4 (1974), 73-75.
19. Clifton, R.J. and L.B. Freund: On the uniqueness of plane elastodynamic solutions for running cracks, J. of Elasticity, 4 (1974), 293-299.
20. Achenbach, J.D. and Z.P. Bazant: Elastodynamic near-tip stress and displacement fields for rapidly propagating cracks in orthotropic materials, J. Applied Mechanics, 42 (1975), 183-189.
21. Maue, A.W.: Die Beugung Elastischer Wellen an der Halbebene, ZAMM, 33 (1953), 1-10.
22. Sih,G.C., Embley, G.T. and R.S. Ravera: Impact response of a finite crack in plane extension, Int. J. Solids and Structures,.7 (1971),731.
23. Thau, S.A. and T.H. Lu: Transient stress intensity factors for a finite crack in an elastic solid caused by a dilatational wave, Int. J. Solids and Structures, 7 (1971), 731-750.
24. Kim, K.S.: Dynamic crack propagation of a finite crack, Int. J. Solids and Structures, 15 (1979), 685-699.
25. Chen, Y.M.: Numerical computation of dynamic stress intensity factors by a Lagrangian finite-difference method(the HEMP code), Eng. Fract. Mech., 7 (1975), 653-660.
26. Aberson, J.A., Anderson, J.M. and W.W. King: Dynamic analysis of cracked structures using singularity finite elements, in: Mechanics of Fracture, 4, (Ed. G.C. Sih) Noordhoff , Leyden 1977.
27. Brickstad, B.: A FEM-analysis of crack arrest experiments, Int. J. Fracture, 21 (1983), 177-191.
28. ABAQUS, User's Manual, version 4.5, Hibbitt, Karlsson and Sorensen Inc., Providence R.I., 1985.
29. Nilsson, F.: Dynamic stress intensity factors for finite strip problems, Int. J. Fract. Mech., 8 (1972), 403-411.
30. Dahlberg, L., Nilsson, F. and B. Brickstad: Influence of specimen geometry on crack propagation and arrest toughness, in: Crack Arrest Methodology and Applications, ASTM STP 711(ed.G.T. Hahn and M.F. Kanninen), ASTM, Philadelphia (1980), 9-108.
31. Willis, J.R.: Self-similar problems in elastodynamics, Phil. Trans Roy. Soc., Series A,.274 (1973), 435-491.
32. Cherepanov, G.P. and E.F. Afas anev: Some dynamic problems of the theory of elasticity- a review, Int. J. Eng. Science, 12 (1974), 665-690.
33. Broberg, K.B.: The propagation of a brittle crack, Arkiv för Fysik, 18 (1960), 739-750.
34. Freund, L.B.: Crack propagation in an elastic solid subjected to general loading-I. Constant rate of extension, J. Mech. Phys. Solids, 20 (1972), 129-140.

35. Eshelby, J.D.: The elastic field of a crack extending non-uniformly under general anti-plane loading, J. Mech.Phys.Solids, 17 (1969), 177-199.
36. Freund, L.B.: Crack propagation in an elastic solid subjected to general loading-II. Non-uniform rate of Extension, J. Mech.Phys.Solids, 20 (1972), 141-152.
37. Freund, L.B.: Crack propagation in an elastic solid subjected to general loading-III. Stress-wave loading, J. Mech.Phys.Solids, 21 (1973), 47-61.
38. Freund, L.B.: Crack propagation in an elastic solid subjected to general loading-IV. Obliquely incident stress pulse, J. Mech.Phys.Solids, 22 (1974), 137-146.
39. Burridge, R.: An influence function for the intensity factor in tensile fracture, Int.J.Eng.Science, 14 (1976), 725-730.
40. Kostrov, B.V.: On the crack propagation with variable velocity, Int. J. of Fracture, 11 (1975), 47-56.
41. Nilsson, F.: Steady crack propagation followed by nonsteady growth-Mode I solution, Int. J. Solids and Structures, 13 (1977), 1133-1139.
42. Brickstad, B. and F. Nilsson: Numerical evaluation by FEM of crack propagation experiments, Int. J. of Fract., 16 (1980), 71-84.
43. Gudmundsson, P.: Validity of asymptotic solutions for plastic materials, To be presented at ICF7, Houston, 1989.
44. Hult, A.J. and F. McClintock: Elastic-plastic stress and strain distributions around sharp notches under repeated shear, Proc. 9th Int. Congr. Appl. Mech., Brussels, 8 1956), 51-58.
45. Rice, J.R.: Mathematical Analysis in the Mechanics of Fracture, in: Fracture: An Advanced Treatise II (Ed. H. Liebowitz), Academic Press, New York 1968,191-308.
46. Hutchinson, J.W.: Singular behaviour at the end of a tensile crack in a hardening material, J. Mech. Phys. Solids, 16 (1968), 13-31.
47. Rice, J.R. and G. Rosengren: Plane strain deformation near a crack tip in a power-law hardening material, J. Mech. Phys. Solids, 16 (1968), 1-12.
48. Slepyan, L.I.: Crack dynamics in an elastic plastic body, Izv. Akad. Nauk SSSR MTT, 11 (1976), 126.
49. Freund, L.B. and A. S. Douglas: The influence of inertia on elastic-plastic antiplane-shear crack growth, J. Mech. Phys. Solids, 30 (1982), 59-74.
50. Drugan, W.J., Rice, J.R. and T.L. Sham: Asymptotic analysis of growing plane strain tensile cracks in elastic-ideally plastic solids, J. Mech. Phys. Solids, 30 (1982), 447-473.
51. Leighton, J.T., Champion C.R. and L.B. Freund: Asymptotic analysis of steady dynamic crack growth in an elastic-plastic material, J. Mech. Phys. Solids 35 (1987), 541-564.
52. Achenbach, J.D. Kanninen, M.F., and C.H. Popelar: Crack-tip fields for fast fracture of an elastic-plastic material, J. Mech. Phys. Solids,.29 (1981), 211-225.
53. Östlund, S. and P.Gudmundson: Asymptotic crack-tip fields for dynamic fracture of linear strain-hardening solids, Department of Strength of Materials, Royal Institute of Technology, Stockholm, Sweden, Report 83, 1987.

54. Nilsson, F. and P. Ståhle: Crack growth criteria and crack tip models, to appear in SM Archives, (1988).
55. Perzyna P.: The constutive equations for rate sensitive plastic materials, Quarterly of Appl. Math., 20 (1963), 321.
56. Hui, C.Y. and H. Riedel The asymptotic stress and strain field near the tip of a growing crack under creep conditions, Int. J. Fract., 17 (1981), 409-425.
57. Lo, K.K.: Dynamic crack tip fields in rate sensitive solids, J. Mech. Phys. Solids, 31 (1983), 287-305.
58. Yang, W. and L.B. Freund: An analysis of anti-plane shear crack growth in a rate sensitive elastic-plastic material, Int. J. Fract., 30 (1986), 157-174.
59. Nilsson, F.: Crack growth initiation and propagation under dynamic loading, Proc. 3rd Conference on Mechanical Properties of High Rates of Strain, Institute of Physics, ser no 70, Oxford, 1984, 185-204.
60. Nilsson, F., Ohlsson, P., Sjöberg, F. and P. Ståhle: A strain-rate dependent strip yield model for rapid loading conditions, to appear in Eng. Fract. Mech., (1989).
61. Wilson, M.L., Hawley, R.H. and. J. Duffy: The effect of loading and temperature on fracture initiation in 1020 hot-rolled steel, Eng. Fract. Mech., 13, 371-385
62. Brickstad, B.: A viscoplastic analysis of rapid crack propagation in steel, J. Mech. Phys. Solids, 31 (1983), 307-327.
63. Freund, L.B. and A.S. Douglas: The influence of inertia on elastic-plastic antiplane - shear crack growth, J. Mech. Phys. Solids, 30 (1982), 50-74.
64. Freund, L.B., Hutchinson, J.W. and P.S. Lam: Analysis of high-strain-rate elastic-plastic crack growth, Eng. Fract. Mech., 23 (1986), 119-129.
65. Mataga, P., Freund L.B. and J.W. Hutchinson: Crack tip plasticity in dynamic fracture, J.Phys. Chem. Solids, 48 (1987), 985-1005.
66. Henshell, R.D. and K.G. Shaw: Crack tip elements are unnecessary, Int J. Num. Meth. Eng., 9 (1975), 495-507.
67. Barsoum,R.S.: On the use of isoparametric finite elements in linear fracture mechanics, Int J. Num. Meth. Eng., 10 (1976), 25-37.
68. Akin, J.E.: The generation of elements with singularities, Int. J. Num. Meth., Eng., 10 (1976), 1249-1259.
69. Thesken, J.C. and P. Gudmundson: Application of a variable order singular element to dynamic fracture mechanics, Comp. Mech., 2 (1987), 307-316.
70. Rydholm, G., Fredriksson, B. and F. Nilsson: Numerical investigation of rapid crack propagation, in: Numerical Methods in Fracture Mechanics, Pineridge Press, 1978, 660-672.
71. Malluck, J.F. and W.W. King: Fast fracture simulated by a finite element analysis which accounts for crack.tip energy dissipation, in: Numerical Methods in Fracture Mechanics, Pineridge Press, 1978, 648-659.
72. Bazant, Z.P., Glazik Jr, J.L. and J.D. Achenbach: Elastodynamic fields near running cracks by finite elements", Comput. and Struct., 8 (1978),193-198.

73. Östlund, S. and P.Gudmundson: The application of moving finite elements for the study of crack propagation in linear elastic solids, Comput. and Struct., 25 (1987),765-774.

EXPERIMENTAL FRACTURE DYNAMICS

J.F. Kalthoff
University of Bochum, Bochum, FRG

CONTENTS

1. INTRODUCTION

This chapter considers the fracture behaviour of propagating and subsequently arresting cracks and of cracks under impact loading. Loading techniques and measuring methodologies for determining the crack arrest toughness K_{Ia} and the impact fracture toughness K_{Id} are discussed. Conventional measuring procedures are outlined; the validity and the applicability range of these procedures are critically examined. Furthermore, the physical processes that control the various dynamic events are analyzed by means of a special optical technique which determines the dynamic stress intensity factor directly from the local stress strain field at the crack tip. On the basis of the established results, improved measuring techniques are developed that are capable of determining true dynamic material strength properties at arbitrary test conditions without any restrictions in crack propagation velocity, loading rate, load duration, etc.

Due to the high loading rates associated with propagating and subsequently arresting cracks and with impacted cracks, the dynamic fracture toughnesses K_{Ia} and K_{Id} in general are lower than the static initiation fracture toughness K_{Ic} (see Fig. 1). This result does not apply in the ductile fracture regime where the material undergoes large plastic deformations before it fails. In these cases, dynamic J-values in general are higher than for equivalent quasistatic situations due to the increase in yield stress with loading rate. Because of this trend in the data, and often also because of the interest in conservative lower bound values, dynamic fracture investigations in the linear-elastic or small-scale yielding range are of particular practical interest and thus are considered predominantly in this chapter.

In Section 2 the shadow optical method of caustics for measuring dynamic stress intensifications is described. An understanding of this technique is necessary to follow the subsequent investigations on dynamic effects influencing the process of crack arrest and impact loading of cracks. Furthermore, the technique, although too sophisticated and expensive for routine material testing, is an appropriate tool for research oriented work in the field of experimental fracture dynamics.

Section 3 considers the problem of crack arrest. The standard procedure for measuring the crack arrest toughness is described. Shadow optical

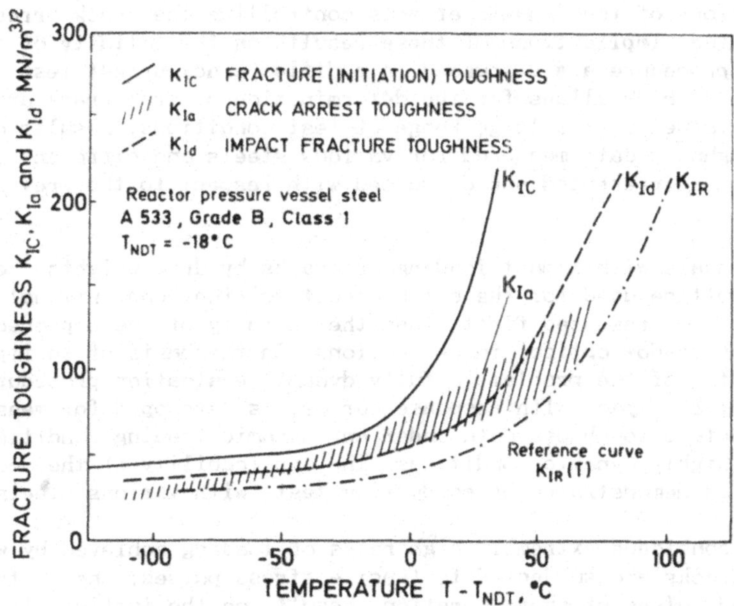

Fig. 1 Dynamic fracture toughness for arresting cracks, K_{Ia}, and for impacted cracks, K_{Id}, compared with the static fracture toughness, K_{Ic}

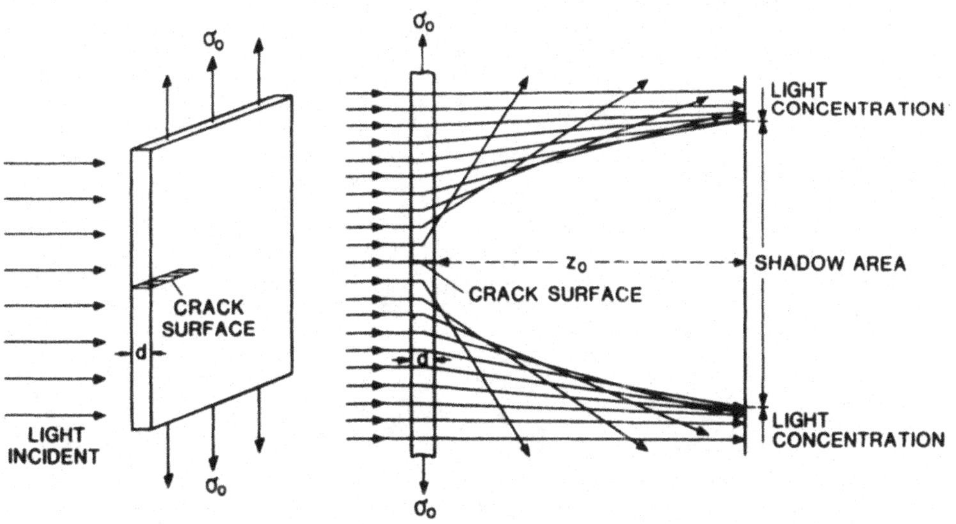

Fig. 2 Principle of the shadow optical method of caustics

investigations of the dynamic effects controlling the crack arrest event are presented. Implications of these results on the validity of the measuring procedure are discussed. A modified crack arrest test specimen is described which allows for the determination of true crack arrest toughness values over a large range of test conditions. Results of crack arrest toughness data measured for various steels and different test conditions are presented and discussed with respect to the previous findings.

Section 4 deals with impact loading of cracks by drop weights. Following the same outline used for the crack arrest section, conventional test procedures are presented first, then the validity of these procedures is analyzed by shadow optical investigations. On the basis of an improved understanding of the problem, a fully dynamic evaluation procedure, i.e. the concept of impact with response curves, is developed for measuring impact fracture toughness data under any dynamic loading condition, including highly dynamic conditions. The applicability of the measuring technique is demonstrated by conducting tests with various steels.

Section 5 considers extremely high rates of loading achieved by wave loading. Cracks are subjected to tensile stress pulses, shear stress pulses, and pulses of short duration. Results on the instability and failure behaviour observed with various materials are discussed. The presented results indicate that the failure behaviour can be very different from that one obtained at lower loading rates. Thus, an effective dynamic material characterization requires a correct simulation of the particular loading rate of interest.

2. SHADOW OPTICAL METHOD OF CAUSTICS

The shadow optical method of caustics is a relatively new experimental technique in stress strain analysis, originally introduced by Manogg [1,2] in 1964. The method is sensitive to stress gradients and therefore is an appropriate tool for measuring stress concentrations, in particular for measuring crack tip stress intensity factors in fracture mechanics applications. The technique was extended later by Theocaris [3-5], Rosakis [6,7] and the author and his colleagues [8-13] to different conditions of loading, material behaviour, in static as well as dynamic situations. This chapter reviews the basic physical principles and the mathematical analyses of the method. Evaluation formulas for the application in fracture mechanics are given.

2.1 Physical Principle of the Shadow Optical Imaging Process

Stresses alter the optical properties of a solid, i.e. the thickness of the body (due to Poisson's effects) and the refractive index of the material. These changes in the optical properties are utilized in the shadow optical method of caustics to make stress distributions in the solid visible.

The physical principle of the method is illustrated in Fig. 2. A specimen with a notch or a crack as a stress riser is subjected to tensile loading. The specimen shall be of a transparent material for the moment. It is illuminated by a parallel light beam. Due to the tensile stress concentration around the crack tip, the thickness of the specimen and the refractive index of the material are steadily reduced when the crack tip is approached. Thus, a light ray traversing the specimen near the crack tip is deflected in a direction away from the crack tip. In this respect, the area surrounding the crack tip acts like a divergent lens. But, the closer the light ray is to the crack tip, the larger is the deflection angle. Consequently, on a screen (image or reference plane) at a distance z_0 from the specimen, the crack tip appears as a shadow area which is surrounded by a region of light concentration, i.e. the caustic. This shadow pattern represents a quantitative decription of the stress intensification around the crack tip in the plate.

Shadow optical light patterns are obtained for tensile as well as compressive stress intensifications (provided notches are utilized instead of cracks in the latter case, such that compressive stress concentrations can build up). The light patterns can be observed with transparent specimens or in reflection with non-transparent specimens. They can be observed as real or as virtual images. Figure 2 represents the most simple case of a tensile stress intensification in a transparent specimen with the observation of a real shadow optical image. The virtual image would be obtained on the opposite side of the specimen where the real image is observed. A consideration of the light rays which are reflected at the front side of the specimen, i.e. the side facing the light source, leads to the formation of shadow images as well, which again can be real or virtual depending on the position of the image plane with respect to the specimen.

Figure 3b illustrates the light ray tracing for the reflection case which is analogous to the corresponding transmission case (Fig. 3a) shown previously in Fig. 2. A steel specimen with a mirrored front surface shall be considered. Due to Poisson's effect, the stress concentration at the crack tip forms a dimple at the specimen surface which acts like a concave mirror. Thus, the light rays are reflected towards the center line. An extension of these light rays into the virtual image space behind the specimen leads to the same or a similar light ray tracing as in the transmission case (see Figs. 3b and 3a for comparison). Thus, the light pattern observed in a virtual image plane with a reflection arrangement is similar to that observed in a real image plane with a transmission arrangement.

In this context it is important to consider one characteristic property of shadow optical light intensity patterns obtained with stress concentration problems. The closer to the crack tip a light ray traverses the specimen, the larger its deflection angle (see Fig. 2). Consequently, for light rays with decreasing distance to the crack tip, the corresponding image points first also approach the image point of the crack tip. But, from a certain distance on, this tendency is reversed. For light rays that are traversing the specimen closer than this distance to the crack tip, the image points move away from the crack tip image point. The shadow optical image, therefore, exhibits a sharp boundary line between the area of darkness and the surrounding area of light concentration. This boundary line between the two areas is called the caustic curve ("caustic" is Greek for "focal line").

These particular light rays which directly hit the caustic curve in the image plane are of special importance in shadow optics; they are called "initial" light rays. The locus of all points where the initial light rays traverse the object plane (specimen) is called the initial curve. Light rays traversing the object plane at points that, compared with the initial curve, are closer to (or farther away from) the crack tip are deflected more steeply (or more shallowly) than the initial light rays. Consequently, in both cases the image points lie outside the caustic curve. The caustic curve in the image plane is the direct image of the initial curve in the object plane. In the mathematical sense of the theory of imaging, the mapping process of the initial curve onto the caustic curve represents

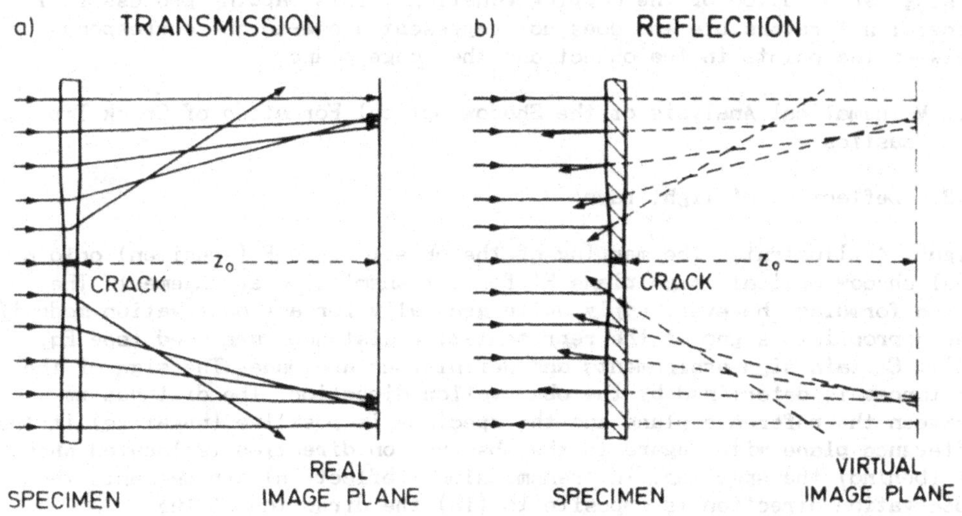

Fig. 3 Light ray tracing for a shadow optical transmission arrangement (a) and a reflection arrangement (b)

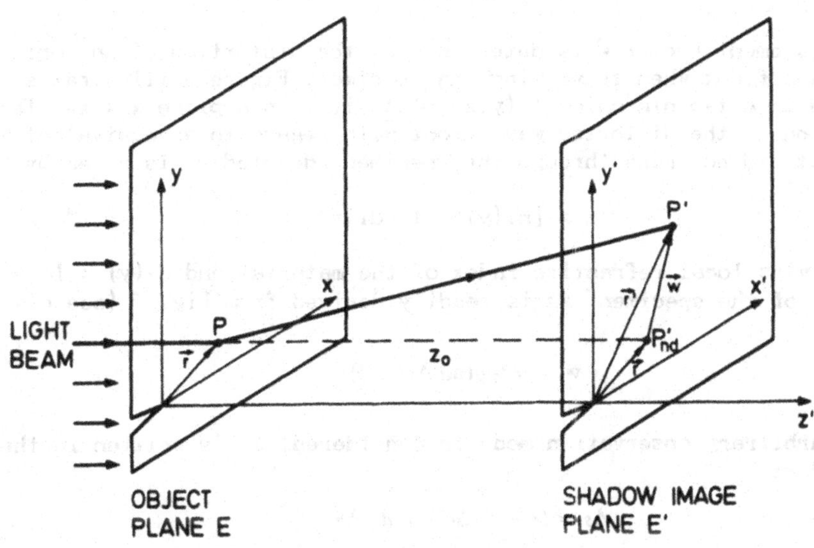

Fig. 4 Mapping of the object plane onto the image plane

a singular solution of the mapping equations. This mapping process is in general not reversible and does not represent a one-to-one correspondence between the points in the object and the image plane.

2.2 Mathematical Analysis of the Shadow Optical Formation of Crack Tip Caustics

2.2.1 Deflection of light rays

Figure 4 illustrates the mapping of the object plane E (specimen) onto a real shadow optical image plane E' for a transmission arrangement. The given formulas, however, apply quite generally for any observation mode if the appropriate signs of the representative distances are used (see Eq. (4)). Certain sign assignments and definitions are made. The sign of the distances is determined by the observation direction. The distance z_0 between the reference plane and the specimen is positive (negative) if the reference plane with regard to the observation direction is located ahead of (behind) the specimen. In transmission (reflection) arrangements the observation direction is opposite to (in) the direction of the illuminating light beam. Tensile stresses are counted positive. According to these definitions the distance z_0 shown in Fig. 4 is positive, $z_0 > 0$.

A light ray traverses the object plane E at the point $P(\vec{r})$, where r is the radial distance from the notch tip. Due to the influence of the stresses in the specimen this light ray is deflected and hits the image plane E' displaced by the vector $\vec{w}$ at the point $P'(\vec{r}')$ with

$$\mathbf{r}' = \mathbf{r} + \mathbf{w} \tag{1}$$

The displacement vector $\vec{w}$ is determined by the distortion of an impinging planar wave front when traversing the specimen. Figure 5 illustrates the situation in a two dimensional (y,z)-plot, i.e. in a plane $x = x_c$. The retardation of the distorted wave front with regard to an equivalent wave front that did not pass through the specimen, denoted s, is given by

$$s = [n_1(y) - 1]\cdot d_1(y) \tag{2}$$

where $n_1(y)$ = local refractive index of the material and $d_1(y)$ = local thickness of the specimen. As is readily derived from Fig. 5 (see also [14])

$$\mathbf{w} = +z_0 \,\mathbf{grad}\, \Delta s(r, \phi) \tag{3}$$

When an arbitrary observation mode is considered, Δs is written in the general form

$$\Delta s = (n - 1)\,\Delta d_{eff} + d_{eff}\,\Delta n \tag{4}$$

with n = refractive index, $d_{eff} = d$ for transmission,
and $n = -1$, $d_{eff} = d/2$ for reflection.

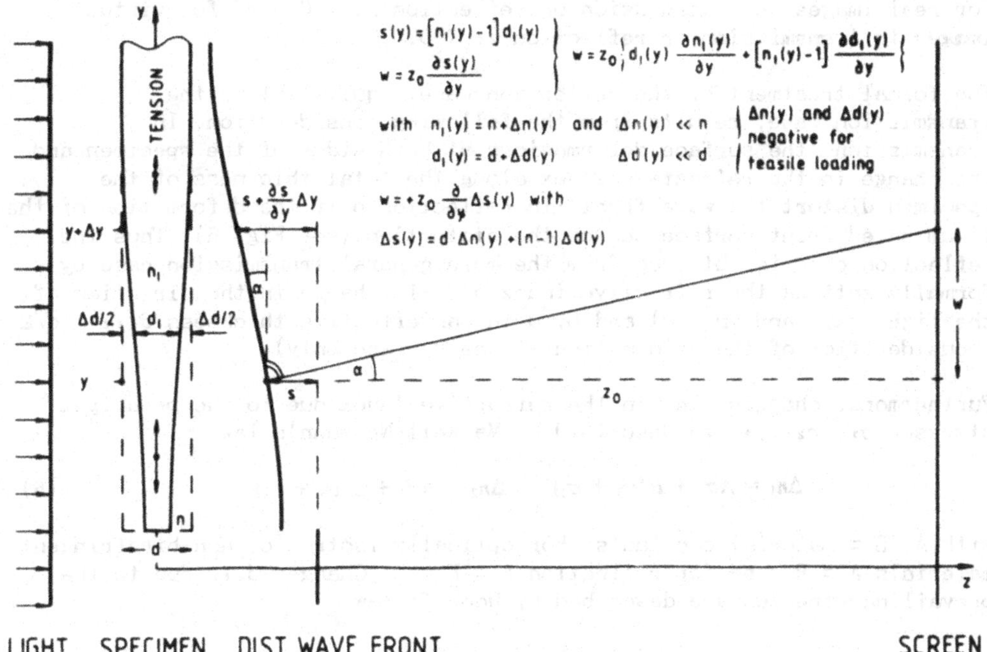

Fig. 5 Deflection of a light ray traversing a specimen with a spatial stress gradient

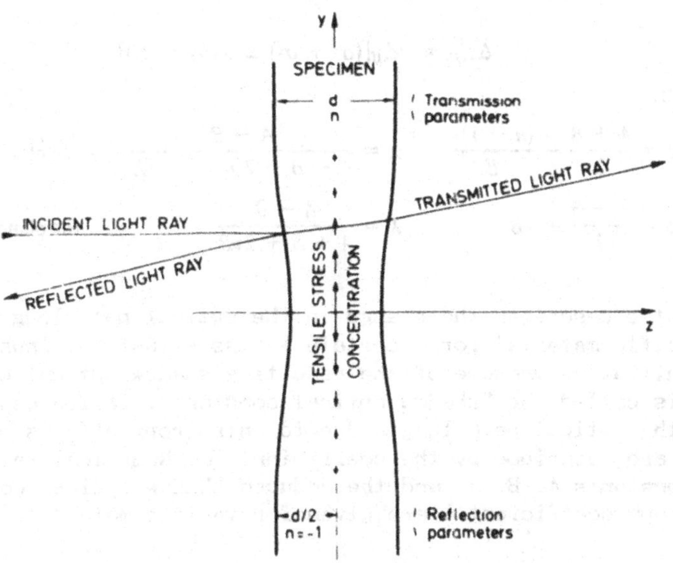

Fig. 6 Relevant parameters for transmission and reflection of light rays

For real images in transmission or reflection $z_0 > 0$, and for virtual images in transmission or reflection $z_0 < 0$.

The formal treatment of the reflection case, equivalent to the transmission case, results from the following consideration. In transmission, the surface deformations at both sides of the specimen and the change in the refractive index along the total thickness of the specimen distort the wave front. In reflection only the deformation of the illuminated front surface causes the distortion (see Fig. 6). Thus the reflection case is obtained from the more general transmission case by formally setting the refractive index n = -1 (change in the direction of the light rays and $\Delta n = 0$) and by using an effective thickness $d_{eff} = d/2$ (consideration of the deformation at one surface only).

Furthermore, changes, Δn, in the refractive index due to the principal stresses σ_1, σ_2, σ_3 are described by Maxwell-Neumann's law

$$\Delta n_1 = A\sigma_1 + B(\sigma_2 + \sigma_3) \qquad \Delta n_2 = A\sigma_2 + B(\sigma_1 + \sigma_3) \tag{5}$$

with A, B = material constants. For optically isotropic, non-birefringent materials A = B, and for reflection A = B = 0. Changes d_{eff} due to the prevailing stresses are described by Hooke's law

$$\Delta d_{eff} = \left[\frac{1}{E}\sigma_3 - \frac{\nu}{E}(\sigma_1 + \sigma_2)\right] d_{eff} \tag{6}$$

with $\sigma_3 = 0$ for plane stress, and $d_{eff} = 0$ for plane strain. With Eqs. (5,6) the Eq. (4) then can be rearranged to

$$\Delta s_{1/2} = c d_{eff}[(\sigma_1 + \sigma_2) \pm \lambda(\sigma_1 - \sigma_2)]$$

where (7)

$$c = \frac{A+B}{2} - \frac{(n-1)\nu}{E} \qquad \lambda = \frac{A-B}{A+B-2(n-1)\nu/E} \qquad \text{for plane stress}$$

$$c = \frac{A+B}{2} + \nu B \qquad \lambda = \frac{A-B}{A+B+2\nu B} \qquad \text{for plane strain}$$

The constant c describes the change in the optical path length obtained with a specific material for a certain stress situation. Thus the constant c is a quantitative measure of the resulting shadow optical effect and therefore is called the "shadow optical constant". Influences on the change of the optical path length due to anisotropy effects of the material ($A \neq B$) are described by the coefficient λ. Numerical values for the material constants A, B, n, and the deduced shadow optical constant c and the anisotropy coefficient λ are given for various materials in Table 1.

Constants for Caustic Evaluation

Material	Elastic Constants		General Optical Constants			Shadow Optical Constants				Effective Thickness
						for Plane Stress		for Plane Strain		
	Young's Modulus MN/m^2	Poisson's Ratio	Refractive Index	A m^2/N	B m^2/N	c m^2/N	λ	c m^2/N	λ	d_{eff}
TRANSMISSION:										
Optically Anisotropic:										
Araldite B	3660**	0.392**	1.592	-0.056×10^{-10}	-0.620×10^{-10}	-0.970×10^{-10}	-0.288	-0.580×10^{-10}	-0.482	d
CR - 39	2580	0.443	1.504	-0.160×10^{-10}	-0.520×10^{-10}	-1.200×10^{-10}	-0.148	-0.560×10^{-10}	-0.317	d
Plate Glass	73900	0.231	1.517	$+0.0032 \times 10^{-10}$	-0.025×10^{-10}	-0.027×10^{-10}	-0.519	-0.017×10^{-10}	-0.849	d
Homalite 100	4820**	0.310**	1.561	-0.444×10^{-10}	-0.672×10^{-10}	-0.920×10^{-10}	-0.121	-0.767×10^{-10}	-0.149	d
Optically Isotropic:										
PMMA	3240	0.350	1.491	-0.530×10^{-10}	-0.570×10^{-10}	-1.080×10^{-10}	~0	-0.750×10^{-10}	~0	d
REFLECTION:										
All materials	E	ν	-1	0	0	2ν/E	0	-	-	d/2

Table 1 Shadow optical constants

2.2.2 Crack tip caustics

Equations (1), (3), and (7) describe the mapping of the object plane onto the shadow optical image plane for arbitrary stress distributions $\sigma_{1,2}(r,\Phi)$ The specific mapping equations for crack problems are obtained by inserting into the general equation (7) the individual formula for the stress distribution around the crack tip.

The mode-I and the mode-II case are considered simultaneously. The left hand equations apply for mode-I, the right hand ones for mode-II. The stress distributions are given by

Mode-I Mode-II

$$\sigma_r = \frac{K_I}{\sqrt{2\pi r}}\frac{1}{4}\left(5\cos\frac{\phi}{2} - \cos\frac{3\phi}{2}\right) \qquad \sigma_r = \frac{K_{II}}{\sqrt{2\pi r}}\frac{1}{4}\left(-5\sin\frac{\phi}{2} + 3\sin\frac{3\phi}{2}\right)$$

$$\sigma_\phi = \frac{K_I}{\sqrt{2\pi r}}\frac{1}{4}\left(3\cos\frac{\phi}{2} + \cos\frac{3\phi}{2}\right) \qquad \sigma_\phi = \frac{K_{II}}{\sqrt{2\pi r}}\frac{1}{4}\left(-3\sin\frac{\phi}{2} - 3\sin\frac{3\phi}{2}\right) \tag{8}$$

$$\tau_{r\phi} = \frac{K_I}{\sqrt{2\pi r}}\frac{1}{4}\left(\sin\frac{\phi}{2} + \sin\frac{3\phi}{2}\right) \qquad \tau_{r\phi} = \frac{K_{II}}{\sqrt{2\pi r}}\frac{1}{4}\left(\cos\frac{\phi}{2} + 3\cos\frac{3\phi}{2}\right)$$

With these stress distributions and Eq. (7) (for $\lambda = 0$; for simplicity only the isotropic case shall be considered in this context) the mapping Eqs. (1) and (3) become

$$x' = r\cos\phi - \frac{K_I}{\sqrt{2\pi}} z_0 c d_{eff} r^{-3/2}\cos\frac{3}{2}\phi \qquad x' = r\cos\phi + \frac{K_{II}}{\sqrt{2\pi}} z_0 c d_{eff} r^{-3/2}\sin\frac{3}{2}\phi$$

$$y' = r\sin\phi - \frac{K_I}{\sqrt{2\pi}} z_0 c d_{eff} r^{-3/2}\sin\frac{3}{2}\phi \qquad y' = r\sin\phi - \frac{K_{II}}{\sqrt{2\pi}} z_0 c d_{eff} r^{-3/2}\cos\frac{3}{2}\phi \tag{9}$$

The complete family of the deflected light rays forms a shadow space behind the object plane. Its surface is an envelope to the light rays and is called the caustic surface. The intersection of this surface with the image plane forms the caustic curve. This caustic curve is a multivalued, singular solution of Eqs. (1) and (3) (i.e., the mapping of points along the caustic curve is not reversible). Thus a necessary and sufficient condition for the existence of the caustic curve is obtained if the Jacobian of Eqs. (1) and (3) vanishes, that is

$$\frac{\partial x'}{\partial r}\frac{\partial y'}{\partial \phi} - \frac{\partial x'}{\partial \phi}\frac{\partial y'}{\partial r} = 0 \tag{10}$$

The coordinates r,Φ of points P that fulfill Eq. (10) form the initial curve in the object plane and the mapping of this curve onto the image plane is the caustic curve.

Consequently, application of Eq. (10) to the mapping Eqs. (9) gives the equation of the initial curve

$$r = \left[\frac{3}{2}\frac{|K_I|}{\sqrt{2\pi}}|z_0|\,|c|\,d_{eff}\right]^{2/5} \equiv r_0 \qquad r = \left[\frac{3}{2}\frac{|K_{II}|}{\sqrt{2\pi}}|z_0|\,|c|\,d_{eff}\right]^{2/5} \equiv r_0 \qquad (11)$$

The initial curves are circles around the crack tip with fixed radii r_0. With the mapping Eqs. (9) the caustic curves are finally obtained as images of the initial curves [Eqs. (11)] and are given as

$$\begin{aligned} x' &= r_0(\cos\phi - \operatorname{sgn}(K_I z_0 c)\tfrac{2}{3}\cos\tfrac{3}{2}\phi) & x' &= r_0(\cos\phi + \operatorname{sgn}(K_{II} z_0 c)\tfrac{2}{3}\sin\tfrac{3}{2}\phi) \\ y' &= r_0(\sin\phi - \operatorname{sgn}(K_I z_0 c)\tfrac{2}{3}\sin\tfrac{3}{2}\phi) & y' &= r_0(\sin\phi - \operatorname{sgn}(K_{II} z_0 c)\tfrac{2}{3}\cos\tfrac{3}{2}\phi) \end{aligned} \qquad (12)$$

Mathematically, the caustic curves are generalized epicycloids. The caustics are shown graphically in Fig. 7 for positive and negative values of the distance z_0 between the specimen and the reference plane, for positive and negative loading, and for transmission and reflection arrangements.

An illustrative picture of the complete distribution of deflected light rays in the image plane is obtained in a simple manner by considering the mapping of rays that traverse the object plane along lines Φ = const. The images of those lines obtained with Eqs. (12) for the two fracture modes considered are shown in Fig. 8. The caustic curves appear as envelopes to the obtained families of image lines (see Fig. 7 for comparison).

For the quantitative evaluation of caustics a length parameter between characteristic points on the caustic curve is defined, e.g., the maximum diameter of the caustic curve given by the distances D in Fig. 7. These distances are related to the radii of the initial curves by the equations

$$D = 3.17 r_0 \qquad D = 3.02 r_0 \qquad (13)$$

With Eqs. (11) and (13) a quantitative formula is obtained in each case relating the size of the shadow optical pattern with the generating load parameter:

$$K_I = \frac{2\sqrt{2\pi}}{3(3.17)^{5/2} z_0 c d_{eff}} D^{5/2} \qquad K_{II} = \frac{2\sqrt{2\pi}}{3(3.02)^{5/2} z_0 c d_{eff}} D^{5/2} \qquad (14)$$

These equations represent the final evaluation formula for determining the stress intensity factor K_I or K_{II} from experimentally observed caustics.

With transparent materials that are optically anisotropic ($\lambda \neq 0$) the caustic curve splits up into a double caustic. Figure 9 shows crack tip caustics for the optically anisotropic material Araldite B in comparison to the corresponding single caustic. Since the coefficient of anisotropy depends on the state of stress ($|\lambda|$ is larger for plane strain than for plane stress; see Table 1) the split-up of the double caustic is larger for plane strain than for plane stress.

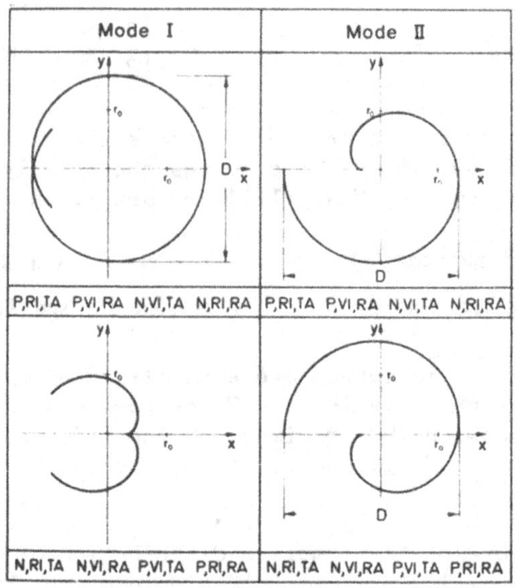

Fig. 7 Crack tip caustic curves for mode-I, and mode-II loading; P, N = positive, negative loading; RI, VI = real, virtual image; TA, RA = transmission, reflection arrangement

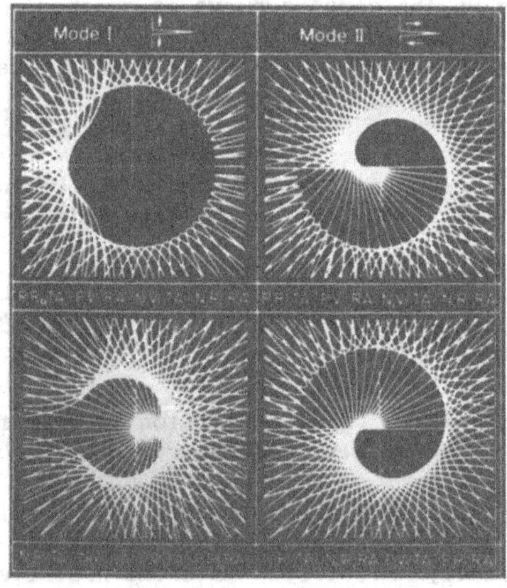

Fig. 8 Shadow optical light intensity distributions (Key see Fig. 7)

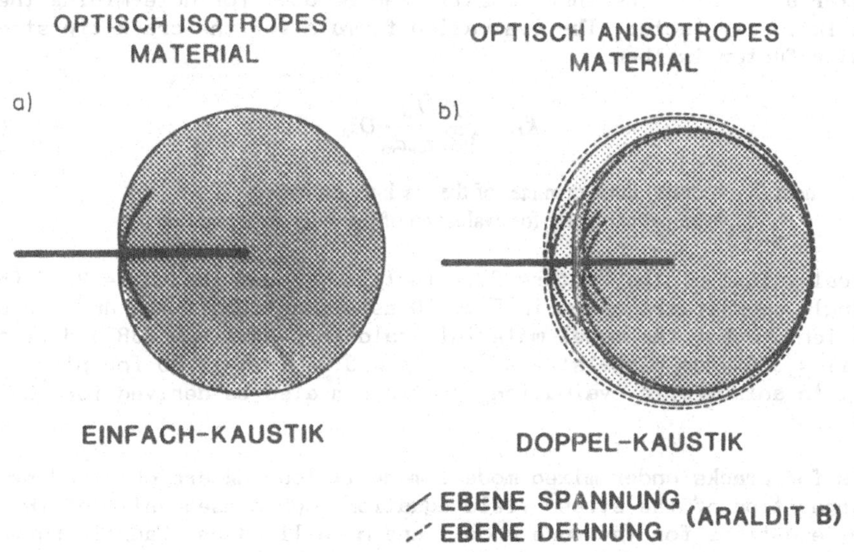

Fig. 9 Single caustic for optically isotropic material (a); **and** double caustic for optically anisotropic material (b)
(optisch isotropes/anisotropes material = optically isotropic/anisotropic material, Einfach-/Doppel-Kaustik = single/double caustic, ebene Spannung/ Dehnung = plane stress/strain)

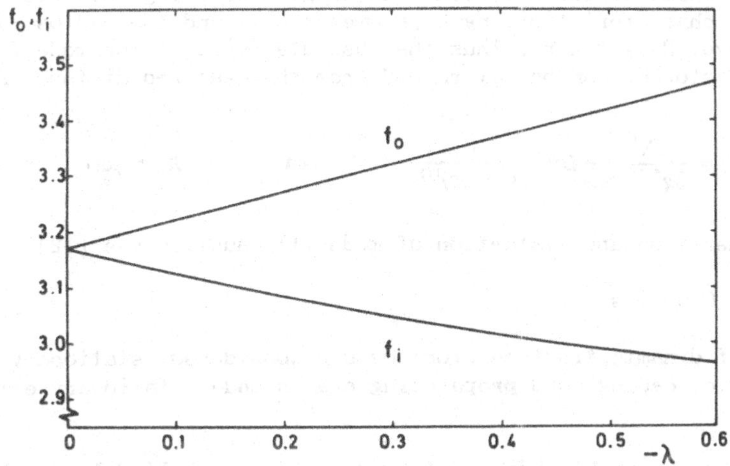

Fig. 10 Numerical factors for evaluating anisotropic crack tip caustics

The outer as well as the inner caustic can be used for determining the stress intensity factor. The evaluation formula for the crack-tip stress intensity factor is then

$$K_I = \frac{2\sqrt{2\pi}}{3f_{o,i}^{5/2} z_0 c d_{eff}} D_{o,i}^{5/2} \tag{15}$$

where $D_{o,i}$ = outer, inner diameter of the crack-tip caustic
$f_{o,i}$ = numerical factor for evaluation of the outer, inner caustic

Numerical values of the factors $f_{o,i}$ that lie around the value 3.17 for the single caustic are given in Fig. 10 as a function of the anisotropy coefficient λ. For the model material Araldite B $\lambda = -0{,}288$ and $f_o = 3.32$, $f_i = 3.05$ for plane stress, and $f_o = 3.34$, $f_i = 2.99$ for plane strain. An anisotropic evaluation formula can also be derived for the mode-II case.

Results for cracks under mixed mode-I-mode-II loading are obtained by a superposition of the stress field equations and consequently of the mapping equations for the pure mode-I and mode-II cases. Caustic curves for different ratios of the mode-II to the mode-I stress intensity factor $\mu = K_{II}/K_I$ are shown in Fig. 11a. Depending on the ratio K_{II}/K_I, all intermediate stages between the pure mode-I and the pure mode-II caustics are possible.

From the two diameters D_{max} and D_{min}, defined in Fig. 11b, the two stress intensity factors K_I and K_{II} are determined [15,16]. With Fig. 12 resulting from Eqs. (12) for the complete mode-I-mode-II caustic curve, the stress intensity factor ratio $\mu = K_{II}/K_I$ is first determined from the measured value $(D_{max} - D_{min})/D_{max}$. According to Fig. 13, this μ value then defines the value of the numerical factor g, describing the relationship between the characteristic length parameter D_{max} and the radius r_o of the initial curve, $D_{max} = g \cdot r_o$. Thus the absolute value of the mode-I stress intensity factor K_I can be determined from the measured distance D_{max} as

$$K_I = \frac{2\sqrt{2\pi}}{3g^{5/2} z_0 c d_{eff}} D_{max}^{5/2} \frac{1}{(1+\mu^2)^{1/2}} \qquad \text{and} \qquad K_{II} = \mu K_I \tag{16}$$

For the generation and evaluation of mode-III caustics see [12].

2.3 Dynamic Caustics

Two cases of dynamic fracture problems are considered: stationary cracks under dynamic loading, and propagating cracks under stationary external loading.

For a stationary crack subjected to a dynamically applied load, the near-field stress distribution is the same as for a statically loaded crack [17]. The stress intensity factor, however, becomes a function of time. Consequently, since the dependencies from the radial distance r and the

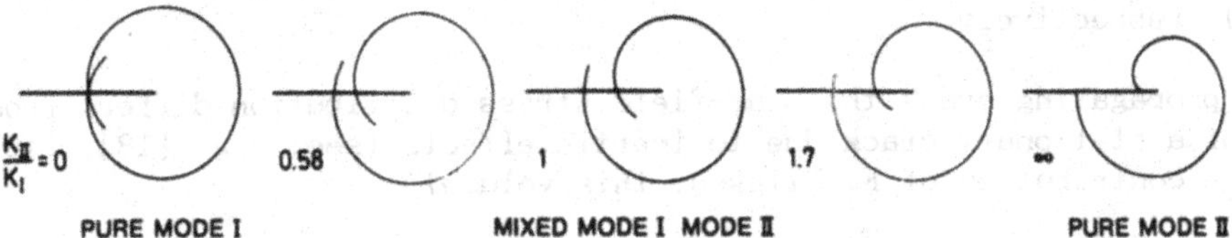

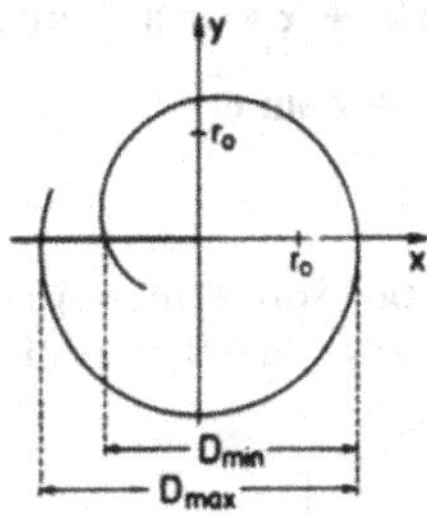

Fig. 11 Mixed-mode crack tip caustics

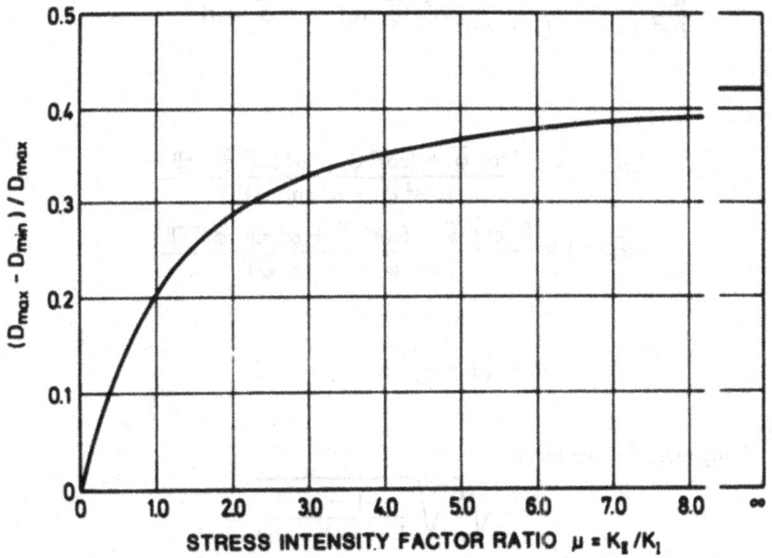

Fig. 12 Determination of the stress intensity factor ratio μ from mixed-mode caustics

angle Φ are identical for the dynamic and static cases, the results presented in the preceding section also apply to dynamically loaded cracks, with the stress intensity factors K_I or K_{II} replaced by $K_I(t)$ or $K_{II}(t)$, respectively.

For a propagating crack the near-field stress distribution differs from that of a stationary crack due to inertia effects (see, e.g. [18], [19], and the contribution of F. Nilsson, this volume).

In a coordinate system (x,y) or (r,Φ) with its origin in the moving crack tip (see Fig. 14)

$$\begin{aligned} x &= a(t) + \bar{x} = a(t) + \bar{r}\cos\bar{\phi} \\ y &= \bar{y} = \bar{r}\sin\bar{\phi} \end{aligned} \tag{17}$$

where a is the crack length

the near-field stresss distribution for a propagating crack with an instantaneous crack velocity v = da/dt is given as

$$\begin{aligned} \sigma_x &= \frac{K_I}{\sqrt{2\pi\bar{r}}}\frac{1+\alpha_2^2}{4\alpha_1\alpha_2-(1+\alpha_2^2)^2}\left[(1+2\alpha_1^2-\alpha_2^2)p(\bar{\phi},\alpha_1)-\frac{4\alpha_1\alpha_2}{1+\alpha_2^2}p(\bar{\phi},\alpha_2)\right] \\ \sigma_y &= \frac{K_I}{\sqrt{2\pi\bar{r}}}\frac{1+\alpha_2^2}{4\alpha_1\alpha_2-(1+\alpha_2^2)^2}\left[-(1+\alpha_2^2)p(\bar{\phi},\alpha_1)+\frac{4\alpha_1\alpha_2}{1+\alpha_2^2}p(\bar{\phi},\alpha_2)\right] \\ \tau_{xy} &= \frac{K_I}{\sqrt{2\pi\bar{r}}}\frac{1+\alpha_2^2}{4\alpha_1\alpha_2-(1+\alpha_2^2)^2}\alpha_1[q(\bar{\phi},\alpha_1)-q(\bar{\phi},\alpha_2)] \end{aligned} \tag{18}$$

where

$$\begin{aligned} p(\bar{\phi},\alpha_j) &= \frac{[\cos\bar{\phi}+(\cos^2\bar{\phi}+\alpha_j^2\sin^2\bar{\phi})^{1/2}]^{1/2}}{(\cos^2\bar{\phi}+\alpha_j^2\sin^2\bar{\phi})^{1/2}} \\ q(\bar{\phi},\alpha_j) &= \frac{[-\cos\bar{\phi}+(\cos^2\bar{\phi}+\alpha_j^2\sin^2\bar{\phi})^{1/2}]^{1/2}}{(\cos^2\bar{\phi}+\alpha_j^2\sin^2\bar{\phi})^{1/2}} \end{aligned} \tag{19}$$

and

$$\alpha_j = \left(1-\frac{v^2}{c^2}\right)^{1/2} \qquad j = 1, 2 \tag{20}$$

with

Longitudinal wave speed:

$$c_1 = \sqrt{\frac{E}{\rho}}\sqrt{\frac{1-\nu}{(1+\nu)(1-2\nu)}}$$

Transverse wave speed:

$$c_2 = \sqrt{\frac{E}{\rho}}\sqrt{\frac{1}{2(1+\nu)}} \tag{21}$$

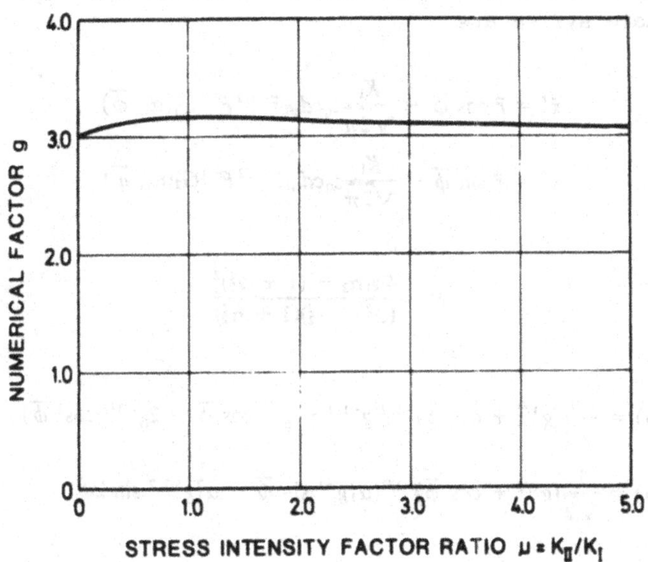

Fig. 13 Numerical factor for stress-intensity-factor-determination for mixed-mode caustics

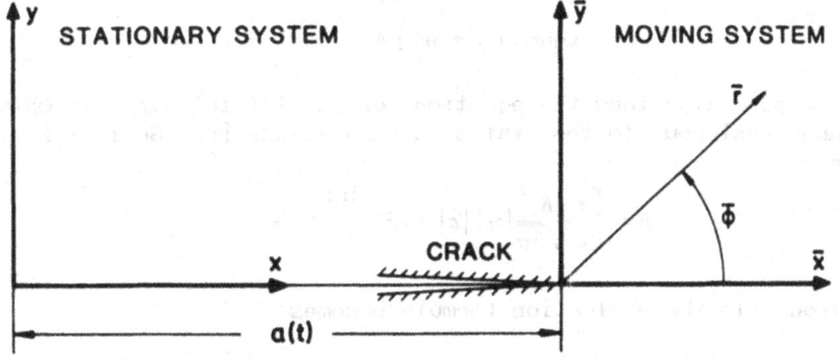

Fig. 14 Coordinate system at the tip of a moving crack

With this stress distribution the mapping equations with regard to the moving coordinate system are

$$\bar{x}' = \bar{r}\cos\bar{\Phi} - \frac{K_I}{\sqrt{2\pi}} z_0 c d_{eff} \bar{r}^{-3/2} F^{-1} G_1(\alpha_1, \bar{\Phi})$$
$$\bar{y}' = \bar{r}\sin\bar{\Phi} - \frac{K_I}{\sqrt{2\pi}} z_0 c d_{eff} \bar{r}^{-3/2} F^{-1} G_2(\alpha_1, \bar{\Phi}) \quad (22)$$

where

$$F = \frac{4\alpha_1\alpha_2 - (1+\alpha_2^2)^2}{(\alpha_1^2 - \alpha_2^2)(1+\alpha_2^2)} \quad (23)$$

and

$$G_1(\alpha_1, \bar{\Phi}) = \frac{-1}{\sqrt{2}}(g^{1/2} + \cos\bar{\Phi})^{-1/2}(g^{-1/2} - g^{-1}\cos\bar{\Phi} - 2g^{-3/2}\cos^2\bar{\Phi})$$
$$G_2(\alpha_1, \bar{\Phi}) = \frac{1}{\sqrt{2}}(g^{1/2} + \cos\bar{\Phi})^{-1/2}(\alpha_1^2 g^{-1}\sin\bar{\Phi} - \alpha_1^2 g^{-3/2}\sin 2\bar{\Phi}) \quad (24)$$

with

$$g = 1 + (\alpha_1^2 - 1)\sin^{-2}\bar{\Phi} \quad (25)$$

It can be easily shown by numerical calculations (see the more detailed discussion in [9]) that the influence of α_1 on the functions $G_1(\alpha_1, \Phi)$ and $G_2(\alpha_1, \Phi)$ is negligibly small for all crack velocities of practical relevance (i.e., $v < 0.3c_1$). In particular, the error made by neglecting the α_1 dependency on G_1 and G_2 is small in comparison to that on the factor F. Thus with an accuracy sufficient for engineering purposes these functions can be approximated by

$$G_1(\alpha_1, \bar{\Phi}) \approx \cos\tfrac{3}{2}\bar{\Phi}$$
$$G_2(\alpha_1, \bar{\Phi}) \approx \sin\tfrac{3}{2}\bar{\Phi} \quad (26)$$

With these approximations the equation for the initial curve is obtained in a manner analogous to the static considerations [see Section 2.2.1, Eq. (11)] as

$$\bar{r} = \left[\frac{3}{2}\frac{|K_I|}{\sqrt{2\pi}}|z_0|\,|c|\,d_{eff}F^{-1}\right]^{2/5} \equiv \bar{r}_0 \quad (27)$$

and consequently the evaluation formula becomes

$$K_I = \frac{2\sqrt{2\pi}\,F}{3(3.17)^{5/2} z_0 c d_{eff}} D^{5/2} \quad (28)$$

Thus the stress intensity evaluation formula for a propagating crack is the same as that for a stationary crack [Eq. (14)] except for a correction factor F(v). This factor accounts for velocity effects on the r- and Φ-distribution of the dynamic stress field for a propoagating crack. The factor F as a function of crack velocity v is given by Fig. 15. F is less than 1, but for pratically relevant crack velocities it is nearly equal to 1.

Figure 16 shows crack-tip caustics for different crack velocities but with a fixed stress intensity factor. The crack velocities are normalized by the Rayleigh wave speed $c_R \approx (0.862 + 1.14\nu)/[\sqrt{2}(1 + \nu)^{3/2}]\,\sqrt{E/\rho}$. The dynamic caustics are shown in comparison to the corresponding caustic for a stationary crack. For all crack velocities the shape of the dynamic caustics is practically the same as for the stationary caustic, but the size of the caustic increases slightly with increasing velocity according to the correction factor F.

For optically anisotropic caustics the situation becomes more complicated due to the additional anisotropy term in the equations and the resulting formation of double caustics. It can be shown, however, that the outer caustic increases with crack velocity in a similar manner as the single caustic for optically isotropic materials [9]. The increase in size of the inner caustic with crack velocity is somewhat larger than that for the single caustic. Thus, the isotropic correction factor, F(v), given by Eq. (23), also provides a good approximation for double caustics if applied to the outer caustic.

2.4 Elastic-Plastic Caustics

For cracks in ductile materials (e.g., structural steels) the linear theory of elasticity is no longer appicable. The crack-tip stress field is described by elastic-plastic relations. The stresses are bounded by the yield stress, and consequently the strains near the crack tip become unbounded when the crack tip is approached. Thus, instead of an elastic stress concentration problem, a plastic strain concentration problem is encountered. But analogous to elastic stress concentrations, caustics also result for plastic strain concentrations [6,7].

With regard to an application to steels, the following consideration applies to the reflection case only. Due to the deformation of the specimen surface the distortion of the wave front of light rays according to Eq. (4) is given by

$$\Delta s = -\Delta d \tag{29}$$

where d is the total thickness of the specimen. The reduction in specimen thickness is obtained from the plastic strain ε as

$$\Delta d = \varepsilon_{zz} d \tag{30}$$

Elastic strains ε^e are neglected relative to the plastic strains ε^p, since $\varepsilon^e \ll \varepsilon^p$. Due to the incompressibility of the material under plastic deformation (constancy of volume),

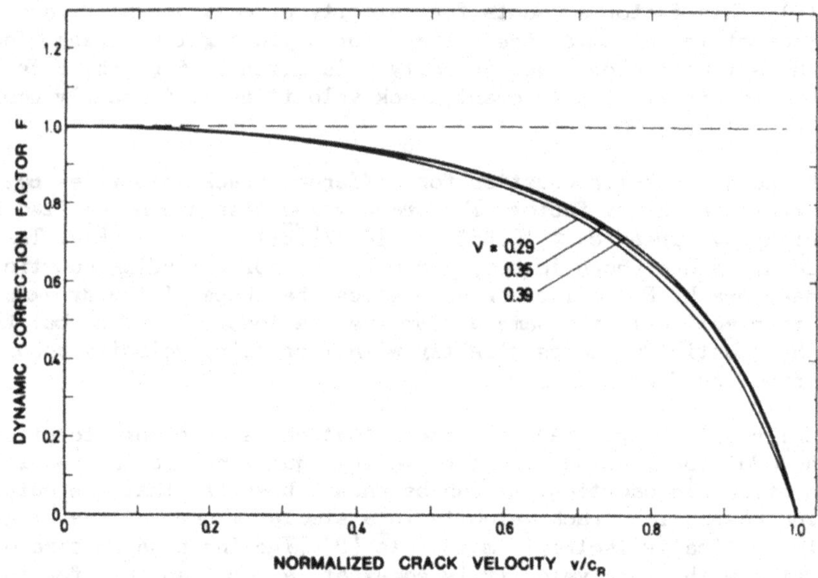

Fig. 15 Correction factor for evaluating dynamic crack tip caustics

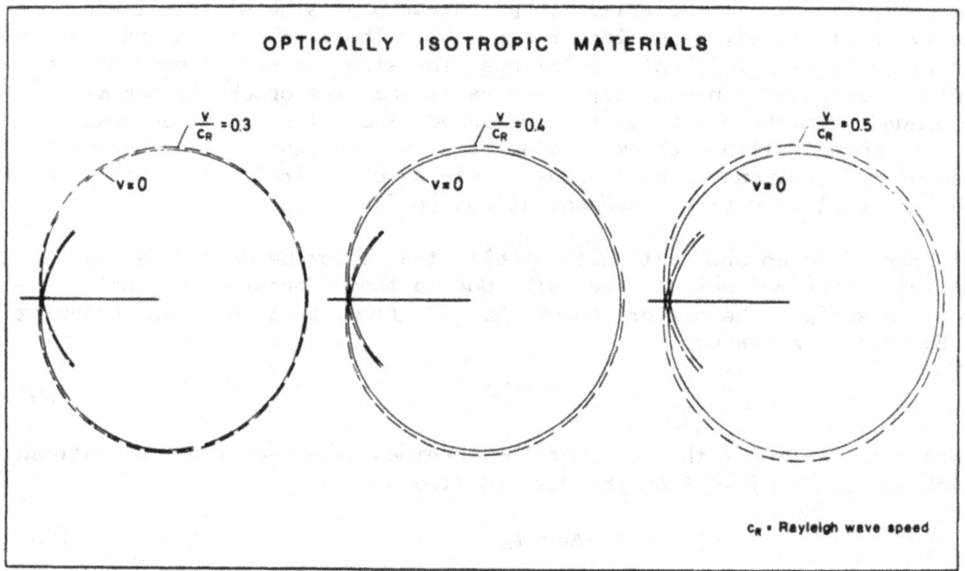

Fig. 16 Caustics for propagating cracks: v = crack propagation velocity

$$\epsilon_{zz} = -(\epsilon_{rr} + \epsilon_{\phi\phi}) \tag{31}$$

Consequently, the mapping equations (1) and (3) take the form

$$\mathbf{r}' = \mathbf{r} + z_0 d \, \mathbf{grad}(\epsilon_{rr} + \epsilon_{\phi\phi}) \tag{32}$$

The strains in the near-field region around a crack tip have been calculated by Hutchinson [20,22], and Rice and Rosengreen [21]. For an elastic-perfectly-plastic material [i.e., a nonhardening material with $n \to \infty$ (see Fig. 17)], the strain values are unbounded in a fan region ($-79.7° < \Phi < +79.7°$) ahead of the crack tip (see Fig. 18). Outside this fan region the strains are bounded by finite values. Quantitatively the strains are given by the following expressions:

$$\left.\begin{aligned} \epsilon_{rr} &= 0 \\ \epsilon_{\phi\phi} &= \frac{J}{2.57\sigma_0 r}\frac{\sqrt{3}}{2}\cos\phi \end{aligned}\right\} -79.7° < \phi < +79.7°$$
$$\left.\begin{aligned} \epsilon_{rr} &= \text{bounded} \\ \epsilon_{\phi\phi} &= \text{bounded} \end{aligned}\right\} \begin{aligned} &-180° < \phi < -79.7° \\ &+79.7° < \phi < +180° \end{aligned} \tag{33}$$

where J is the so-called J integral and σ_0 is the tensile yield stress. J is assumed to be positive. With Eqs. (33) the mapping equations (32) become

$$\left.\begin{aligned} x' &= r\cos\phi - \frac{Jz_0 d\sqrt{3}}{2.57\sigma_0(2)} r^{-2}\cos\phi \\ y' &= r\sin\phi - \frac{Jz_0 d\sqrt{3}}{2.57\sigma_0(2)} r^{-2}\sin\phi \end{aligned}\right\} -79.7° < \phi < +79.7° \tag{34}$$

The bounded strains outside the fan region $-79.7° < \Phi < +79.7°$ cannot contribute to the caustic. Application of Eq. (10) yields the initial curve

$$r = \left[\frac{\sqrt{3}|J|\,|z_0|d}{2.57\sigma_0}\right]^{1/3} \equiv r_0 \tag{35}$$

which again is a circle of fixed radius r_0 around the crack tip (but which is only defined for segment $-79.7° < \Phi < +79.7°$). The mapping of the initial curve (35) onto the reference plane by Eqs. (34) gives the equations of the caustic curve:

$$\left.\begin{aligned} x' &= r_0(\cos\phi + \operatorname{sgn}(z_0)\tfrac{1}{2}\cos 2\phi) \\ y' &= r_0(\sin\phi + \operatorname{sgn}(z_0)\tfrac{1}{2}\sin 2\phi) \end{aligned}\right\} -79.7° < \phi < +79.7° \tag{36}$$

The caustic curve in a reference plane $z_0 < 0$ (virtual image plane) is graphically shown in Fig. 19. Analogous to the elastic case, the characteristic length parameter is defined as the maximum diameter D of the caustic in the y-direction, which is related to the radius of the

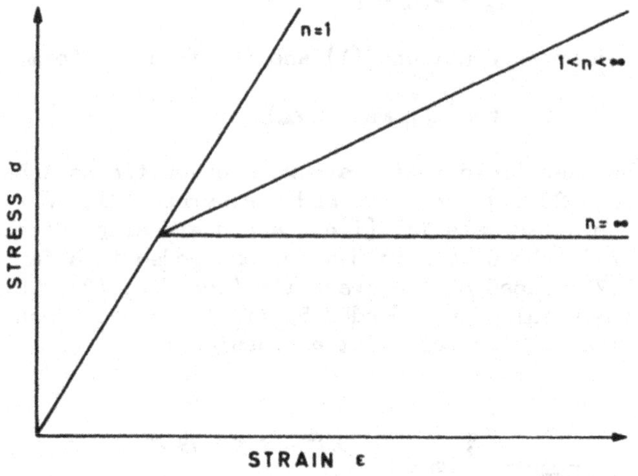

Fig. 17 Elastic-plastic stress-strain behaviour

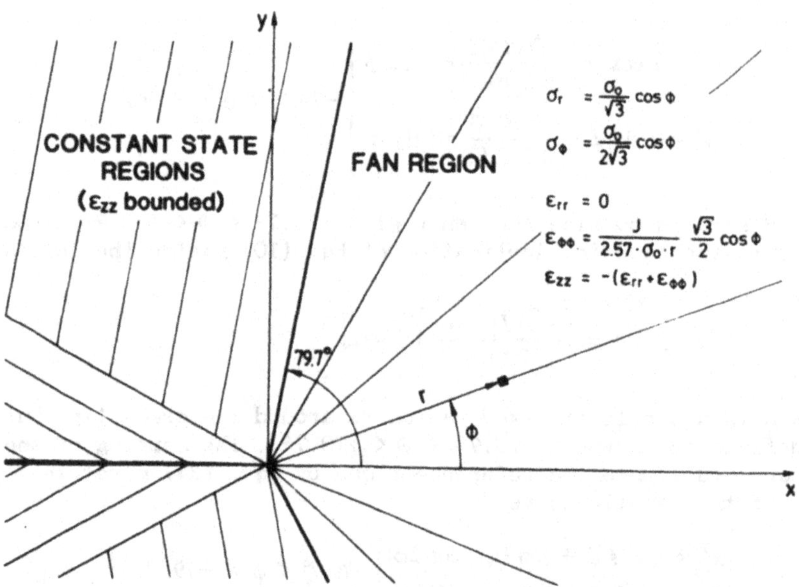

Fig. 18 Elastic-plastic stress-strain field at the tip of a crack, HRR field

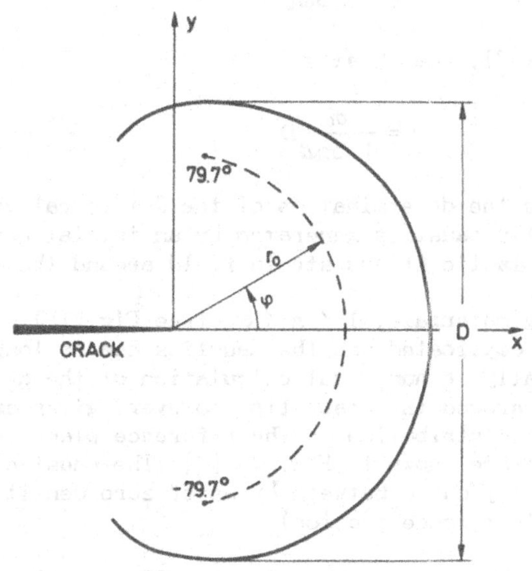

Fig. 19 Elastic-plastic crack tip caustic, n = ∞ (from [6])

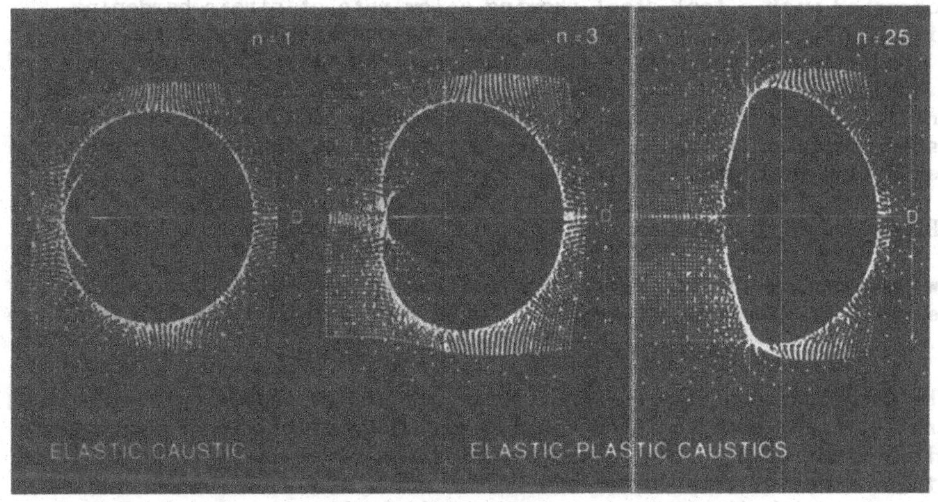

Fig. 20 Shadow optical light distributions for a crack in an elastic-plastic material (from [7])

initial curve by

$$D = 2.38 r_0 \tag{37}$$

With Eqs. (35) and (37), one obtains

$$J = \frac{\sigma_0}{13.5 z_0 d} D^3 \tag{38}$$

This equation allows the determination of the J-integral value from elastic-plastic crack-tip caustics generated by an initial curve located within the elastic-plastic stress strain field around the crack tip.

For strain-hardening materials, $1 < n < \infty$ (see Fig. 17), the strain fields become more complicated and the caustics can no longer be calculated analytically. A numerical calculation of the deflections of an array of light rays around the crack tip, however, gives an illustrative picture of the light distribution in the reference plane. Results obtained for n = 25, 3, and 1 are shown in Fig. 20 [7]. The caustic curve becomes visible as the boundary curve between areas of zero density (shadow area) and high density (light concentration).

It can be seen that the caustic curve for n = 25 resembles the analytically determined caustic curve for nonhardening materials, $n = \infty$. Furthermore, for the limiting case n = 1 (i.e., an elastic material), the caustic curve obtained is identical to the elastic crack tip caustic curve. The caustic curve for n = 3 represents an intermediate stage between these two limiting cases.

Figure 21 shows an experimentally observed elastic-plastic shadow pattern photographed with a tool steel showing a low rate of strain hardening. Despite the formation of slip lines which necessarily disturb the reflection conditions of the polished specimen surface, the shadow pattern is of high quality. The resulting caustic curve is in good agreement with the theoretically derived caustic curve and demonstrates the applicability of the shadow optical method of caustics for investigating plastic strain concentration problems.

2.5 Experimental Techniques

A few comments on the laboratory application of the shadow optical technique shall be given in this review, for a more detailed consideration see [12].

In the description of the physical principle of the shadow optical method, a parallel light beam has been considered. One possibility of realising a parallel light beam in the laboratory is illustrated in Fig. 22. In practice, parallel light beams are seldom used, however. Shadow optical arrangements with divergent or convergent light beams are preferred, shown schematically in Fig. 22 too. The illustrations show the transmission case, but they apply in an analogous manner to the reflection case also. Regardless of what arrangement is utilized, the light beam has to fulfill

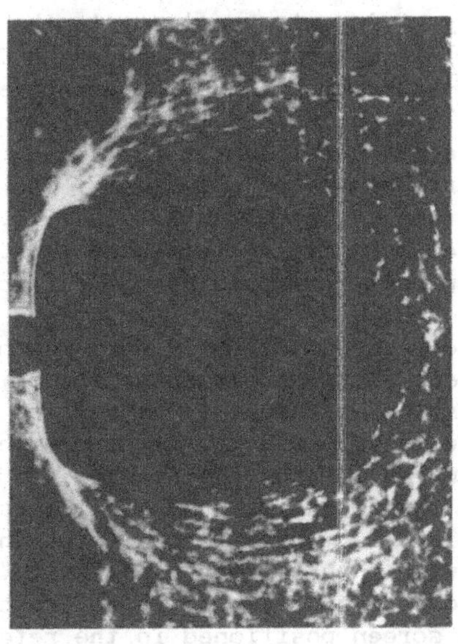

Fig. 21 **Experimentally observed elastic-plastic shadow pattern (from [6])**

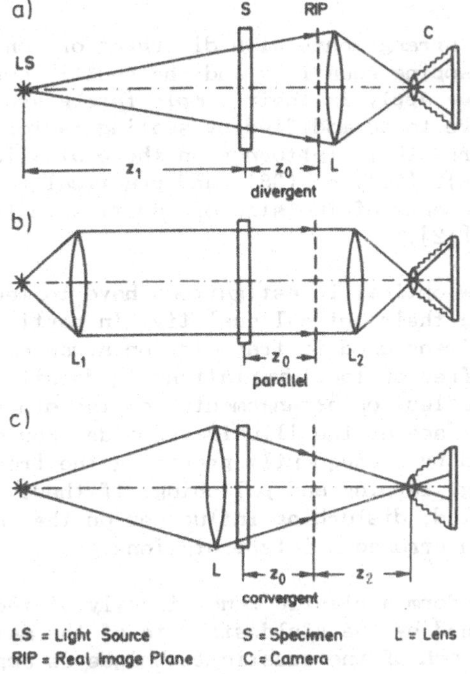

Fig. 22 **Shadow optical arrangements with (a) divergent, (b) parallel, or (c) convergent light beams**

only one, but a very stringent, requirement: the light beam has to be exceptionally parallel, divergent, or convergent. To achieve this property, the light source must have the essential feature of a point source, i.e., a small aperture and a large distance from the object. If these conditions are not sufficiently fulfilled, the shadow optical pattern gets a blurry "nonsharp" appearance. The boundary line between the regions of darkness and light concentration (i.e., the caustic curve) is no longer represented by a distincly marked line. Thus, the quantitative evaluation of shadow patterns would become difficult or even erroneous.

Real caustics can be recorded directly on a photographic film positioned in the reference plane. This direct recording of shadow patterns is possible in transmission arrangements as well as in reflection arrangements. In the latter case the light beam has to be slightly tilted with regard to the normal of the specimen to get a separation of the reflected beam from the impinging beam. The direct recording of shadow patterns, however, is seldom used in practice. Shadow patterns are generally recorded with a photographic camera (e.g., a conventional 35-mm camera) which is focused on the specific reference plane under consideration. The use of a screen positioned in the reference plane is not needed. The recording of shadow patterns with a photographic camera has an advantage in that not only real but also virtual shadow patterns can be registered. In the latter case it is only necessary to focus the camera on the respective virtual reference plane. (For reasons of simplicity, the virtual image planes are not marked in Fig. 22).

When shadow optical arrangements with divergent or convergent light beams are utilized, the mapping equations and the caustic curves for parallel light beams no longer apply in their simple form given by Eqs. (9), (22) or (34) but they have to be modified by scaling factors. A derivation of these scaling factors, their influence on the evaluation formulas, i.e. Eqs. (14), (15), (16), (28), or (38), and practical recommendations for quantitative measurements of the size of caustics in laboratory applications are given in [12].

Specimens for shadow optical investigations have to meet relatively high standards concerning their optical qualitiy. In particular, care must be taken that the specimens used in transmission arrangements are made from materials that are free of local variations in density and in thickness. Specimens used in reflection arrangements, on the other hand, must have an optically planar surface at the illuminated side. For steel specimens best results are obtained by subsequently preparing the front surface of the specimen by grinding, lapping and polishing. If these conditions are not sufficiently fulfilled, disturbing influences on the shadow patterns may appear, resulting in erroneous interpretations.

Every material will form a plastic zone directly at the crack tip the size of which is determined by the yield strength of the material. If the linear-elastic approach of the caustic technique is applied it is therefore necessary to assure that these plasticity effects do not have a disturbing influence on the formation of the caustic. In practice, this is usually achieved by utilizing shadow optical arrangements with an initial curve being larger and lying outside the plastically deformed region. The caustic is then generated from areas where the linear-elastic stress-strain relations remain valid. This requirement is realized experimentally by choosing a sufficiently large distance $|z_0|$ between the specimen and the reference plane so that

$$r_0 > r_{pl} \tag{39}$$

where r_{pl} is the size of the plastic zone. On the other hand, when the elastic-plastic approach of the caustic technique is applied, one has to make sure that the initial curve lies within the plastic zone, i.e.

$$r_0 < r_{pl} \tag{40}$$

Furthermore, it is important to recognize that the values of the shadow optical constant, c, and of the anisotropy coefficient, λ, are different for plane stress and plane strain (see Table 1). The state of stress around the crack tip, however, varies with the distance from the crack tip (plane strain conditions apply very near to the crack tip, plane stress conditions further away from the crack tip, and a mixed state of stress in between). In order to avoid difficulties, the distance $|z_0|$ between the specimen and the reference plane must again be chosen sufficiently large so that the initial curve lies outside this region of mixed state of stress,

$$r_0 > r_{p\sigma} \tag{41}$$

where $r_{p\sigma}$ is the smallest radius around the crack tip outside which a state of plane stress applies.

Application of the shadow optical method for investigating dynamic processes requires the recording of time-dependent shadow patterns with high-speed cameras, e.g. a Cranz-Schardin 24-spark high-speed camera. Basically, the Cranz-Schardin system consists of a photographic recording unit which contains 24 individual cameras with optical axes that meet in the object. The light for illuminating the specimen is generated by 24 sparks which are assembled in a special light unit. The light emitted from one individual spark is focused into the lens of the corresponding individual camera via a field lens or a field mirror. The sparks are triggered at different predetermined times (minimum picture interval time = 1 μs) via an electronic control unit.

For further details on experimenmtal techniques in shadow optical applications see [12,13].

2.6 Summary

An overview on the shadow optical method of caustics for measuring crack tip stress or strain intensifications has been given. It has been shown that the shadow optical method of caustics is based on the deflection of

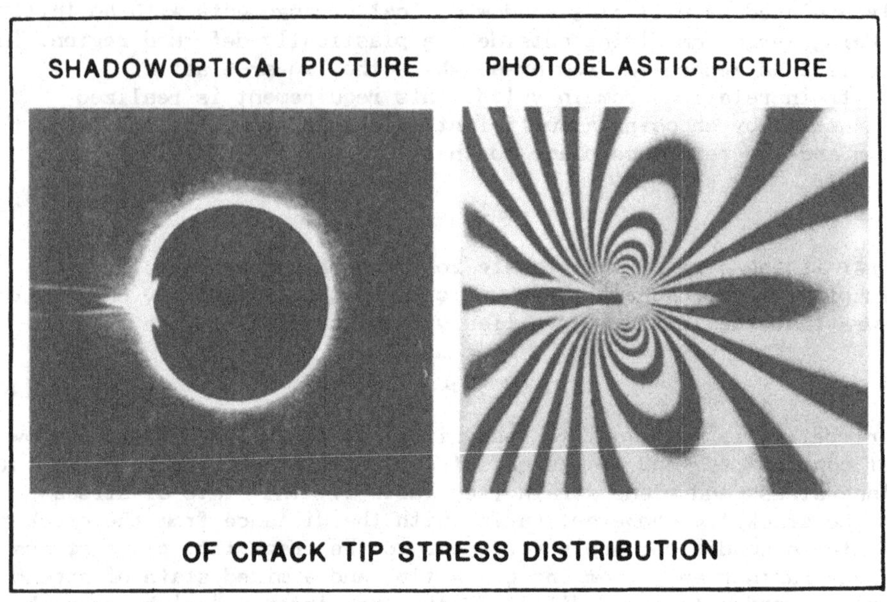

Fig. 23 Comparison of (a) shadow optical and (b) photoelastic crack tip patterns

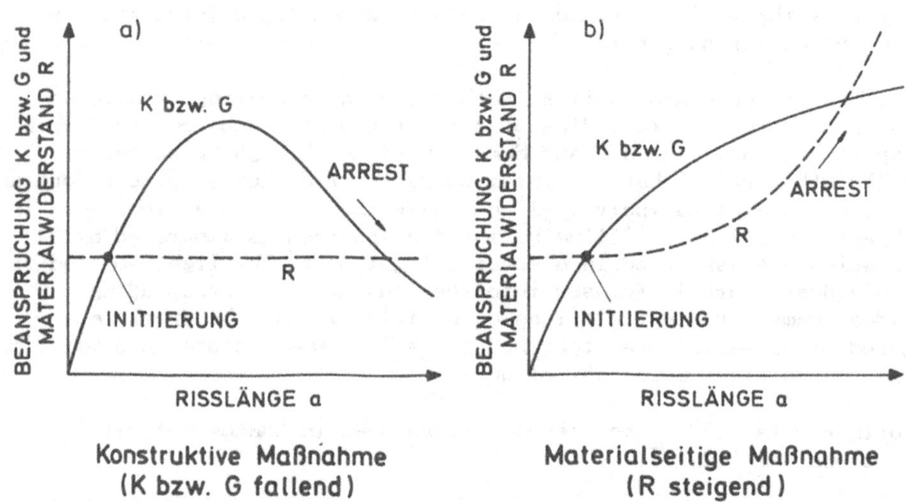

Fig. 24 Techniques for arresting a crack (Beanspruchung = loading, Widerstand = resistance, Rißlänge = crack length, Konstruktive/Materialseitige Maßnahme = design/material provision)

light rays due to gradients in stresses or strains. Most other experimental techniques of stress analysis are based on effects that are directly proportional to stresses or strains. The specific advantages and disadvantages of the shadow optical method of caustics due to these differences in the dependence from the stress become evident by a comparison of a shadow optical pattern and a photoelastic pattern obtained for the same crack-tip stress field shown in Fig. 23. The shadow spot represents a direct quantitative measure of the stress intensification at the crack tip. The photoelastic pattern is more complicated than the shadow optical pattern due to the large number of isochromatic fringes. Thus the evaluation of photoelastic stress concentration patterns naturally becomes a laborious task. In particular, in the near-field region around the crack tip the reduced resolution makes an evaluation difficult, and the derived results are necessarily of reduced accuracy. Furthermore, data on the loading condition in the direct vicinity of the crack tip can be obtained only by extrapolation of data measured at distances farther away from the crack tip. The shadow optical picture does not suffer from these disadvantages. The simplicity and clearness of the shadow optical picture result from the fact that stress distributions with small variations of stress do not become visible. This, however, also implies certain disadvantages. The far-field distributions around the crack tip does not result in a shadow optical effect because of the small stress gradients and therefore cannot be investigated. The photoelastic pattern, however, yields accurate information in particular in the far-field range. The shadow optical method of caustics and the photoelastic method of isochromatic fringes are therefore not to be considered as competitive experimental tools that can be used with the same success and efficiency in obtaining the same kind of data. Depending on the specific property of interest and the problem to be investigated, one or the other measuring technique is more appropriate. Both techniques have their specific ranges of applicability and complement each other. Crack tip singularities, however, are easier visualized and quantified by the caustic technique than by the photoelastic technique.

3. CRACK ARREST

3.1 Introduction

The crack initiation concept of fracture mechanics represents a well accepted methodology for quantifying the safety of structures or components that may exhibit inherent flaws. Test procedures have been established and standardized for measuring material strength values quantifying the resistance of the material against crack initiation under different conditions of loading and material behaviour. Fracture mechanics safety analyses provide both predictions of tolerable crack sizes for specific service loads, and values for the maximum allowable load a structure with a specific detected flaw may safely bear.

Complete safety of a structure, however, cannot be assured by the crack initiation concept, since it is not possible to consider with 100 % certainty all the parameters that may influence fracture. For example, in inspection procedures a crack may be overlooked due to human errors. Or, unforseen overloads caused by local impact or thermal shocks may initiate a crack that would not have been initiated under normal loading conditions. The safe operation of structures is increased if such cracks, that may become unstable, are arrested before they cause complete catastrophic failure of the structure. It is the goal of crack arrest safety analyses to arrest such a crack early enough that the structure can still function to the degree that it is removed from service without loss of human life or damage to the entire structure. Thus, crack arrest safety analyses represent a safety reserve. They are applied in particular to structures which have to meet very high safety standards.

There are two principally different ways to arrest a crack, illustrated in Fig. 24 [23,24]. First, the stress intensity factor K or the energy release rate G is decreased when the crack extends, i.e. the crack propagates into a decreasing K-field (see Fig. 24a). Secondly, the resistance of the material against crack propagation (i.e. the fracture toughness K_{Ic} or G_{Ic} - in Fig. 24 denoted by the symbol R) is increased, i.e. the crack propagates into an area of increasing toughness (see Fig. 24b).

Both techniques are used in practice. Figure 25 shows the reduction of the stress intensity factor for an expanding crack in a decreasing K-field

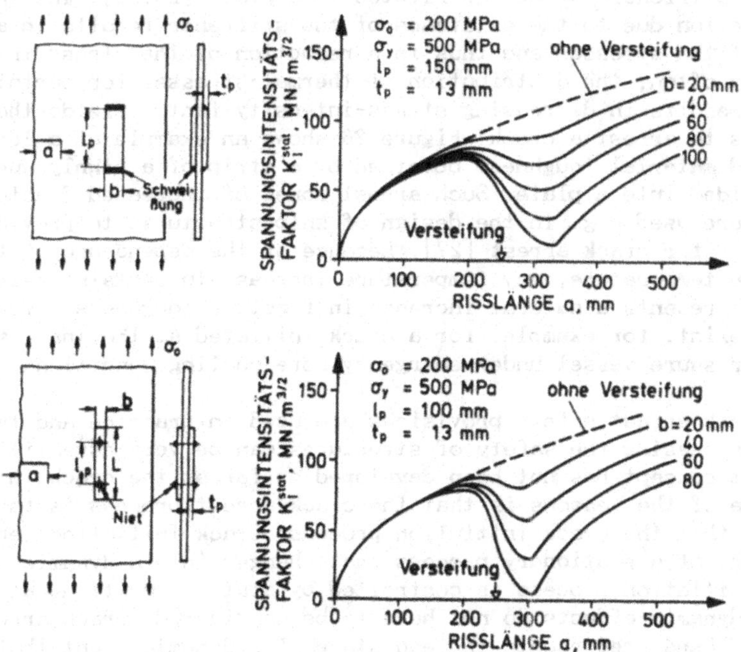

Fig. 25 Decreasing stress-intensity-factor field acheived by stiffeners (after [24,25])
(Spannungsintensitätsfaktor = stress intensity factor, ohne Versteifung = without stiffener, Rißlänge = crack length)

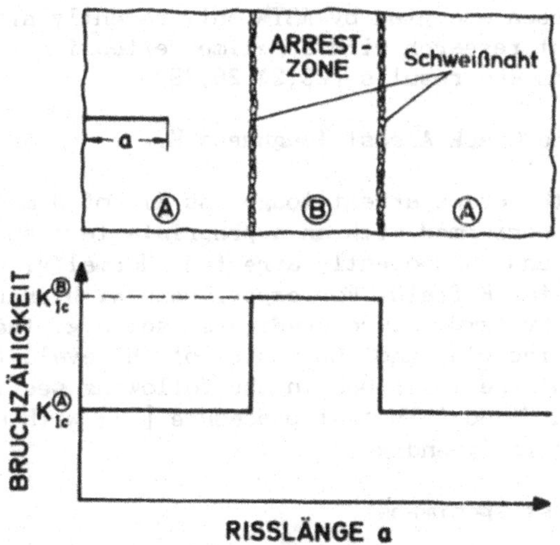

Fig. 26 Increasing fracture toughness by arrest zone
(Schweißnaht = weldment, Bruchzähigkeit = fracture toughness, Rißlänge = crack length)

achieved by stiffeners welded or riveted to a plate [25,26]. The increase in cross section due to the existence of the stiffener results in a reduction of the stresses and thus in a reduction of the stress intensity factor. Very often, the distribution of thermal-stresses for certain crack paths also results in decreasing stress-intensity-factor fields that provide ways to arrest a crack. Figure 26 shows an example of a local zone of increased material toughness obtained by a strip of a highly ductile material welded into a plate. Such arrest zones of increased fracture resistance are used e.g. in the design of ship structures to provide possibilities for crack arrest [27]. Because of the dependence of fracture toughness on temperature, any temperature increase in certain areas of a structure represents a natural increase in fracture toughness. These conditions exist, for example, for a crack initiated at the inner wall of a reactor pressure vessel under emergency core cooling conditions.

Although crack arrest safety provisions are used in practice and their effect in increasing the safety of structures can be very essential, the crack arrest concept has not been developed as far as the crack initiation concept. One of the reasons is that the crack arrest process is far more complicated than the crack initiation process. Crack initiation represents the end stage of a stationary process that changes into a dynamic one. Thus, the initiation process is controlled by static conditions at the crack tip; dynamic effects do not have to be considered. Crack arrest, on the other hand, represents the end stage of a dynamic event that changes into a static stage. Thus, the possible influences of dynamic effects on the crack arrest process associated with the previous crack propagation event have in principle to be considered. These dynamic effects may influence the mechanics of the crack arrest process, i.e. the stress intensity factor at arrest. Furthermore, they will influence the material behaviour itself. A procedure for measuring the crack arrest toughness, K_{Ia}, has been accepted by ASTM only recently after a long period of crack arrest research with sometimes extensive discussions of controversial opinions and results [23,24,28,29].

3.2 Measurement of the Crack Arrest Toughness K_{Ia}

In order to measure the crack arrest toughness K_{Ia} of a material, laboratory experiments are performed with an appropriate test specimen in which a crack is initiated and subsequently arrested. Normally, crack arrest is achieved by a decreasing K-field. For experiments with increasing material toughnesses obtained by temperature gradients, see e.g. the Robertson test [30]. The physical principles and the basics of the evaluation procedure of a crack arrest test are described in the following sections (see also [23, 24]). The standardized ASTM test procedure [31] with all experimental details is reproduced in Appendix I.

3.2.1 Crack arrest test specimens

Double-cantilever-beam (DCB) specimens (rectangular or tapered) and compact (C) specimens are usually used in crack arrest tests. The geometry of the rectangular DCB specimen and the C specimen, are given in Figs. 27

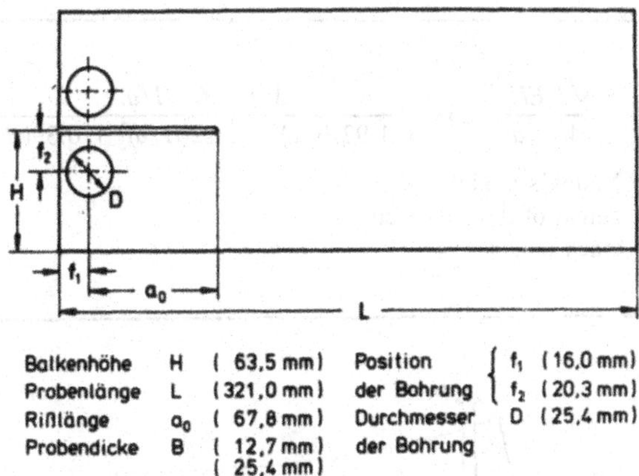

Fig. 27 Rectangular double-cantilever-beam (DCB) specimen (Balkenhöhe = beam height, Probenlänge = specimen length, Rißlänge = crack length, Probendicke = specimen thickness, Position/Durchmesser der Bohrung = center/diameter of hole)

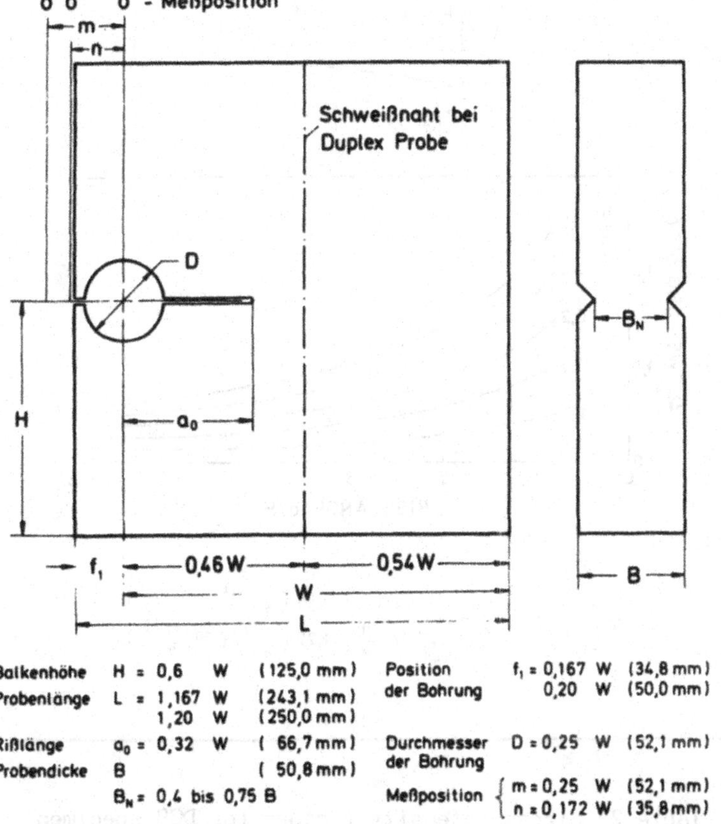

Fig. 28 Compact (C) specimen (see above, Meßposition = measuring point)

$$K_I = \frac{\sqrt{3}}{4}\frac{EH^{3/2}}{a^2}2\delta\frac{1 + 0.64(H/a)}{1 + 1.92(H/a) + 1.22(H/a)^2 + 0.39(H/a)^3}$$

where E = Young's modulus
$2H$ = height of the specimen
a = crack length

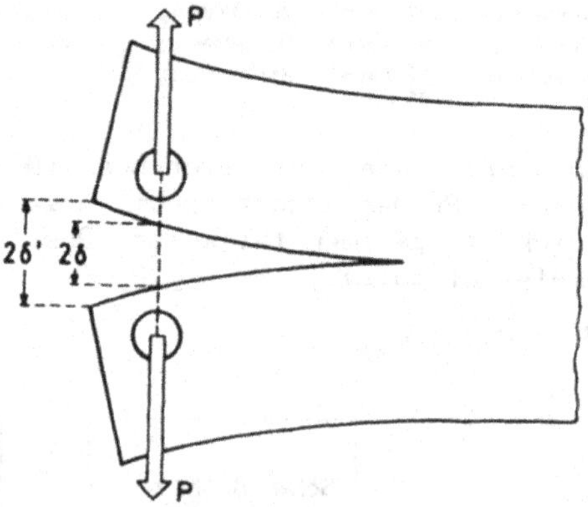

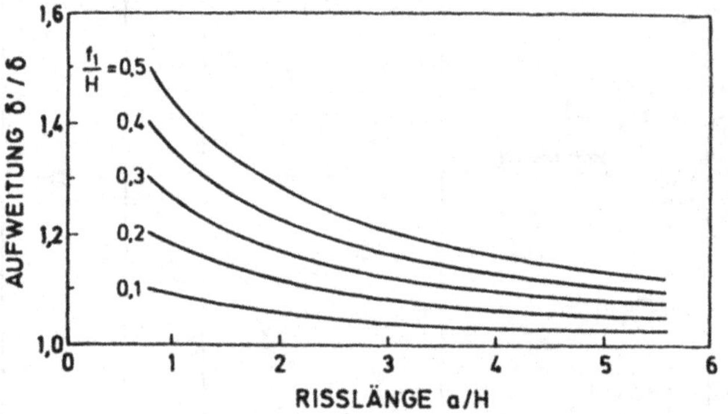

$$\frac{\delta'}{\delta} = 1 + \frac{3}{2}\frac{f_1}{a}\left[\frac{\left(1 + 0{,}64\left(\frac{H}{a}\right)\right)^2}{1 + 1{,}92\left(\frac{H}{a}\right) + 1{,}22\left(\frac{H}{a}\right)^2 + 0{,}39\left(\frac{H}{a}\right)^3}\right]$$

Table 2 Stress intensity factors for DCB specimen (Aufweitung = displacement; Rißlänge = crack length)

$\frac{a}{W}$	$\frac{K_I^{stat}\sqrt{W}}{E\,2\delta}$	$\frac{K_I^{stat}\sqrt{W}}{E\,2\delta'}$	$\frac{K_I^{stat}\sqrt{W}}{E\,2\delta''}$
.30	.369	.275	.219
.31	.363	.271	.218
.32	.357	.267	.217
.33	.351	.263	.216
.34	.345	.259	.215
.35	.339	.256	.214
.36	.333	.252	.212
.37	.327	.248	.210
.38	.321	.245	.209
.39	.316	.240	.207
.40	.310	.237	.205
.41	.304	.233	.202
.42	.299	.229	.200
.43	.293	.226	.198
.44	.288	.223	.196
.45	.283	.219	.194
.46	.278	.215	.191
.47	.272	.212	.189
.48	.267	.209	.186
.49	.262	.205	.184
.50	.257	.202	.182
.51	.253	.199	.179
.52	.248	.195	.177
.53	.243	.192	.174
.54	.239	.189	.172
.55	.234	.186	.170
.56	.230	.183	.167
.57	.225	.180	.165
.58	.221	.177	.162
.59	.217	.174	.160
.60	.213	.171	.157
.61	.209	.168	.155
.62	.205	.165	.153
.63	.201	.163	.150
.64	.197	.160	.148
.65	.194	.158	.145
.66	.190	.155	.143
.67	.186	.152	.140
.68	.183	.150	.138
.69	.179	.147	.135
.70	.175	.144	.133
.71	.171	.142	.130
.72	.168	.139	.127
.73	.164	.136	.125
.74	.160	.133	.122
.75	.156	.130	.119
.76	.151	.127	.116
.77	.147	.123	.113
.78	.142	.120	.110
.79	.138	.116	.107

Table 3 Stress intensity factors for C specimens

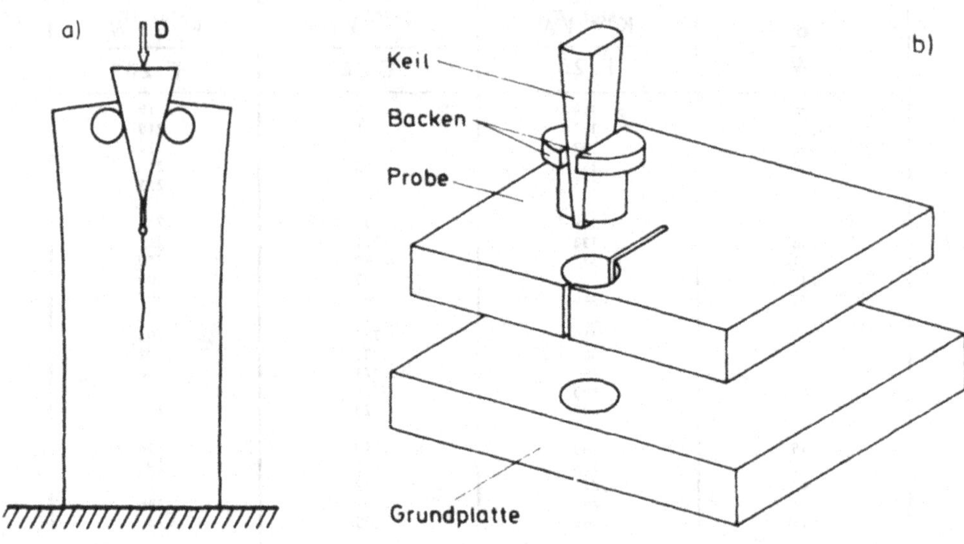

Fig. 29 Longitudinal (a) and transverse (b) wedge loading (Keil = wedge, Backen = split-D, Probe = specimen, Grundplatte = base plate)

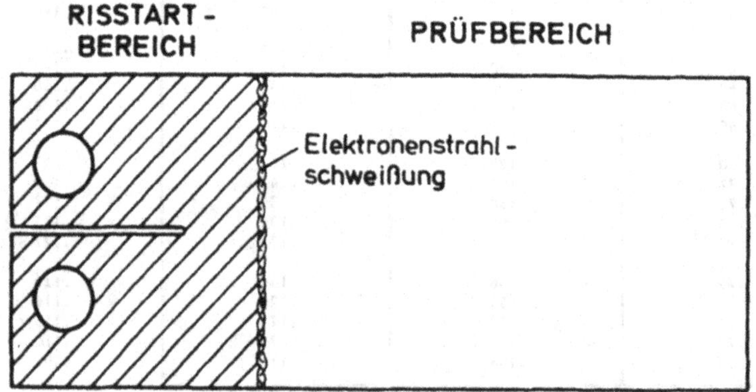

Fig. 30 Duplex specimen (Elektronenstrahlschweißung = electron beam weldment, Rißstartbereich = crack initiation section, Prüfbereich = test section)

consequently the higher the crack propagtion velocity. With materials that show ductile behaviour, the cracks are initiated from locally embrittled zones, e.g. from brittle welds at the root of the starter notch (similar to the Pellini-test). Or, so called duplex specimens are utilized, which consist of a brittle starter section electron beam welded to the arrest section of the specimen, made from the material to be tested (see Figs. 30, 31, and 28). A fast propagating crack is initiated in the brittle starter area and then penetrates into the test material where it is arrested later on.

3.2.2 Test procedure and data evaluation

A crack is initiated from an initial notch of length a_0 at a displacement-value $2\delta_c$. The crack length a_0 and the displacement $2\delta_c$ determine the crack initiation stress intensity factor, K_{Io} or K_{Iq}. The crack propagates into a decreasing K-field and comes to arrest at an arrest crack length, a_a. The arrest crack length is determined by inspection of the fracture surface after the test. The fracture surface is first marked by heat tinting, then the specimen is broken open completely to allow for the crack length measurement.

The crack arrest toughness value K_{Ia} is derived from these data on the basis of the following concept [31-33] (see also Fig. 32). It is assumed that a static stress intensity factor prevails at the moment of arrest, i.e. crack arrest is assumed to occur if the quasistatically determined stress intensity factor curve $K_I(2\delta_c, a)$ falls below the material resistance curve (denoted R in Fig. 32). In a more precise formulation, the concept admits that a static stress condition will not apply directly at the moment of arrest, but that this will be the case at a short time after arrest (1 ms). In this modified version of the concept this stress intensity factor at a short time after arrest is denoted K_{Ia} and is assumed to be a good estimate of the stress intensity factor prevailing exactly at the moment of arrest.

Thus, the crack arrest process is basically considered as a crack initiation process, but with the time scale being reversed. Consequently, the crack arrest toughness K_{Ia} is determined from the measured data δ_c and a_a by the relationship

$$K_{Ia} = K_I(\delta_c, a_a) \quad (43)$$

where K_I is the quasistatic stress intensity factor calculated by the relationships given in Tables 2 and 3. For details concerning validity conditions, limitations in the use of test parameters, and practical recommendations concerning test specimens, loading fixtures, and operational procedures, see the ASTM test procedure given in Appendix I.

and 28 respectively. Typical specimen dimensions for testing a reactor pressure vessel steel are given in brackets in the inserts of Figs. 27 and 28. A decreasing function of the stress intensity factor for the expanding crack is obtained by wedge loading. The usual arrangement for longitudinal wedge loading is shown in Fig. 29a, and for transverse wedge loading in Fig. 29b. The wedge is quasistatically driven into the specimen, thus loading the crack by an opening displacement 2δ (measured at the load line or $2\delta'$ if measured at the end of the specimen, see Table 2). Because of the stiffness of the wedge loading system, the displacement stays practically constant during crack propagation. Thus, due to the increasing compliance of the specimen, the stress intensity factor is reduced with increasing crack length. Quantitative relationships for the stress intensity factor as a function of crack length are given by Table 2 for the DCB specimen and by Table 3 for the C specimen.

In order to assure straight crack propagation, side grooves (45°, 0.25 mm root radius) are usually used with crack arrest specimens (see Fig. 28). The side grooves have another advantage in that the multiaxiality of the crack tip stress field in the near surface area of the specimen is increased. Thus, with an appropriate choice of the depths of the side grooves, it is possible to achieve a state of stress that resembles that in the inner part of the specimen. When side grooves are used, the stress intensity factor is increased with regard to the usual stress intensity factor. On the basis of energy considerations it is concluded that

$$K^{sg} = \sqrt{B/B_N}\, K \qquad (42)$$

where K^{sg} is the stress intensity factor value with side grooves and K without. For the definition of B and B_N see Fig. 28.

As with procedures for measuring the crack initiation toughness, K_{Ic}, crack arrest test specimens must fulfill certain minimum size requirements (in particular with regard to thickness) to result in geometry independent material toughness values. The size of the plastic zone at a crack tip is given in terms of $(K_I/\sigma_Y)^2$, with σ_Y = yield stress. In dynamic cases, i.e. for propagating and subsequently arresting cracks, $\sigma_Y^{dyn} > \sigma_Y^{stat}$. Consequently, the size of plastic zones for dynamic problems in general is smaller than for the equivalent static case. Thus, the minimum size of crack arrest test specimens in general is smaller (up to about a factor of 10) than for usual K_{Ic}-specimens. Thus, the linear elastic regime of fracture mechanics in crack arrest tests is valid over a considerably larger range than for crack initiation investigations.

Depending on the material to be tested, different procedures are used to initiate a fast propagating crack. With specimens that show brittle behaviour, cracks are initiated from blunted notches. The stress intensity factor at initiation (denoted K_{Iq} or K_{Io}) thus becomes larger than the fracture toughness K_{Ic} of the material. As a consequence, considerable elastic energy can be stored in the specimen which then is used by the crack to propagate at a high velocity through the specimen. The larger the crack tip bluntness the larger the initiation stress intensity factor and

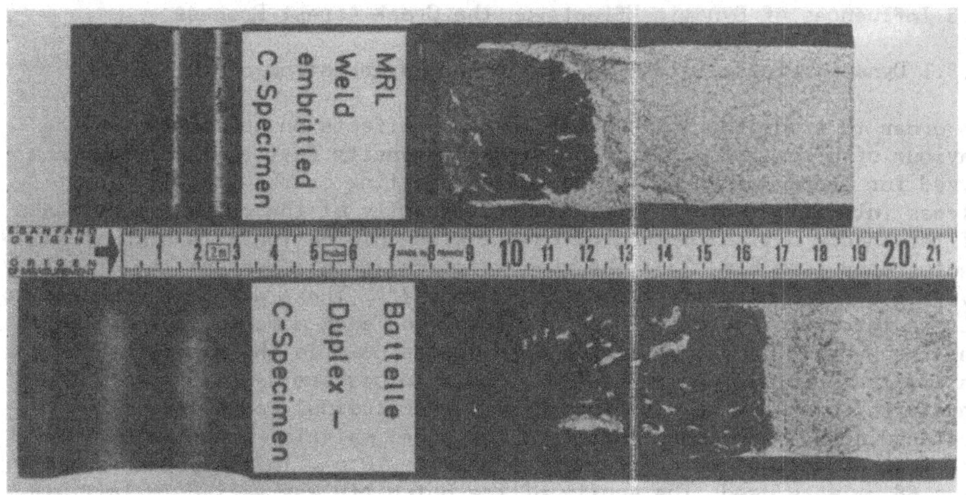

Fig. 31 Fracture surfaces of crack arrest specimens

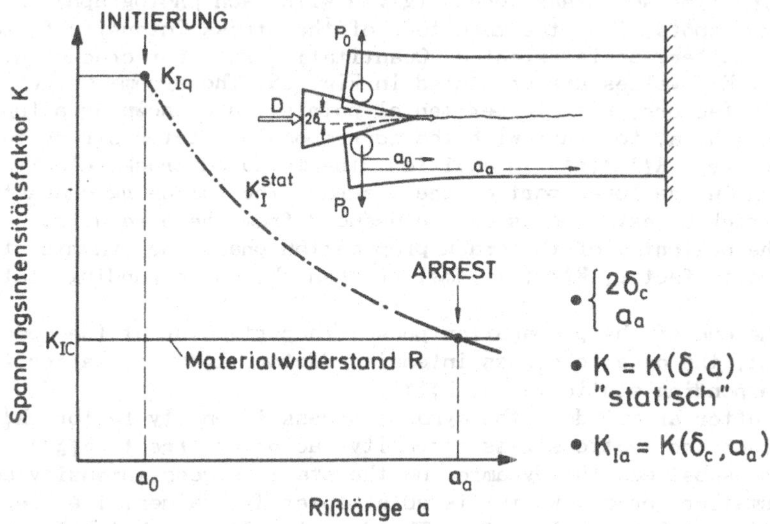

Fig. 32 Determination of the crack arrest toughness $K_{Ia} = K_{Ia}^{stat}$ (Spannungsintensitätsfaktor = stress intensity factor, Widerstand = resistance, Initiierung = initiation, Rißlänge = crack length, statisch = static)

3.3 Influences of Dynamic Effects on the Crack Arrest Process

3.3.1 Dynamic stress intensity factor for an arresting crack

In order to study the influence of dynamic effects on the mechanical behaviour of arresting cracks, the stress intensity factor history was analyzed for propagating and subsequently arresting cracks [34,35]. The stress intensity factors were measured directly at the crack tip by means of the shadow optical method of caustics in combination with high speed photography (see Fig. 33). The shadow optical data represent the actual dynamic stress intensity factors, denoted K_I^{dyn}. These data are compared to stress intensity factor values calculated for each momentary crack length and the prevailing crack opening displacement δ_c via the static stress intensity formula given in Table 2. These values are denoted K_I^{stat}. The investigations have been performed with longitudinal wedge loaded rectangular DCB specimens made from the model material Araldite B. The specimen dimensions were 321 mm by 127 mm by 10 mm. Initial notches with a_0 = 66 mm were used. The acuity of the notch tip was varied to initiate cracks at different K_{Iq}-values. With the shadow optical method of caustics applied in transmission, real shadow patterns were recorded in a reference plane $z_0 > 0$ with a Cranz-Schardin 24-spark high speed camera. The camera was triggered by the crack itself which, at the onset of propagation, interrupted a laser beam to provide the trigger pulse.

A series of six shadow optical photographs is shown in Fig. 34. With increasing time and crack length (given with each photograph) the size of the shadow spots, i.e. the magnitude of the stress intensity factors, becomes smaller, as is expected. Quantitative data for cracks initiated at different K_{Iq}-values are presented in Fig. 35. The dynamic stress intensity factors, K_I^{dyn} (experimental points), are shown as a function of crack length, a, together with the corresponding static stress intensity factor curves, $K_I^{stat}(a)$. In addition, the measured crack velocities, v, are given in the lower part of the diagram. The following characteristics of the crack arrest process can be deduced from these results:

- At the beginning of the crack propagation phase the dynamic stress intensity factor, K_I^{dyn}, is smaller than the corresponding static value, K_I^{stat}.
- At the end of the propagation phase, in particular at the moment of arrest, the dynamic stress intensity factor, K_I^{dyn}, is larger than the corresponding static value, K_I^{stat}.
- Only after arrest does the dynamic stress intensity factor, K_I^{dyn}, approach the static stress intensity factor at arrest, K_{Ia}^{stat}.

Differences between the dynamic and the static stress intensity curves become smaller for cracks initiated at lower K_{Iq}-values, i.e. for cracks propagating at lower velocities. The dynamic effects obviously decrease with decreasing velocity as one might expect. Furthermore, it is seen from Fig. 35 that the higher the crack initiation stress intensity factor, K_{Iq}, the larger the resulting crack velocity, v, and the larger the resulting crack jump distance, Δa.

Implications of this behaviour on the determination of the crack arrest

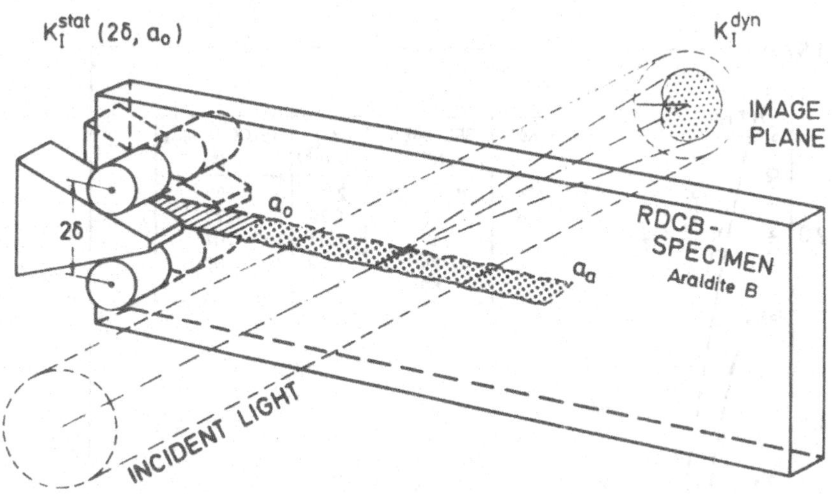

Fig. 33 Experimental set-up (schematically)

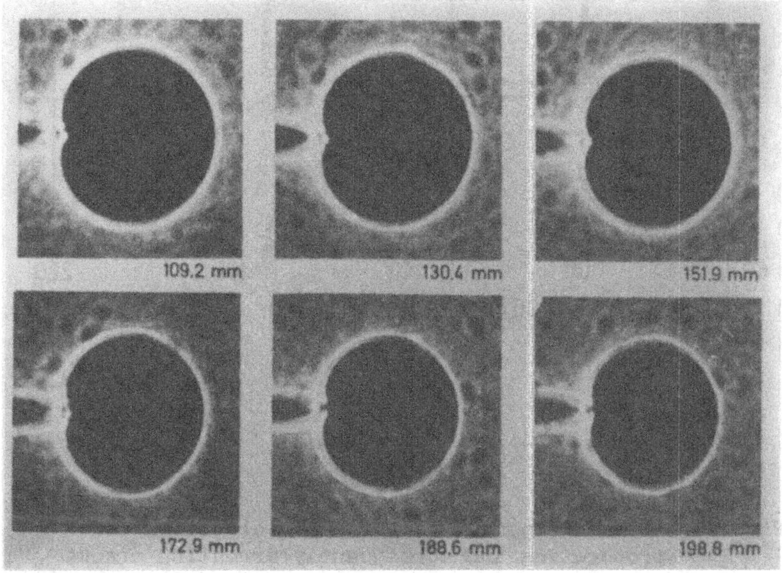

Fig. 34 Caustics for a fast propagating and subsequently arresting crack (Araldite B, real image, z_o > 0, transmission)

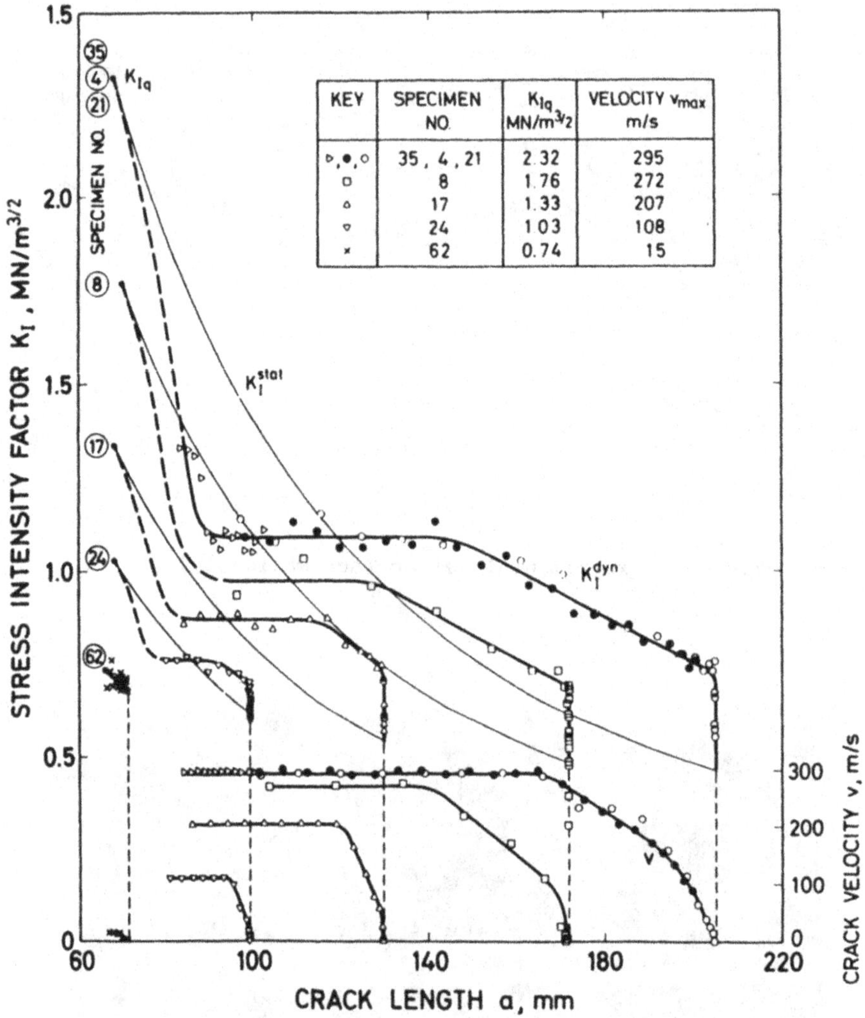

Fig. 35 Stress intensity factors and velocities for propagating and subsequently arresting cracks

toughness are illustrated in Fig. 36. The value of the K_I^{stat}-curve at the arrest crack length a_a, denoted K_{Ia}^{stat} for clarity in this context, represents the crack arrest toughness K_{Ia}, as determined by the ASTM test procedure. The value of the dynamic stress intensity factor curve, K_I^{dyn}, at the arrest crack length, a_a, represents the actual stress intensity factor at the moment of arrest, i.e., this value represents the real crack arrest toughness, denoted K_{Ia}^{dyn} for clarity in this context. The ASTM procedure considers $K_{Ia} = K_{Ia}^{stat}$ to be a reasonable approximation of the real crack arrest toughness K_{Ia}^{dyn}. It is recognized from Fig. 36, however, that K_{Ia}^{stat} values decrease with increasing crack jump distance, Δa, i.e., with the crack velocity prior to arrest, whereas K_{Ia}^{dyn} scatters around a horizontal line. The differences between K_{Ia}^{dyn} and K_{Ia}^{stat} are remarkable. Thus, it must be concluded that the quasistatically determined crack arrest toughness does not represent a good estimate of the true arrest property, but does depend on the particular test conditions. The dynamically determined toughness, however, obviously does not depend on the choice of the test conditions.

Such behaviour was predicted by Hahn et al. [23,24,28,29,36,37]. The authors analyzed the crack arrest process using an energy balance approach. According to their theoretical results, as the crack accelerates in the initial phase of crack propagation, kinetic energy in the form of stress waves is built up in the specimen which is subsequently available to contribute to the stress intensity factor at arrest. Thus, as illustrated in Fig. 37, at the beginning of the crack propagation phase, the dynamic stress intensity factor, K_I^{dyn}, should be smaller than the corresponding static value, K_I^{stat}, while at arrest, it should be larger. This result implies that a crack can propagate even if K_I^{stat} becomes less than the material restistance. Consequently, an arrest length a_a^{**} larger than a_a^* is obtained, where a_a^* is the arrest crack length that would result from the quasistatic crack arrest concept. Furthermore, a dynamically determined crack arrest toughness, K_{Ia}^{dyn}, should be larger than a quasistatically determined value, K_{Ia}^{stat}. The reported experimental findings based on shadow optical investigations confirm this dynamic concept of recovered kinetic energy.

The stress intensity factor behaviour directly before, at, and after arrest has been studied in more detail. The results obtained from caustic investigations are shown in Fig. 38. The dynamic stress intensity factor, K_I^{dyn}, is plotted as a function of time. K_I^{dyn} oscillates around the value of the static stress intensity factor at arrest, K_{Ia}^{stat}, via an oscillation with damped amplitude. The initial amplitude of this oscillation, i.e. $K_{Ia}^{dyn} - K_{Ia}^{stat}$, is a measure of the influence of dynamic effects on the crack arrest process. Only at a large time after arrest does the dynamic stress intensity factor approach this static value.

Experiments similar to those performed with the rectangular DCB specimens have been performed with specimens of other geometries. Figure 39 shows results for the tapered DCB specimen and the C specicmen in comparison to data obtained for the DCB specimen. For all three types of specimens the dynamic stress intensity factor, K_I^{dyn}, after arrest shows an oscillitary

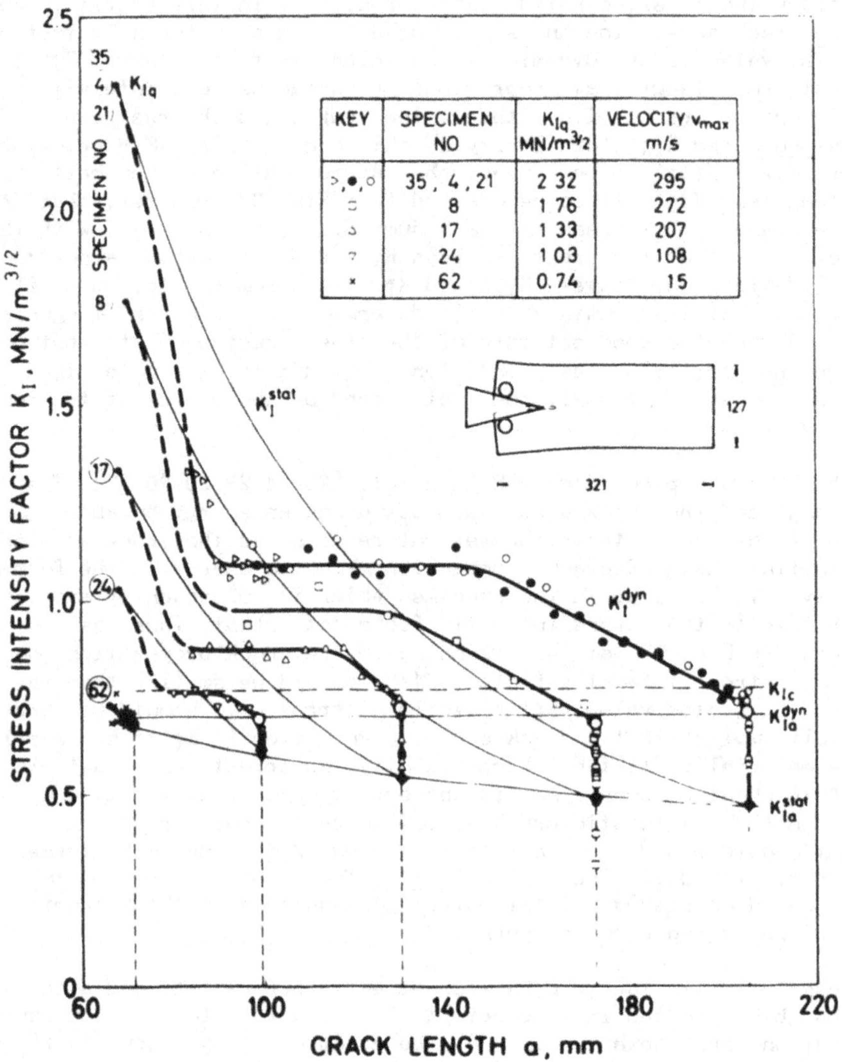

Fig. 36 Crack arrest toughness values determined quasistatically, K_{Ia}^{stat}, and dynamically, K_{Ia}^{dyn}

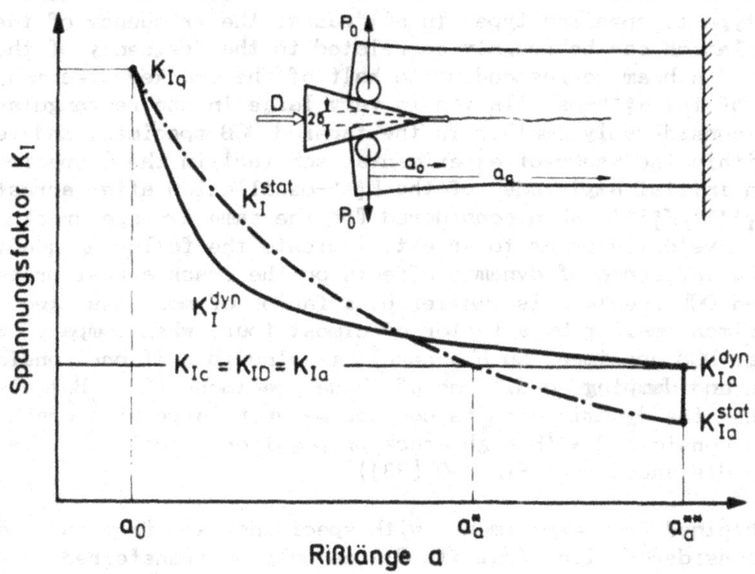

Fig. 37 Influence of recovered kinetic energy on the crack arrest process
(Spannungsfaktor = stress intensity factor, Rißlänge = crack length

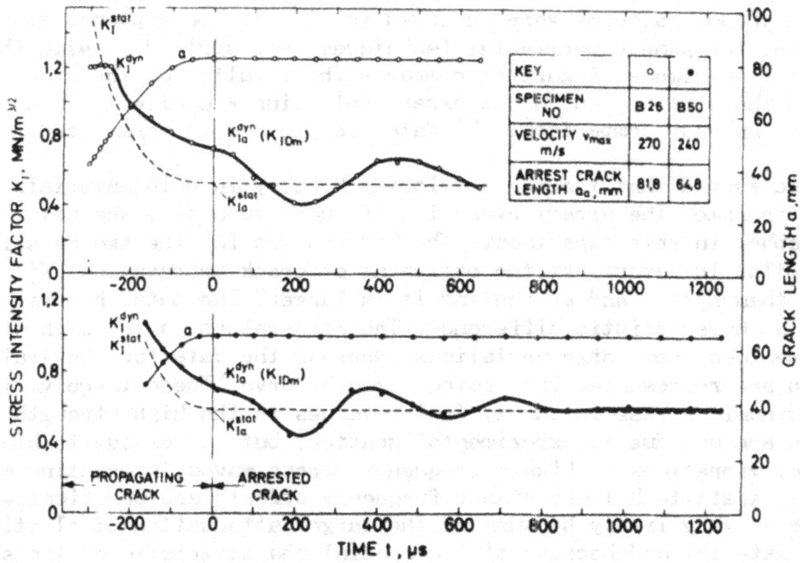

Fig. 38 Stress intensity factors after crack arrest

behaviour. The amplitude and the frequency of these oscillations vary from specimen type to specimen type. In all cases, the frequency of the K_I^{dyn}-oscillation can be roughly correlated to the frequency of the vibration of a beam corresponding to half of the cracked specimen. The amplitude of the K_I^{dyn}-oscillation is very large in the rectangular DCB specimen, considerably smaller in the tapered DCB specimen, and very low (almost within the range of experimental scatter) in the C specimen. The normalized initial amplitudes of the K_I^{dyn}-oscillation after arrest, $(K_I^{dyn} - K_I^{stat})/K_I^{stat}$, when considered for the same average crack propagation velocity prior to arrest, indicate the following quantitative result. The influence of dynamic effects on the crack arrest process in the tapered DCB specimen is smaller by a factor of more than two, and in the C specimen smaller by a factor of almost four, when compared with the rectangular DCB specimen. Such a result is plausible if one considers the reflection and damping conditions of these specimens [23]. But also with C specimens, the dynamic effects can become very large when crack arrest events are considered with high crack propagation velocities, i.e., large crack jump distances (see Fig. 40 [38]).

Results obtained from experiments with specimens made from the model material considered, i.e. Araldite B, can only be transferred to steel specimens with caution, since the velocities and the damping of elastic waves in the materials, the velocities of the propagating cracks, etc. are very different. Therefore, experiments similar to those performed with Araldite B specimens have been performed with specimens made from steel. The high strength maraging steel X2 NiCoNo 18 9 5 (similar to the American designation - 18 Ni maraging grade 300) was used. Because of its linear elastic material behaviour, this steel is very well suited for shadow optical investigations. The shadow optical technique was applied in reflection and shadow patterns were observed in virtual image planes $z_0 < 0$. Otherwise, the same experimental techniques were applied as with the Araldite B specimens. Figure 41 compares the results for the steel specimen with those of an equivalent experiment using Araldite B. In order to allow for an easy comparison, the data are plotted in normalized form.

It is recognized that the crack velocity history in both materials is almost the same. The arrest event itself, unfortunately, was not photographed in this experiment. The K_I^{dyn}-curves for the two materials are also similar in nature. At the beginning of crack propagation, K_I^{dyn} is smaller than K_I^{stat}, and at the end it is larger. The data, however, exhibit a characteristic difference. The K_I^{dyn}-values in the high strength steel specimen show large variations, whereas the data for the Araldite B specimen are represented by a rather smooth curve. These large variations of the dynamic stress intensity factor values in the high strength steel specimen are not due to experimental scatter, but rather due to higher frequency vibrations or higher frequency stress waves interacting with the crack. In Araldite B these higher frequency effects are practically not noticeable, very likely because of the larger attenuation of elastic waves in this material and because of the special characteristic of the shadow optical method of caustics, being less sensitive to surface deformations in transmission than in reflection.

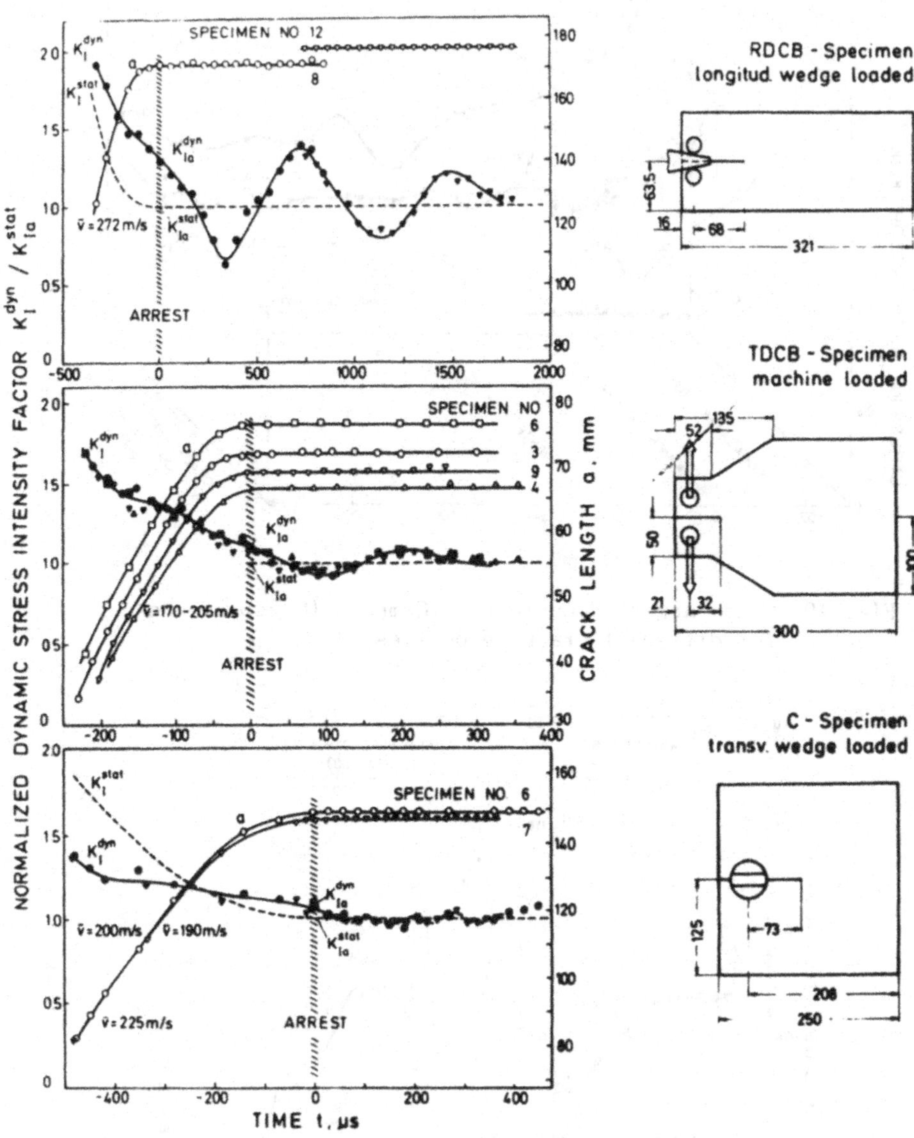

Fig. 39 Influences of dynamic effects on the crack arrest process measured with various crack arrest test specimens

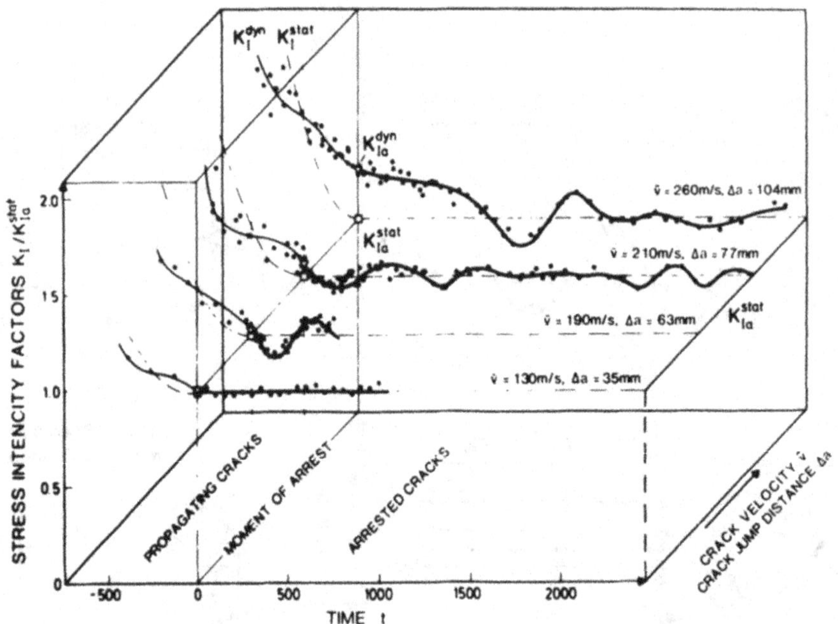

Fig. 40 Crack arrest behaviour in Compact (C) specimens for different crack velocities

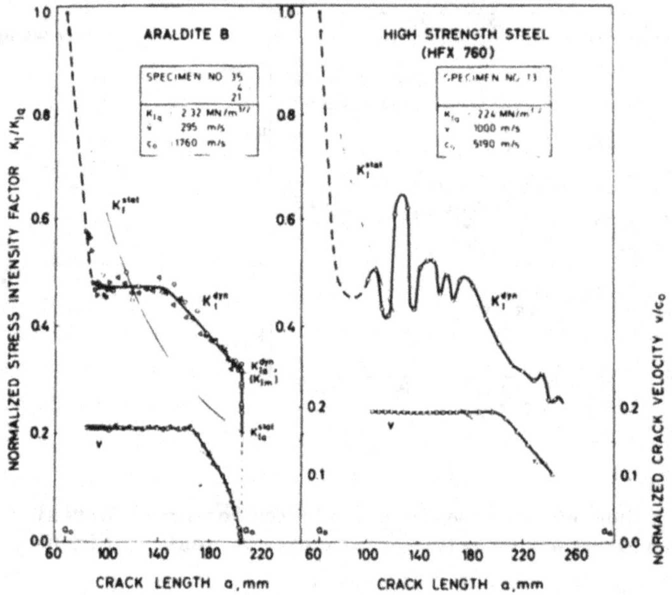

Fig. 41 Stress intensity factors for propagating and subsequently arresting cracks in steel and Araldite B specimens

The experimental findings obtained by the shadow optical investigations are summarized in simplified form in Fig. 42. The results indicate that dynamic effects do have an influence on the crack arrest process. Visible evidence of these dynamic effects is given in Fig. 43. A fast propagating crack (1000 m/s) in a high strength steel (X2 NiCoMo 18 9 5) specimen was photographed in a reflection arrangement with $z_0 < 0$. In addition to the shadow spot at the crack tip, the picture shows the generation of elastic waves at the tip of the propagating crack, their reflection at the finite boundaries of the specimen, and the later interaction of these waves with the crack tip.

Consequently, the stress condition at arrest is not static but still dynamic although the crack velocity is zero at the moment of arrest. Crack arrest toughness values which are determined on the basis of static evaluation procedures (as with the ASTM procedure) in principle, therefore, cannot represent a true material property. Such values, K_{Ia}^{stat}, must be smaller than the true crack arrest toughness, K_{Ia}^{dyn}. The differences will vary with specimen type and crack jump distance. The larger the crack jump distance, the smaller the statically determined crack arrest toughness, K_{Ia}^{stat}, in comparison to the dynamically determined value, K_{Ia}^{dyn}. Only with small crack jump distances, i.e., small crack propagation velocities, can the true crack arrest toughness, K_{Ia}^{dyn}, be approximated by $K_{Ia}^{stat} = K_{Ia}$.

Because of this dependence, the influence of dynamic effects on the crack arrest process can also be deduced from measurements of the crack arrest toughness, K_{Ia}^{stat} (identical to K_{Ia}, measured according to the ASTM procedure), when plotted as a function of the crack jump distance, Δa, or the arrest crack length, a_a. If dynamic effects do influence the crack arrest process, the crack arrest toughness, K_{Ia}, should show a decreasing tendency with increasing Δa or a_a (similar to that shown in Fig. 36). The steeper the slope of this curve, the larger the influence of dynamic effects. This is discussed in more detail in the next paragraph.

3.3.2 Dependence of crack arrest toughness from prior history

The influence of dynamic effects on the crack arrest process is checked on the basis of the previous speculation by a consideration of K_{Ia}-data measured for various steels. Most crack arrest data have been obtained within an ASTM Round-Robin test program [42]. 25 institutions from different countries took part in the program. As specified in the ASTM test procedure, compact specimens locally embrittled at the starter notch were used. Four sets of data were measured with two bridge steels, A 514 and A 588 at -30 °C, and with the reactor pressure vessel steel, A 533 B at +10 °C and +25 °C. Each participant tested three specimens under each of the above specified four conditions. Figure 44 represents the measured crack arrest toughness data, K_{Ia}, as a function of the resulting crack arrest length, a_a. Some data show too small or too large arrest crack lengths that do not fulfill the validity criteria set forth in the test procedure. The validity ranges, therefore, are given in Fig. 44. But even the consideration of valid data only shows a strongly decreasing trend of

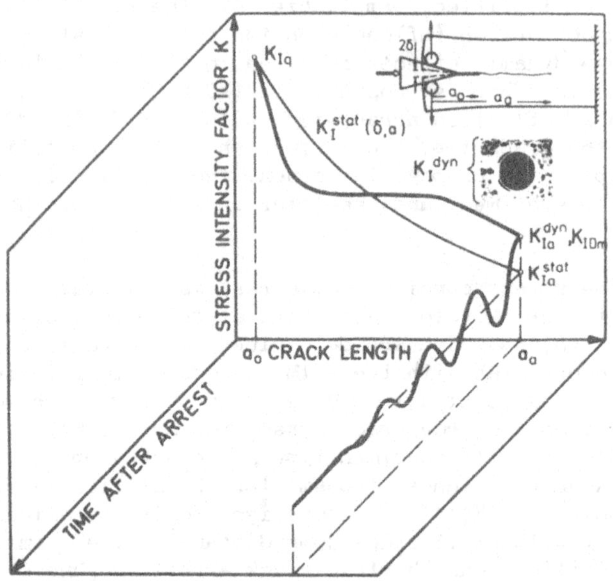

Fig. 42 Crack arrest behaviour (schematically)

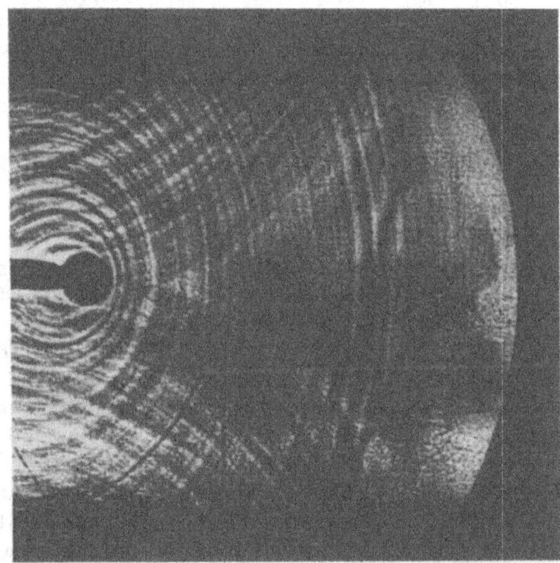

Fig. 43 Shadow optical photograph of a fast propagating crack in a steel specimen (recording of a virtual image, $z_o < 0$, in a reflection arrangement)

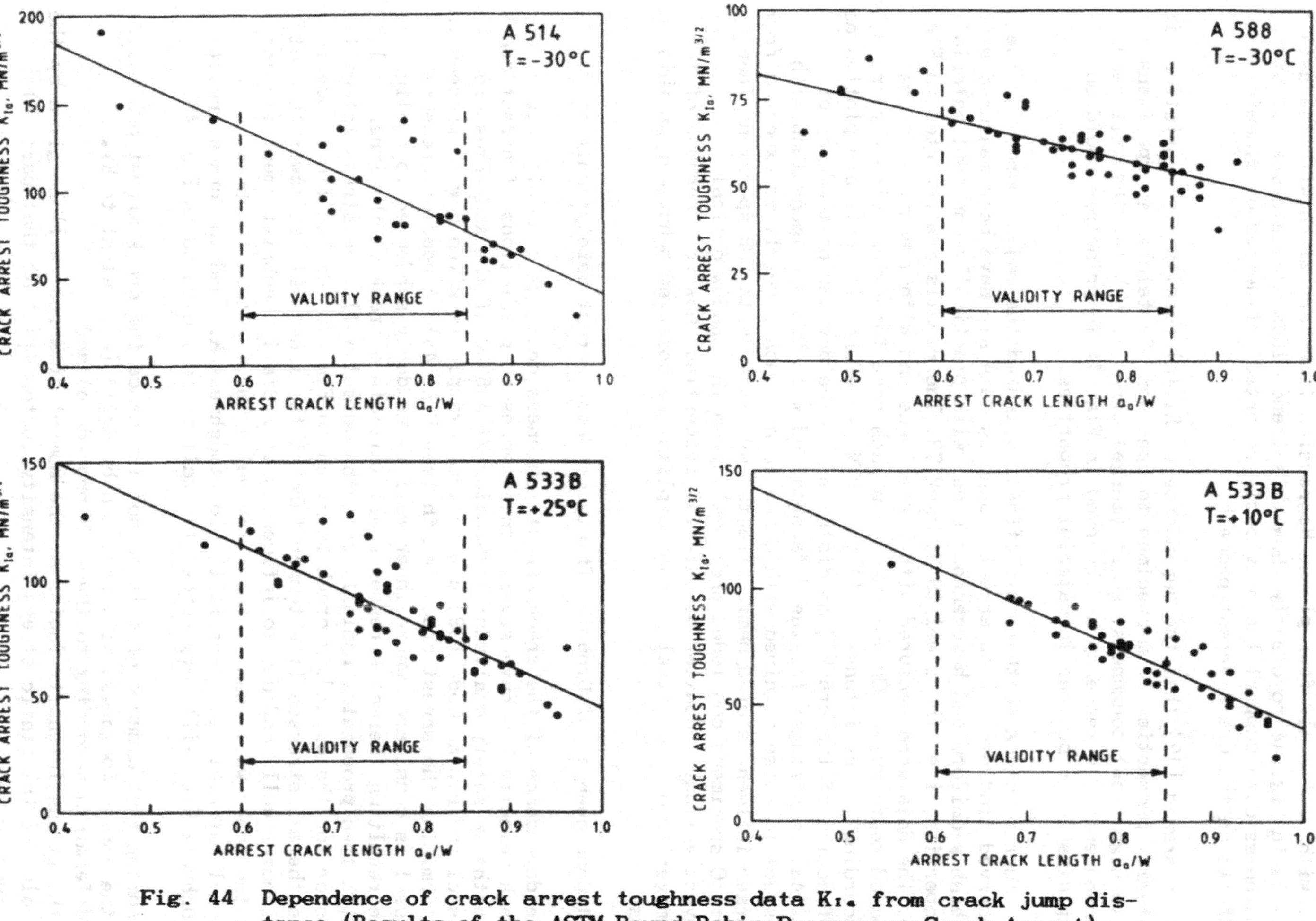

Fig. 44 Dependence of crack arrest toughness data K_{Ia} from crack jump distance (Results of the ASTM Round Robin Program on Crack Arrest)

K_{Ia} with a_a for all four plots. Regression lines are given with the data points in Fig. 44 to clearly show this trend. Thus, these crack arrest toughness data support the previously reported influences of dynamic effects on the crack arrest process.

It is argued [42] that the dependence of $K_{Ia}(a)$ results from variations in material properties from specimen to specimen. Certainly, in specimens with high (low) toughness, short (large) arrest crack lengths will result. Consequently, a tendency as observed in Fig. 44 in principle could be explainable by scatter in material properties.

In order to check whether this effect is indeed the major source for the observed findings, crack arrest toughness data K_{Ia} have been measured with a highly uniform and homogeneous steel with practically constant material properties, i.e. the steel 38 NiCrMoV 73. The results are reported in Fig. 45. The data were measured with C specimens and with transverse wedge loaded rectangular DCB specimens, in each case following (directly or according to the intension of) the ASTM procedure. The data are plotted as a function of the crack jump distance Δa. The decreasing tendency of $K_{Ia}(\Delta a)$ is stronger for the data obtained with DCB specimens than with respect to those obtained with C specimens. Such a result is expected from larger influences of dynamic effects associated with DCB specimens than with C specimens concluded previously from the shadow optical investigations. Scatter of material properties, however, because of the homogenity of this steel, cannot explain the observed behaviour in this case.

3.4 Consequences of Dynamic Effects on Crack Arrest Safety Analyses

The dependence of the crack arrest toughness on crack jump distance or on crack velocity can have severe consequences on predictions of arrest crack lengths in safety analyses: In Section XI A 5300 of the ASME Pressure Vessel and Piping Code [39] a practical example is given how to proceed in determining the arrest crack length (see Fig. 46). A reactor pressure vessel is considered whose inner wall is suddenly subjected to cooling. The resulting thermal stresses shall initiate a crack at the inner wall which then propagates radially into the vessel wall. The stress intensity factor for the expanding crack is given in Fig. 46 (dashed line). Due to the thermal stresses (the temperature at the inner wall is lower than at the outer wall) and due to influences of radiation embrittlement (larger at the inner wall than at the outer wall) the fracture toughness of the vessel material, i.e. the initiation toughness K_{Ic} and the crack arrest toughness K_{Ia}, will vary across the wall as is indicated in Fig. 46.

Assuming that dynamic effects do not influence the crack arrest process,

- the arrest toughness will be given by K_{Ia}^{stat} (equivalent to K_{Ia} determined according to the ASTM procedure) and
- the stress intensity factor at the moment of arrest will be given by the value of the static stress intensity factor K_{I}^{stat} at that particular crack length.

Thus, crack arrest is given by the cross-over of the lines K_{I}^{stat} and K_{Ia}^{stat}

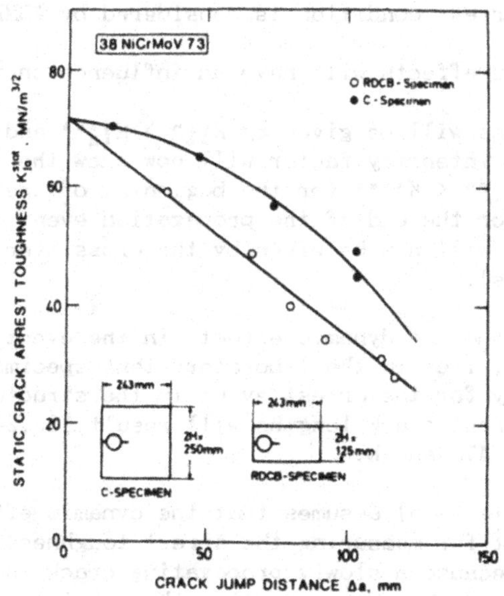

Fig. 45 Dependence of crack arrest data K_{Ia} on crack jump distance for DCB and C specimens

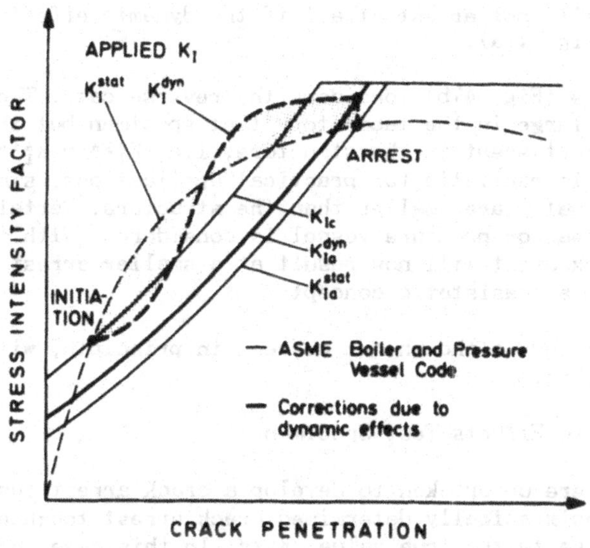

Fig. 46 Initiation and arrest of a crack in a pressure vessel wall (schematically)

(thin lines). This arrest condition is considered by ASME.

Assuming that dynamic effects will have an influence on the crack arrest process,

- the arrest toughness will be given by $K_{Ia}^{dyn} > K_{Ia}^{stat}$ and
- the dynamic stress intensity factor will now show the typical characteristics: $K_I^{dyn} < K_I^{stat}$ for the beginning of the propagation event and $K_I^{dyn} > K_I^{stat}$ for the end of the propagation event.

Consequently, arrest will now be given by the cross-over of the lines K_I^{dyn} and K_{Ia}^{dyn} (thick lines).

Thus, when the magnitude of dynamic effects in the event for determining the arrest toughness, i.e. in the laboratory test specimen, are different from those that apply for the arrest event in the structure, e.g. the vessel, different arrest crack lengths will result [23,24,43]. This is demonstrated by Fig. 47a and b.

The first example (Fig. 47a) assumes that the dynamic effects associated with the arrest event for measuring the arrest toughness are small and can be neglected, e.g. because a slowly propagating crack in a rather large test specimen is considered. Consequently, the quasistatically determined toughness, K_{Ia}^{stat}, is a good approximation of the true toughness, K_{Ia}^{dyn}. The dynamic effects in the structure, however, shall be large, e.g. because the crack shall propagate at a relatively high velocity in a relatively small structure with nearby boundaries. Consequently, for the crack approaching arrest, $K_I^{dyn} > K_I^{stat}$. This crack, then, arrests only at a larger arrest crack length than predicted according to a quasistatic concept. Or, it will not arrest at all if the dynamic effects should be very large (see Fig. 47a).

The second example (Fig. 47b) considers the reverse case. The dynamic effects shall be large in the laboratory test specimen but they shall be small for the arrest event in the structure, i.e. $K_I^{dyn} = K_I^{stat}$ but $K_{Ia}^{dyn} > K_{Ia}^{stat}$. This case is realistic for practical applications, since laboratory test specimens usually are smaller than the structure. Certainly this is the case when a reactor pressure vessel is considered. With these assumptions, crack arest will now result at a smaller arrest crack length than predicted by a quasistatic concept.

Thus, static analyses, although not correct in principle, will yield conservative results.

3.5 Reduced-Dynamic-Effects Test Specimen

Special efforts were undertaken to develop a crack arrest test specimen for which the quasistatically determined crack arrest toughness, K_{Ia}, would be very close to the true value, K_{Ia}^{dyn}. In this case, a K_{Ia}-value determined according to the ASTM procedure would indeed represent an acceptable estimate of the true crack arrest toughness. Such a specimen would have to exhibit very small dynamic effects, i.e. the recovery of kinetic energy due to reflected stress waves at the finite boundaries of

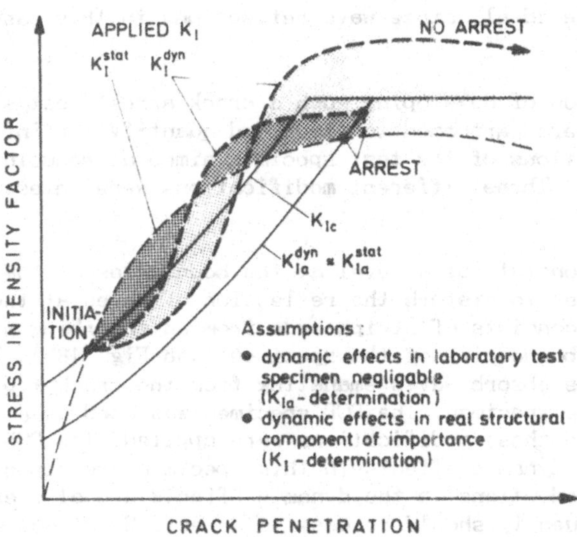

Fig. 47a Influences of dynamic effects on crack arrest safety analyses (schematically)

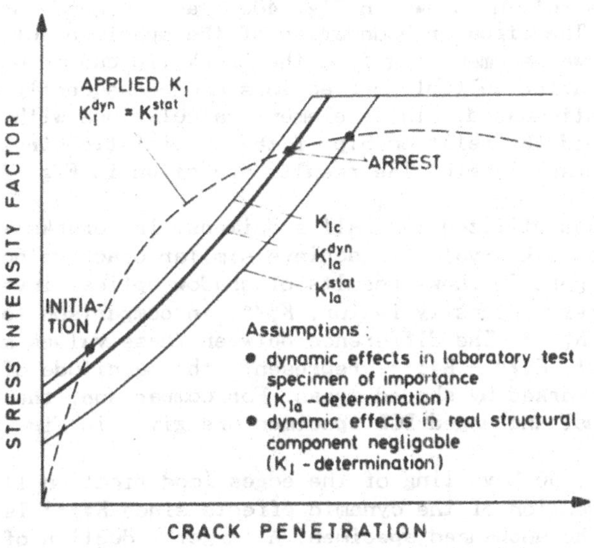

Fig. 47b Influences of dynamic effects on crack arrest safety analyses (schematically)

the specimen would have to be very small. Trivially, an infinitely large specimen would be ideal, since wave reflections in this case could not occur at all.

With the intention of developing such a crack arrest test specimen, investigations were performed to study and quantify the influence of certain modifications of the test specimen aimed at reducing wave reflections [41]. Three different modifications were investigated (see Fig. 48).

Modification I consists of a bevel at the boundaries of the specimen (see Fig. 48a) in order to disturb the reflection of waves at the boundaries. Modification II consists of strips made from an absorbing material attached to the boundaries of the specimen (see Fig. 48b). It is expected that these strips absorb waves emanating from the crack tip and damp vibrations of the specimen. The DCB specimen was used as a reference specimen to which these modifications were applied. The DCB specimen was chosen since the dynamic effects in this specimen are especially large; therefore, the reductions in the dynamic effects are also expected to be large and consequently should be easy to detect. Specimens made from the model material Araldite B were utilized in these investigations. As a wave absorbing medium for the modification type II specimens, a hard polyurethane rubber, Vocollan Shore 90, was used. The acoustic impedance of this material is about the same as that of Araldite B. Thus, practically all incident waves are transmitted through the interphase.

Modification III represents a completely different type of specimen. The shape of this specimen, shown in Fig. 48c, was designed to defocus reflected waves. The circular boundaries of the specimen act like diverging mirrors. Thus, waves emanating from the crack tip can interfere with the crack tip only after multiple reflections and consequently after they have been strongly attenuated. Finite element calculations with this specimen geometry provided the relationship of the quasistatic stress intensity factor versus crack length. The results are given in Fig. 49.

Wedge loading was utilized with all specimens. The cracks were initiated at almost the same K_{Iq}-value to achieve similar crack velocities in all experiments. Figure 50 shows results of shadow optical investigations on the dynamic stress intensity factor, K_I^{dyn}, in comparison to the equivalent static values, K_I^{stat}. The difference between these values at the moment of arrest, $K_{Ia}^{dyn} - K_{Ia}^{stat}$, represents the magnitude of the dynamic effects and is marked by shaded areas. For comparison, the corresponding data for a normal unchanged DCB specimen are given in Fig. 50 as well.

As shown in Fig. 50 bevelling of the edges (modification I) leads to a significant reduction of the dynamic effects since K_{Ia}^{stat} is now closer to K_{Ia}^{dyn} than for the unchanged specimen. A larger reduction of the specimen dynamics is caused by the damping elements (modification II). More detailed investigations (not described here) show that the dynamic effects decrease with increasing widths of the absorbing strips, as it is expected (Fig. 50 shows the result for a width b = 144 mm and a specimen height 2H

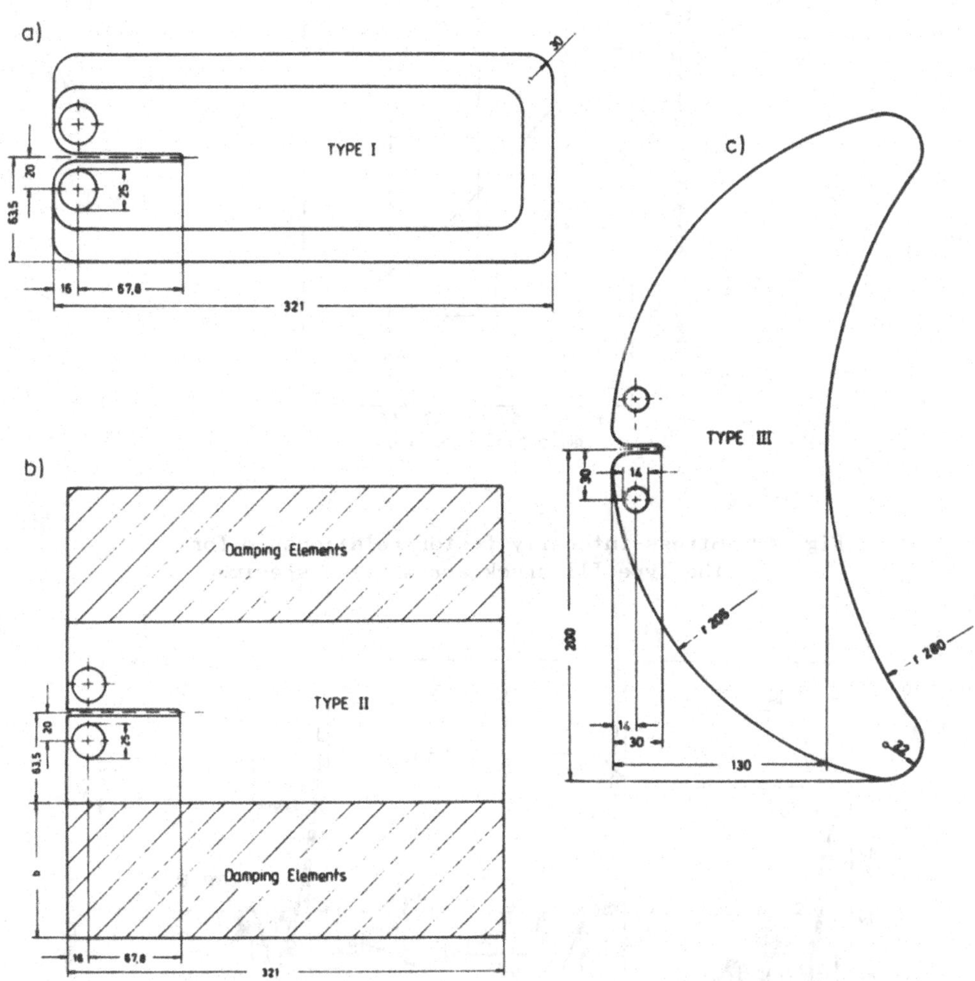

Fig. 48 Modifications of crack arrest test specimens to reduce dynamic effects

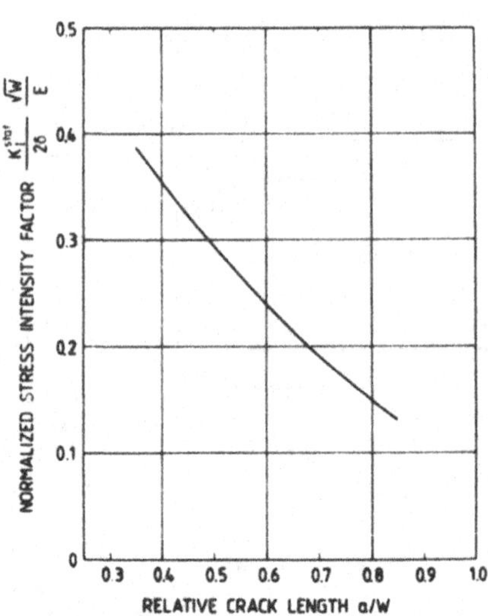

Fig. 49 Stress-intensity-factor relationship for the Type III crack arrest test specimen

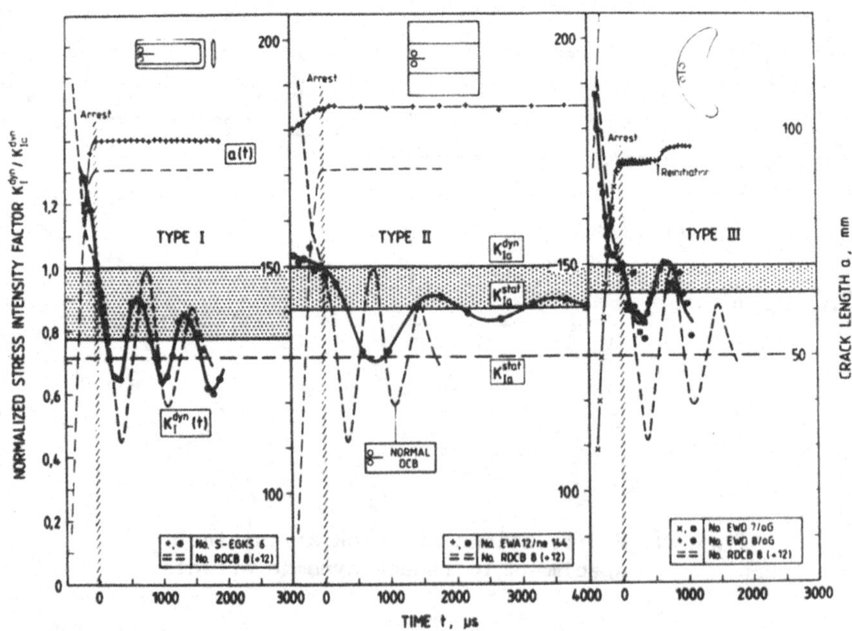

Fig. 50 Reduction of dynamic effects for modified crack arrest test specimens

= 127 mm). The strongest reduction of dynamic effects, however, is observed with the modification type III specimen. Compared with the DCB specimen the dynamic effects are reduced by about a factor of three.

All modifications were combined and slightly altered for practical reasons to design a test specimen with reduced dynamic effects that can be applied in laboratory tests with steel specimens. This specimen, called the reduced-dynamic-effects (RDE) specimen, is shown in Fig. 51. Its geometry is similar to that of the modification type III specimen, but the shape of the boundary has been approximated by straight lines in order to allow for easier machining. For the same reason, nonsymmetric bevelling is used. In order to damp waves and vibrations, damping elements were attached (not at the edges but) on the surface of the specimen. Rectangular plates made of lead are used for this purpose. This special attachment of lead plates to the specimen has the additional advantage that the damping elements becomes even more effective and that the manufacturing process becomes more simple. The lead plates are glued onto the specimen with an epoxy adhesive and can be removed after the experiment by heating the specimen. The lead plates together with additional weights attached to the specimen serve another function; they reduce the eigen frequency of the specimen and thus increase the natural time base. Thus the crack comes to arrest before it can significantly be influenced by vibrational effects. Figure 52 shows a photograph of the RDE specimen without (above) and with damping elements and additional weights (below).

In order to check the effectiveness of this specimen design in reducing dynamic effects, fracture toughness values K_{Ia} have been measured with this specimen following the ASTM test procedure. Test specimens were made from the steel 38 NiCrMoV 73 used before in experiments with DCB and C specimens, allowing for a comparison of results.

Figure 53 shows the obtained K_{Ia}-values plotted versus crack jump distance, Δa. The data points lie pratically on a straight line of only slightly decreasing slope. Figure 54 compares the results of the RDE specimen with those obtained for the C and the DCB specimen. The data points are omitted in this plot for clearness, only the least square fit lines are shown in Fig. 54. The dependence on crack jump distance is about 3 or 4 times smaller for the RDE specimen than for the C specimen.

Thus with the RDE-specimen, crack arrest toughness data can be determined by applying a quasistatic evaluation procedure which yields an arrest toughness value K_{Ia} which represents a reasonably close approximation of the true crack arrest toughness. Such an arrest toughness value would be sufficiently accurate for engineering purposes and would avoid overconservatism in predictions of arrest crack lengths.

3.6 Dynamic Correction Formula for C Specimens

The RDE specimen has the disadvantage of being expensive to machine. The C specimen is certainly much easier to machine. But also, when C specimens are utilized, the significance and the accuracy of the measured crack

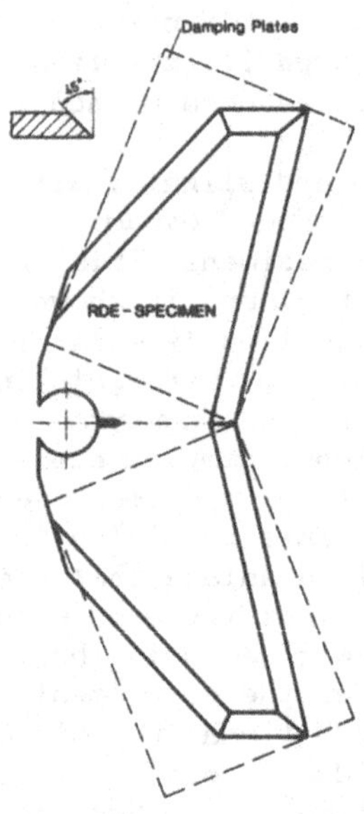

Fig. 51 Reduced-dynamic-effects (RDE) specimen

Fig. 52 Photographs of reduced-dynamic-effects (RDE) specimen

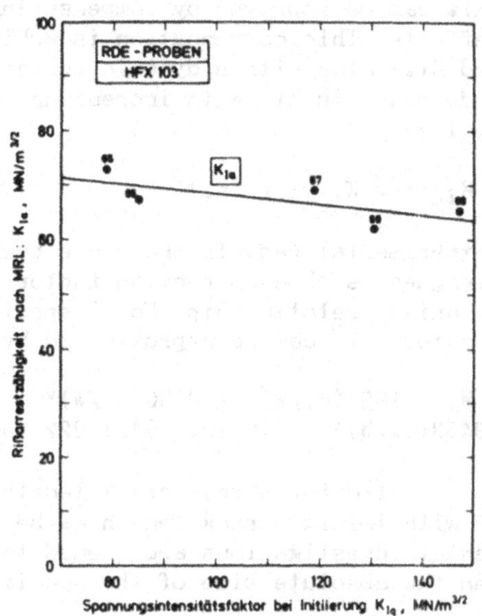

Fig. 53 Dependence of crack arrest toughness values K_{Ia} from crack jump distance measured with RDE test specimens (Spannungsintensitätsfaktor bei Initiierung = Crack initiation stress intensity factor, Rißarrestzähigkeit = crack arrest toughness)

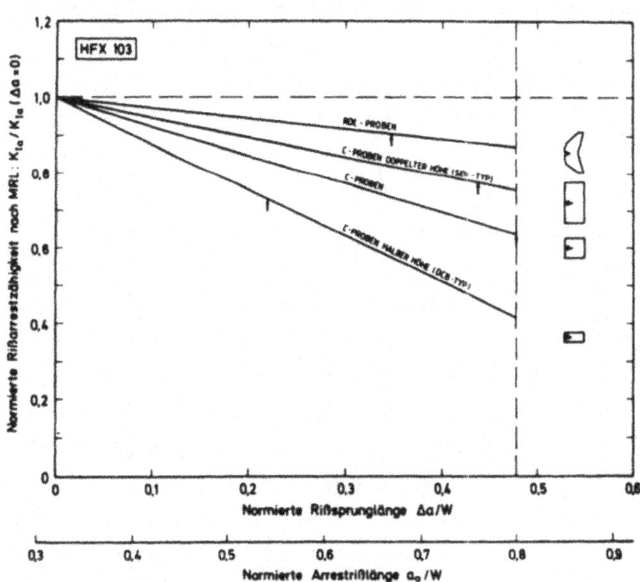

Fig. 54 Comparison of crack arrest toughness data K_{Ia} measured with different crack arrest test specimens (Normierte Rißarrestzähigkeit = normalized crack arrest toughness, Normierte Rißsprunglänge/Arrestlänge = normalized crack jump distance/ arrest length)

arrest toughness data K_{Ia} can be improved by compensating for the influences of dynamic effects. This compensation is achieved by multiplying the obtained K_{Ia}-value with a dynamic correction factor, f_c, which accounts for the decrease in K_{Ia} with increasing crack jump distance or crack arrest length, i.e.

$$K_{Ia}^{true} = K_{Ia} \cdot f_c(a_a) \tag{44a}$$

Based on the numerous experimental results that have been established by the author and his colleagues, such a correction factor was found to represent a practically unique relationship. For C specimens with W = 200 mm the correction factor, f_c, can be approximated by the relationship

$$f_c = -21.72 + 319.2(a_a/W) - 1894(a_a/W)^2 + 6150(a_a/W)^3 \\ -11818(a_a/W)^4 + 13453(a_a/W)^5 - 8406(a_a/W)^6 + 2227(a_a/W)^7 \tag{44b}$$

This dynamic correction is valid for arrest crack length a_a varying between 0.3 W and 0.8 W with initial crack length a_0 between 0.3 W and 0.4 W. Further experimental investigations are needed to check how the results are changed when the absolute size of the specimen is altered.

This correction factor approach, although unsatisfactory from a fundamental point of view, helps to minimize errors that are inherent with the quasistatic crack arrest procedure. The decreasing trend of the measured K_{Ia}-data with increasing arrest length resulting from dynamic effects is compensated for in an approximative manner.

4. IMPACT LOADING OF CRACKS BY DROP WEIGHTS

4.1 Introduction

Gravity activated masses in the form of pendulums or drop weights are usually utilized to investigate the strength of steels under impact loading. The V-notched Charpy or the Izod-specimen (measuring 55 mm by 10 mm by 10 mm) is the most widely used specimen type tested in a bending arrangement with a Charpy pendulum device (see Fig. 55). The impact velocity is usually 5 m/s according to an initial drop height of 1.25 m (sometimes the impact velocity can be varied by reducing the initial drop height). Depending on the mass of the hammer, pendulums with energy capacities in the range of 150 J, 300 J, or 750 J are used.

In the simplest form of the test, the energy required to break the specimen is measured (Section 4.1.1). The instrumented impact test (Section 4.1.2) was introduced to discriminate between partial energies associated with different phases of the fracture process. In the context of this review, however, mainly the measurement of the impact fracture toughness value K_{Id} will be considered (Sections 4.2 - 4.4).

4.1.1 Energy measurement

The energy to break a specimen, denoted impact or Charpy energy, is determined from the difference of the position of the pendulum before and after the test. This impact energy when plotted as a function of test temperature shows a characteristic shape (see Fig. 55) which is quantified by three properties. The lower and the upper shelf energy and the transition temperature, i.e. the temperature where the energy changes rapidly. These three properties characterize the fracture behaviour of steels under dynamic conditions of loading. In the lower shelf regime the specimen breaks by a brittle cleavage crack and in the upper shelf regime the specimen fails by a ductile shear type of failure which is proceeded by large plastic deformations. Thus, the lower the transition temperature, the less the danger of brittle fracture and, the higher the upper shelf energy, the higher the capability of the material to deform. The upper and lower shelf energy and the transition temperature, however, only represent relative material properties. They do not represent absolute strength values that can be used in the design of structures. This is demonstrated by Fig. 56 showing the shift of the transition temperature

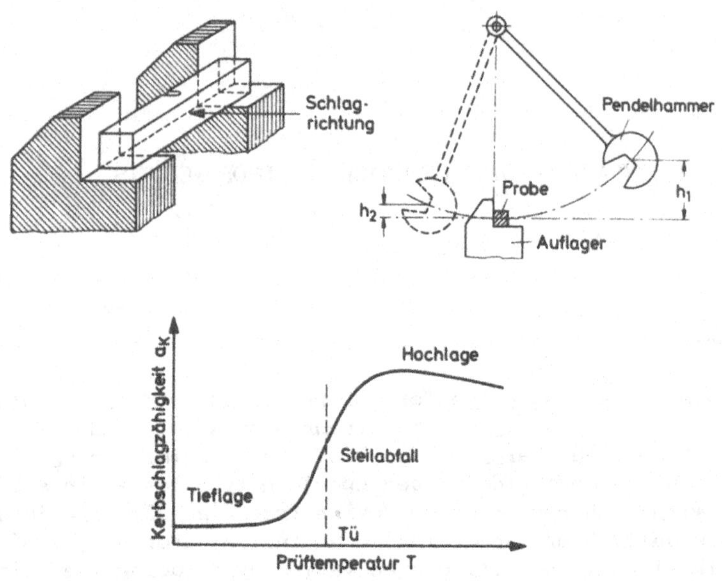

Fig. 55 Conventional impact test technique
(Schlagrichtung = striking direction, Pendelhammer = pendulum hammer, Probe = specimen, Auflager = support, Kerbschlagzähigkeit = Charpy energy, Prüftemperatur = test temperature, Tieflage = lower shelf regime, Steilabfall = transition regime, Hochlage = upper shelf regime)

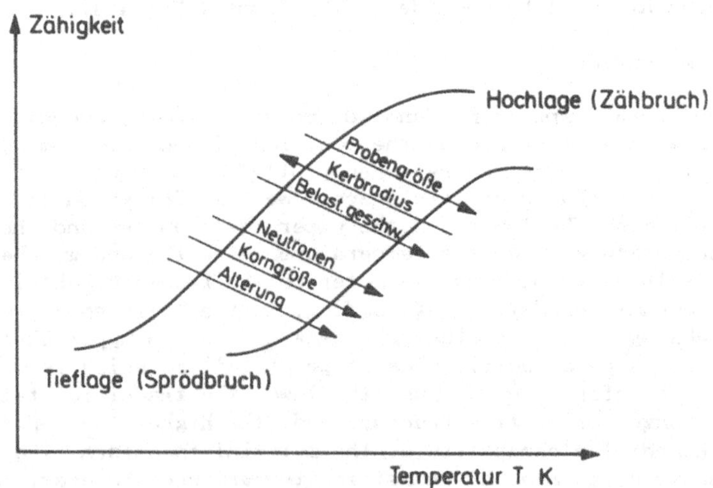

Fig. 56 Dependence of transition temperature on test parameters
Zähigkeit = toughness, Probengröße = specimen size, Kerbradius = notch tip radius, Belastungsgeschwindigkeit = loading rate, Neutronen = neutron irradiation, Korngröße = grain size, Alterung = aging, see also Fig. 55)

when certain test parameters are altered. In particular, the transition temperature is increased when the size of the specimen is increased. Thus a small specimen may fail in a ductile manner, whereas a large structure at the same temperature may fail by brittle cleavage cracks. This is usually taken care of by design rules requiring that the service temperature of a structure must be conconsiderably higher than the transition temperature measured with the Charpy test.

Furthermore, the Charpy impact energy represents an integral value which is determined by the energy to initiate a crack, the energy to propagate the crack through the specimen, and the energy to pull the specimen through the anvils of the test machine (in case that the specimen does not break completely during the impact process). These partial energies for crack initiation and for crack propagation can be very different for different materials, furthermore, they vary with test temperature. Consequently, the total energy in general will vary too, but it is also possible (and actually observed) that the total energy may practically stay the same although the partial energies are different. Thus, the Charpy energy does not allow to discriminate between the resistance of the material against crack initiation or crack propagation. But, depending on the objective and the purpose a structure or a component of a structure has to fulfill, an optimum design requires detailed knowledge about these specific quantities.

4.1.2 Instrumented impact test

The intention of measuring the energy associated with different phases of the fracture process led to the development of the instrumented impact test. With strain gages at the tup of the striking hammer, the load is measured as a function of time during the impact process (similar to the load being measured in a quasistatic test during loading). Furthermore, the displacement of the specimen is determined, either by directly measuring the displacement during the test or by calculating it from the measured load-time-curve via Newton's law, i.e.

$$v(t) = v_0 - 1/m \int P(t)dt$$
$$s(t) = \int v(t)dt. \qquad (45)$$

The energy the specimen absorbs is then given as

$$E(s) = \int P(s)ds. \qquad (46)$$

A typical load-displacement trace is given in a schematic representation in Fig. 57 [44]. Brittle cleavage crack propagation which is associated with a high propagation velocity is indicated by the sharp load drop. On the fracture surface it is recognized by a crystalline appearance. At higher test temperatures the material fails by a ductile or a shear type fracture, and the load-displacement trace shows a bell shaped curve without any load drop (see also Fig. 59). The fracture surface assumes a fibrous appearance. The energy up to fast crack initiation or to maximum

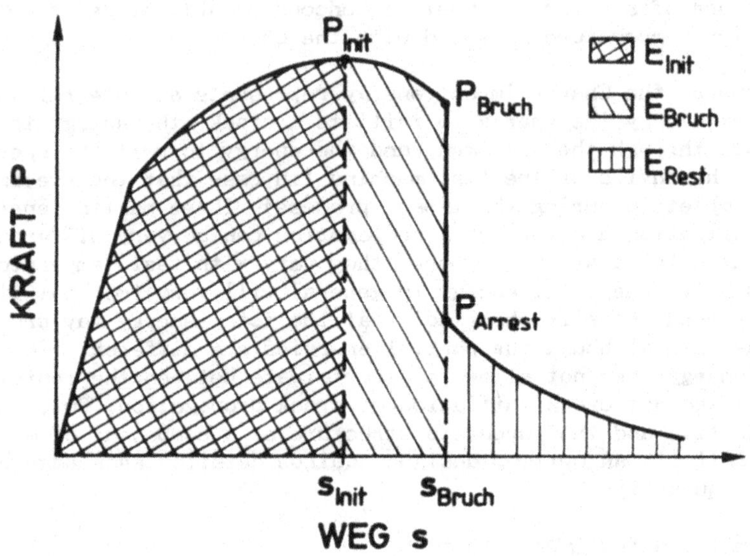

Fig. 57 Results of an instrumented impact test: load-displacement-trace, partial energies
(Kraft = load, Weg = displacement, Bruch = fracture)

load defines the crack initiation energy, the subsequent energy represents the crack propagation energy. Real load-displacement traces measured with the steel 26 NiCrMoV 85 at different test temperatures are shown in Fig. 58. At low temperatures a completely brittle failure of the specimen is obtained showing a crystalline fracture surface. The crack propagation energy is negligible in comparison to the crack initiation energy. With increasing temperature the sharp load drop becomes less pronounced and the ratio of fibrous to crystalline failure surface increases. The high temperature traces are an indication of a ductile shear type failure with a completely fibrous fracture appearance. The crack propagation energy is remarkably high in comparison to the crack initiation energy.

The load-time or load-displacement traces exhibit characteristic oscillations which result from specimen vibrations that are excited by the impinging hammer. These oscillations are damped with time, but they can only be neglected after a considerably large period of time. These oscillations make it difficult to quantify a load value at a specific time or displacement during the early time range of the impact event. They are of almost no influence, however, for the determination of the displacement or the energy, since the integration procedure averages over these oscillations. Thus the determination of energy values is practically unaffected by these oscillations.

This short review of the conventional and the instrumented Charpy impact test gives only the basis of the test techniques and the evaluation procedure. Detailed recommendations on the measuring methodology and the procedure to determine material characteristic properties have been worked out by Task Groups of the Deutscher Verband fuer Materialpruefung (DVM) [45] and the Verein Deutscher Eisenhuettenleute (VDEh) [46]. These recommendations are reproduced in Appendix II. For further details on the standard Charpy test see also the contribution of F. Klepaczko, this volume. In the context of this review, the determination of energy values is not considered further, it is the instrumented impact test with precracked specimens which is of interest. These specimens allow for the determination of fracture toughness values, K_{Id}. In contrast to the Charpy energy or partial energy quantities, the fracture toughness values have the advantage that they can be transferred to the structure - at least in principle, i.e. when certain requirements are fulfilled. Thus they can be used as design criteria or they can be used in safety analyses to quantify the maximum load a structure with a pre-existing flaw can be subjected to

4.2 Measurement of the Impact Fracture Toughness K_{Id}

The impact fracture toughness K_{Id} is measured with precracked specimens. Large 3-point-bend specimens are utilized with drop weight towers or Charpy specimens are used with pendulum test devices. The drop weight or the pendulum is instrumented with strain gages at the tup of the striking hammer to allow for the measurement of the load the specimen is subjected to during the impact event. The following discussion applies to precracked Charpy specimens, in particular since a test procedure has been developed and recommended to ASTM for measuring K_{Id} with the instrumented Charpy test [47]. But the test technqiue and the evaluation procedure applies in

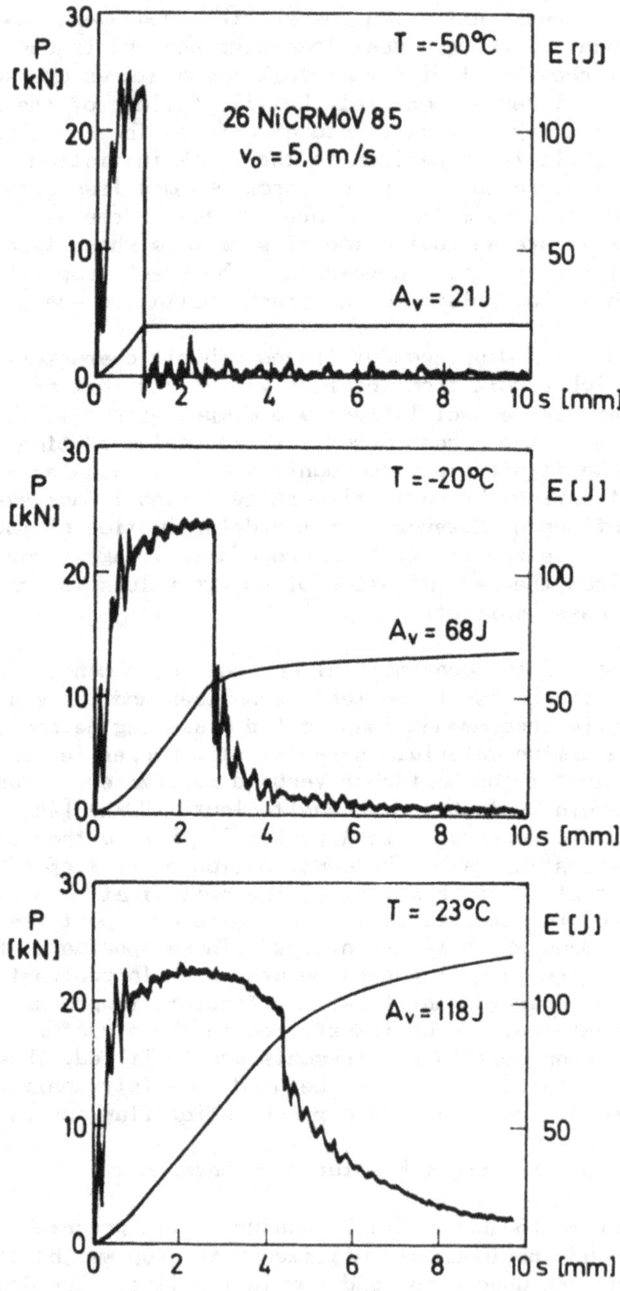

Fig. 58 Load-displacement traces and energy curves obtained for different test temperatures

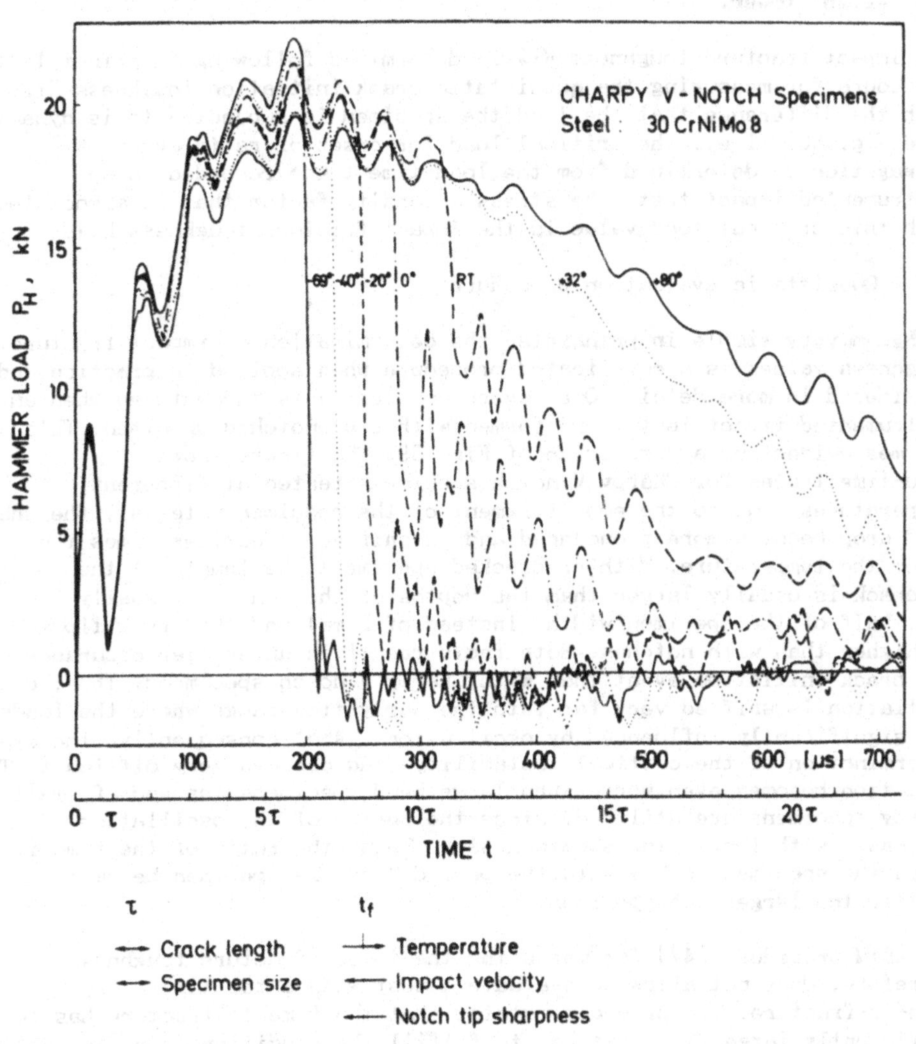

Fig. 59 **Characteristic load-time traces for different test temperatures**

an analogous form also for larger bend specimens that are impacted in a drop weight tower.

The impact fracture toughness K_{Id} is determined following in principle the procedure for measuring the quasistatic crack initiation toughness K_{Ic}, with the difference that the load the specimen is subjected to is dynamic (see Fig. 60): i.e., the critical load for onset of rapid crack propagation is determined from the load-time trace recorded in an instrumented impact test. The stress intensity factor that is associated with this critical load value is the impact fracture toughness K_{Id}.

4.2.1 Quasistatic evaluation procedure

Although very simple in principle, the determination of impact fracture toughness values is a complicated procedure when applied in practice and considered in more detail. One severe complexity is encountered when an instrumented impact test is performed with a prenotched specimen. This becomes evident by a discussion of Fig. 59. The figure shows load-time traces for Charpy V-notch specimens tested at different temperatures. Due to the embrittlement of the specimen material, the sharp load drop becomes more pronounced and is shifted to earlier times the lower the temperature. With precracked specimens the length of the precrack is usually larger than the depths of the V-notch (usually 5 mm, i.e. half of the specimen width, instead of 2 mm) and the crack tip acuity is higher than with notches. Both facts result in an earlier occurance of the crack initiation event than with Charpy V-notch specimens. Thus, crack initiation is shifted very far into the early time range where the loads are significantly influenced by oscillations. And, consequently, the exact determination of the critical instability load becomes very difficult. The situation becomes even worse when large bend specimens instead of small Charpy specimens are utilized, since the period of the oscillation increases with increasing specimen size. Thus, the ratio of the time it takes the specimen to break to the period τ of the specimen becomes smaller the larger the specimen.

The ASTM procedure [47] for measuring the impact fracture toughness, therefore, does not allow to evaluate a test with a too low time-to-fracture. The procedure states that the time-to-fracture has to be sufficiently large, i.e., it has to fulfill the condition (see Fig. 61)

$$t_f > 3\tau, \qquad (47)$$

where τ is the period of the oscillation in the load signal caused by the vibrating specimen. With Eq. (47) the following assumptions are believed to apply:

1. The oscillations are sufficiently damped out that the crack initiation load can be determined with sufficient accuracy.
2. A quasistatic state of stress applies in the specimen, i.e. a quasi-static formula for the stress intensity factor as a function of load and crack length can be utilized to determine the fracture toughness value (see equation given in Fig. 60).

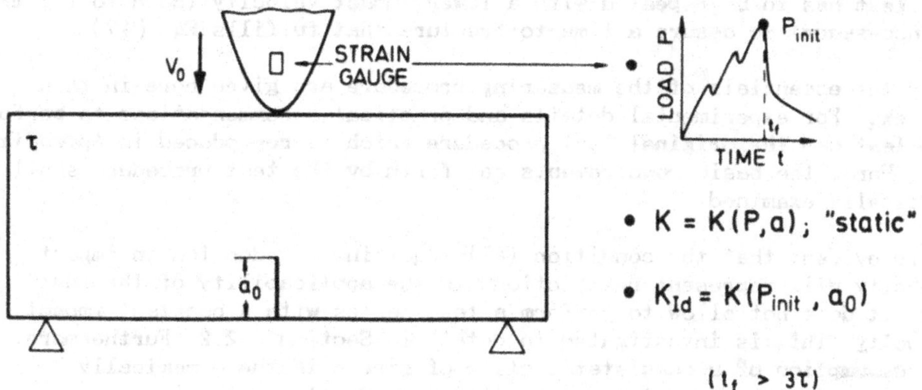

Fig. 60 Determination of the impact fracture toughness K_{Id}

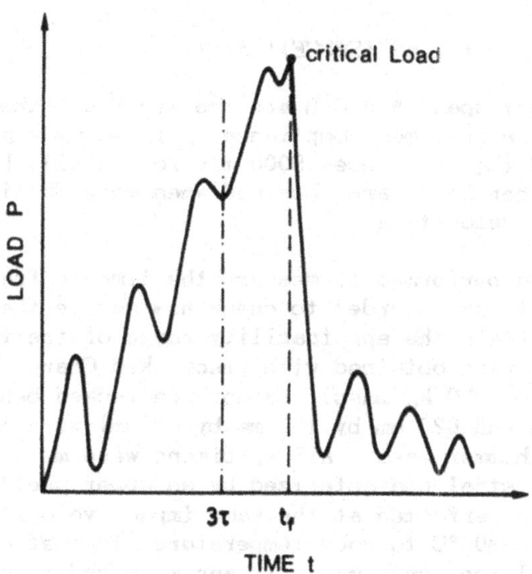

Fig. 61 The 3τ - criterion

If a test should not result in a time-to-fracture being larger than 3τ, the test has to be repeated with a lower impact velocity (down to 1.2 m/s, if necessary) to assure a time-to-fracture that fulfills Eq. (47).

Only the essentials of the measuring procedure are given here in this context. For experimental details and practical recommendations to perform the test see the original test procedure which is reproduced in Appendix III. Here, the basic requirements set forth by the test procedure shall be critically examined.

It is evident that the condition (47) requiring a reduction in impact velocity will represent restrictions in the applicability of the test: e.g. it does not allow to perform a test series with a constant impact velocity. This is investigated in detail in Section 4.2.2. Furthermore, the assumption of a quasistatic state of stress in the dynamically impacted specimen is critically analyzed in Section 4.3.

4.2.2 Applicability range of the quasistatic evaluation procedure

In order to fulfill the criterion $t_f > 3\tau$, it is favourable to utilize specimens with a short period, τ, and test conditions which yield a large time-to-fracture, t_f. Short periods, τ, result for small specimens. For Charpy specimens the 3τ-value is in the range of about 0.1 to 0.2 ms, depending on the crack length. Ireland [48] gives an empirical formula for determining the period τ.

$$\tau = 1.68\ (SWBCE)^{1/2}/c_0 \qquad (48)$$

where S is the support span, W and B are the width and the thickness of the specimen, C is the specimen compliance, E is Young's modulus and c_0 is the sound wave speed (Eq. (48) uses 5000 m/s for steel). Large fracture times, t_f, on the other hand, are obtained when more ductile materials are tested at low impact velocities.

Experiments have been performed to measure the time-to-fracture under different test conditions in order to check how severe the validity conditions (47) restricts the applicability range of the test. Figure 62 summarizes data that were obtained with precracked Charpy specimens impacted by a pendulum (20 kg mass). Larger precracked bend specimens (220 mm by 40 mm by 20 mm and 620 mm by 150 mm by 40 mm) were tested in a drop weight tower (90 kg hammer mass). All specimens were machined from the same pressure vessel steel characterized by an upper shelf energy of 200 Joule. The tests were performed at the same impact velocity of 5 m/s in a temperature range of -80 °C to room temperature. In most cases, in particular with the larger specimens, t_f was measured to be far below the 3τ-value of the specimen. Only with Charpy specimens tested at temperatures higher than 10 °C could the requirement $t_f > 3\tau$ be met. Data from further experiments with Charpy specimens machined from another pressure vessel steel with an upper shelf energy of only 40 Joule tested at different impact velocities are shown in Fig. 63. The requirement $t_f > 3\tau$ could only be fulfilled at lower temperatures by reducing the

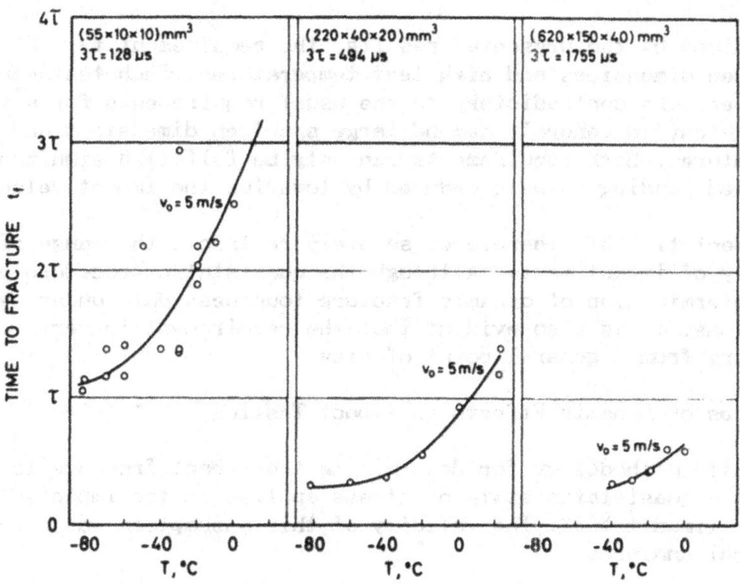

Fig. 62 Times-to-fracture obtained with specimens of different sizes

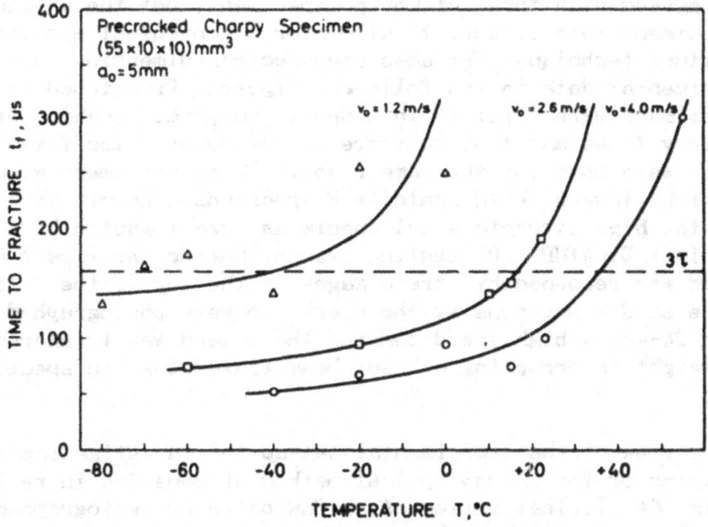

Fig. 63 Times-to-fracture obtained for different impact velocities

impact velocity.

~~As is recognized~~ by the presented results, the requirement $t_f > 3\tau$ demands small specimen dimensions and high test temperatures. Such test conditions, however, are contradictory to the usual requirements for a valid LEFM-test, which, in general, demand large specimen dimensions and low test temperatures. Both requirements can only be fulfilled simultaneously if the applied loading rate is reduced by lowering the impact velocity.

The requirement $t_f > 3\tau$, therefore, severely restricts the range of applicability of impact tests. Although the quasistatic procedure will allow the determination of dynamic fracture toughness data under certain test conditions, it is also evident that the requirement is very unsatisfactory from a general point of view.

4.3 Influences of Dynamic Effects in Impact Testing

The quasistatic methodology for determining the impact fracture toughness assumes that a quasistatic state of stress applies in the impacted test specimen for times $t > 3\tau$. The validity of this assumption was checked by shadow optical analyses.

4.3.1 Dynamic stress intensity factors for an impacted crack

Experiments were performed to investigate the influence of dynamic effects on the hammer load and the actual stress intensification at the crack tip [49]. Prenotched bend specimens were utilized for these investigations made from two materials, the epoxy resin, Araldite B, and the high strength maraging steel, X2 NiCoMo 18 9 5. The main relative dimensions of the specimens agreed with those of Charpy specimens, but the absolute sizes of the specimens were larger, to allow for a convenient application of the shadow optical technique. The absolute specimen dimensions are given with the experimental data in the following figures. Prenotched instead of precracked specimens were utilized in order to suppress breaking of the specimen at early times and thus to increase the observation time. Low impact velocities were used for the same reaon. These specimens were impacted in drop weight towers. With Araldite B specimens a hammer of 4.9 kg was utilized, the high strength steel specimens were impacted by a hammer of 90 kg mass in a DYNATUP 8100 testing system. During the impact event the hammer load was recorded by strain gages at the tup of the striking hammer, and the shadow patterns at the crack tip were photographed using a Cranz-Schardin 24-spark high speed camera. The camera was triggered by the falling drop weight interrupting a laser beam traversing the specimen near its rear edge.

An illustrative view of the experimental set-up for investigating steel specimens by means of the shadow optical method of caustics in reflection is given in Fig. 64. Typical series of shadow patterns photographed with Araldite B specimens and with high strength steel specimens are given in Figures 65 and 66, respectively. Only a few of the total 24 pictures are reproduced. Each picture shows the central part of the impacted specimen

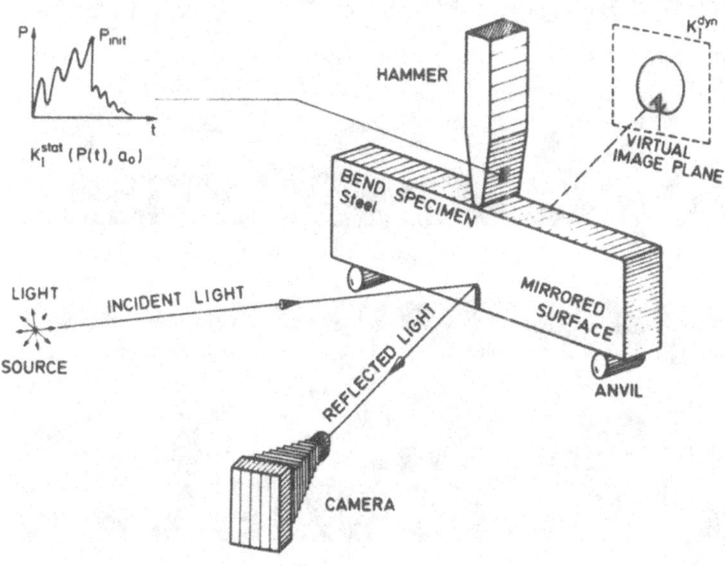

Fig. 64 Experimental set-up (schematically)

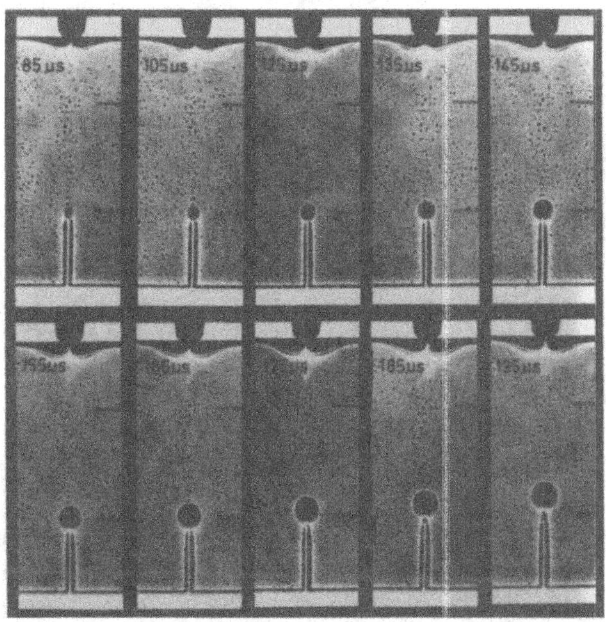

Fig. 65 Shadow optical photographs of a precracked bend specimen during impact loading (Araldite B, real image, $z_0 > 0$, transmission)

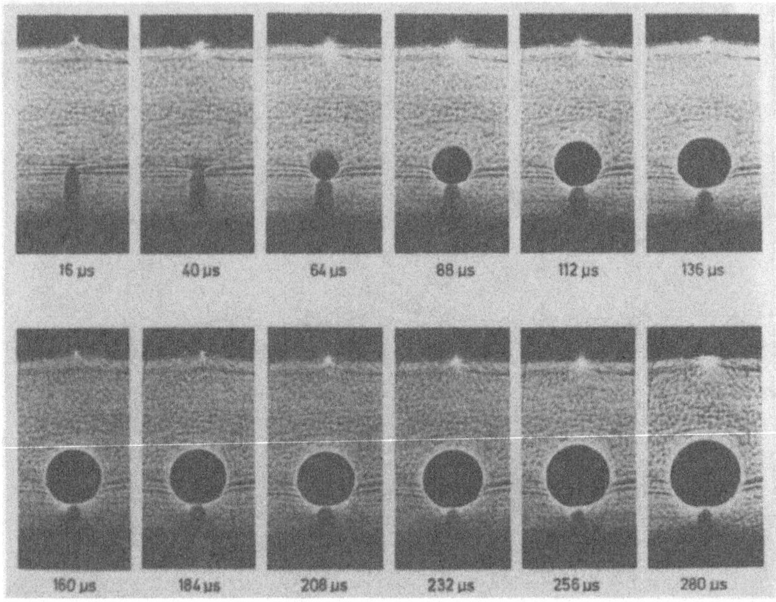

Fig. 66 Shadow optical photographs of a precracked bend specimen during impact loading (Steel, virtual image, z₀ < 0, reflection)

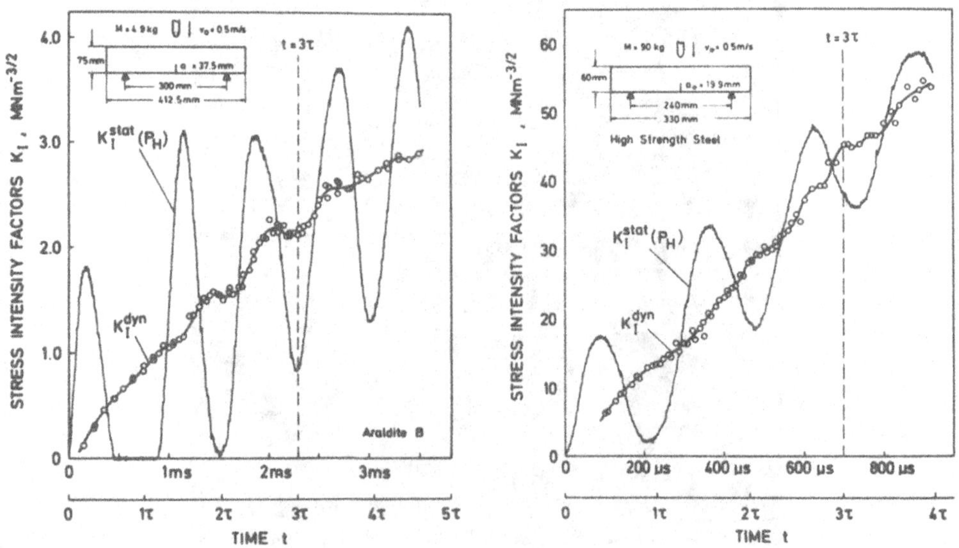

Fig. 67 and 68 Stress intensity factors during impact loading: (67) Araldite B, (68) high strength steel

at subsequent times.

Quantitative data of an Araldite B specimen are given in Fig. 67. The shadow optically determined dynamic stress intensity factors K_I^{dyn} are plotted as functions of time. In addition, the stress intensity factors $K_I^{stat}(P_H)$ are shown, determined from the hammer load signal P_H utilizing conventional static stress intensity factor formulas from ASTM E 399. The times are given in absolute units and also in relative units by normalization with the τ-value of the specimen. The K_I^{stat}-values show a strongly oscillating behaviour, whereas the actual dynamic stress intensity factor values K_I^{dyn} show a more steadily increasing tendency. In particular, it is recognized that for times larger than 3τ the influences of dynamic effects obviously have not vanished. Marked differences are still measured between K_I^{stat} and K_I^{dyn}. As is expected, in the small time range, $t < 3\tau$, these differences are very pronounced. One observation shall especially be addressed. After the inertial peak, the hammer load decreases to zero, indicating a loss of contact between the specimen and the hammer, as was also observed by Krisch [50]. During the loss of contact between the hammer and the specimen, the shadow optical measurements, however, indicate a steadily increasing crack tip loading.

Results obtained with steel specimens are shown in Fig. 68. The results are similar to those obtained with Araldite B specimens. The oscillations in the $K_I^{stat}(P_H)$-curve are somewhat smaller than those obtained with the Araldite B specimens, however. This is due to the different contact stiffnesses between the tup of the hammer and the specimen when Araldite B or steel specimens are impacted by the same striking hammer. In summary, the differences between K_I^{dyn} and K_I^{stat} are remarkable, they become smaller with increasing time, but still are unacceptably high at times $t_f > 3\tau$.

4.3.2 Loss-of-contact effect

Further experiments have been performed that revealed another dynamic effect influencing the behaviour of impacted specimens.

Figure 69 shows a result similar to the one reported in Fig. 67, but obtained at an impact velocity of 1.0 ms. This 1-m/s-experiment has been repeated under identical experimental conditions as given in Figure 69 but utilizing a specimen of reduced length: 412 mm instead of 550 mm. The support span, however, was kept the same. Data obtained with this specimen of reduced length are shown in Fig. 70. From a quasistatic point of view a reduction in specimen length should not affect the data. A comparison of these two sets of data (Figs. 69 and 70), however, shows a remarkable difference. The specimen with reduced length shows a more oscillatory behaviour for both the K_I^{stat}- and the K_I^{dyn}-curve. The pronounced oscillations in the K_I^{dyn}-curve also indicate that a mean load line of the hammer load record (as proposed by Turner et al. [51]) must not be a good representation of the actual loading condition at the crack tip.

In order to understand the very different behaviour of these two specimens of different length, further experiments were performed in which

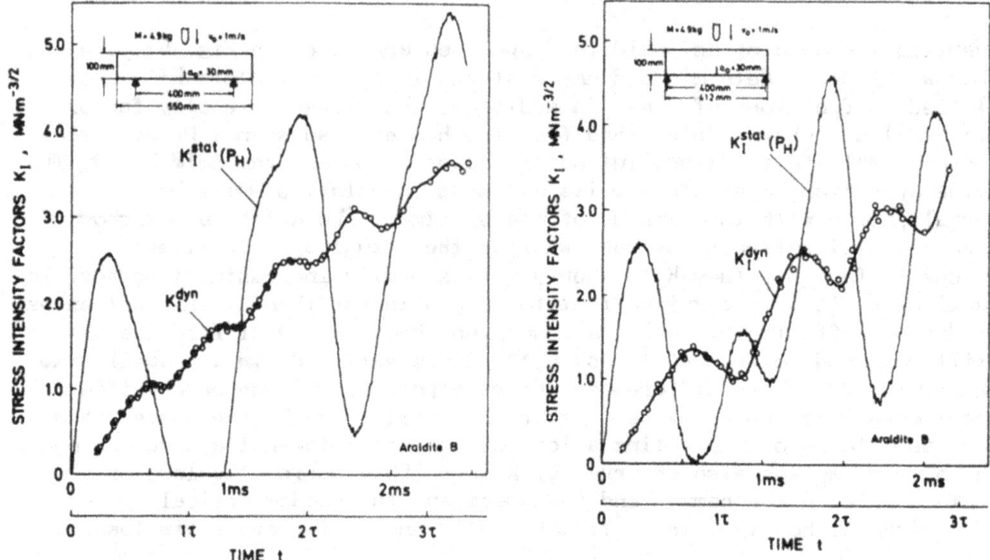

Fig. 69 and 70 Stress intensity factors during impact loading for specimens of normal (69) and reduced (70) length

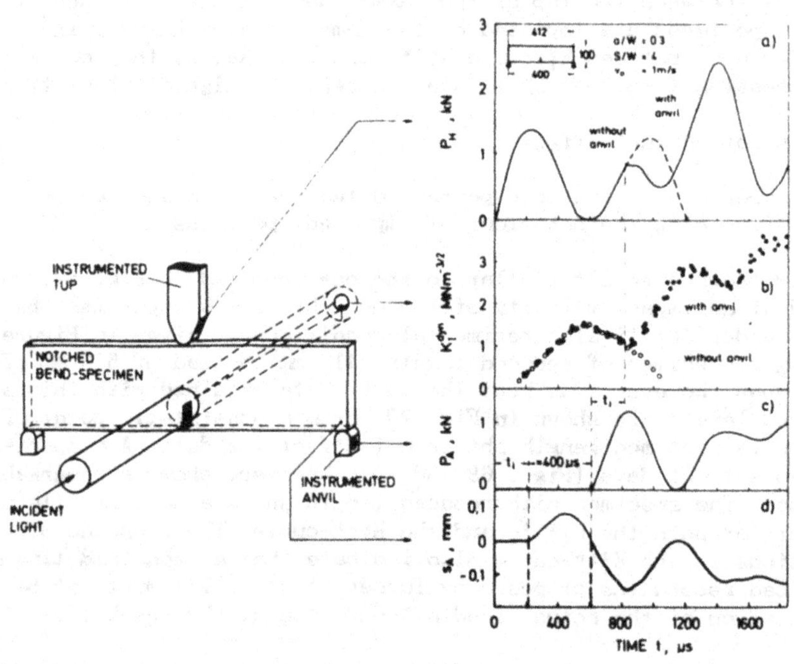

Fig. 71 Extended measuring technique: experimental set-up and results

additional measurements were taken [52]. The loads P_A at the anvils were recorded and also the movements s_A at the specimen ends were registered by an inertia free electro-optical clip gage. Figure 71 compares the load measured at the striking hammer (a), the stress intensity factor measured at the crack tip (b), the load measured at the anvils (c), and the position of the specimen ends with regard to the anvils (d). The data were obtained with Araldite B specimens impacted at 1 m/s. The τ-value of the specimens utilized for these investigations is about 700 μs. A comparison of the four signals indicates that non-zero loads at the anvils are registered only after a rather long time of about 600 μs. This time is about three times larger than the time it would take the slowest wave, i.e. a transverse wave, to travel from the point of impact to the anvils. This unexpected behaviour is explained by diagram (d) in Fig. 71. A loss of contact is observed between the specimen ends and the anvils. The loss of contact starts at about 200 μs. This time is in agreement with the above consideration of wave propagation times. For about the next 400 μs the specimen is completely free and only after this time, i.e. at a time of about 600 μs total (i.e. a time which corresponds to about 1τ), do the specimen ends come into contact with the anvils again. In accordance with this observation, loads are then recorded at the anvils.

Since in these experiments the anvils were obviously of no influence during the early phase of the impact process (i.e. for about 1τ); additional experiments were performed with unsupported, i.e.completely free specimens. The results are represented by the dashed curves and open data points in Fig. 71. In accordance with speculation, the early ($t < \tau$) specimen reaction is the same for both, the supported and the unsupported specimen.

Another series of experiments has been performed to investigate the influence of specimen stiffness on the loss-of-contact behaviour at the anvils. Fig. 72 summarizes the results obtained with specimens of varying width. The loss-of-contact effect is largest for the most compliant, slender specimen. This specimen looses contact even two times. The effect is reduced when the specimen width is enlarged. Almost no loss-of-contact is observed for the stiffest specimen. Similar results were obtained when the stiffness was varied by decreasing the length of the initial notch instead of increasing the width of the specimen.

The observed loss-of-contact between specimen and anvil contradicts the boundary condition of a constant zero displacement at the support which is usually assumed for simply supported beams under impact loading [53-55]. Rather, a vibration of a nonsupported specimen is excited directly after impact. For this type of vibration the nodal points of motion in general are not identical with the support points. It is believed that the loss-of-contact is larger the further away the nodal points are from the support points. For a specimen with a length in accordance with the original dimensions of the Charpy specimen, the nodal points are "by chance" close to the support points, thus, little loss-of-contact effects are expected for this specimen type, but larger effects are expected when the specimen length is reduced (see Fig. 72). This behaviour explains the

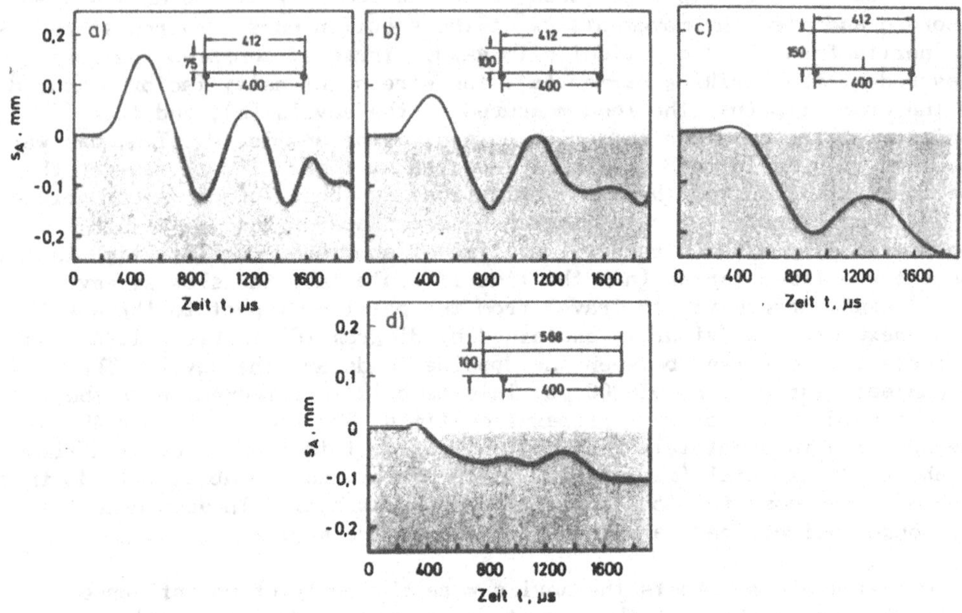

Fig. 72 Loss-of-contact effects for different specimen geometries (Zeit = time)

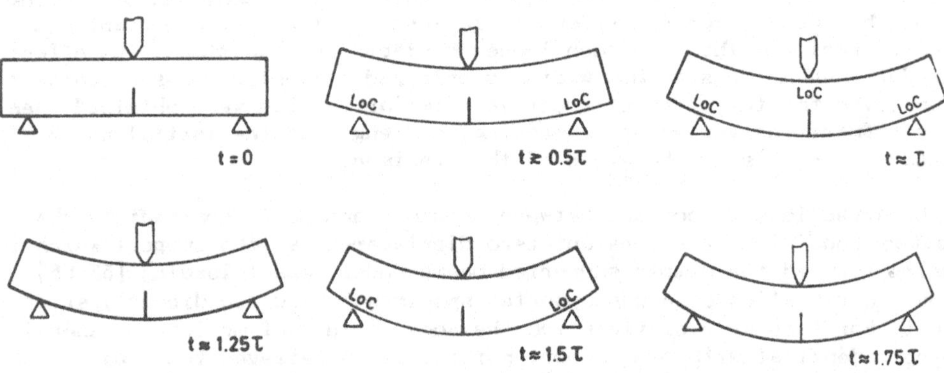

Fig. 73 Loss-of-contact effects observed with prenotched bend specimens under drop weight loading

different results in the K_I^{stat} and K_I^{dyn} behaviour observed in Figs. 69 and 70.

A schematic overview of all the loss-of-contact (LOC) effects observed between the hammer and the specimen and the specimen and the anvils together with the respective time ranges are given in Fig. 73. Also, on the basis of these results, it will be very unlikely that from a time $t \approx 3\tau$ on, quasistatic behaviour of the specimen will apply.

4.4 Method of Impact Response Curves

As was shown before, the quasistatic procedure for determining the impact fracture toughness K_{Id} (Section 4.2.1) has two severe shortcomings:

1) Requirement (47) obviously is not sufficient to guaranty a quasistatic state of stress in the impacted specimen. Such a condition obviously will only be reached later on.
2) Requirement (47) and the associated requirement for reduced impact velocities restricts the applicability range of the method.

In order to overcome these difficulties a fully dynamic measuring procedure has been developed by the author and his colleagues [56-58], called the impact response curve technique.

4.4.1 Basic principle of the measuring procedure

For fixed test conditions, such as specimen geometry (in particular specimen size and crack length) and impact velocity, the history of the dynamic crack tip stress intensity factor, $K_I^{dyn}(t)$, is established. This $K_I^{dyn}(t)$-curve quantitatively relates the response of the specimen to that particular impact event and therefore is called "impact response curve". This curve depends only on the elastic reaction of the specimen/striker system and therefore is unique for the system considered and, thus, applies to all steel specimens tested under the same impact conditions. In particular, this one relationship applies to steels of different toughnesses as long as the elastic properties of the steels, i.e. the elastic modulus and Poisson's ratio, are the same and the conditions for linear-elastic fracture mechanics or small-scale yielding are fulfilled.

The dynamic fracture toughness for a given structural steel is then determined by performing an impact experiment under the same test conditions as before but with a specimen made from the steel under study. In this experiment only the resulting time-to-fracture is measured. The dynamic fracture toughness value K_{Id} is then obtained from the pre-established impact response curve and the measured time-to-fracture, t_f, (see Fig. 74) by using the relation:

$$K_{Id} = K_I^{dyn}(t = t_f) \tag{49}$$

This fully dynamic procedure for measuring the impact fracture toughness does not have the previously discussed restriction that the time-to-fracture must be larger than a certain minimum value (see Eq. (47)). Therefore, the procedure can be applied for any experimental test

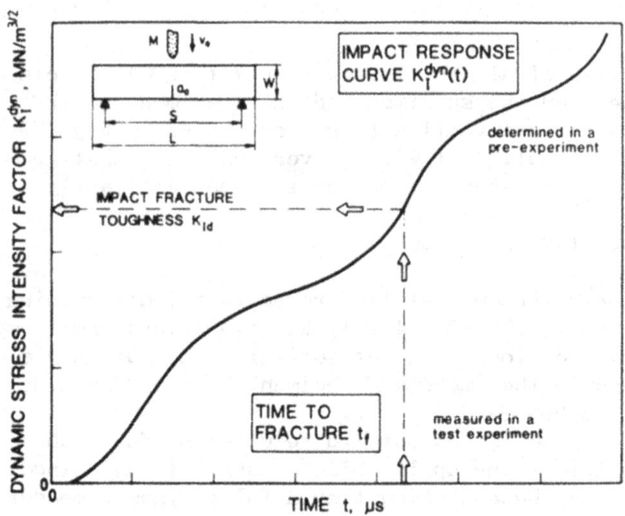

Fig. 74 Determination of the dynamic fracture toughness K_{Id} using the method of impact response curves

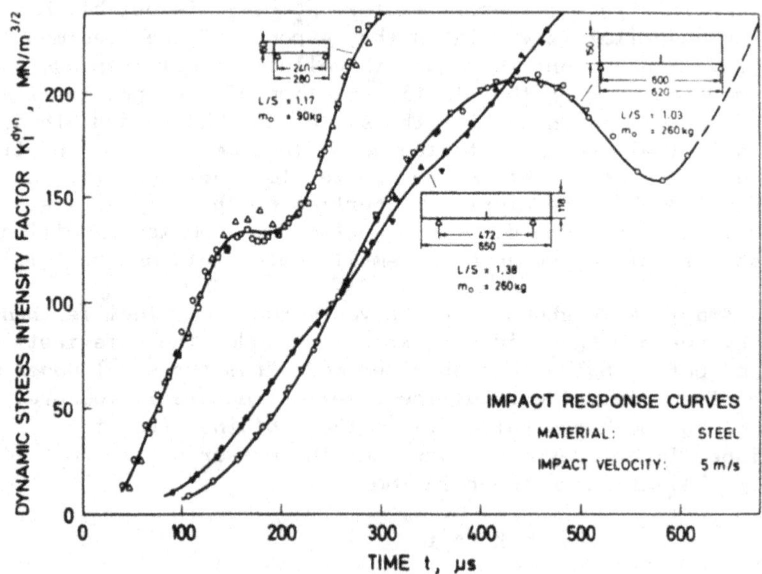

Fig. 75 Impact response curves for precracked bend specimens

condition, particularly also in the brittle fracture and high-velocity impact range, as long as the usual conditions for small-scale yielding are fulfilled.

4.4.2 Impact response curves

With large bend specimens tested in drop weight towers the impact response curves were determined by means of the shadow optical method of caustics. Three specimen sizes have been considered.

	Type I	Type II	Type III
a) Specimen length	280 mm	620 mm	650 mm
b) Specimen width	60 mm	150 mm	118 mm
c) Crack length	30 mm	75 mm	59 mm
d) Hammer mass	90 kg	260 kg	260 kg

Specimens of these types made from the high strength maraging steel X2 NiCoMo 18 9 5 were prepared with a mirrored front surface and have been impacted in a DYNATUP 8100 drop weight tower at a velocity of 5 m/s. The tip of the notch was blunted to increase the load carrying capacity of the specimen and thus to allow the measurement of impact response curves for stress intensity factors exceeding the fracture toughness of the steel (i.e. about 80 $MNm^{-3/2}$). The shadow optical method was applied in reflection. Shadow graphs similar to those shown in Fig. 66 were obtained. The resulting stress intensity factor histories, i.e. the impact response curves, are given in Fig. 75 for the three types of specimens considered. Several tests have been performed with one specimen type to generate more than 24 data points and to cover a larger time range.

Strain gage instrumentation of the specimen near the crack tip provides the simplest way of establishing impact response curves with precracked Charpy specimens (see e.g. [59]). In a static pre-experiment, the signal obtained from a strain gage at a specific location near the crack tip is calibrated in terms of the stress intensity factor. Because of the location of the strain gage near to the crack tip, the strain gage signal obtained in an impact event can be assumed to represent a good measurement of the dynamic crack tip stress intensity factor as well.

Figure 76 illustrates impact response curves for steel Charpy specimens with an initial crack length of a = 5 mm tested at different impact velocities of v_0 = 2, 3.8 and 5 m/s. The data were obtained with a 300 J pendulum test device (tup radius 2 mm, machine compliance $C_M = 8.1 \times 10^{-9}$ m/N). The strain gages were positioned about 2 mm to the side of the crack tip. To increase the load carrying capacity of the specimen, blunted notches instead of fatigue-sharpened initial cracks were again utilized in the experiments.

Because of the great practical importance of Charpy specimens, special efforts were undertaken to present the established impact response curves in a convenient form for practical applications. It is recognized from

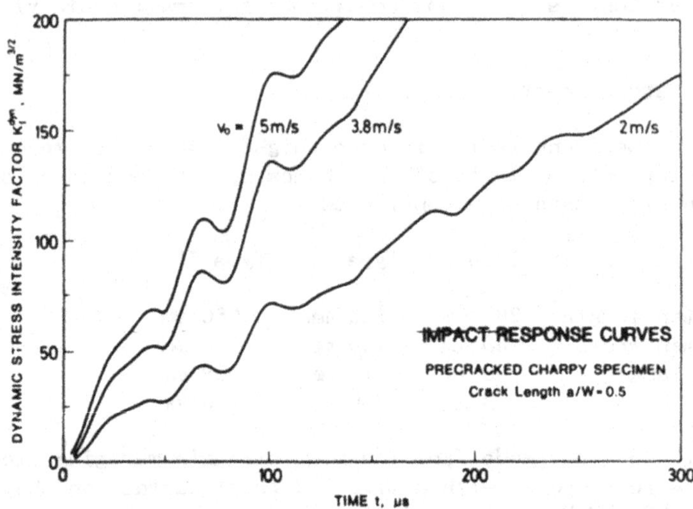

Fig. 76 Impact response curves for precracked Charpy specimens

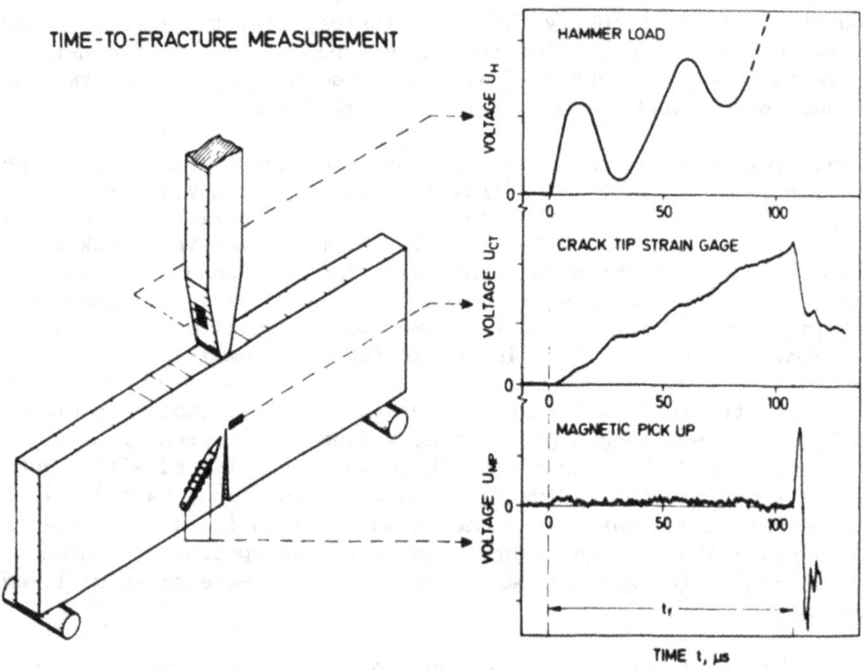

Fig. 77 Time-to-fracture measurement

Fig. 76 that the impact response curves for different impact velocities are similar. The stress intensity factors K_I^{dyn} depend linearly on the impact velocity v_0. Furthermore, it is evident that the data are independent of the hammer mass m_0, provided the energy used for breaking the specimen is small compared to the total impact energy [60]. This condition is usually fulfilled with conventional pendulum impact test devices, if the specimen breaks in the early time range. The impact response curves for arbitrary impact velocities can therefore be represented by a single relationship. It can be seen from Fig. 76 that this relationship can be mathematically described by a linear dependence of K on time with superimposed dynamic corrections. This relationship differs with crack length, but for the small variations in crack length used in bend specimens ($0.45 < a/W < 0.55$, as recommended in ASTM E 399), the resulting differences are only modest and can be compensated for in an approximate manner.

For practical applications, it is convenient to use the expression:

$$K^{dyn} = R \cdot v_0 \cdot t'' \tag{50}$$

where $t'' = f(t')$ given by Table 4

$$t' = g(t) = t\{1 - 0.62(a/W - 0.5) \pm 4.8(a/W - 0.5)^2\}$$

with $R = 301\ GN/m^{5/2}$. v_0 is the impact velocity, a is the crack length, W is the specimen width, t" and t'are modified times, and t is the measured physical time. The functions f and g account for the dynamic corrections and variations of crack length, respectively. The above value for R applies for stiff test devices with a machine compliance $C_M = 8.1 \times 10^{-9}$ m/N. If the compliance of a test device should be different, the resulting influence can be taken into account by multiplying the given value of R by the first-order correction factor $1.276/(1 + 0.276\ C_M/8.1 \times 10^{-9}\ m/N)$. Procedures for determining the machine compliance of impact test devices are described in [61].

This approximate formula shows good agreement when compared with semianalytical results [60]. Equation (50) describes the impact response curves for all practically relevant test conditions with an accuracy sufficient for engineering purposes.

4.4.3 Time-to-fracture measurements

The time-to-fracture of a precracked Charpy specimen or a large bend specimen tested in a drop tower can be obtained from signals of two uncalibrated strain gages, one of which is located on the tup of the striking hammer and the other on the specimen to the side of the crack tip. The leading edge signal from the hammer strain gage marks the beginning of the impact event. The onset of crack propagation, on the other hand, is indicated by the rapid drop in load registered by the crack tip strain gage. The time-to-fracture, t_f, is the interval between the two signals. Typical oscillograms of time-to-fracture measurements are shown

t' [μs]	t"= f(t') [μs]	t' [μs]	t"= f(t') [μs]	t' [μs]	t"= f(t') [μs]
0	0	100	118	200	198
2	0	102	119	202	202
4	2	104	118	204	204
6	4	106	117	206	207
8	6	108	115	208	210
10	9	110	115	210	212
12	13	112	115	212	212
14	17	114	115	214	213
16	20	116	116	216	213
18	24	118	118	218	214
20	28	120	120	220	216
22	30	122	122	222	219
24	33	124	124	224	222
26	35	126	126	226	225
28	36	128	128	228	230
30	38	130	129	230	233
32	39	132	130	232	236
34	40	134	131	234	239
36	42	136	132	236	241
38	43	138	134	238	243
40	45	140	136	240	244
42	46	142	138	242	245
44	47	144	141	244	245
46	46	146	145	246	245
48	45	148	148	248	245
50	45	150	152	250	245
52	46	152	155	252	245
54	49	154	157	254	245
56	53	156	159	256	246
58	57	158	161	258	249
60	61	160	164	260	251
62	65	162	166	262	253
64	69	164	169	264	255
66	72	166	172	266	257
68	73	168	175	268	258
70	73	170	177	270	260
72	72	172	180	272	261
74	70	174	183	274	262
76	69	176	185	276	265
78	68	178	187	278	267
80	69	180	188	280	269
82	70	182	188	282	272
84	75	184	187	284	275
86	81	186	186	286	277
88	88	188	185	288	280
90	94	190	186	290	282
92	100	192	187	292	284
94	106	194	189	294	286
96	111	196	192	296	288
98	116	198	195	298	289

Table 4 Time functions

in Fig. 77.

An indication of the time at which the crack becomes unstable can also be obtained by another procedure which does not require specimen instrumentation. At the onset of rapid crack propagation, a magnetic signal is generated by the accelerating crack tip if the specimen has been slightly magnetized before testing, e.g. with a permanent magnet [62]. This signal is picked up by a magnetic sensor (e.g. a coil) located at the crack tip near (but not in contact with) the specimen surface. Very good results have been obtained with commercially available magnetic pickups that are used in tape recorders.

The signal thus obtained has a short rise time and gives a clear indication of the time at the moment of fracture instability. Figure 77 shows a magnetic crack initiation signal compared to a crack tip strain gage signal. This magnetic measuring procedure is inexpensive and highly advantageous for routine testing, because it does not require extensive or costly specimen instrumentation. Fig. 78 shows a magnetic pickup (in a sheltering case) attached to the hammer of a Charpy pendulum to allow for a convenient measurement.

4.4.4 Application of the impact response curve technique

Impact fracture toughness data K_{Id} have been measured by means of the impact response curve technique and time-to-fracture measurements. Three steels, 30 CrNiMo 8, 15 MnNi 63, and St E 460, were tested at different temperatures. The geometrical and impact conditions were chosen to be identical to those for which impact response curves have been developed before. Two specimen sizes were utilized, i.e. the Charpy specimen and the Type I geometry. The Type I specimens were impacted at 5 m/s, the Charpy specimens at 5 m/s or at 2.5 m/s. Times-to-fracture in the range of 18 μs to 95 μs were measured. The resulting K_{Id} values are shown in Figs. 79a, b, and c. Static initiation fracture toughness values K_{Ic} (as far as they were available from literature data) are given for comparison in the figures as well. The observed results are plausible. First, the K_{Id} values are lower than K_{Ic}. Second, since the high toughness steel 15 MnNi 63 and the mild steel ST E 460 are more strain rate sensitive than the high strength steel 30 CrNiMo 8, the differences in K_{Ic} and K_{Id} are larger with these steels than with the latter one. The times-to-fracture at which the specimen broke, lie in the range of 0.3τ to 1.9τ; i.e. they are considerably lower than the limit value 3τ set forth in the quasistatic ASTM procedure. Thus, the ASTM procedure would not have been able to establish valid toughness values at all. The impact response curve technique, however, despite the small times-to-fracture, yields reliable K_{Id}-data, since it is a fully dynamic measuring technique.

4.4.5 Advantages and limitations of the impact reponse curve technique

The technique of measuring the impact fracture toughness K_{Id} with impact response curves and time-to-fracture measurements has several advantages over the conventional quasistatic ASTM procedure [47]. The impact response

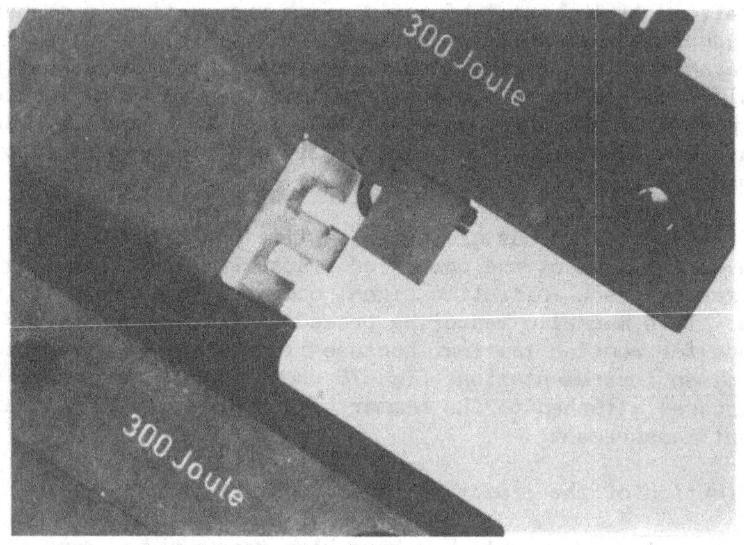

Fig. 78 Charpy pendulum hammer with magnetic pickup

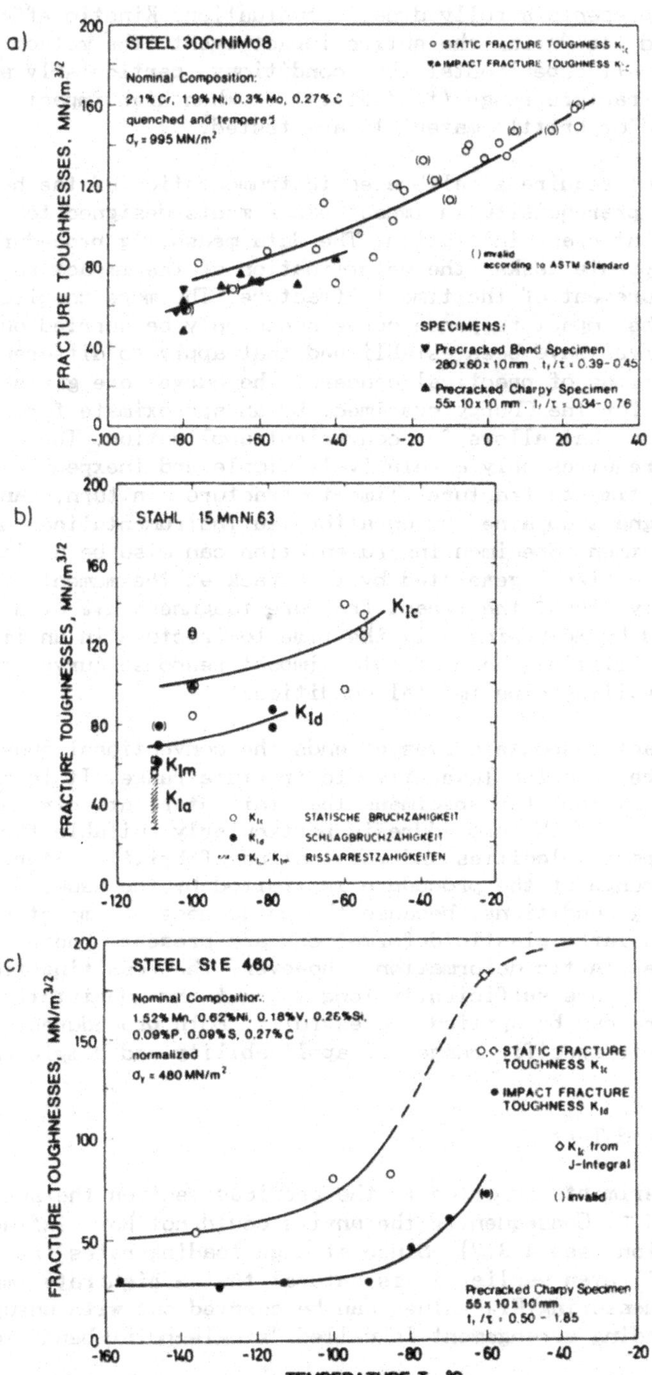

Fig. 79 Measured impact fracture toughness data for low-alloy steels a) 30 CrNiMo 8, b) 15 MnNi 63, c) St E 460

curve technique represents a fully dynamic evaluation. Kinetic effects are correctly accounted for during the entire impact event. The method can thus be applied to all experimental test conditions, particularly also in the small time-to-fracture range ($t_f < 3\tau$), i.e. when high impact velocities are used or brittle materials are tested.

This method does not require a calibrated instrumentation of the hammer, which is usually a prerequisite in impact experiments designed to determine the load at crack initiation. The data measuring procedure consists of two separate tasks: the determination of the impact response curve and the measurement of the time-to-fracture. The more complicated determination of the impact response curve needs only be carried out once. Impact response curves have been established that apply to different experimental conditions of practical concern. The curves are given in graphical form or, for the Charpy specimen, by an approximate formula and a tabulated function that allows for convenient application. The actual K_{Id} determination requires only a relatively simple and inexpensive measurement of the time-to-fracture. Time-to-fracture, in turn, can be determined from signals obtained by uncalibrated instrumentations of the hammer and the specimen. Specimen instrumentation can also be avoided by sensing the magnetic signal generated by the crack at the moment of fracture instability. Thus, the impact fracture toughness K_{Id} of a given steel is determined by measuring only the time-to-fracture in an impact experiment and by utilizing the particular impact response curve that applies to the prevailing experimental conditions.

The concept of impact response curves extends the conventional quasistatic evaluation procedure into the lower time-to-fracture range. It is an appropriate measuring tool for specimens that fail at relatively short times after impact, i.e. the procedure is particularly suitable for testing at high impact velocities and for testing of brittle materials. The applicability range of the procedure is limited by the usual small-scale yielding conditions, because the uniqueness of impact response curves is lost when large plastic deformations are present. Under the conditions of large plastic deformations, however, the resulting times-to-fracture generally are sufficiently long so that the quasistatic evaluation procedure can be applied successfully. Both procedures, therefore, have their specific ranges of applicability and complement each other.

4.5 Single-Point-Bend Test

In many of the experiments reported in the previous section the specimens broke at times $t_f < \tau$. Consequently the anvils could not have influenced the specimen reaction (see 4.3.2). Since at high loading rates the specimens would fail even earlier it is natural that a high rate impact test ($\gg 5$ m/s) for measuring K_{Id}-values can be carried out with unsupported specimens. This loading arrangement is called "single-point-bend" loading.

Typical results of "single-point-bend" experiments [63] are shown in Figs. 80 and 81. Steel specimens 270 mm long and 50 mm wide were utilized in

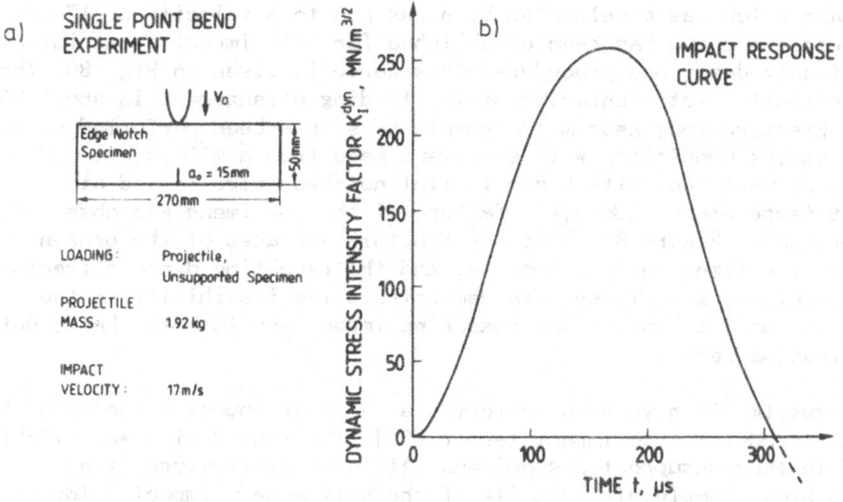

Fig. 80 Single-point-bend technique: a) test conditions, b) impact response curve

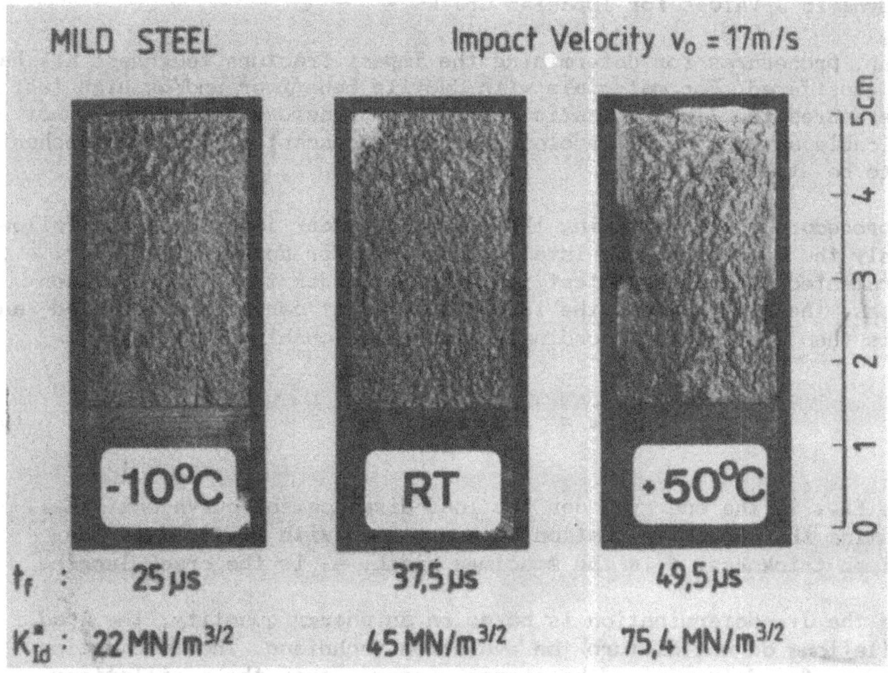

Fig. 81 Results of one-point-bend test

these investigations. Impact loading was achieved using a projectile of 1.9 kg mass which was accelerated by a gas gun to a velocity of 17 m/s. An impact response curve has been established for this impact event following the previously described procedure. This curve is given in Fig. 80. The stress intensity factor capacity of the loading arrangement is about 260 $MN/m^{3/2}$. Fracture toughness measurements have then been performed under the same impact conditions with specimens made from a mild steel. Specimens 20 mm thick, with sharp initial notches, were tested at different temperatures. Complete failure of the specimens was observed in all experiments. Figure 81 shows the fracture surfaces of the broken specimens, the times-to-fracture, t_f, and the resulting dynamic fracture toughness values, K_{Id}. These data demonstrate the feasibility of the single-point-bend technique for measuring impact fracture toughness data at high loading rates.

Similar experiments have been performed at a lower impact velocity of 5 m/s with a pendulum type impact tester [63]. The test device was slightly modified to load unsupported specimens. At lower temperatures the specimens broke completely in spite of the only modest impact velocity.

It is evident that a quasistatic evaluation procedure would not be able to describe these experiments. The fully dynamic method of impact response curves, however, can successfully be applied even under these highly dynamic test conditions.

4.6 Dynamic J-Values for Impacted Cracks

So far, procedures for determining the impact fracture toughness K_{Id} have been considered. For materials with ductile bahaviour and/or high test temperatures the linear-elastic approach of fracture mechanics is not applicable anymore, and the elastic-plastic concept of fracture mechanics has to be utilized.

The procedure for determining the J_{Id}-value under impact loading follows closely the equivalent quasistatic procedure for determining J_{Ic}: i.e., an instrumented 3-point-bend test is performed under impact loading conditions, the energy under the load-displacement curve is calculated, and J_{Id} is then determined according to the relationship

$$J_{Id} = \frac{2\ E_{max}}{d\ (W-a_0)} \tag{51}$$

Where E_{max} is the energy under the load-displacement-curve until E_{max} (assuming that crack initiation is associated with E_{max}), d is the specimen thickness, W is the specimen width, a_0 is the crack length.

Since the J_{Id}-determination is based on an energy quantity, the load oscillations do not disturb the evalution technique. The integration procedure for determining the energy averages over the oscillations. Furthermore, the whole time scale, in particular the time at which E_{max} occurs. is shifted to high values where the oscillations are sufficiently damped already. Consequently, dynamic effects on the mechanics of the test and the evaluation procedure (although of great importance in K_{Id}-determination) are of almost no influence in the J_{Id}-procedure. In this respect an elastic-plastic J_{Id}-test is much simpler than a linear elastic K_{Id}-determination and consequently is not considered further in this review. It shall be mentioned, however, that another problem is of significance in this context. It is well known that first stable crack growth occurs before maximum load (in contrast to Eq. (51)), i.e. a J_{Id}-value calculated from E_{max} may not be correct and yield nonconservative results. Several research efforts are concentrated on finding appropriate techniques for detecting first stable crack growth under impact loading conditions.

5. WAVE LOADING OF CRACKS

In practice, dynamic material strength values are usually measured in drop weight experiments. By convenience, drop heights in the range of one or a few meters are utilized resulting in impact velocities of about 5 m/s to less than 10 m/s. The corresponding crack tip loading rate $\dot{K}_I$ is of the order of 10^5 $MNm^{-3/2}s^{-1}$. It is evident that one single strength value, measured at such a loading rate, cannot characterize the material property for all loading rates, in particular not for very high rates. This section considers the fracture behaviour at extremely high rates of loading.

In principle there are three possibilities for the dependence of the impact fracture toughness from loading rate (see Fig. 82): (a) The impact fracture toughness continues to decrease with increasing loading rate. (b) The impact fracture toughness reaches a certain minimum level and remains at this level at higher loading rates. (c) The impact fracture toughness reaches a minimum and increases again if the loading rate exceeds a certain limit.

Behaviour (a) is not very realistic since this would imply that the impact fracture toughness ultimately becomes zero. The existence of a minimum fracture toughness as in case (b) is more likely. Dislocation movements reduce the stress concentration at the crack tip and lead to the formation of plastic zones around the crack tip. These plastic zones absorb energy before failure occurs and thus control the impact-fracture-toughness value. With increasing loading rate however, the dislocation processes become less effective because of the finite velocities associated with dislocation movements. Consequently, at sufficiently high loading rates the dislocation processes cannot be activated at all, and a constant behaviour should result from then on for all higher loading rates. It is of course also possible that material processes of a completely different nature become effective at very high loading rates, causing an increase in impact fracture toughness with increasing loading rate, as in case (c). A determination as to which behaviour applies can only be made on the basis of experimental investigations.

In one of the earliest publications on this subject Eftis and Krafft [64] reported on dynamic-fracture-toughnesss data. The authors measured the impact fracture toughness K_{Id} of mild steels in drop-weight experiments but they did not make any direct measurements at higher loading rates.

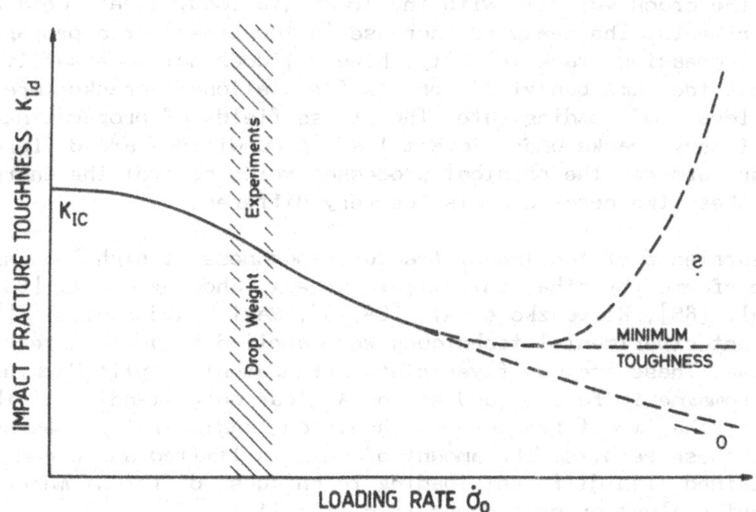

Fig. 82 Possible behaviour of impact fracture toughness on loading rate

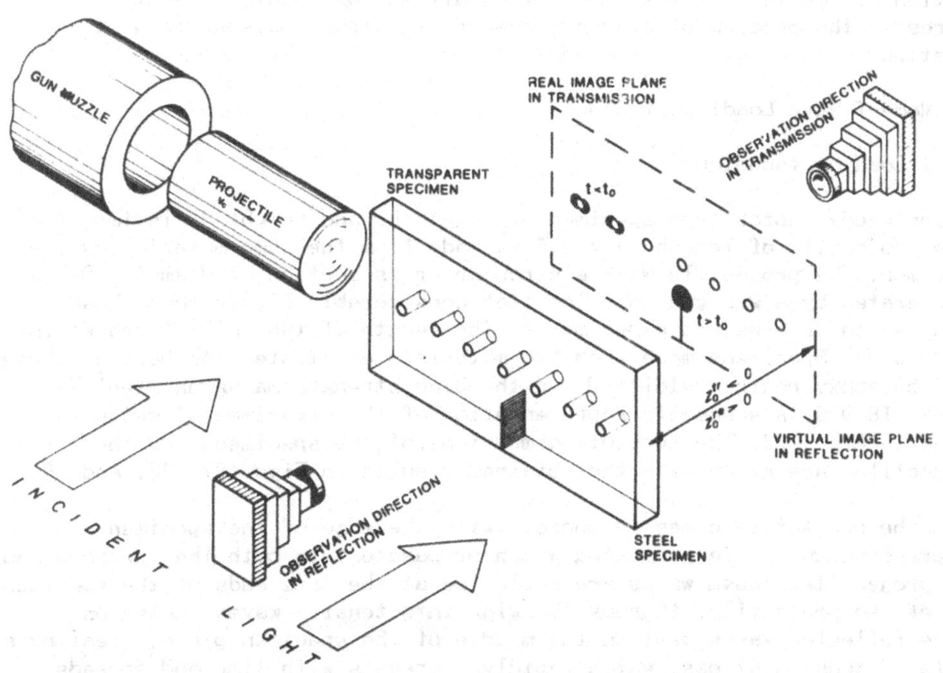

Fig. 83 Mode-I wave loading of cracks: experimental set-up and shadow optical arrangement (schematically)

They measured the fracture toughness of propagating cracks instead and correlated the crack velocity with the crack-tip loading rate obtained in impact experiments. The measured increase in toughness for a propagating crack with increasing crack velocity, however, does not necessarily indicate that the same behaviour applies for stationary cracks impacted at increasing levels of loading rate. The stress fields of propagating cracks and of stationary cracks under impact loading conditions are different [18,65]. Furthermore, the physical processes which control the energy balance in these two cases can also be very different.

Direct measurements of the impact fracture toughness at high loading rates have been performed by other investigators, e.g. Shockey et al. [66,67], Costin et al. [68], Klepaczko et al. [69,70], Ravi-Chandar et al. [71], etc. Different experimental techniques were applied to achieve very high loading rates. These include flyer-plate arrangements, split Hopkinson-bars, electromagnetic forces, and so on. A clear understanding of the dependence of the impact fracture toughness on loading rates does not emerge from these results. The amount of data is limited and a comparison of data obtained with different loading techniques, different materials, and different evaluation procedures is difficult.

The following sections present techniques to subject cracks to extremely high rates of loading by impinging stress waves and discuss the measured fracture behaviour. Mode-I loading is considered in Section 5.1, whereas Section 5.2 deals with fast mode-II conditions of loading. Section 5.3 addresses the problem of cracks subjected to stress pulses of short duration.

5.1 Mode-I Wave Loading of Cracks

5.1.1 Loading technique

A single-edge-notch type specimen of length L_S and height H_S is impacted by a projectile of length $L_P = 0.5 \cdot L_S$ made from the same material as the specimen. The projectile with a circular cross section of diameter D_P is accelerated by a gas gun to velocities considerably higher than those obtained in drop-weight experiments. The length of the initial cracks is $a = 0.5 \cdot H_S$. Specimens made from two materials are tested. As before, these are the epoxy resin Araldite B and the high strength maraging steel X2 NiCoMo 18 9 5. A schematic representation of the experimental setup is given in Fig. 83. The absolute dimensions of the specimens and the projectiles are given with the obtained results in Figs. 87, 88, and 89.

When the projectile comes in contact with the edge of the specimen compressive waves are generated which propagate into both the specimen and the projectile. These waves are reflected at the free ends of the specimen and of the projectile, thereby changing into tensile waves. Later on, these reflected waves meet in the middle of the specimen plate, creating a state of tensile stress, which rapidly increases with time and spreads across the entire specimen. Due to subsequent reflection processes, compressive stresses build up again at later times. The total duration of the resulting tensile-stress phase is equal to the time for the stress waves to make a round trip along the length of the projectile. The amplitude of the stress pulse is controlled by the impact velocity. The cracks are brought to instability within the early phase of the tensile stress state when the stresses increase rapidly with time.

In order to examine whether the stress history in the specimen follows the above described expectations, the shadow optical method of caustics is applied in pre-experiments using small holes drilled into the specimen as stress indicators. Analogous to crack tip stress intensifications, stress concentrations around holes also generate shadow patterns. Figure 84 shows caustics as they result for real shadow images $z_0 > 0$ in a transmission arrangement for tensile and for compressive loading. Tensile stresses generate a shadow pattern with its intersection pointing in the direction of the applied stress, whereas for compressive stresses the intersection is oriented perpendicular to the direction of the applied stress. The larger the caustic, the larger the applied stress (for details see ([12]).

Experiments have been performed with an Araldite B specimen containing a series of 11 holes of 1 mm diameter and 10 mm distance drilled into the specimen along the longitudinal direction of the specimen (see schematic representation in Fig. 83). A series of 24 shadow optical pictures photographed at successive times is reproduced in Fig. 85a. The recording times are given with each frame. The time at which compression changes into tension has been set equal to zero in the figure.

The first pictures show the propagation of the compressive wave into the specimen (frames 1-5) creating a rather uniformly distributed compressive stress field (frames 6-11). The compressive stresses then become smaller and change into tensile stresses, as is indicated by the change in the direction of the shadow patterns (frames 12-14). Within a relatively short time of about 50 µs the shadow patterns reach their maximum size. Uniformly distributed tensile stresses are observed for the following time interval (frames 15-22); later on the stresses change into compression again (frames 23-24). Further investigations with a row of holes oriented along the transverse direction of the specimen also showed rather uniformly distributed stresses in this direction, particularly at later times during the event, i.e. during the tension phase.

The formation of crack tip caustics under these loading conditions has been investigated in another experiment (see Fig. 85b). A single edge notch instead of a sharp crack has been utilized in this investigation to allow for compressive stress concentrations to build up at the notch tip. According to expectations (compare with Fig. 7) a compressive notch-tip shadow pattern is observed during loading in the compressive stress phase ($t < 0$) which then changes into tensile shadow patterns after the applied stress had changed into tension ($t > 0$). Thus, the measured data fully support the theoretically predicted stress history behaviour.

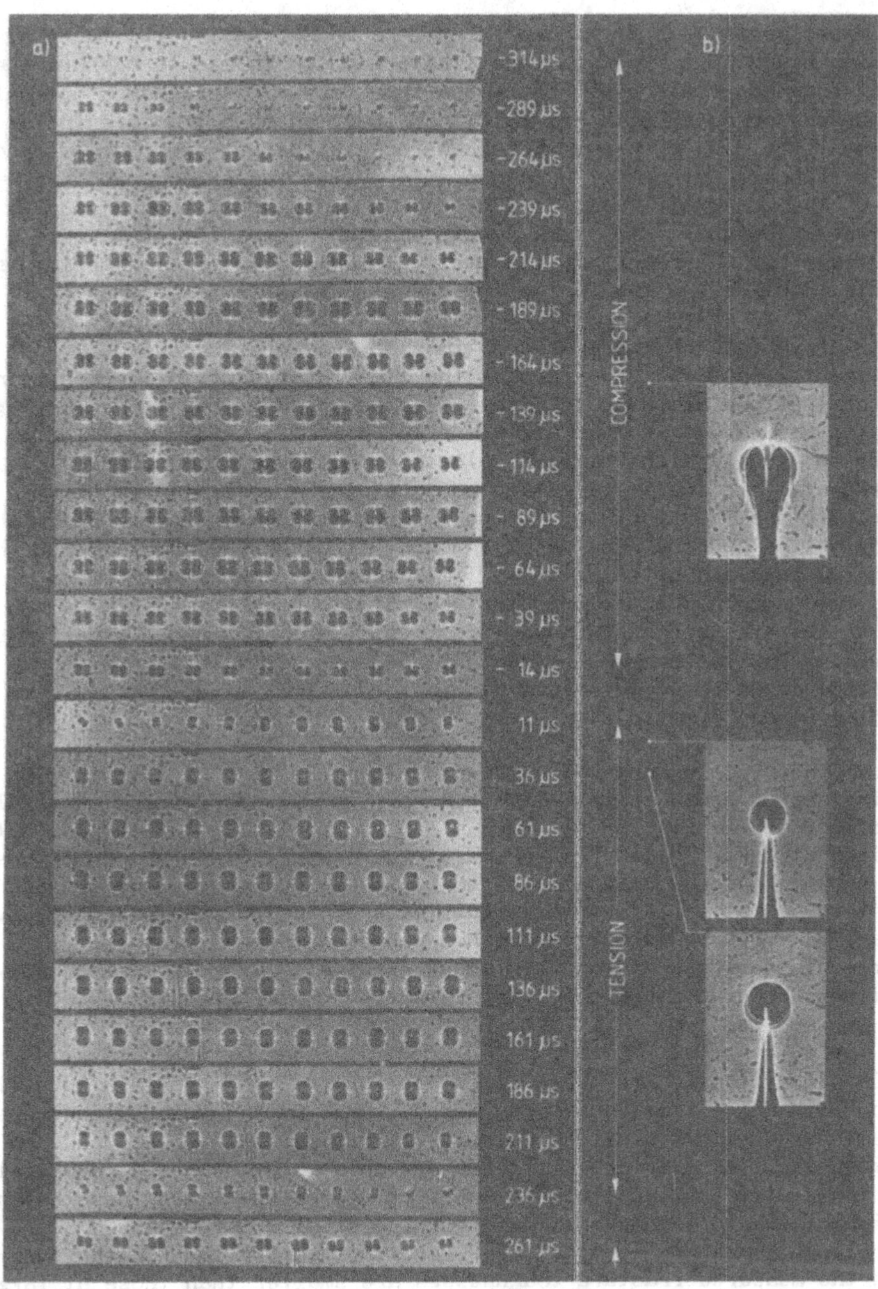

Fig. 85 High speed series of shadow optical photographs: a) row of holes, b) notch tip

5.1.2 Fracture behaviour

Main experiments with precracked specimens made from the two materials, Araldite B and high strength steel, have been performed to measure the impact fracture toughness. Figure 86a shows a series of shadow optical photographs observed with an Araldite B specimen in a transmission arrangement for $z_0 > 0$. Figure 86b shows equivalent photographs observed with a high-strength-steel specimen in a reflection arrangement for $z_0 < 0$. The crack-tip caustics indicate an undisturbed mode I-loading of the crack. The growth in crack length and the generation of waves emanating from the crack-tip (in particular with the steel experiments, see Fig. 86b) define the instant of crack instability.

Quantitative data on the measured times-to-fracture are shown in Fig. 87. The time-to-fracture is defined as the time interval from the beginning of the tensile-loading phase to onset of rapid crack propagation. The impact velocity was varied from about 10 m/s to 40 m/s for Araldite B and to 80 m/s for steel. Higher impact velocities could not be used due to experimental difficulties: damage occured at the contact area between the projectile and the specimen. Furthermore, disturbances showed up in the shadow optical photographs resulting from stress waves generated by the impact process. In spite of these limitations in impact velocity, times-to-fracture down to 7 μs in the Araldite B experiments and down to 6 μs in the high-strenght-steel experiments were obtained. These times are extremely short as becomes evident by a comparison with the distances that longitudinal waves would travel within these times, i.e., about 18 mm in Araldite B and 35 mm in steel.

Impact fracture toughness data measured from the size of the caustics at the instant of crack instability are shown in Fig. 88 for the material Araldite B and in Fig. 89 for high-strength steel. For comparison, data obtained at lower loading rates and at quasi-static loading conditions are also shown. These data were measured by the author and his colleagues with fast tension tests [73], drop weight tests [74], and precracked Charpy tests [73]. The shadow optical method was used in most of these experiments for determining the impact fracture toughness. With Charpy specimens, however, the concept of impact response curves (see Section 4.4) was applied.

The obtained results are quite different for the two materials investigated. Within the time range considered, the impact fracture toughness data measured with Araldite B do not show a significant dependence on loading rate. The data are scattered around a value which is identical with the static fracture toughness K_{Ic} of the material. On the other hand, the impact fracture toughness data for the steel X2 NiCoMo 18 9 5 are strongly dependent on loading rate. For times-to-fracture in the range of 10 μs the measured fracture toughnesses are smaller than those at larger times-to-fracture. Furthermore, data measured at times-to-fracture below 10 μs show a sharply increasing trend. Impact fracture toughness values measured at 6 μs become even larger than the static fracture toughness value K_{Ic}. Although the data necessarily show a large scatter due to the

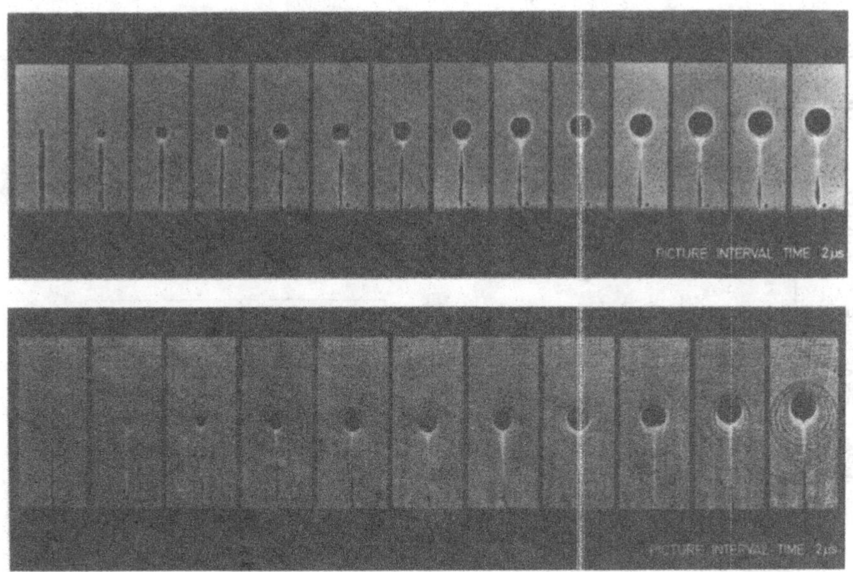

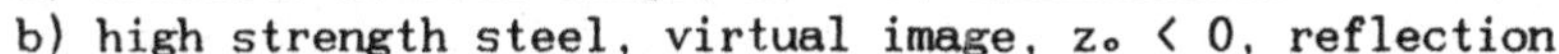

Fig. 86 High speed series of shadow optical photographs:
a) Araldite B, real image, $z_0 > 0$, transmission
b) high strength steel, virtual image, $z_0 < 0$, reflection

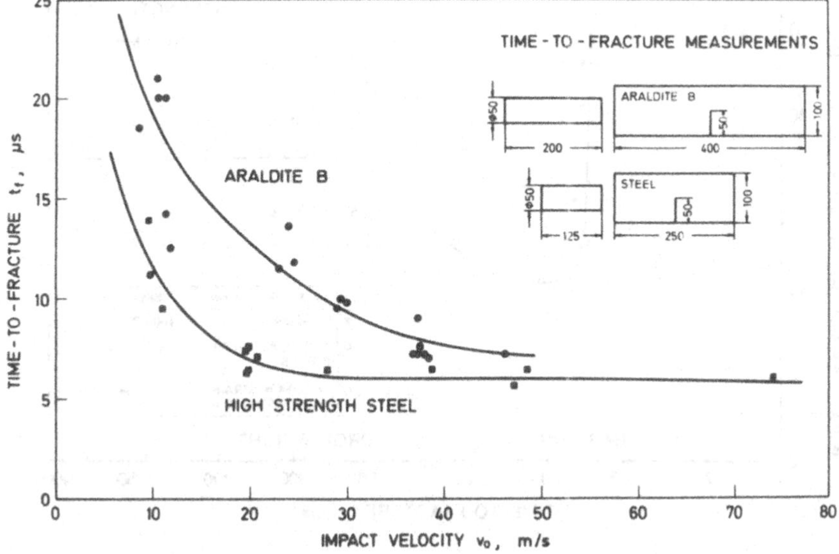

Fig. 87 Times-to-fracture measured in wave loading experiments

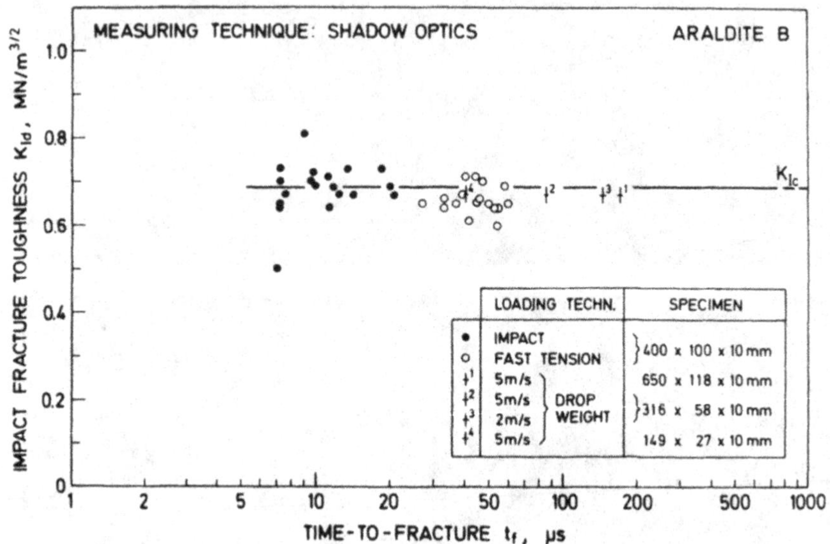

Fig. 88 Dependence of impact fracture toughness on time-to-fracture for Araldite B

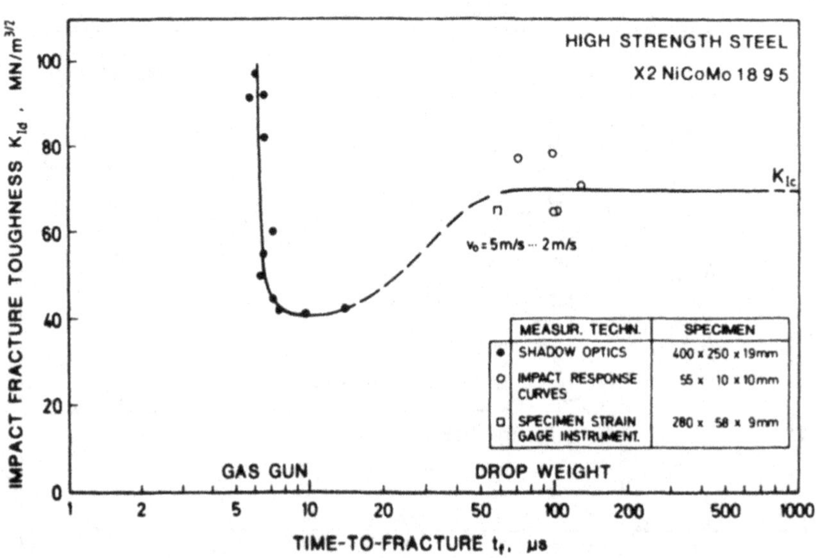

Fig. 89 Dependence of impact fracture toughness on time-to-fracture for a high strength steel

limited time resolution of the high speed camera (i.e. 1 μs) the trend in the shadow optical data is very pronounced and far beyond any uncertainties due to experimental scatter. At loading rates above a certain limit the impact fracture toughness of this steel obviously increases with increasing loading rate.

5.1.3 Incubation time

Such increase of the impact fracture toughness with decreasing time-to-fracture can be explained if the usual static instability criterion is modified by assuming the existence of an incubation time, as is illustrated in Fig. 90. According to this assumption, the crack tip would have to experience a supercritical stress intensity factor $K_I > K_{crit}$ for a certain minimum period of time before onset of rapid crack propagation can occur. This time interval is called the incubation time. It is very likely that this time is related to certain material processes and, therefore, is different for different materials. If the stress intensity factor K_I^{dyn} increases slowly with time, the delay in instability due to this incubation time is negligible. Consequently, the critical stress intensity factor for the onset of rapid crack propagation, K_{Id}, would be practically the same as the value K_{crit}. However, if the slope of the $K_I^{dyn}(t)$ curve is very steep then the stress intensity factor will increase significantly within the incubation time. Thus, the critical stress intensity factor for the onset of rapid crack propagation, K_{Id}, would be considerably larger than the value K_{crit}. Consequently, impact fracture toughness values determined from measurements taken at the onset of rapid crack propagation would increase with decreasing time-to-fracture, i.e. with increasing loading rate. The data presented in Figs. 88 and 89 for the two materials investigated appear to indicate that the incubation time for Araldite B is considerably smaller than that for steel.

It should be emphasized that although the concept of the incubation time explains the experimental observations presented here, this does not necessarily imply that it is the only explanation. The existence of the incubation time is a hypothetical assumption. An explanation of the physical background of the incubation time is not presented. The validity of the concept must be tested through further experiments with a variety of materials and through theoretical modeling and analysis of the material and mechanical processes that may occur near the crack tip at such high loading rates.

5.2 Mode-II Wave Loading of Cracks

5.2.1 Loading technique

A specimen with two parallel edge cracks or edge notches is impacted by a projectile having a diameter d equal to the distance h of the two cracks, as is schematically shown in Fig. 91. The impinging projectile initiates a compressive wave in the middle part of the specimen that generates a mode-II loading at the crack tips. The sign of loading is different for

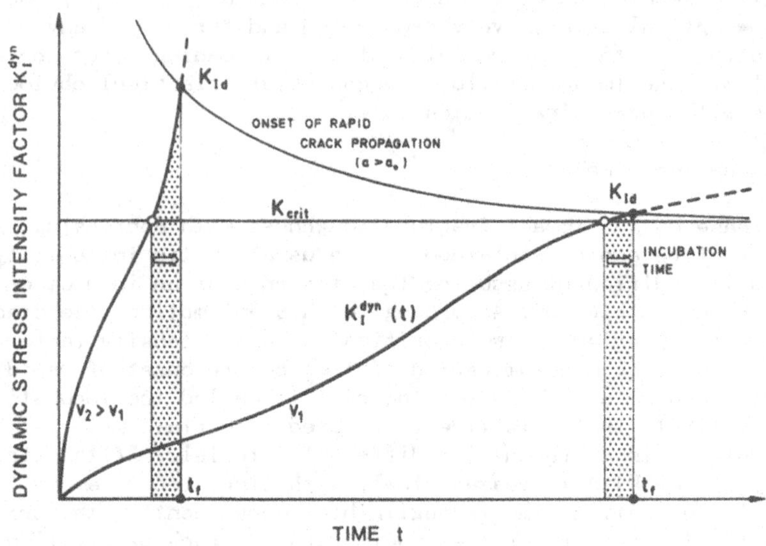

Fig. 90 Influence of the incubation time on the instability behaviour of cracks

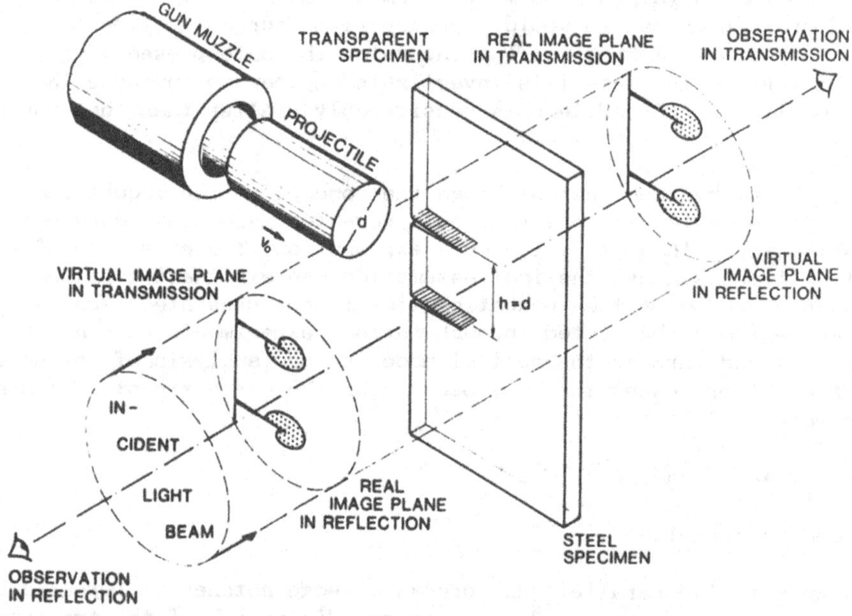

Fig. 91 Mode-II wave loading of cracks: experimental set-up and shadow optical arrangement (schematically)

the two notches. The mode-II loading condition should apply for the early time range of the impact event, i.e. until waves that are reflected at the finite boundaries of the specimen will interfere with the crack tip. The specimen is completely free: it is not held in a special loading fixture. But a correct alignment of the specimen perpendicular to the impact direction is necessary. A cylindrical steel projectile of 50 mm diameter is used to impact the specimen. For specimens with crack distances $h < 50$ mm a projectile with a front end of reduced diameter is utilized. The projectile is accelerated by an air gun to velocities v_0 ranging from 10 m/s to about 100 m/s. Typical specimen dimensions are about 100 mm x 200 mm. The usual crack lengths are about half the specimen width. (For further details see [75,76]).

An overview on the loading condition was obtained from experiments with specimens made from the material PMMA. This specimen material was chosen because it yields single caustics that can easily be compared to the theoretically predicted caustic curves. For quantitative analyses other materials have later been used since PMMA suffers from viscoelastic effects that cause difficulties in the exact evaluation of dynamic stress intensity factors. Experiments were performed with symmetric double-edge-notch specimens as well as with unsymmetric single-edge-notch specimens.

Figure 92 shows a series of high speed photographs for a symmetric specimen with a double edge notch configuration. The distance between the two notches is 30 mm. The picture interval time is 3 µs. Figure 93 shows an equivalent series of high speed photographs for an unsymmetric specimen with a single edge notch. The picture interval time is 2 µs. All photographs were recorded with a shadow optical transmission arrangement in a virtual reference plane $z_0 < 0$. A comparison of the results for the "upper" notch in both experiments shows identical behavior. With increasing time an almost undisturbed pure mode-II loading is obtained (see Figs. 7 and 8 for comparison). For the very early time range of the impact event, however, a superimposed compressive mode-I loading is observed (see Fig. 11). Note that with the chosen shadow optical observation mode a compressive stress intensification now generates a shadow spot instead of the light concentration pattern (see Fig. 8). The recorded superimposed compressive stress intensification results from the lateral compressive strain associated with the initiated compressive stress pulse. With notches, as utilized in the experiments under consideration, a compressive stress intensification can build up at the notch tip. With cracks, the compressive stresses may cause uncontrollable friction effects between the fracture surfaces. Although these compressive stress intensifications are very large during the early phase of loading, they become negligible as time proceeds and as the absolute magnitude of the mode-II loading increases. The last frames in Figs. 92 and 93 indicate the beginning of instability of one of the notches. The pure mode-II caustic tends to change into a tensile mode-I caustic, which for the chosen observation mode shows up as the light concentration pattern now (see Fig. 8). Since the unsymmetric single-edge-notch specimen obviously yields the same results as the symmetric double-edge-notch specimen, single-edge-notch specimens have been utilized in all subsequent experiments because they

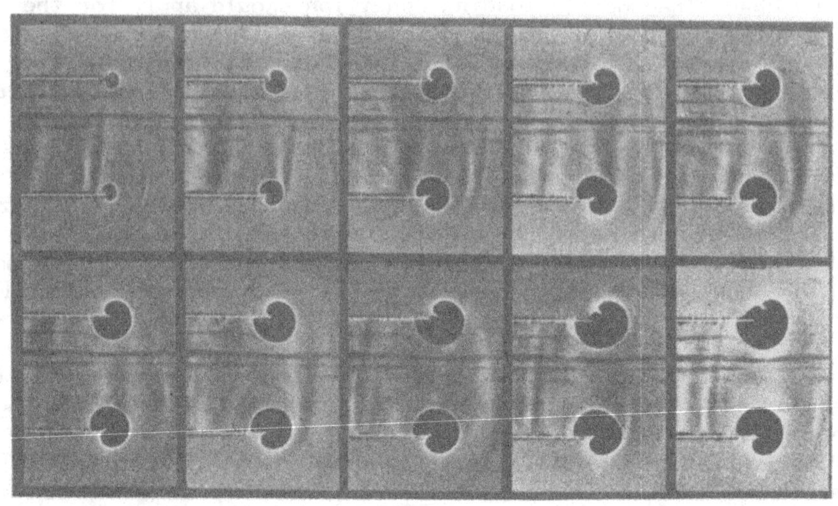

Fig. 92 High speed series of shadow optical photographs (Double-edge-notch configuration, PMMA, virtual image, $z_0 < 0$, transmission, 3 μs)

Fig. 93 High speed series of shadow optical photographs (Single-edge-notch configuration, PMMA, virtual image, $z_0 < 0$, transmission, 2 μs)

are easier to manufacture.

Quantitative data on the stress intensification rate $\dot{K}_{II}$ are shown in Fig. 94 for experiments performed with specimens made from the model material Araldite B and the high strength maraging steel X2 NiCoMo 18 9 5. The impact velocities have been 12 m/s or 13 and 33 m/s, respectively. The maximum crack tip stress intensification rates $\dot{K}_{II}$ are about 2 x 10^5 $MNm^{-3/2}s^{-1}$ for Araldite B and 2 x 10^7 $MNm^{-3/2}s^{-1}$ for steel. These rates are extremely high. They are by one or two orders of magnitude higher than those obtained with drop weight experiments, they are of the same order or even higher than those obtained with Hopkinson-bars.

5.2.2 Fracture behaviour

Main experiments have been performed to study the failure behaviour of steel specimens subjected to different rates of shear loading [76]. Two steels have been utilized: the high strength maraging steel, X2 NiCoMo 18 9 5, and the Chromium-Molybdenum steel, 42 CrMo 4. In the experiments, the notch tip acuity and the impact velocity were varied. Notches with tip radii ranging from ρ = 0.8 mm to precracked notches were utilized and impact velocities from 12 m/s to 72 m/s were applied. The strain at the tip of the notch increases with the acuity of the notch (i.e. with decreasing values of the notch tip radius). Furthermore, the rate at which this notch strain builds up increases with the impact velocity. Both parameters thus control the local strain rate at the tip of the notch - the higher the acuity and the higher the impact velocity are, the higher the notch tip strain rate becomes.

Damage paths observed with the steel X2 NiCoMo 18 9 5 are graphically shown in Fig. 95. The following characteristics of damage behaviour are recognized:

I) In the case where a notch of low acuity (ρ = 0.85 mm) is impacted at a low velocity (v_0 = 13 m/s) no damage is observed. The notch does not become unstable. The maximum strain resulting for this impact event is not sufficient to cause instability.

II) With increasing notch tip strain rate, i.e. for combinations of high notch tip acuity but low impact velocities, or for low notch tip acuities but high impact velocities damage is obtained in a direction that forms an angle of about 70° with respect to the ligament. Only the beginning of the damage path is shown in Fig. 95. In all experiments complete failure of the specimen is observed, i.e. damage spreads across the entire specimen thus dividing the specimen into two halves. A photograph of the damage surface is reproduced in Fig. 96a. It shows the usual characteristics of a fracture surface: roughness as it is typical for this steel and shear lips at the edges of the surface. A shadow optical picture taken a few microseconds after instability (Fig. 97) shows a characteristic tensile (mode-I) crack tip caustic. These observations indicate that obviously this damage is controlled by a tensile fracture process, as it is expected from fracture mechanics considerations.

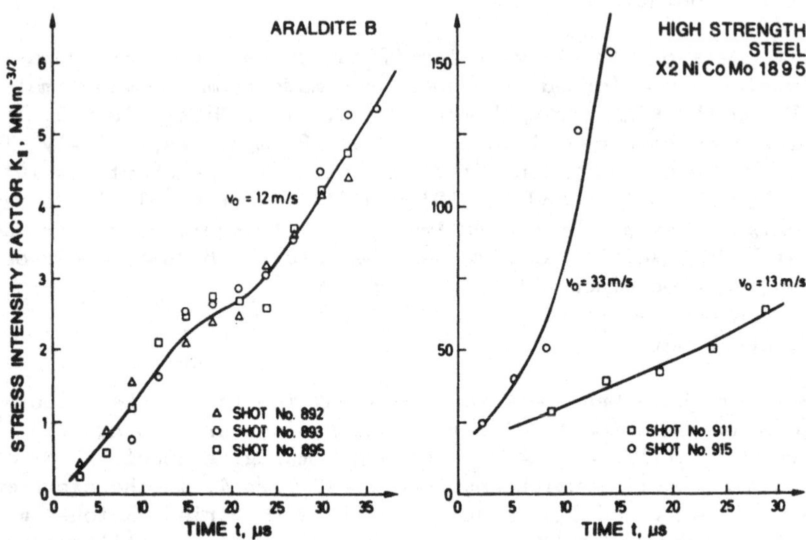

Fig. 94 Dynamic in-plane-shear (Mode-II) stress intensity factors K_{II}

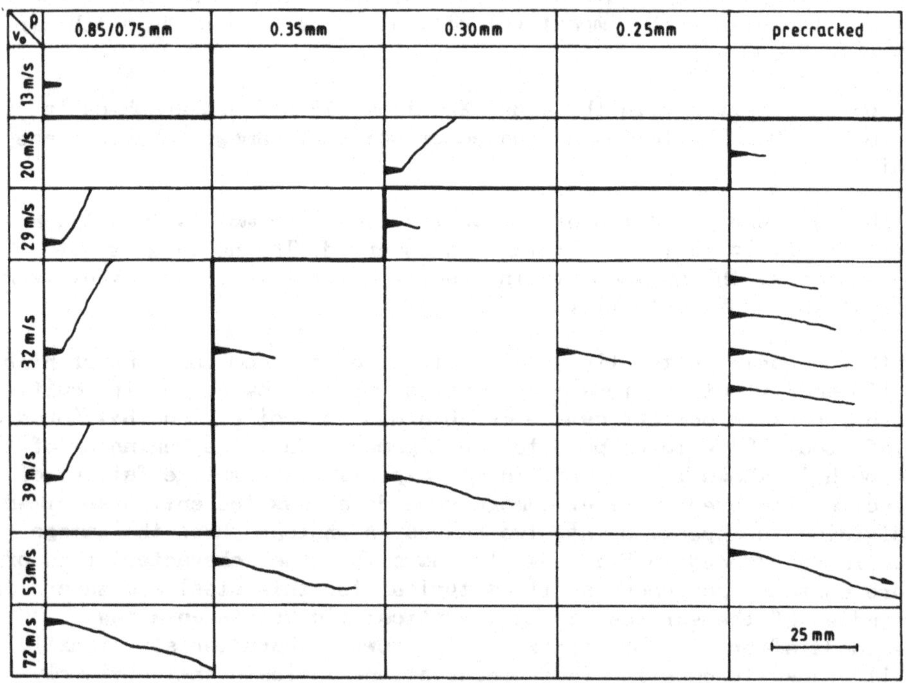

Fig. 95 Damage paths

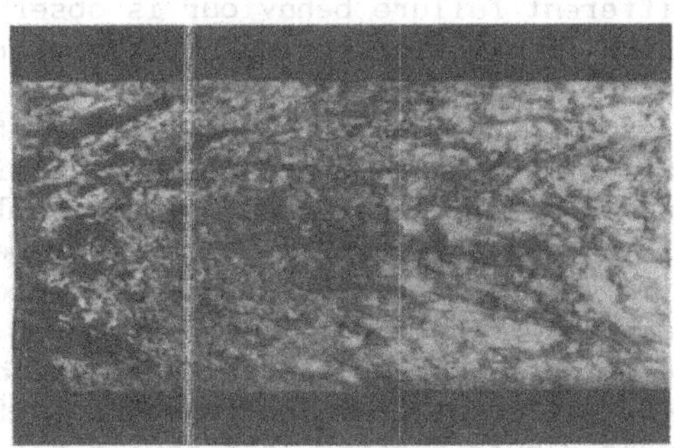

Fig. 96 Appearance of (a) fracture surface obtained at low loading rates and (b) damage surface obtained at high loading rates

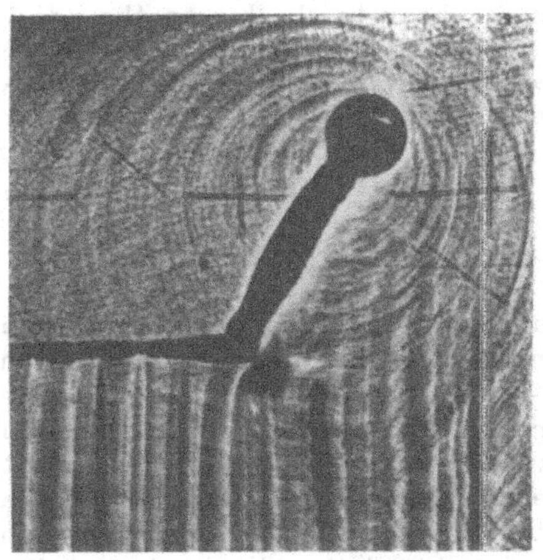

Fig. 97 Shadow optical photograph after instability: steel, virtual image, $z_0 < 0$, reflection

III) When the notch tip strain rate is increased further, i.e. when notches of high acuity are impacted at high velocities a completely different failure behaviour is observed. Damage develops in a direction which is almost identical with the original direction of the notch. Actually the damage path is inclined by a small angle with respect to the ligament, but on the opposite side of where the 70°-cracks are observed. Furthermore, damage extends over a limited length only and comes to arrest in almost all cases. But the damage length increases with increasing notch tip strain rate. In one experiment performed at a very high notch tip strain rate, (precracked notch, impact velocity v_0 = 53 m/s) complete failure of the specimen was observed. A typical photograph of the resulting damage surfaces is given in Fig. 96b. This damage surface looks very different from the fracture surface shown in Fig. 96a. It shows large smeared-over regions that have a shiny mirror-like appearance. Furthermore, there is no indication of any shear lips. Damage extends straight to the very end of the cross section of the specimen. Metallographic micrographs of a cut through the specimen perpendicular to the damage surface are presented in Fig. 98 for the two steels investigated. Heavily sheared localized bands are visible. The shear band photographed with the steel X2 NiCoMo 18 9 5 shows a continuously increasing concentration of shear deformation towards the central region of the shear zone. With the steel 42 CrMo 4, a white edging band is visible indicating that a phase transformation has taken place with this material in the central region of the shear band. In particular the white edging band is associated with a strong increase of hardness with respect to the base material, as is indicated by Fig. 99.

Details of the formation of shear bands due to the adiabatic conditions and softening processes that control this damage process shall not be discussed here (for details see (77,78)). In this context it is important to recognize that damage at high levels of notch tip strain rate is obviously not caused by tensile fracture processes but by localized shear mechanisms.

5.2.3 Discussion of results

In the previous presentation of the observed experimental findings the parameter "notch tip strain rate" has been defined. Quantitatively the strain at the notch tip varies proportional to $1/\sqrt{\rho}$, where ρ is the notch tip radius (79). Furthermore, the rate at which the strain builds up increases proportionally with v_0, where v_0 is the impact velocity. Thus the term $v_0/\sqrt{\rho}$ should be an appropriate quantitative measure of the notch tip strain rate. Notch tip strain rate factors, $v_0/\sqrt{\rho}$, have been calculated for each experiment and the observed experimental findings have been plotted as a function of this parameter in Fig. 100. The plot allows for a joint presentation of all the observed experimental findings in one diagram.

I) For notch tip strain rates less than $(v_0/\sqrt{\rho})_{TF}$ damage does not occur. The initial notches stay stable.

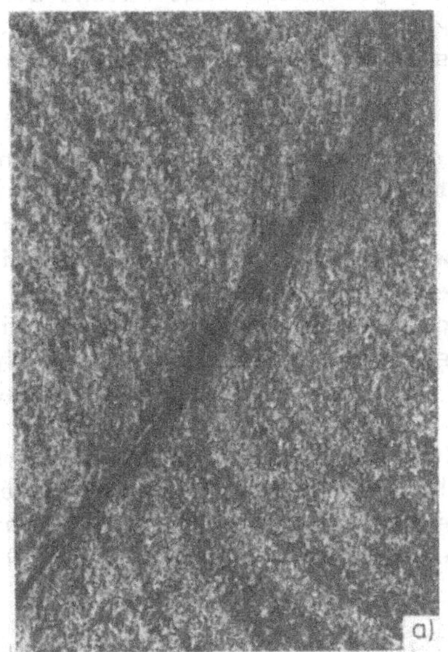

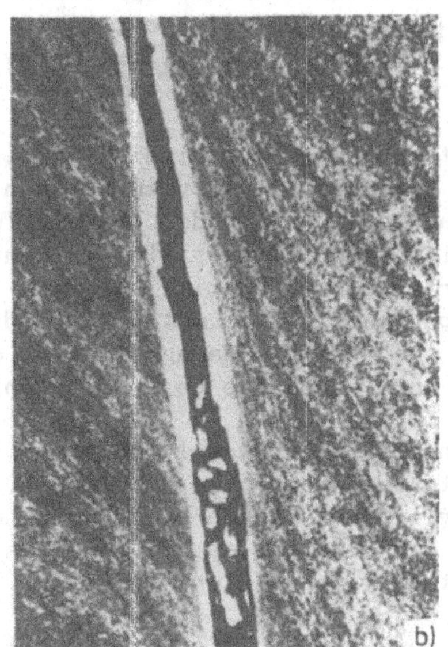

Fig. 98 Micrographs of shear bands: a) deformed band, Steel X2 NiCoMo 18 9 5; b) transformed band, Steel 42 CrMo 4

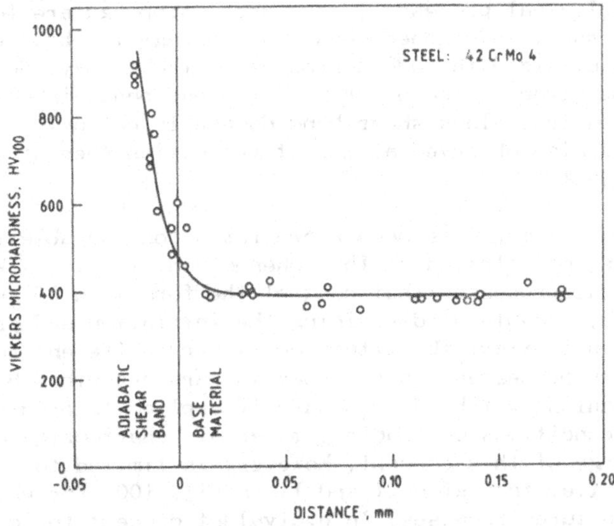

Fig. 99 Hardness profile across shear band

II) For notch tip strain rates between the threshold values $(v_0/\sqrt{\rho})_{TF}$ and $(v_0/\sqrt{\rho})_{SB}$ failure occurs by tensile fracture processes that extend at an angle of about 70° with respect to the ligament, causing complete failure of the specimen. This failure behaviour is in accordance with fracture mechanics considerations.

III) At the notch tip strain rate $(v_0/\sqrt{\rho})_{SB}$ failure mode transition takes place. For notch tip strain rates higher than this value, failure occurs by localized shear bands. This failure is limited in length and it extends in a direction which is almost identical with the direction of the original starter notch. Despite the very different test parameters utilized in the experiments (i.e. notch tip radii, impact velocities, specimen thickness), the observed damage lengths obviously follow the same curve. This curve, thus, seems to be able to uniquely describe the experimental data.

One conclusion is readily obtained from this plot. The transition from complete failure of the specimen by tensile fracture processes to damage of only limited length by shear band processes indicates that considerably higher energies are associated with the generation and growth of shear bands than with fracture processes.

Shear bands are one of the major processes that control damage at high rates of loading. Thus shear band processes are of great practical relevance in high rate applications. Usually high impact velocities are necessary to generate shear bands. In the described experiments, failure mode transition from fracture to shear band processes, however, is observed at rather modest impact velocities (of the order of 10 m/s) only. This is due to the fact that notches are used as stress-strain-risers in the presented experiments. Similar stress-strain-rising effects apply for many other technological processes, e.g. when chippings are formed in metal cutting processes; and indeed shear band damage is observed with fast cutting tools as utilized in modern production technology. Several other examples could be given on the occurance of shear bands in common industrial applications. Thus shear band damage is not only a high rate phenomenon but is also observed at modest velocities when strain concentrations are present.

Although shear band damage has been known for a considerable time and much research work has concentrated on this phenomenon, little is known about the actual physical processes that control the formation of shear bands. A complete analysis, capable of describing the initiation and propagation of shear bands does not exist. The situation is very different for failure caused by fracture processes. These processes are described by the concept of fracture mechanics, which is successfully applied to predict failure under different conditions of loading, material behaviour, etc. The applicability range of this concept, however, is limited to low or modest rates of strain, i.e. to regions I and II in Fig. 100, for which damage is controlled by fracture processes. An equivalent concept for describing shear band damage, i.e. region III in Fig. 100, is still missing.

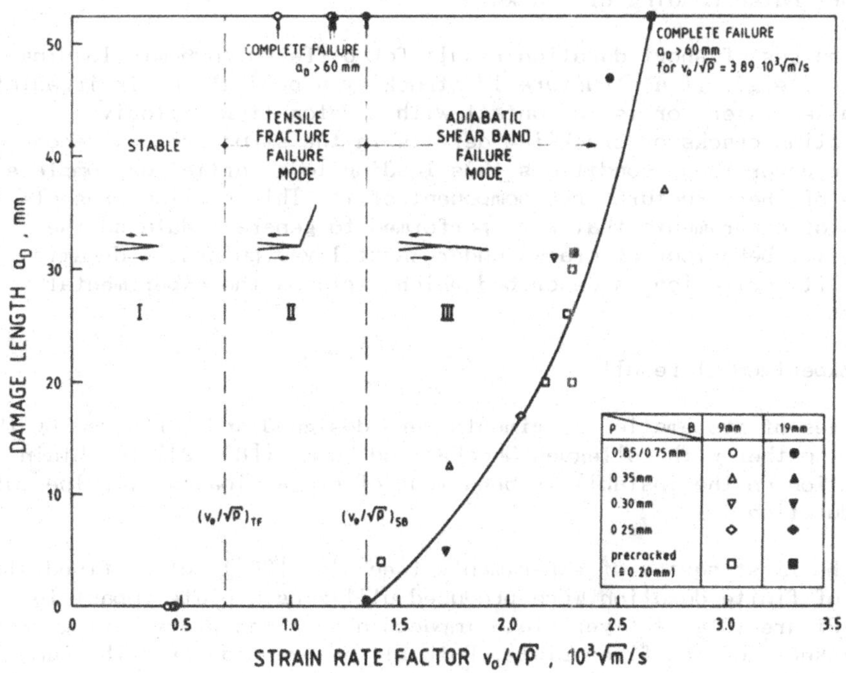

Fig. 100 Dynamic shear failure diagram

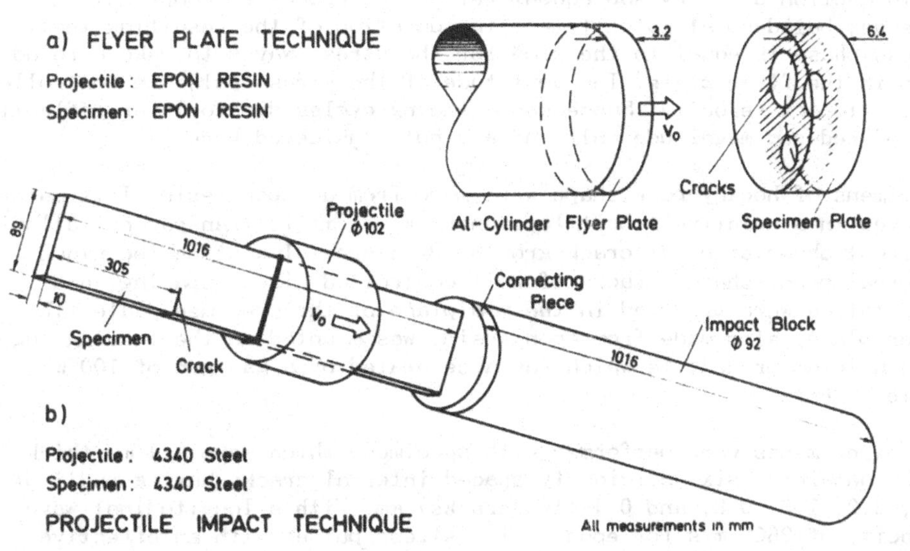

Fig. 101 Techniques for short pulse loading of cracks

5.3 Short Pulse Loading of Cracks

Stress pulses of short duration result for different dynamic loading processes, e.g., if a structure is struck by a projectile, is irradiated by a pulsed laser, or is in contact with a detonating explosive. Pre-existing cracks or cracklike defects in the structure can become unstable under these conditions thus leading to a partial or complete failure of the structure or a component of it. This section presents the results of experiments that were performed to generate data on the instability behaviour of cracks under short lived pulses. A dynamic instability criterion is described which explains the experimental findings.

5.3.1 Experimental results

Two series of systematic experiments were designed and performed by the author together with collegues Shockey and Homma ([80-82]) to obtain information on the instability behaviour of cracks loaded by pulses of short duration.

A) In the first series of experiments (see also [80]) well defined stress pulses of finite duration were produced utilizing a plate impact technique, Figure 101a. A flyer plate impacts a specimen plate having twice the thickness of the flyer plate. Both plates are made from the same material. Due to the impact process, waves of uniaxial compressive strain are generated which propagate into both the flyer plate and the specimen plate. These waves are reflected at the free ends of the two plates, thereby changing sign. Later on, the reflected waves must meet in the middle of the specimen plate creating a tensile state of stress (similar as in Section 5.1). By subsequent reflection processes compressive stresses build up at later times. The duration of the resulting tensile stress phase is equal to the time for the stress waves to make a round trip in the flyer plate. The amplitude of the stress pulse is controlled by the impact velocity. Subsequent loading cycles due to later reflections are of reduced magnitude only and are not considered here.

Specimens of hockey-puck shape were made from an epon resin. This material behaves in a relatively brittle manner. Also, it is transparent, allowing a direct observation of crack growth. By a special casting technique internal penny-shaped cracks of well controlled sizes, spacing, and orientation were produced in the mid-plane of the specimen plate. The flyer plate, also made from epon resin, was mounted on the leading edge of an aluminium projectile which was accelerated by a gas gun of 100 mm diameter bore.

Six experiments were performed with specimens which were 6.4 mm thick, each containing six judiciously spaced internal cracks having radii of 6.4, 3.2, 1.6, 0.8, and 0.4 (two cracks) mm. With a longitudinal wave velocity of 2600 m/s for epon resin, stress pulses with an effective duration of about 2 μs result for the chosen specimen thickness. Impact velocities ranging from 110 to 275 m/s, corresponding to peak tensile stresses ranging from 14.9 to 37.3 MPa (calculated by a one dimensional wave propagation code), were chosen in an attempt to make different crack sizes grow in different experiments.

The results are shown in Fig. 102. The solid points, indicating crack instability, and the open circles, denoting no crack growth, define the regimes of crack instability and stability, respectively. The boundary between the two regimes (see shaded band in Fig. 102) represents the threshold conditions for crack instability. Some scatter exists in the data. Nevertheless, the band provides a reliable indication of the shape and average location of the instability curve. For short cracks the instability stress decreases with increasing crack length, as in static fracture mechanics. For cracks above a certain length, however, the instability stress does not decrease further but stays at the same constant level.

B) In the second series of experiments (see also [81,82]) single-edge-notch specimens were loaded by a special projectile impact technique, Fig. 101b. The specimens were 1026 mm long, 89 mm wide and 10 mm thick, made from 4340 steel, oil quenched and tempered for 1h at 205 °C to obtain a hardness of R_c51. Initial cracks of lengths varying from 20 % to 80 % of the specimen width were located at 305 mm from one end of the specimen. At the end furthest from the crack the specimen was connected to a cylindrical block of solution treated 4340 steel measuring 1016 mm in length and 92 mm in diameter. A connecting piece was welded to the specimen and attached to the block by a bolt. The projectiles were 102 mm diameter steel cylinders of 4340 steel (R_c51) containing an axial slot large enough to fit around the specimen.

At the instant of impact, compressive waves are generated in the projectile and in the impact block, whereas a tensile wave is produced in the specimen. These waves travel the length of the members and are reflected upon reaching the free ends, changing sign and traveling in the opposite directions. Since the projectile is much shorter than the specimen and the block, the compressive wave in the projectile reflects from its free end and arrives back at the original impact surface before the waves in the specimen and impact block reach their free ends. When the tensile wave in the projectile reaches the original impact surface, the projectile separates from the impact block and unloading waves follow the initial loading waves down the specimen and the impact block. Consequently, the duration of the resulting first tensile stress pulse in the specimen again is equal to the time for the stress wave to make a round trip in the projectile. And similar to the previous experiments, the impact velocity controls the amplitude of stress pulse. The reflected waves in the specimen and in the impact block cause subsequent stress pulses which, however, are of reduced amplitudes only. In particular, the tensile wave from the free end of the impact block cannot be effectively transmitted to the specimen because the block tends to detach from the specimen. Only partial transmission occurs because the bolt prevents complete detachment. The subsequent stress pulses, therefore, are not considered in this context.

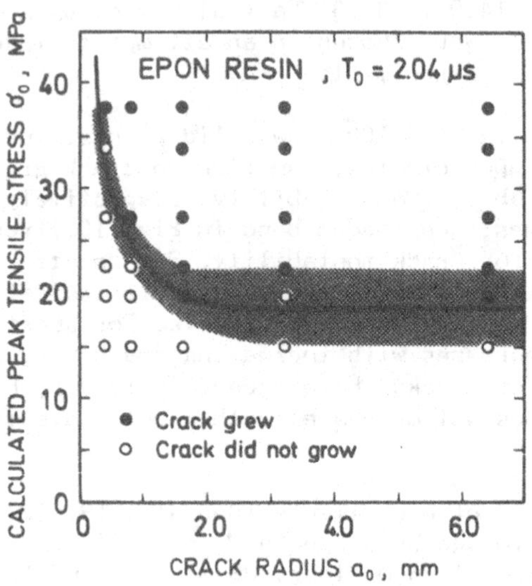

Fig. 102 Instability stresses under short pulse loading

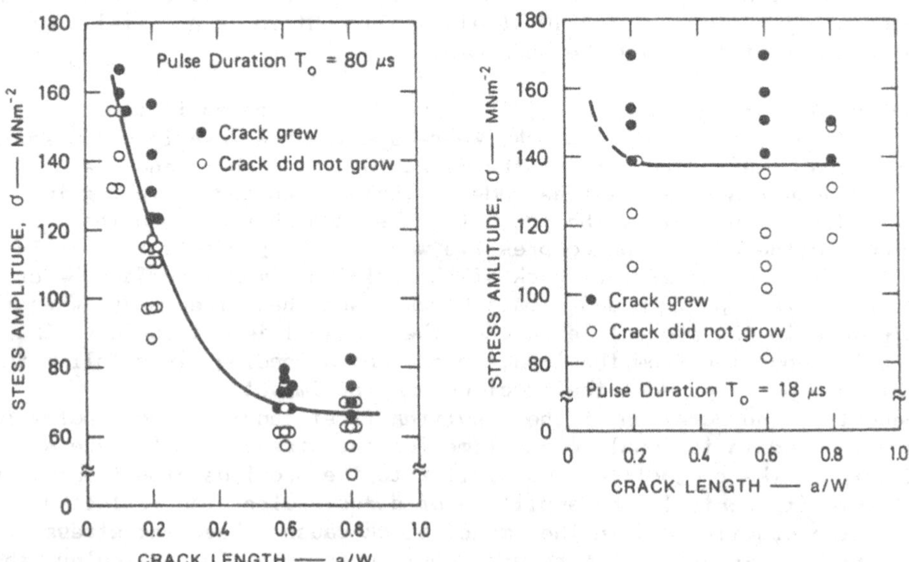

Fig. 103 Instability stresses under short pulse loading

Projectiles 203 mm and 46 mm long were utilized in the experiments producing effective pulse durations of 80 μs and 18 μs, respectively. The projectiles were accelerated in a special launching device. Several experiments were performed at increasing impact velocities, i.e. increasing stress amplitudes, until incremental crack growth could be detected by replicating the crack.

The results are shown in Fig. 103. The trends in the variation of the critical stress with crack length are similar to those observed in the epon resin experiments. For short crack lengths the stress for crack instability decreases with increasing crack length, but it tends to stay at a constant level for larger crack lengths. The shorter the pulse duration, the smaller the critical crack length after which constant behaviour is observed, and the larger the instability stress.

5.3.2 Theoretical considerations

The observed experimental findings demonstrate that a static instability criterion fails to describe the mechanical behaviour of cracks under short lived pulses. Obviously a stress intensity factor following the simple $\sigma_0\sqrt{\pi a_0}$ relationship does not apply anymore for short pulse fracture experiments if the crack length exceeds a certain limit. The actual loading conditions at the crack tip are indeed controlled by a rather complex stress intensity factor history.

The behaviour of cracks under step function loads (see Fig. 104) has theoretically been investigated by Achenbach [83], Freund [84], Sih [85], and others. The principal result is shown in Figs. 104c and 105a. The dynamic stress intensity factor K_I^{dyn} is plotted as a function of the dimensionless time $(c_1/a_0)t$, where c_1 is the longitudinal wave velocity. The stress intensification first increases with time according to a square root of time relationship, overshoots the equivalent static stress intensity factor by a considerable amount, and then reaches a constant static value after damped oscillations. This behaviour is plausible. Wave diffraction processes take place (see Fig. 104a) which can be interpreted the following way. For very early times, $t < a_0/c$ only a certain part of the crack can contribute to the stress intensification at the crack tip due to finite "information" velocities. Consequently, not the real crack length a_0 but an effective crack length a_{eff}, with $a_{eff} = ct < a_0$, determines the stress intensity factor $K_I^{dyn} = \sigma_0\sqrt{\pi a_{eff}} = \sigma_0\sqrt{\pi ct}$, where c is is the "information velocity", i.e. an average wave propagation velocity (see Fig. 104b). After the waves have travelled several times along the crack, $t \gtrsim 10a_0/c_1$, a static situation is reached and the real crack length a_0 determines the stress intensification, $K^{dyn} = \sigma_0\sqrt{\pi a_0}$.

Based on these results, the stress intensity histories were derived [86-88] for cracks of increasing lengths subjected to a rectangular stress pulse of finite duration T_0. The results are shown in Fig. 105b. The dynamic stress intensity factors K_I^{dyn} (full curves), are compared to the equivalent static stress intensity factors, K_I^{stat} (dashed lines). For convenience the crack lengths a_0 are measured in units c_1T_0.

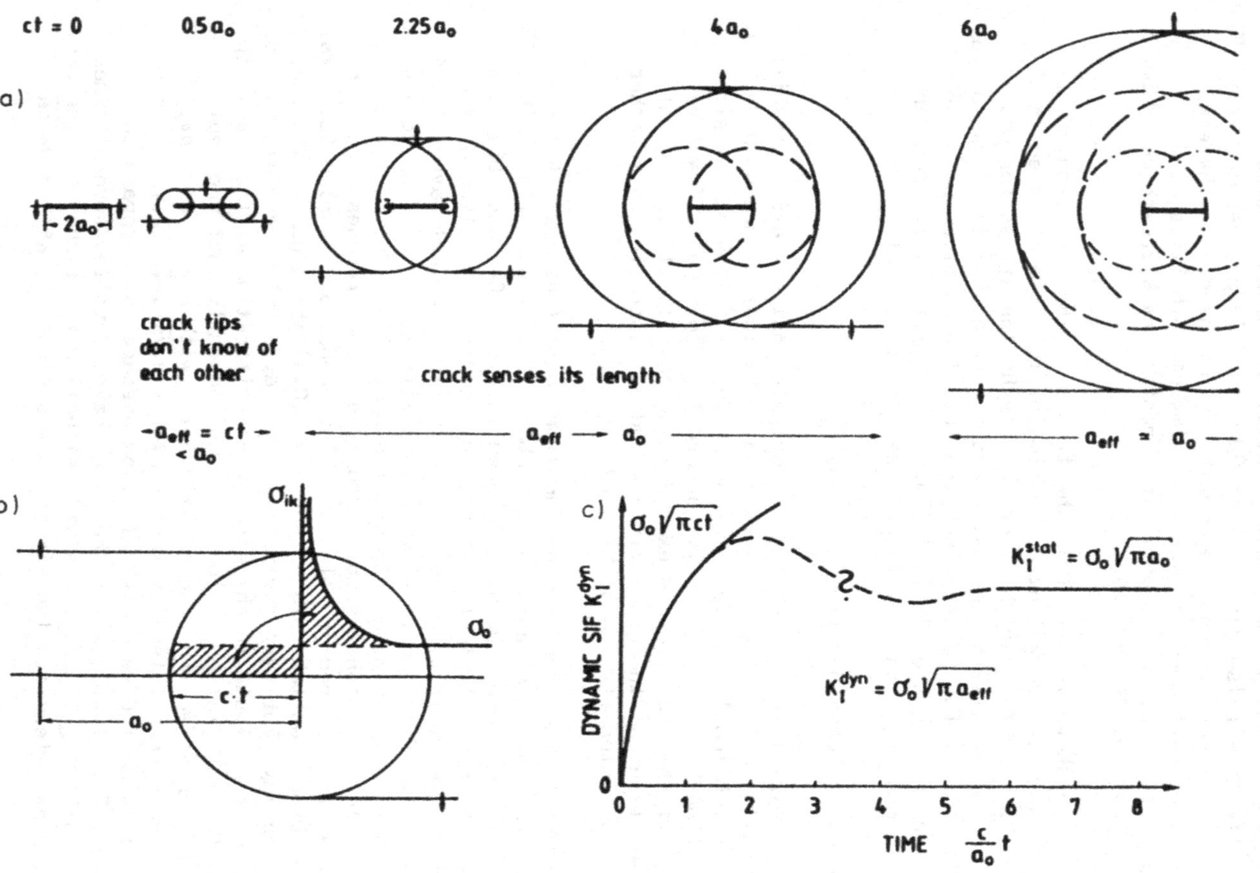

Fig. 104 Stress pulse loading of a crack: diffraction phenomena

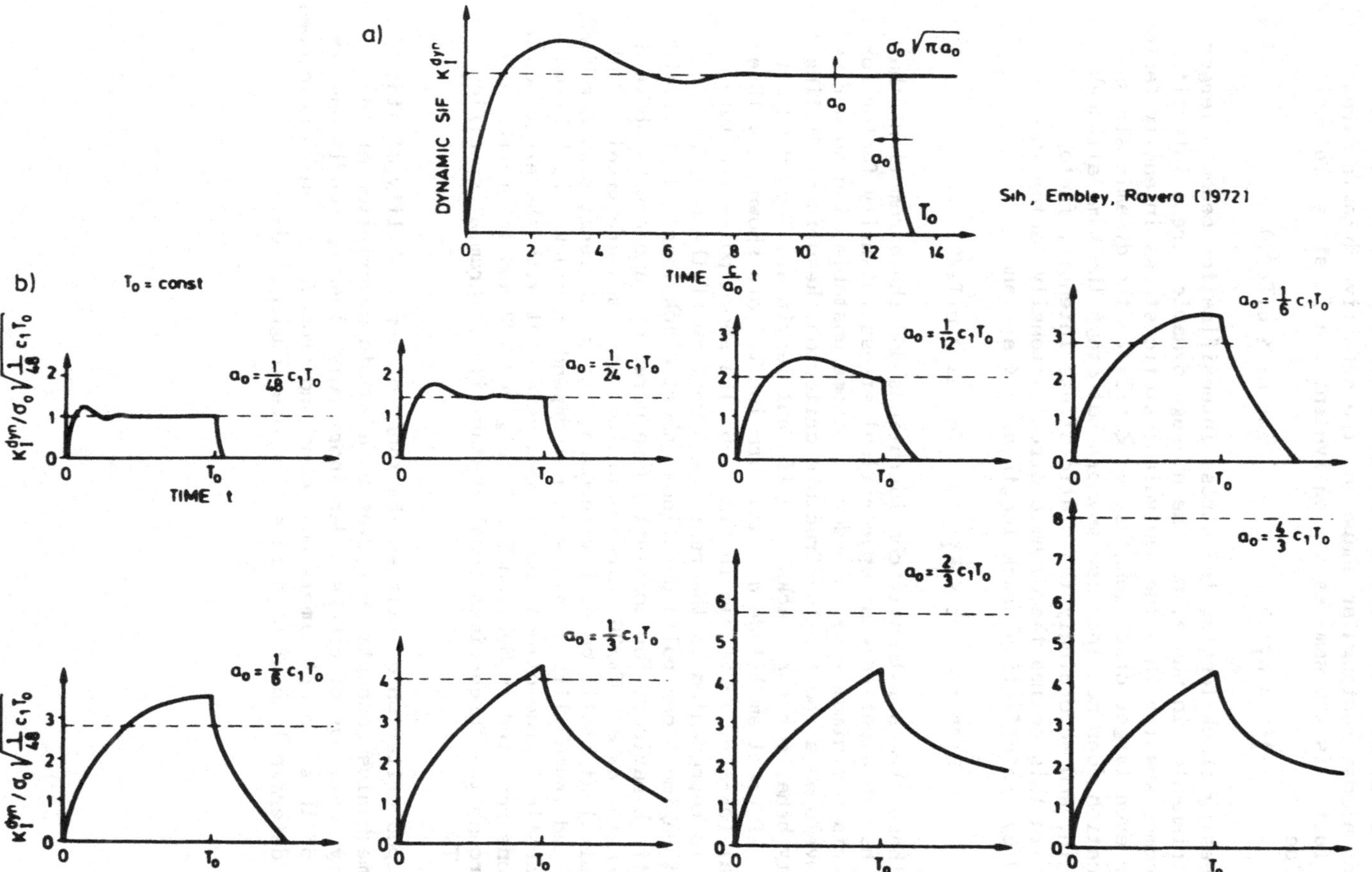

Fig. 105 Stress intensity histories for cracks under short pulse loading

For short cracks, $a_0 \lesssim c_1T_0/30$, the stress intensity history is characterized by an almost rectangular shape and the effective dynamic stress intensity factor is the same as the equivalent static stress intensity factor value

$$K_I^{dyn} = K_I^{stat} \qquad \text{for } a_0 \lesssim c_1T_0/30. \qquad (52)$$

With increasing crack length, the stress intensification becomes larger and more triangular in shape, but the average dynamic stress intensity factor becomes smaller than the equivalent static stress intensity factor value. For even larger crack lengths, $a_0 \gtrsim c_1T_0/2$, the dynamic stress intensification does not increase anymore but stays the same although the crack length, and accordingly the static stress intensity factor, increases. In this regime the dynamic stress intensity factor is controlled by the effective crack length $a_{eff} < a_0$, and

$$K_I^{dyn} = \text{const} < K_I^{stat} \qquad \text{for } a_0 \gtrsim c_1T_0/2. \qquad (53)$$

Based on these stress intensity considerations and the assumptions that the crack has to experience a supercritical stress intensity factor for at least a certain minimum time in order to become unstable, the author and Shockey developed a short pulse fracture criterion. Results predicting the instability behaviour for cracks of different length a_0, loaded by stress pulses of different amplitude σ_0, and duration T_0, are shown in a three dimensional (σ_0-a_0-T_0)-diagram in Fig. 106. The short pulse fracture behaviour is represented in the rear right section of the diagram, the front left regime (long pulse durations, short crack lengths) shows the usual static behaviour. For constant pulse durations and cracks above a certain length, $a_0 \gtrsim c_1T_0/30$, higher critical stresses are predicted than in the equivalent static case. Furthermore, for crack lengths $a_0 \gtrsim c_1T_0/2$ the predicted instability stresses do not depend on crack length anymore. These theoretical predictions are in good agreement with the experimental observations reported in Section 5.3.2. Figures 102 and 103 represent slices through the three-dimensional instability diagram at specific durations T_0.

The developed criterion quantifies the range of applicability of static fracture mechanics concepts and gives an adequate description of the instability behaviour of cracks under short pulse loading conditions. On the basis of this concept simple equipment and procedures can be developed for valid determination of high rate fracture toughness data.

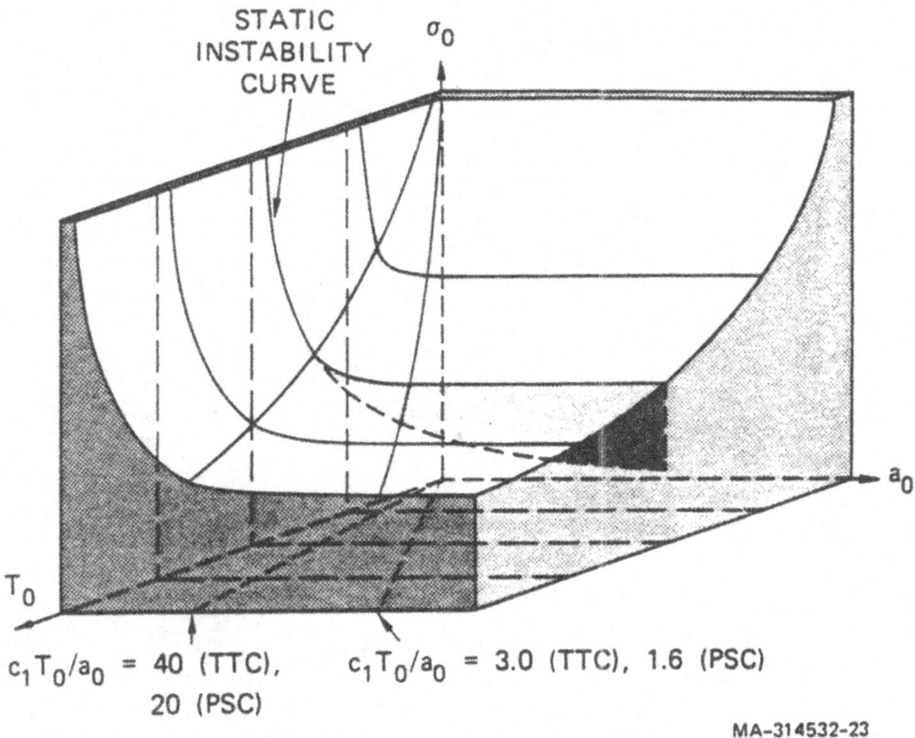

Fig. 106 Instability behaviour of cracks under short pulse loading

APPENDIX I

Designation: E 1221 – 88

Standard Test Method for Determining Plane-Strain Crack-Arrest Fracture Toughness, K_{Ia}, of Ferritic Steels[1]

This standard is issued under the fixed designation E 1221; the number immediately following the designation indicates the year of original adoption or, in the case of revision, the year of last revision. A number in parentheses indicates the year of last reapproval. A superscript epsilon (ε) indicates an editorial change since the last revision or reapproval.

1. Scope

1.1 This test method employs a side-grooved, crack-line-wedge-loaded specimen to obtain a rapid run-arrest segment of flat-tensile separation with a nearly straight crack front. The test method provides a static analysis determination of the stress intensity factor at a short time after crack arrest. The estimate is denoted K_a. When certain size requirements are met, the test result provides an estimate, termed K_{Ia}, of the plane-strain crack-arrest toughness of the material.

1.2 The specimen size requirements, discussed later, provide for in-plane dimensions large enough to allow the specimen to be modelled by linear elastic analysis. For conditions of plane-strain, a minimum specimen thickness is also required. Both requirements depend upon the crack arrest toughness and the yield strength of the material. A range of specimen sizes may therefore be needed, as specified in this test method.

1.3 If the specimen does not exhibit rapid crack propagation and arrest, K_a, cannot be determined.

1.4 Values stated in inch-pound units are to be regarded as the standards. SI units are provided for information only.

1.5 *This standard may involve hazardous materials, operations, and equipment. This standard does not purport to address all of the safety problems associated with its use. It is the responsibility of the user of this standard to establish appropriate safety and health practices and determine the applicability of regulatory limitations prior to use.*

2. Referenced Documents

2.1 *ASTM Standards:*

E 8 Test Methods of Tension Testing of Metallic Materials[2]

E 23 Methods for Notched Bar Impact Testing of Metallic Materials[3]

E 208 Test Method for Conducting Drop-Weight Test to Determine Nil-Ductility Transition Temperature of Ferritic Steels[3]

E 399 Test Method for Plane-Strain Fracture Toughness of Metallic Materials[3]

E 616 Terminology Relating to Fracture Testing[3]

3. Terminology

3.1 *Definitions:*

3.1.1 Definitions in Terminology E 616 are applicable to this test method.

3.2 *Descriptions of Terms Specific to This Standard:*

3.2.1 *crack-arrest fracture toughness,* K_a ($FL^{-3/2}$)—the value of the stress intensity factor shortly after crack arrest.

NOTE 1—The in-plane specimen dimensions must be large enough for adequate enclosure of the crack-tip plastic zone by a linear-elastic stress field.

NOTE 2—In this test method, side-grooved specimens are used. The calculation of K_a is based upon measurements of both the arrested crack length and of the crack-mouth opening displacements prior to initiation of a fast-running crack and shortly after crack arrest.

3.2.2 *plane-strain crack-arrest fracture toughness,* K_{Ia} ($FL^{-3/2}$)—the value of crack-arrest fracture toughness, K_a, for a crack that arrests under conditions of crack-front plane-strain.

NOTE 3—The requirements for attaining conditions of crack-front plane-strain are specified in the procedures of this test method.

3.2.3 *stress intensity factor at crack initiation,* K_o ($FL^{-3/2}$)—the value of K at the onset of rapid fracturing.

NOTE 4—In this test method, only a nominal estimate of the initial driving force is needed. For this reason, K_o is calculated on the basis of the original (machined) crack (or notch) length and the crack-mouth opening displacement at the initiation of a fast-running crack.

4. Summary of Test Method

4.1 This test method estimates the value of the stress intensity factor, K, at which a fast running crack will arrest. The test is made by forcing a wedge into a split-pin, which applies an opening force across the crack starter notch in a modified compact specimen, causing a run-arrest segment of crack extension. The rapid run-arrest event suggests need for a dynamic analysis of test results. However, experimental observations **(1, 2)**[4] indicate that, for the test method described here, an adjusted static analysis of test results provides a useful estimate of the value of the stress intensity factor at the time of crack arrest.

4.2 Calculation of a nominal stress intensity at initiation, K_o, is based on measurements of the machined notch length and the crack-mouth opening displacement at initiation. The value of K_a is based on measurements of the arrested crack

[1] This test method is under the jurisdiction of ASTM Committee E-24 on Fracture Testing and is the direct responsibility of Subcommittee E24.01 on Fracture Mechanics Test Methods.

Current edition approved Feb. 1, 1988. Published April 1988.

[2] *Annual Book of ASTM Standards*, Vols 01.02, 02.01–02.03, and 03.01.

[3] *Annual Book of ASTM Standards*, Vol 03.01.

[4] The boldface numbers in parentheses refer to the list of references at the end of this test method.

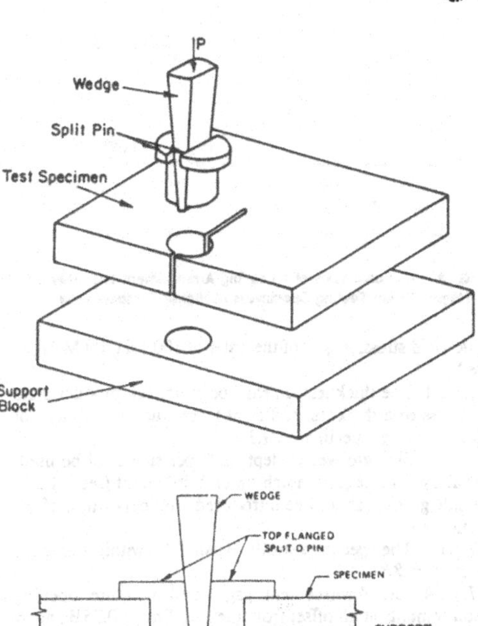

FIG. 1 Schematic Pictorial and Sectional Views Showing the Standard Arrangement of the Wedge and Split-Pin Assembly, the Test Specimen, and the Support Block.

length and the crack-mouth opening displacements prior to initiation and shortly after crack arrest.

5. Significance and Use

5.1 In structures containing gradients in either toughness or stress, a crack may initiate in a region of either low toughness, or high stress, or both, and arrest in another region of either higher toughness, or lower stress, or both. The value of the stress intensity factor during the short time interval in which a fast-running crack arrests is a measure of the ability of the material to arrest such a crack. Values of the stress intensity factor of this kind, which are determined using dynamic methods of analysis, provide a value for the crack-arrest fracture toughness which will be termed K_A in this discussion. Static methods of analysis, which are much less complex, can often be used to determine K at a short time (1 to 2 ms) after crack arrest. The estimate of the crack-arrest fracture toughness obtained in this fashion is termed K_a. When macroscopic dynamic effects are relatively small, the difference between K_A and K_a is also small **(1–4)**. For cracks propagating under conditions of crack-front plane-strain, in situations where the dynamic effects are also known to be small, K_{Ia} determinations using laboratory-sized specimens have been used successfully to estimate whether, and at what point, a crack will arrest in a structure **(5, 6)**. Depending upon component design, loading compliance, and the crack jump length, a dynamic analysis of a fast-running crack propagation event may be necessary in order to predict whether crack arrest will occur and the arrest position. In such cases, values of K_{Ia} determined by this test method can be used to identify those values of K below which the crack speed is zero. More details on the use of dynamic analyses can be found in Ref **(4)**.

5.2 This test method can serve at least the following additional purposes:

5.2.1 In materials research and development, to establish in quantitative terms significant to service performance, the effects of metallurgical variables (such as composition or heat treatment) or fabrication operations (such as welding or forming) on the ability of a new or existing material to arrest running cracks.

5.2.2 In design, to assist in selection of materials for, and determine locations and sizes of, stiffeners and arrestor plates.

6. Apparatus

6.1 The procedure involves testing of modified compact specimens that have been notched by machining. To minimize the introduction of additional energy into the specimen during the run-arrest event, the loading system must have a low compliance compared with the test specimen. For this reason a wedge and split-pin assembly is used to apply a load on the crack line. This loading arrangement does not permit easy measurement of opening loads. Consequently, opening displacement measurements in conjunction with crack size and compliance calibrations are used for calculating K_o and K_a.

6.2 *Loading Arrangement:*

6.2.1 A typical loading arrangement is shown in Fig. 1. The specimen is placed on a support block whose thickness should be adequate to allow completion of the test without interference between the wedge and the lower crosshead of the testing machine. The support block should contain a hole that is aligned with the specimen hole, and whose diameter should be between 1.05 and 1.15 times the diameter of the hole in the specimen. The load that forces the wedge into the split-pin is transmitted through a load cell.

6.2.1.1 The surfaces of the wedge, split-pin, support block, and specimen hole should be lubricated. Lubricant in the form of thin (0.005 in. or 0.13 mm) strips of TFE-fluorocarbon is preferred. Molybdenum disulfide (both dry and in a grease vehicle) and high-temperature lubricants can also be used.

6.2.1.2 A low-taper-angle wedge and split-pin arrangement is used. If grease or dry lubricants are used, a matte finish (grit blasted) on the sliding surfaces may be helpful in avoiding galling. The split-pin must be long enough to contact the full specimen thickness, and the radius must be large enough to avoid plastic indentations of the test specimen. In all cases it is recommended that the diameter of the split-pin should be 0.005 in. (0.13 mm) less than the diameter of the specimen hole. The wedge must be long enough to develop the maximum expected opening displacement. Any air or oil hardening tool steel is suitable for making the wedge and split-pins. A hardness in the range from R_C45 to R_C55 has been used successfully. With the

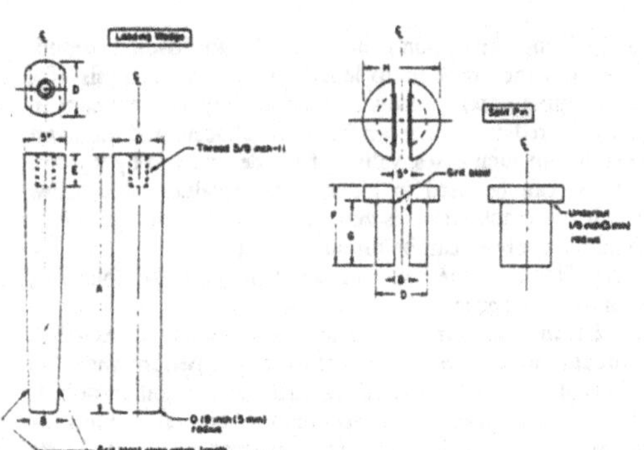

	in.	mm
A	8.00	203
B	0.33	8.4
D	0.99	25.1
E	1.00	25.4
F	2.25	57.2
G	2.00	50.8
H	1.50	38.1

NOTE—The dimensions given are suitable for use with a 1.0 in. (25.4 mm) diameter loading hole in a 2.0 in. (50.8 mm) thick test specimen. These dimensions should be scaled appropriately when other hole diameters and specimen thicknesses are used.

FIG. 2 Suggested Geometry and Dimensions of a Wedge and Split-Pin Assembly

recommended wedge angle and proper lubrication, a loading machine producing 1/5 to 1/10 the expected maximum opening load is adequate. The dimensions of a wedge and split-pin assembly suitable for use with a 1.0 in. (25.4 mm) diameter loading hole are shown in Fig. 2. The dimensions should be scaled when other hole diameters are used. A hole diameter of 1.0 in. (25.4 mm) has been found satisfactory for specimens having $5 < W < 6.7$ in. ($125 < W < 170$ mm).

NOTE 5—Specimens tested with the arrangement shown in Fig. 1 may not exhibit an adequate segment of run-arrest fracturing, for example, at testing temperatures well above the NDT temperature. In these circumstances, the use of the loading arrangement shown in Fig. 3 has been found to be helpful (2, 7) and may be employed.

6.3 *Displacement Gages*—Displacement gages are used to measure the crack-mouth opening displacement at $0.25W$ from the load-line. Accuracy within 2 % over the working range is required. Either the gage recommended in Test Method E 399 or a similar gage modified to accommodate conical seats is satisfactory. It is necessary to attach the gage in a fashion such that seating contact with the specimen is not altered by the jump of the crack. Two methods that have proven satisfactory for doing this are shown in Fig. 4. Other gages can be used so long as their accuracy is within 2 %.

7. Specimen Configuration, Dimensions, and Preparation

7.1 *Standard Specimen:*

7.1.1 The configuration of a compact-crack-arrest (CCA) specimen that is satisfactory for low and intermediate strength steels is shown in Fig. 5. (In this context, an intermediate strength steel is considered to be one whose static yield stress, σ_{YS}, is of the order of 100 ksi (700 MPa) or less.)

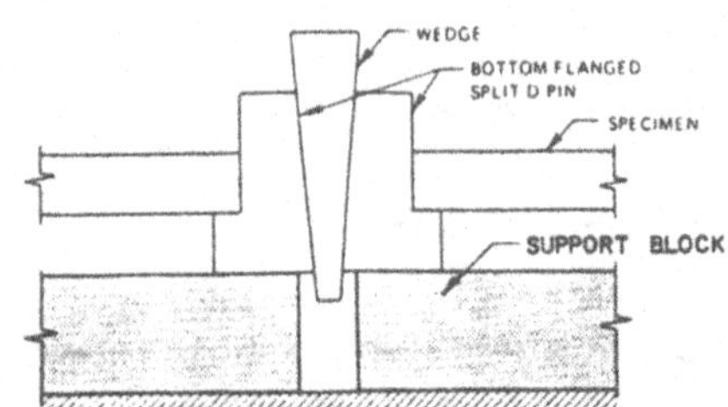

FIG. 3 Sectional View of a Loading Arrangement That May Be Helpful When Testing Specimens at Higher Temperatures

7.1.1.1 The thickness, B, shall be either full product plate thickness or a thickness sufficient to produce a condition of plane-strain, as specified in 9.3.3.

7.1.1.2 Side grooves of depth $B/8$ per side shall be used. For alloys that require notch-tip embrittlement (see 7.1.3.2) the side grooves should be introduced after deposition of the brittle weld.

7.1.1.3 The specimen width, W, shall be within the range $2B \leq W \leq 8B$.

7.1.1.4 The displacement gage shall measure opening displacements at an offset from the load line of $0.25W$, away from the crack tip.

7.1.2 *Specimen Dimensions:*

7.1.2.1 In order to limit the extent of plastic deformation in the specimen prior to crack initiation, certain size requirements must be met. These requirements depend upon the material yield strength. They also depend upon K_a, and therefore the K_o needed to achieve an appropriate run-arrest event.

7.1.2.2 The in-plane specimen dimensions must be large enough to allow for the linear elastic analysis employed by this test method. These requirements are given in 9.3.2 and 9.3.4, in terms of allowable crack jump lengths.

7.1.2.3 For a test result to be termed plane-strain (K_{Ia}) by this test method, the specimen thickness, B, should meet the requirement given in 9.3.3.

7.1.3 *Starting Notch:*

7.1.3.1 The function of the starting notch is to produce crack initiation at an opening displacement (or wedging force) that will permit an appropriate length of crack extension prior to crack arrest. It is convenient to express the crack initiation condition in terms of a calculated stress intensity factor, K_o (see 3.2.3). Different materials require different starter notch preparation procedures to produce an acceptable value of K_o.

7.1.3.2 The recommended starter notch for low and intermediate strength steels is a notched brittle weld, as shown in Fig. 6. It is produced by depositing a weld across the specimen thickness. Guidelines on welding procedures are given in Appendix X1.

7.1.3.3 Alternative crack starter configurations and embrittlement methods may also be used. One example, which has been used successfully with high strength steel (**8**), is a simple machined notch. Another example of an alternative crack starter configuration is the duplex compact spec-

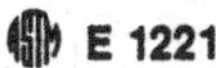
E 1221

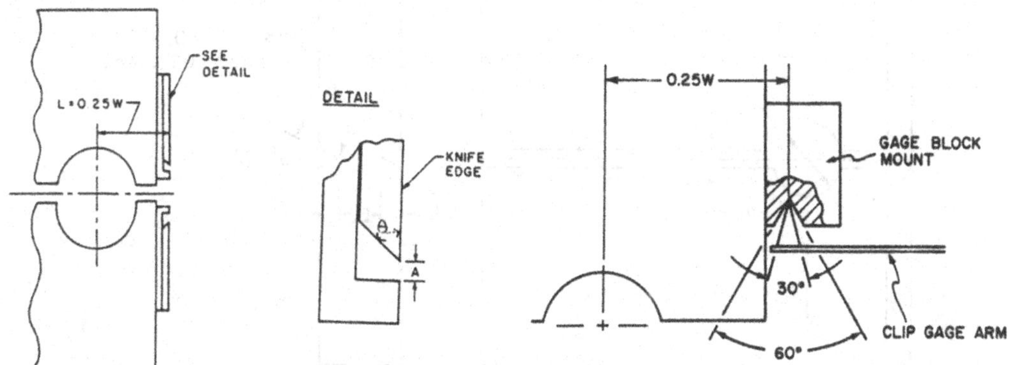

NOTE 1— Dimension A should be 0.002–0.010 in. (0.05–0.25 mm) less than the thickness of the clip gage arm.
NOTE 2—The knife edge can be attached to the specimen with mechanical fasteners or adhesives.
NOTE 3—The clip gage is installed by sliding it into the gap.

FIG. 4 Two Alternative Clip Gage Seating Arrangements Using Knife Edges and Using Conical Mounts

imen (1), which consists of a block of high-strength, low-toughness alloy containing the starter notch, electron-beam welded to the alloy being evaluated.

8. Procedure

8.1 *Number of Tests*—It is recommended that at least three valid test results be obtained at a single test temperature.

8.2 *Specimen Measurement*—Measure the specimen thickness, B, and the crack plane thickness, B_N, to ±1 % of B. Measure the specimen width, W, to ±1 % of W.

8.3 *Temperature Control and Measurement:*

8.3.1 Specimens may be heated or cooled to the selected test temperature by any appropriate method. A method that has been used successfully for elevated-temperature tests employs electric-resistance heating tapes in combination with a variable power source. Tests at subambient temperatures have been conducted using cooling coils embedded in the specimen support block (see Fig. 7); a controlled flow of liquid nitrogen or other suitable coolant through the cooling coils permits low temperatures to be reached without difficulty. To minimize temperature gradients through the specimen thickness, it is necessary to surround the specimen with a good thermal insulator. Prior to starting the test, the specimen should be held at the test temperature for a time sufficient to allow the specimen to attain a uniform temperature (to within 5°F (3°C)).

8.3.2 Measure the specimen temperature with a thermocouple welded to the top surface of the specimen at a location near the side groove, about 1 in. (25 mm) ahead of the starter notch. In reporting the test results, the test temperature shall be the temperature measured on the specimen at the time of the rapid run-arrest event.

8.4 *Loading Procedure:*

8.4.1 The test method calls for the use of a cyclic loading technique. In this technique, load is applied to the wedge until a rapid crack initiates, or until the crack-mouth opening displacement (measured by the clip gage) reaches a predetermined value. If a rapid fracture has not initiated prior to the recommended maximum displacement being reached, the specimen is unloaded until the wedge loses contact with the split-pin. The specimen is then reloaded in the same manner as before and load application is once again terminated either by initiation of a rapid crack or upon the opening displacement reaching a specified value. Successively higher values of the recommended maximum opening displacement are allowed on each loading cycle, until a rapid crack initiates or until the test is discontinued.

8.4.2 The loading technique of this test method does not allow direct measurement of the opening loads applied to the specimen by the wedge and split-pin assembly. The load applied to the specimen is therefore obtained from measurements of the crack-mouth opening displacement. Components of the opening displacement that do not contribute to the opening load can occur. These have their origin primarily in seating of the load train and clip gage, local cracking in the brittle weld, and interference with crack closure due to incomplete strain reversal in the plastic zone near the root of the starter notch. The purpose of the cyclic loading technique is to identify and estimate the magnitude of these contributions.

8.4.3 The effects of load train seating and weld bead cracking can essentially be limited to the first loading cycle by an appropriate limit on the maximum opening displacement imposed in that cycle. This limit is designed to keep the first loading cycle linear elastic in a global sense. These influences can then be eliminated, with some degree of conservatism, by excluding the zero-load displacement offset recorded at the end of the first loading cycle from the displacement used to calculate K_a.

8.4.4 The second undesirable contribution to the total measured crack opening displacement is due to the local yielding that occurs around the root of the starter notch prior to initiation of a rapid fracture. The formation and growth of this plastic zone can be regarded as being mainly responsible for the zero-load displacement offsets that are recorded after

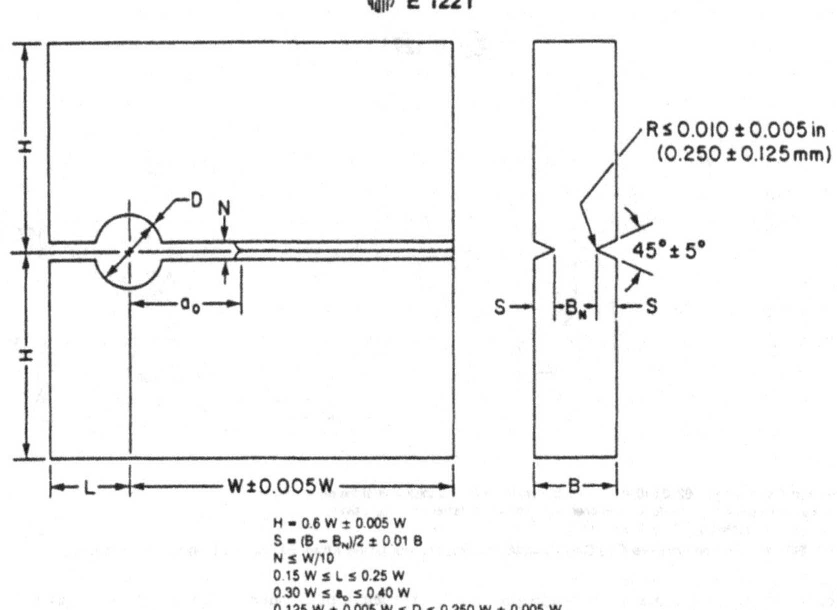

FIG. 5 Geometry and Dimensions of a Crack-Line-Wedge-Loaded Compact-Crack-Arrest (CCA) Test Specimen that is Satisfactory for Low and Medium Strength Steels

completion of the first loading cycle, that is, between cycles 2 and 3, 3 and 4, etc. The influence of this effect could be eliminated in its entirety by excluding all of the zero-load offset in opening displacement measured prior to the start of the loading cycle during which the run-arrest event occurs. However, there is evidence to suggest that such a step may be overly conservative. Model tests have shown that, when the plastic zone is well enclosed by the linear elastic stress field in the specimen, nearly all of the offset in the zero-load displacement is recovered if the plastic zone is severed by a saw cut or by a brittle crack (**9**). The degree to which this component of the strain energy stored in the specimen is recovered in time to influence the run-arrest behavior of the rapid crack is unclear at the present time. The *K*-calculation procedure of this test method therefore avoids the extremes of excluding all or none of the zero-load displacement offsets which accumulate in the second and subsequent loading cycles, and excludes one half of these effects.

NOTE—Dimension *N* must be large enough to allow entry of the welding electrode being used.

FIG. 6 Details of the Notched Brittle Weld that is Recommended for Use as a Crack Starter for Low and Medium Strength Steels

8.4.5 An autographic record of wedge-load versus crack-mouth-opening-displacement should be obtained. The recorder should not be re-zeroed between loading cycles since knowledge of the accumulated zero-load displacement offset is desired. It could also be useful to obtain information about the final segment of the opening-displacement versus time record on an oscillograph or other high-rate recording device. This would provide additional information about the nature of the run-arrest event.

8.4.6 Apply load to the wedge until the crack-mouth-opening-displacement measured by the clip gage reaches the recommended maximum value given by:

$$[(\delta_o)_1]_{max} = \frac{0.69\,\sigma_{YS}\, W\, \sqrt{B_N/B}}{E f(a_o/W)} \qquad (1)$$

where:

σ_{YS} = static yield strength of the specimen material (or, in the case of the duplex specimen, of the crack-starter-section material).

The other terms are as defined in 9.2. The testing machine should be operated in displacement control, with a free-running crosshead speed of 0.1 to 0.5 in./min (2 to 12 mm/min).

NOTE 6—Rapid fracture initiation on the first loading cycle is unlikely in the brittle weld CCA specimen. However, if a run-arrest event does occur, proceed with the calculations of K_o and K_a in the same manner as if sequential load-unload cycling had been used. In subse-

E 1221

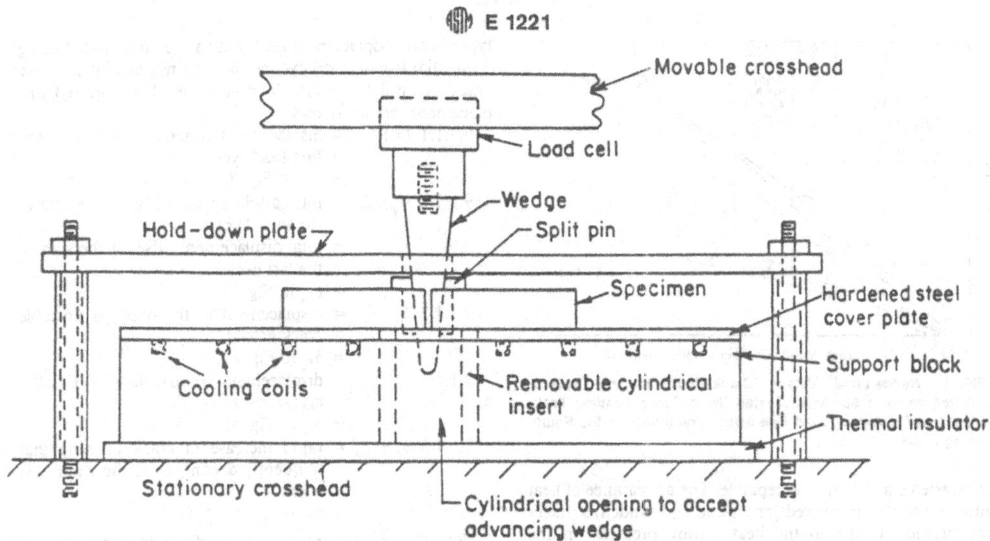

FIG. 7 Schematic Illustration of a Loading Arrangement that Facilitates Wedge Extraction When Using the Cyclic Loading Technique

quent tests of replicate specimens, the first cycle displacement limit should be reduced sufficiently so that the first loading cycle can be completed without intervention of a rapid fracture.

8.4.7 Unload the specimen by extracting the wedge in preparation for a second loading cycle. The clip gage should remain in place during unloading and wedge removal to maintain a record of the displacement offset that occurs upon a return to zero load.

NOTE 7—Wedge extraction and cyclic loading can be simplified greatly by the use of the arrangement shown schematically in Fig. 7. Key features include a hold-down plate and a wedge that is fastened to the loading ram. However, the hold-down plate may not be required when using lubricant in the form of TFE-fluorocarbon strips (see 6.2.1.1).

8.4.8 Without re-zeroing the recorder, reinsert and apply load to the wedge at the same displacement rate as on the first cycle. Continue loading until a rapid crack jump occurs or until the displacement measured with the clip gage reaches a predetermined value. The recommended maximum opening displacement on the second and subsequent cycles can be calculated from

$$[(\delta_o)_n]_{max} = [1.0 + 0.25(n - 1)]\left[\frac{0.69\ \sigma_{YS}\ W\ \sqrt{B_N/B}}{E\ f(a_o/W)}\right] \quad (2)$$

where:

n = the cycle number.

The other terms are the same as in Eq (1). If an unstable crack is not initiated upon reaching the prescribed displacement limit, again unload and extract the wedge as specified in 8.4.7. Label the load/displacement record with the appropriate cycle number and repeat 8.4.8.

NOTE 8—If a large number of load/unload cycles are required, it may be necessary to relubricate the wedge and split-pin assembly. Increased friction will be indicated by an increased slope in the load-displacement record and the need for large reverse loads to extract the wedge.

8.4.9 To measure K_a a segment of unstable crack extension must occur. The occurrence of unstable crack extension will normally be apparent to the operator, both audibly and as an abrupt load drop on the test record. (In the brittle weld CCA specimen, a load drop of 50 to 60 % has been found to indicate that a sufficient length of unstable fracturing has occurred.) After the event, the operator should remove the load on the wedge to avoid further crack propagation.

8.4.10 If on subsequent loading cycles, it is observed that attempts to increase the opening displacement are accompanied by a decrease in the applied wedge load, that is, stable tearing is occurring, it is unlikely that the specimen will exhibit rapid run-arrest fracturing. It is recommended that under these circumstances, the test be discontinued. It may be helpful at this point to remachine the specimen to remove the weld bead and the material ahead of the starter notch that has been subjected to plastic deformation. A fresh starter notch can then be prepared and the specimen retested at a lower temperature (35 to 70°F (20 to 40°C lower)) in an attempt to obtain useful data from the specimen.

NOTE 9—A displacement limit beyond which the specimen is unlikely to give successful results can be estimated from:

$$[\delta_o]_{limit} = \frac{1.50\ \sigma_{YS}\ W\ \sqrt{B_N/B}}{E\ f(a_o/W)} \quad (3)$$

which is approximately twice the quantity calculated from Eq (1).

NOTE 10—The quantity of material that must be removed from an unsuccessful specimen can be approximated by the radius of the plastic zone surrounding the starter notch under plane strain conditions, and calculated from $(K_o/\sigma_{YS})^2/6\pi$. A sufficient quantity of material must be machined out to remove any stable tearing that may have occurred.

8.5 *Marking the Arrested Crack:*

8.5.1 The position of the arrested crack can be marked by heat tinting. Heating at temperatures in the range 500 to 700°F (260 to 370°C) for 10 to 90 min has proved successful. Any time and temperature combination that clearly marks

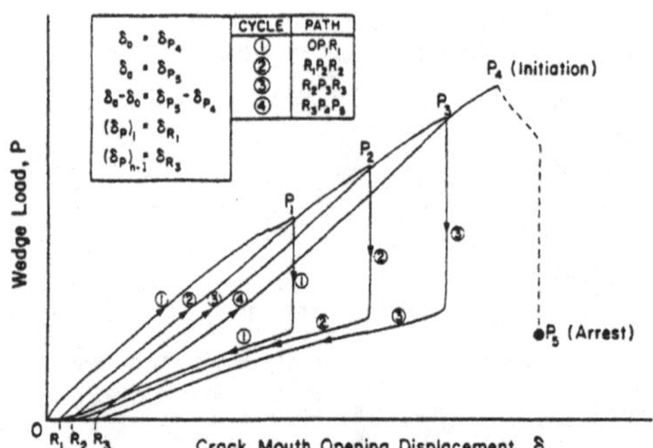

FIG. 8 Wedge-Load Versus Crack-Mouth-Opening-Displacement Record for a Specimen Tested Using Cyclic Loading Techniques, that Displayed Rapid Run-Arrest Fracturing on the Fourth Loading Cycle

the arrested crack front is acceptable. The appearance of heat tinting on freshly machined (or ground and sanded) surfaces may provide a clue to the heat tinting progress on the fracture surfaces. If a fractographic examination of the fracture surfaces is to be performed, the use of lower heat tinting temperatures or the marking of the arrested crack front by means of fatigue may be desirable.

8.5.2 After marking the crack front the specimen is broken completely in two. This can usually be done with the wedging apparatus used in testing the specimen. The breaking open of structural steel specimens is greatly facilitated by cooling them in dry ice or liquid nitrogen.

8.6 *Measurement of Arrested Crack Length:*

8.6.1 The heat-tinted fracture surface should first be examined to determine whether it displays irregularities serious enough to warrant exclusion of the test result. The occurrence of tunnelling, a failure to follow the side grooves on one or both sides, and the presence of large, unbroken ligaments on the fracture surface are all behaviors that may give erroneous results for K_a. Annex A1 provides more detailed information on this subject.

8.6.2 The average of three measurements defines the arrested crack length, a_a. These measurements are to be made on the heat-tinted fracture surface, to within 1 %, at the following positions: at the center (mid-thickness) of the specimen, and midway between the center and the bottom of the side groove on each side. Since crack front irregularities may make it difficult to determine the crack length at the specified locations, it is suggested that the measurement be taken as a visual average across a strip of width, $B_N/4$, centered at each measurement location. Examples of sample crack length determinations using this technique are also provided in Annex A1.

NOTE 11—It is recommended that a photographic record of the heat-tinted fracture surface be made a part of the test report, particularly if there are any unusual perturbations in the crack front contours. Descriptive comments may also be helpful.

9. Calculation and Interpretation of Results

9.1 *Displacement Measurement:*

9.1.1 From the autographic load-displacement record, several displacement values should be determined. Fig. 8 is a typical load-displacement record for a specimen tested using sequential load-unload cycling that did not exhibit unstable cracking until the fourth loading cycle. The required displacements are as follows:

9.1.1.1 $(\delta_p)_1$ = displacement offset at the end of the first load cycle
= δ_{R_1} in Fig. 8;

9.1.1.2 $(\delta_p)_{n-1}$ = total displacement offset at the end of the $(n-1)$ cycle
= total displacement offset at the start of the last cycle
= δ_{R_3} in Fig. 8;

9.1.1.3 δ_o = displacement at the onset of unstable crack growth
= δ_{P_4} in Fig. 8;

9.1.1.4 δ_a = displacement approximately 0.1 s after crack arrest
= δ_{P_5} in Fig. 8;

9.1.1.5 $\delta_a - \delta_o$ = rapid increase in crack opening that frequently accompanies the run-arrest event
= $\delta_{P_5} - \delta_{P_4}$ in Fig. 8.

NOTE 12—The preferred interpretation of δ_a is the opening displacement at about 2 ms after crack arrest. However, this measurement may not be possible with the instrumentation used. This testing practice assumes that δ_a at about 100 ms after crack arrest does not differ significantly from δ_a at 2 ms.

9.1.2 In the brittle weld CCA specimen, a load drop of 50 to 60 % has been found to indicate that a sufficient length of unstable fracture has occurred and that δ_a is a usable arrest displacement value. See 9.3.2 and 9.3.4 for limitations on the length of the run-arrest segment.

9.2 *Calculation of K_o and K_a:*

9.2.1 Calculate K_o and K_a from:

$$K = E\,\delta\, f(x)\,\sqrt{B/B_N}/\sqrt{W}\ \text{ksi}\sqrt{\text{in.}}\ (\text{MPa}\sqrt{\text{m}}) \tag{4}$$

where:

$$f(x) = \frac{2.24(1.72 - 0.9x + x^2)\sqrt{1-x}}{(9.85 - 0.17x + 11x^2)},\ \text{and} \tag{5}$$

$x = a/W$.

9.2.2 The expression for $f(x)$ used here was obtained by an experimental compliance calibration procedure followed by a curve-fitting process, the details of which can be found in Ref **(10)**. Values of $f(x)$ computed from Eq (5) for various values of x are given in Table 1. The other terms in Eq (4) are as follows:

E = Young's modulus, ksi (MPa),
a = initial slot length, a_o, or final crack length, a_a, as determined in 8.6, in. (m),
W = specimen width, in. (m),
B = specimen thickness as shown in Fig. 5, in. (m),
B_N = specimen thickness at crack plane as shown in Fig. 5, in. (m), and
δ = crack mouth opening displacement, in. (m).

9.2.3 To calculate K_o, use $a = a_o$ and $\delta = d_o$. To calculate K_a, use $a = a_a$ and $\delta = d_a$. The quantities d_o and d_a are given by:

$$d_o = \delta_o - (\delta_p)_{n-1},\ \text{and} \tag{6a}$$

$$d_a = \delta_o - (\delta_p)_1 - 0.5[(\delta_p)_{n-1} - (\delta_p)_1] + 0.5[\delta_a - \delta_o] \tag{6b}$$

$$= 0.5[\delta_o + \delta_a - (\delta_p)_1 - (\delta_p)_{n-1}] \tag{6c}$$

E 1221

TABLE 1 Values of $f(x)$ for use in Eq (4)

x	$f(x)$	x	$f(x)$	x	$f(x)$
0.30	0.268	0.50	0.192	0.70	0.128
0.31	0.263	0.51	0.189	0.71	0.125
0.32	0.260	0.52	0.185	0.72	0.122
0.33	0.256	0.53	0.182	0.73	0.119
0.34	0.252	0.54	0.179	0.74	0.116
0.35	0.247	0.55	0.175	0.75	0.113
0.36	0.244	0.56	0.172	0.76	0.110
0.37	0.240	0.57	0.169	0.77	0.107
0.38	0.236	0.58	0.165	0.78	0.104
0.39	0.232	0.59	0.162	0.79	0.101
0.40	0.228	0.60	0.159	0.80	0.098
0.41	0.225	0.61	0.156	0.81	0.095
0.42	0.221	0.62	0.153	0.82	0.092
0.43	0.217	0.63	0.150	0.83	0.089
0.44	0.213	0.64	0.146	0.84	0.086
0.45	0.210	0.65	0.143	0.85	0.082
0.46	0.206	0.66	0.140		
0.47	0.203	0.67	0.137		
0.48	0.199	0.68	0.134		
0.49	0.196	0.69	0.131		

NOTE 13—The quantities in brackets in Eq (6b) both represent displacement components whose exact contribution to the energy available to drive the running crack is unclear at the present time. The rationale for the selection of the premultiplier of 0.5 for each of these quantities is discussed in Ref (2). Equation (6c) is simply a mathematical simplification of Eq (6b) and may be more convenient to use from a computational standpoint.

NOTE 14—If a rapid run-arrest event occurs on the first loading cycle, Eqs (6a–c) should be used with $(\delta_p)_{n-1}$ and $(\delta_p)_1$ set equal to zero.

9.3 *Validity Requirements:*

9.3.1 The value of K_a calculated from Eq (4) can be considered a linear-elastic plane-strain value, K_{Ia}, provided the criteria described in 9.3.2 through 9.3.4 and summarized in Table 2 are satisfied. Comments on the adequacy of these requirements can be found in Refs **(2)** and **(11)**.

9.3.1.1 Use is made in the following of σ_{Yd}, a formal dynamic yield strength estimate for appropriate loading times at the test temperature. For structural steels, it is being assumed here that σ_{Yd} is 30 ksi (205 MPa) greater than the yield strength, σ_{YS}, measured by Test Methods E 8.

NOTE 15—The extremely high strain rates associated with yielding near the tip of a fast running crack and the abrupt nature of crack arrest suggest that the trueebelevation of σ_{Yd} over σ_{YS} should be much greater. The value of σ_{Yd} that is being used here is therefore thought to substantially underestimate the actual effective resistance to plastic flow at crack arrest **(12)**.

9.3.2 The unbroken ligament, $W - a_a$, must equal or exceed both $0.15W$ and $1.25\,(K_a/\sigma_{Yd})^2$.

9.3.3 The thickness, B, must equal or exceed $1.0\,(K_a/\sigma_{Yd})^2$.

9.3.4 The minimum crack jump, $a_a - a_o$, must be at least twice the slot width, N, defined in Fig. 5, and greater than the plane-stress plastic zone radius associated with the initial loading, $(K_o/\sigma_{YS})^2/2\pi$.

NOTE 16—If a duplex specimen is used, the alternative requirement is that the crack penetrate a distance equal to or greater than B_N into the test section.

10. Report

10.1 Report the following information:

10.1.1 *Test Identification:*

10.1.1.1 Date,

10.1.1.2 Specimen number, and

TABLE 2 Summary of Criteria Used to Ensure That K_a Is a Linear Elastic, Plane-Strain Value

Feature	Criterion
Unbroken ligament	(A) $W - a_a \geq 0.15W$
Unbroken ligament	(B) $W - a_a \geq 1.25\,(K_a/\sigma_{Yd})^2$
Thickness	(C) $B \geq 1.0\,(K_a/\sigma_{Yd})^2$
Crack-jump length	(D) $a_a - a_o \geq 2N$
Crack-jump length	(E) $a_a - a_o \geq (K_o/\sigma_{YS})^2/2\pi$

10.1.1.3 Crack plane orientation.

10.1.2 *Material:*

10.1.2.1 Material type,

10.1.2.2 Young's modulus,

10.1.2.3 Yield strength (offset − 0.2 %) as determined by Test Methods E 8, and

10.1.2.4 Dynamic yield strength used in 9.3.2 and 9.3.3.

10.1.3 *Test Temperature:*

10.1.4 *Starter Notch:*

10.1.4.1 Type of brittle weld, if any,

10.1.4.2 Notch width, N, and

10.1.4.3 Notch root radius, ρ.

10.1.5 *Specimen Dimensions:*

10.1.5.1 Specimen thickness, B,

10.1.5.2 Specimen thickness at crack plane, B_N,

10.1.5.3 Thickness ratio, B_N/B, and

10.1.5.4 Width, W.

10.1.6 *Crack Length Measurements:*

10.1.6.1 Method used for marking the arrested crack front,

10.1.6.2 Crack length at machined notch, a_o, and

10.1.6.3 Crack length at arrest,

(*a*) At mid-thickness, a_2,

(*b*) At ¼ points of net thickness, a_1 and a_3, and

(*c*) Average crack length at arrest, $a_a = (a_1 + a_2 + a_3)/3$.

10.1.7 *Test Record:*

10.1.7.1 Load and displacement records and associated calculations,

10.1.7.2 First cycle limiting displacement, $[(\delta_o)_1]_{max}$,

10.1.7.3 Opening displacement increment for subsequent cycles, $(\delta_o)_{inc} = 0.25\,[(\delta_o)_1]_{max}$,

10.1.7.4 Number of load/unload cycles, n,

10.1.7.5 Displacements measured from load-displacement records,

(*a*) Displacement offset at end of first loading cycle, $(\delta_p)_1$,

(*b*) Total displacement offset accumulated prior to start of last loading cycle, $(\delta_p)_{n-1}$,

(*c*) Displacement at onset of unstable crack growth, δ_o,

(*d*) Displacement at crack arrest, δ_a, and

(*e*) Displacement increase accompanying the run-arrest event, $\delta_a - \delta_o$,

10.1.7.6 Displacements used to calculate K_o and K_a,

(*a*) Displacement used to calculate K_o,

$$d_o = \delta_o - (\delta_p)_{n-1}, \text{ and}$$

(*b*) Displacement used to calculate K_a,

$$\begin{aligned} d_a &= \delta_o - (\delta_p)_1 - 0.5[(\delta_p)_{n-1} - (\delta_p)_1] + 0.5[\delta_a - \delta_o] \\ &= 0.5[\delta_o + \delta_a - (\delta_p)_1 - (\delta_p)_{n-1}] \end{aligned}$$

10.1.7.7 Load drop as a percentage of P_{max}.

10.1.8 *Calculated Values of K_o and K_a (K_{Ia}):*

10.1.8.1 K_o, and

E 1221

10.1.8.2 K_a (K_{Ia}).
10.1.9 *Validity Requirements (see Table 2):*
10.1.9.1 Uncracked ligament length,
(*a*) Compared to $0.15W$, and
(*b*) Compared to $1.25\ (K_a/\sigma_{Yd})^2$,
10.1.9.2 Thickness, compared to $1.0\ (K_a/\sigma_{Yd})^2$,
10.1.9.3 Crack jump length,
(*a*) Compared to $2N$, and
(*b*) Compared to $(K_o/\sigma_{YS})^2/2\pi$.
10.1.10 *Photographic Record of Fracture Surfaces and Descriptive Comments (Optional):*

11. Precision and Bias

11.1 *Precision:*

11.1.1 The precision of a K_{Ia} determination by this test method is a function of the precision and bias of the various measurements of linear dimensions of the specimen and testing fixtures, the precision of the displacement measurements, the precision and bias of the recording devices used to produce the load displacement record, and the precision and bias of the measurements of the arrested crack length. It is not possible to make meaningful statements concerning precision and bias for all of these measurements. However, it is possible to derive useful information concerning the precision of a K_{Ia} measurement in a global sense from the results of an interlaboratory round robin test program that was conducted to evaluate the originally proposed test method on which this document is based **(2)**.[5]

[5] Information on K_{Ia} round robin data and the round robin program is available in Ref (2), a copy of which is available from ASTM Headquarters. Request RR: E 24-1014.

TABLE 3 Grand Means and Standard Deviations for K_{Ia} for Three Steels as Obtained From a Large Interlaboratory Round Robin Test Program[A]

Material Tested	A514[B]	A588[C]	A533B[D]	A533B[D]
Test Temperature	−30°C	−30°C	10°C	25°C
No. of Test Results	12	40	30	28
Mean K_{Ia}, MPa$\sqrt{m}$	88.4	61.5	78.2	83.4
Standard Deviation, MPa$\sqrt{m}$ and percent	10.2 (12 %)	6.4 (10 %)	9.7 (12 %)	10.6 (13 %)

[A] A total of 21 laboratories reported test results from the program.
[B] Specimens were cut from 2-in. (50.8 mm) thick rolled plate and tested full thickness in an L-T orientation; σ_{YS} = 129 ksi (890 MPa); NDT = −58°F (−50°C); RT_{NDT} = 10°F (−12°C).
[C] Specimens were cut from 2-in. (50.8-mm) thick rolled plate and tested full thickness in an L-T orientation; σ_{YS} = 48 ksi (330 MPa); NDT = 14°F (−10°C); RT_{NDT} = 16°F (−9°C).
[D] Specimens of 2-in. (50.8-mm) thickness were cut from 10-in. thick rolled plate and tested in an L-S orientation; σ_{YS} = 70 ksi (480 MPa); NDT = 10°F (−12°C); RT_{NDT} = 28°F (−2°C).

11.1.2 The results from this program are summarized in Table 3. It should be appreciated that the measures of precision shown in Table 3 apply to tests conducted with materials that exhibit strong transitional behavior in terms of temperature. A larger degree of scatter in K_{Ia} measurements could therefore be expected in tests conducted higher in the transition range, although the coefficient of variation appears to be temperature insensitive for some materials **(6)**. The materials tested also exhibit significant inhomogeneity and a size effect may be apparent when testing specimens of different sizes, with tests on smaller specimens being characterized by larger scatter than will tests on larger specimens.

11.2 *Bias*—There is no accepted standard value for the plane-strain crack-arrest fracture toughness of any material. In the absence of such a true value, any statement concerning bias is not meaningful.

ANNEX

(Mandatory Information)

A1. GUIDELINES FOR ASSESSING FRACTURE SURFACE ACCEPTABILITY AND PROCEDURES FOR DETERMINING THE ARRESTED CRACK LENGTH

A1.1 *Introduction:*

A1.1.1 The idealized fracture surface of a crack arrest specimen is flat, continuous, and straight-fronted. This idealization can be closely approached in practice, as evidenced by Figs. A1.1 and A1.2. However, the fracture surfaces of crack arrest specimens can be complicated by features that, when present in excess, can lead to questionable results for the crack arrest fracture toughness of the material being tested. This annex provides guidelines for identifying, from the fracture surface appearance, test results that are probably not representative of the bulk of the material being tested. It also provides guidelines for determining the arrested crack length, a_a, for a specimen with an irregular crack front.

A1.1.2 Deviations from the ideal fracture surface appearance generally fall into three broad categories. These are: the presence of remaining ligaments, a lack of crack front straightness, and crack propagation out of the plane of the side grooves. The extent to which one or more of these behaviors can occur without adversely affecting the test result cannot be easily quantified at the present time. The purpose of this annex is to provide a basis for the decision-making process that is required in assessing fracture surface acceptability and it is intended to serve as a set of guidelines for the personnel involved in obtaining crack arrest toughness data.

A1.1.3 The final decision as to which fracture surfaces will be classified as unacceptable presently rests primarily on the judgment of the individual (or individuals) performing and evaluating the test. This judgement should be based on experience, or on some knowledge of what is representative for the particular steel and test temperature combination under consideration. Judgmental decisions of this nature are clearly undesirable from a standards viewpoint. However, the alternative would be to disregard test results from all specimens with any degree of fracture surface irregularity.

E 1221

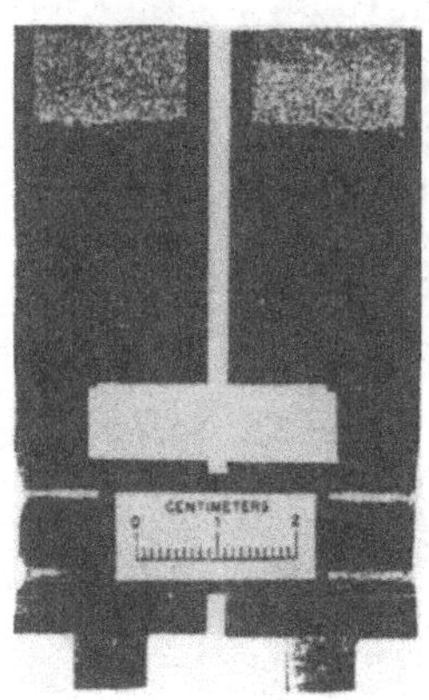

FIG. A1.1 An Example of an Ideal Fracture Surface of a CCA Specimen

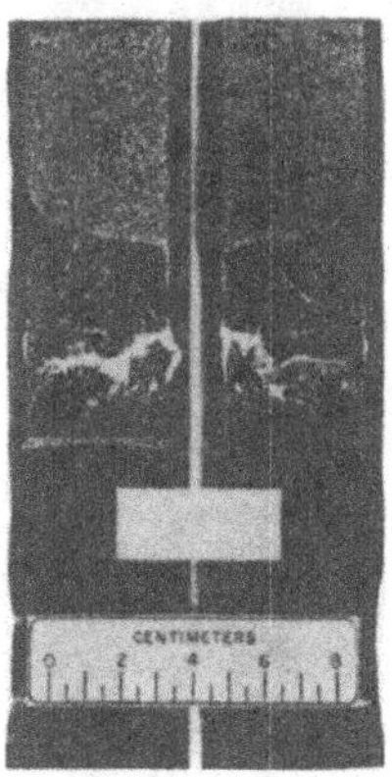

FIG. A1.3 Fracture Surface of a CCA Specimen Showing a Large, Transverse Remaining Ligament

This would not only reduce the percentage of successful, valid tests to a very small number, but would also eliminate virtually all tests performed under certain conditions, for example, at temperatures well into the transition range for a given material. As the testing technology involved here matures, it is anticipated that this annex will be updated to a more quantitative level.

A1.2 *Fracture Surface Acceptability:*

A1.2.1 *Remaining Ligaments:*

A1.2.1.1 In a number of steels, portions of the surface formed by a rapid fracture frequently remain unbroken. These unbroken regions, termed ligaments, are more frequently observed at higher temperatures and higher toughness levels. These unbroken ligaments have been observed in large structural tests as well as in the laboratory sized specimens employed by this test method (2, 5, 6). Since ligaments are commonly observed, their presence on the fracture surfaces of crack arrest specimens tested in the laboratory using this test method may be representative of the fracture of a structural member fabricated from the test material, in service at the test temperature.

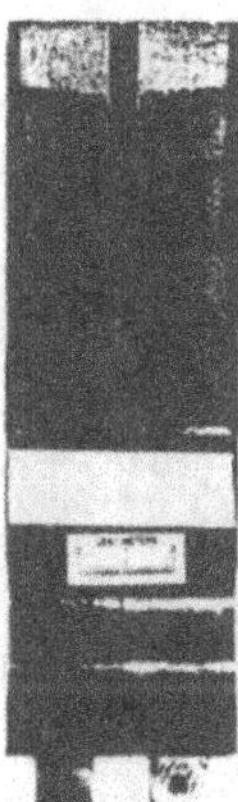

FIG. A1.2 Another Example of an Ideal Fracture Surface of a CCA Specimen

A1.2.1.2 In the transition temperature range, ligamentation of the fracture surface is due to connections between non-coplanar regions of cleavage, which subsequently break by hole-joining. However, the strain required for their fracture may not occur prior to crack arrest and the result is a ligamented fracture surface. These ligaments are often undercut and do not always exert excessive closing forces. The degree of influence of remaining ligaments on the running crack is uncertain. However, their tendency to produce an earlier crack arrest position (and consequently, a larger K_{Ia}) might be greater in a moderate sized laboratory test specimen than for a large crack in a service component.

A1.2.1.3 In assessing the influence of remaining ligaments on the fracture behavior of a CCA specimen tested using this test method, it is necessary to distinguish between the presence of a number of remaining ligaments distributed over the entire fracture surface and one or more large ligaments that may sometimes be present. Specimens that display excessively large remaining ligaments should be disregarded when using this test method. Two examples of single large unbroken ligaments (the light areas on the heat-tinted fracture surfaces) are shown in Figs. A1.3 and A1.4. In both cases, the test results obtained from these specimens were high in comparison to other test results for the same steels tested under the same conditions (2).

A1.2.1.4 When distributed ligaments are present, the total area of the remaining ligaments may be so large as to give a nonrepresentative value of K_{Ia}. The only method of evaluating whether the degree of remaining ligaments on a given specimen is truly representative of the material behavior at the test temperature of interest, is to perform a large number of tests under identical conditions, thereby establishing a baseline for observations. By comparing the results of an individual specimen with the baseline data set, it can be

FIG. A1.4 Fracture Surface of a CCA Specimen Showing a Large Remaining Ligament Aligned with the Direction of Crack Propagation

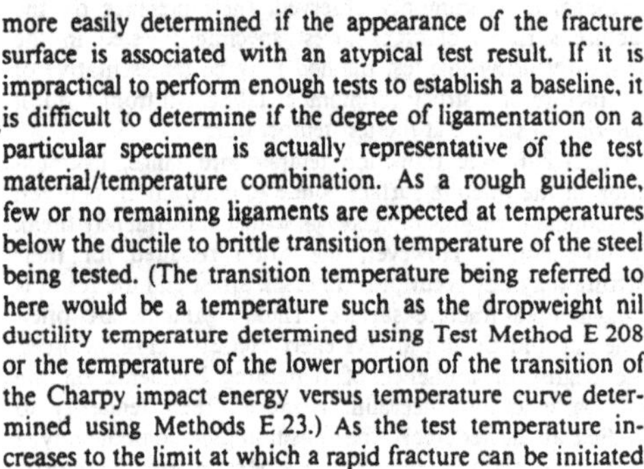

more easily determined if the appearance of the fracture surface is associated with an atypical test result. If it is impractical to perform enough tests to establish a baseline, it is difficult to determine if the degree of ligamentation on a particular specimen is actually representative of the test material/temperature combination. As a rough guideline, few or no remaining ligaments are expected at temperatures below the ductile to brittle transition temperature of the steel being tested. (The transition temperature being referred to here would be a temperature such as the dropweight nil ductility temperature determined using Test Method E 208 or the temperature of the lower portion of the transition of the Charpy impact energy versus temperature curve determined using Methods E 23.) As the test temperature increases to the limit at which a rapid fracture can be initiated

FIG. A1.6 Fracture Surface of a CCA Specimen Showing a Moderate Degree of Ligamentation

using the brittle-weld CCA specimen described in this standard (typically from 55 to 70°F (30 to 40°C) above the dropweight NDT) the portion of the fracture surface composed of remaining unbroken ligaments may approach 50 % and still give acceptable results for K_{Ia}. For the purposes of this standard, 50 % will be taken as the upper limit on the area of the fracture surface that can be covered by the remaining ligaments before the test result is disregarded.

A1.2.1.5 Examples of CCA specimen fracture surfaces displaying varying degrees of distributed remaining ligaments are shown in Figs. A1.5, A1.6, and A1.7. The extent of ligamentation in Fig. A1.5 is small. Figure A1.6 shows a more strongly ligamented fracture surface, but the degree of ligamentation is not severe enough to justify discarding the data. The fracture surface shown in Fig. A1.7 borders on the limits of what could be considered acceptable and its retention would have to be justified through comparison

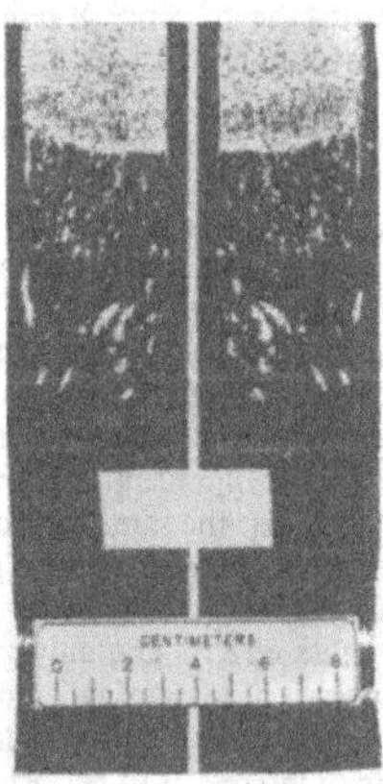

FIG. A1.5 Fracture Surface of a CCA Specimen Showing Very Few Remaining Ligaments

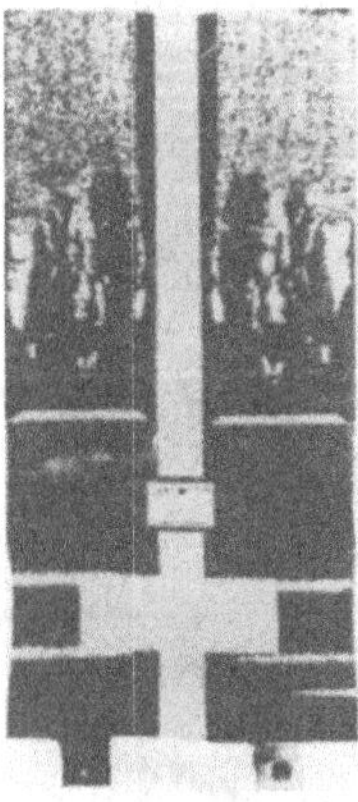

FIG. A1.7 Fracture Surface of a CCA Specimen Showing Extensive Ligamentation

E 1221

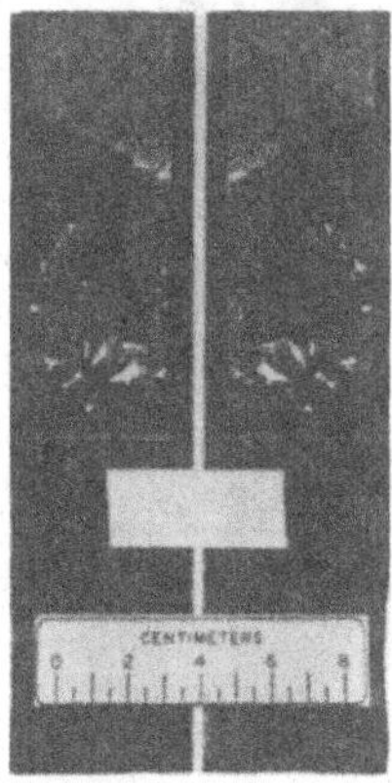

FIG. A1.8 Fracture Surface of a CCA Specimen with a Slanted and Straight Crack Front

with a baseline data set as discussed in A1.2.1.4.

A1.2.2 *Lack of Crack Front Straightness:*

A1.2.2.1 Nonideal behaviors that fall into this classification can be divided into two areas: slanted crack fronts and crack front curvature (or tunneling).

A1.2.2.2 Slanted crack fronts by themselves are not generally considered to be a problem in crack arrest testing as long as they are not extreme. Specimens with crack fronts inclined at angles of up to 45° from the side of the specimen have been seen to produce results that agree well with large data sets (**2**), see for example Figs. A1.8 and A1.9. Results from specimens for which the angle of the crack front exceeds 45° should be disregarded. (In testing CCA specimens, consistently slanted crack fronts can be caused by improper machining, alignment, lubrication, or seating of the load train, specimen, and support block. These should all be checked and corrected if necessary.)

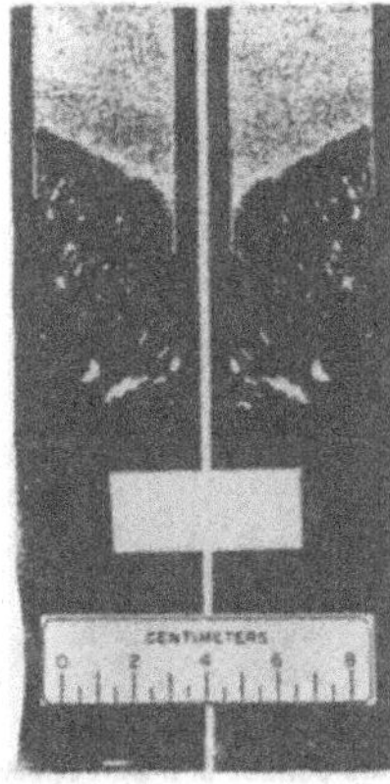

FIG. A1.9 Fracture Surface of a CCA Specimen with a Slanted and Nearly Straight Crack Front

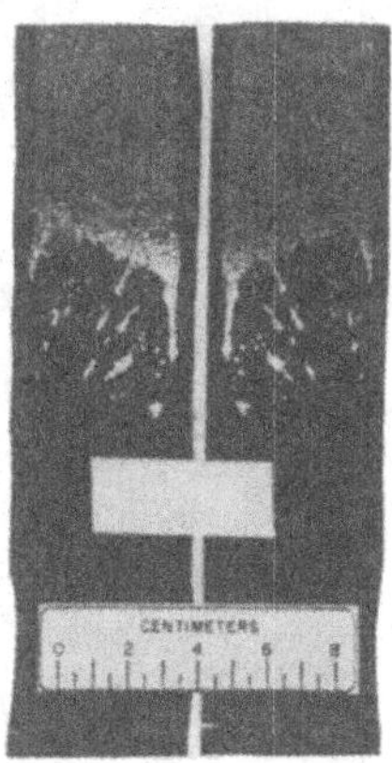

FIG. A1.10 Fracture Surface of a CCA Specimen with a Tunneled but Acceptable Crack Front

A1.2.2.3 Excessive crack front curvature is another reason to discard a test result on the basis of the appearance of the fracture surface. The major problem occurs when the crack propagates only in the center of the specimen (tunnels) and leaves substantial unbroken ligaments at the edges of the fracture surface, but the crack plane remains within the plane of the side grooves, that is, still within the net specimen thickness, B_N. (The situation in which the crack tunnels and propagates out of the plane of the side grooves is a separate case and is discussed in A1.2.3.) Such behavior, when it occurs to a limited extent, does not seem to significantly affect the value of the crack arrest toughness calculated for the specimen, in the same sense as a limited degree of ligamentation distributed over the bulk of the fracture surface does not affect the test result. The principal concern with this type of behavior is that, when it is present to a significant degree, it is not representative of the fracture of a structure and is caused by the lower triaxial constraint that exists near the edges of the crack plane even at the base of the side grooves. Specimens with excessive tunneling in which substantial amounts of the edges of the fracture surface remain unbroken have been found to produce abnormally high values of arrest toughness and should be disregarded (**2**).

A1.2.2.4 Figures A1.10 and A1.11 show two examples of cracks that haae tunnelled but for which the crack plane has remained within the plane of the side grooves. The fracture surface shown in Fig. A1.10 would be acceptable, while the example shown in Fig. A1.11 would not.

A1.2.3 *Out-of-Plane Crack Propagation:*

A1.2.3.1 Test results should also be disregarded if the plane of crack propagation either makes an angle greater than 10° with the plane of the side grooves or if the crack plane lies entirely outside the midplane of the specimen.

A1.2.3.2 In cases where the crack moves only slightly out of the midplane and the portion that is outside the net thickness of the specimen does not fracture during the run-arrest event, a fracture surface is produced that appears

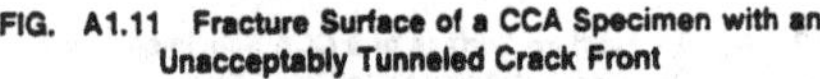

FIG. A1.11 Fracture Surface of a CCA Specimen with an Unacceptably Tunneled Crack Front

tunnelled, as for example in Figs. A1.12 and A1.13. Such behavior is not generally considered to be a problem from the standpoint of either tunneling or out of plane propagation since the unbroken ligaments at the edges of the fracture surface forming the tunnel are not restrained by the material that was removed to form the side grooves. They are consequently able to bend during the run-arrest event, thus allowing it to progress more or less normally.

A1.2.3.3 In the extreme situation, when the crack has moved almost entirely out of the side groove into the gross thickness of the plate, the test result should be discarded, regardless of whether or not the portion of the fracture surface outside the net thickness has fractured. Examples of fracture surfaces that would be unacceptable for this reason are shown in Figs. A1.14 and A1.15.

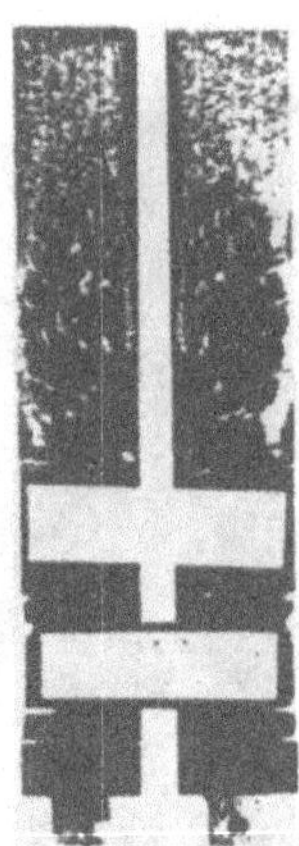

FIG. A1.13 Fracture Surface of a CCA Specimen for Which the Crack has Grown Out of the Midplane of the Specimen but has Remained Mostly Within the Net Specimen Thickness, B_N

A1.3 *Determination of the Arrested Crack Length, a_a:*

A1.3.1 As described in 8.6 of this test method, the arrested crack length, a_a, is taken to be the average of three measurements made on the heat-tinted fracture surface of the specimen. These measurements are to be made at the center of the specimen ($B_N/2$) and halfway between the center of the specimen and the edge of the side groove, on each side ($B_N/2 \pm B_N/4$). It is further stated that, at each measurement location, the measurement should be taken as a visual average of the position of the crack front across a strip of width $B_N/4$, centered at the measurement location.

A1.3.2 The visual average is specified to avoid taking a measurement at a point that may not accurately represent

FIG. A1.12 Fracture Surface of a CCA Specimen for Which the Crack has Started to Grow Out of the Midplane of the Specimen but has Remained Within the Net Specimen Thickness, B_N

FIG. A1.14 Fracture Surface of a CCA Specimen for Which the Crack has Grown Out of the Midplane of the Specimen and Broken Through the Gross Thickness of the Specimen on One Side

E 1221

FIG. A1.15 Fracture Surface of a CCA Specimen for Which the Crack has Tunneled and Run Out of the Midplane of the Specimen as Well as Almost Entirely Out of the Side Grooved Section

the local average position of the crack front in the vicinity of the measurement location. For purposes of analysis, it is assumed that the crack front is both straight and smooth. The visual averaging technique is intended to provide a crack tip location with locally as much unbroken material ahead of it as there is broken material behind it, when averaged across the width of the measurement strip. The visual averaging technique may thus provide some degree of conservatism, since the calculated stress intensity factor decreases as the crack length increases in the calculational procedure employed in this test method.

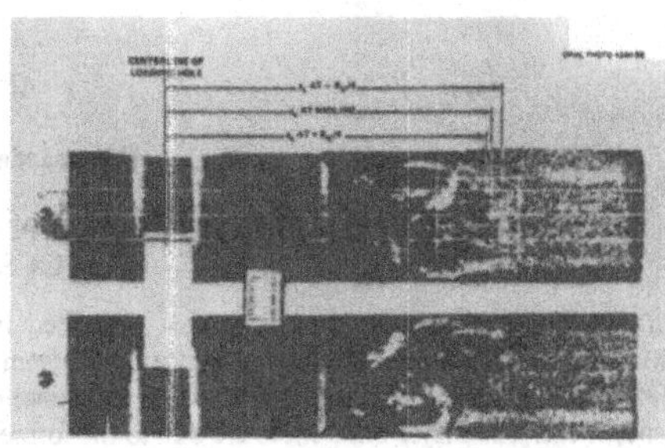

FIG. A1.17 Example of Arrested Crack Length Determination for a CCA Specimen with an Irregular and Slanted Crack Front and a Moderate Degree of Ligamentation

A1.3.3 Examples of local irregularities in arrested crack fronts can be readily seen in Figs. A1.1 through A1.15, which have been presented previously. Three examples of arrested crack length determinations for different degrees of measurement complexity are shown in Figs. A1.16 through A1.18. In each case, strips of width $B_N/4$ are shown, each centered on one of the three lines along which the crack length determinations are to be made. A visual average of the position of the crack front has then been taken within each individual strip. The distance of each of the three visually averaged crack front locations from the center of the specimen loading hole has also been indicated. The resulting three measurements are then averaged to calculate the arrested crack length, a_a, for use in subsequent calculations.

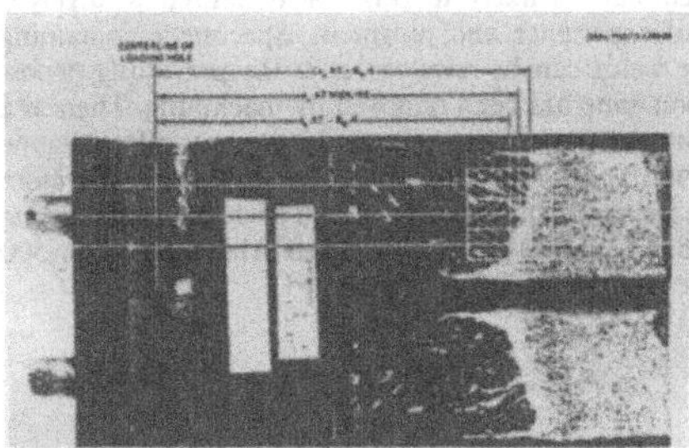

FIG. A1.16 Example of Arrested Crack Length Determination for a CCA Specimen with a Slanted Crack Front and a Small Amount of Ligamentation on the Fracture Surface

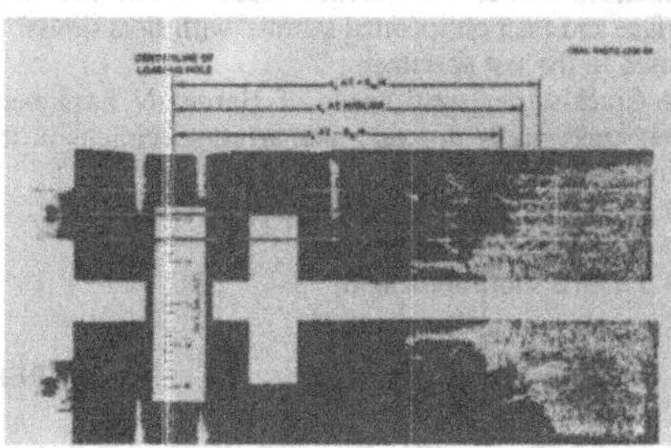

FIG. A1.18 Example of Arrested Crack Length Determination for a CCA Specimen with a Highly Irregular Crack Front and a Heavily Ligamented Fracture Surface

E 1221

APPENDIX

(Nonmandatory Information)

X1. GUIDELINES ON STARTER NOTCH PREPARATION TECHNIQUES FOR BRITTLE-WELD CCA SPECIMENS

X1.1 A welding procedure similar to that described in Test Method E 208 has been found satisfactory, with some modification, for starter notch preparation of brittle-weld CCA specimens. The major differences are due to the former having a flat surface, while the latter has a radiused slot.

X1.2 The technique requires a crack starter notch of sufficient width for the electrode to reach the bottom of the slot. The notch can be fabricated by drilling a hole with its center at the desired location of a_o, with the sides of the notch being produced by saw cutting to the hole. Weld starter and runout blocks are used. The finished single-pass weld should have a relatively flat surface to facilitate subsequent notching. Experience has shown that introduction of a brittle weld does not introduce serious distortion. A light surface grind operation may however be necessary. A notch is machined in the weld as shown in Fig. 6. A notch root radius, ρ, of 0.010 to 0.015 in. (0.25 to 0.38 mm) or even larger can be used.

X1.3 The Murex-Hardex-N electrodes specified in Test Method E 208 have been found to perform satisfactorily, as have Foxdur-500 electrodes (2). Embrittling the notch root by autogenous welding procedures such as TIG or EB welding has also proven to be successful. For welders having little experience with the recommended electrodes, it is recommended that several practice beads be laid, first on a flat surface and then on a slotted sample, with slots similar to those used in the test specimen.

X1.4 Crack-starter weld beads of Hardex-N have been deposited successfully in steel crack arrest specimens in the following manner:

X1.4.1 Demagnetize the specimen prior to welding,

X1.4.2 Use DC, reverse polarity,

X1.4.3 Use a short arc-length,

X1.4.4 Use a welding current of 200 to 225 A,

X1.4.5 Use a welding speed of approximately 4 in./min (100 mm/min),

X1.4.6 Use starter tabs and runout tabs, approximately 0.25 in. (6 mm), thick and 1.0 in. (25 mm) square,

X1.4.7 Complete the weld in one pass, and

X1.4.8 Weld with a slight weaving motion of the electrode. Similar procedures may be used with the Foxdur-500 electrodes. However, drying of the Foxdur-500 electrodes for 2 h at 450 to 650°F (250 to 350°C) is recommended. The Hardex-N electrode appears to perform best as received, but may be dried for 1 h at 200 to 300°F (100 to 150°C).

X1.5 The appearance of cracks in the weld bead does not necessarily indicate that the bead will not perform satisfactorily as a crack starter. The introduction of a notch into the weld bead may remove the cracks. However, even if the notching operation does not remove the weld cracks, rapidly propagating cracks have been initiated successfully in many instances from cracked Hardex-N weld beads. In fact, it has been observed frequently that initially sound brittle welds develop cracks visible to the naked eye under load, well before initiation of an unstable crack. This is presumably because the hardened portion of the heat-affected-zone beneath the weld bead primarily controls initiation, rather than the notched weld bead itself. The decision as to whether a cracked weld is likely to perform as desired is largely a matter of experience and judgment. Specimens containing defective welds can be rewelded after the previously deposited fusion zone has been removed by machining. There is a possibility that cracks may appear in the heat-affected-zone of the specimen. Preheating of the specimens may therefore be necessary and temperatures of 300°F (150°C) when using the Hardex-N and 210°F (100°C) when using the Foxdur-500 electrodes may be helpful.

REFERENCES

(1) Crosley, P.B., Fourney, W.L., Hahn, G.T., Hoagland, R.G., Irwin, G.R., and Ripling, E.J., "Final Report on Cooperative Test Program on Crack Arrest Toughness Measurements," *NUREG/CR-3261*, University of Maryland, College Park, MD (April 1983).

(2) Barker, D.B., Chona, R., Fourney, W.L., and Irwin, G.R., "A Report on the Round Robin Program Conducted to Evaluate the Proposed ASTM Test Method for Determining the Crack Arrest Fracture Toughness, K_{Ia}, of Ferritic Materials," *NUREG/CR-4996 (ORNL/Sub/79-7778/4)*, University of Maryland, College Park, MD (January 1988).

(3) Rosenfield, A.R., "Validation of Compact-Specimen Crack-Arrest Data," *Journal of Engineering Materials Technology*, Vol 106, pp. 207–208 (1984).

(4) Kanninen, M.F., and Popelar, C.H., *Advanced Fracture Mechanics*, Oxford University Press, NY (1985).

(5) Cheverton, R.D., Ball, D.G., Bolt, S.E., Iskander, S.K., and Nanstad, R.K., "Pressure Vessel Fracture Studies Pertaining to the PWR Thermal Shock Issue: Experiments TSE-5, TSE-5A, and TSE-6," *NUREG/CR-4249 (ORNL-6163)*, Oak Ridge National Laboratory, Oak Ridge, TN (June 1985).

(6) Bryan, R.H., Bass, B.R., Bolt, S.E., Bryson, J.W., Edmonds, D.P., McCulloch, R.W., Merkle, J.G., Nanstad, R.K., Robinson, G.C., Thoms, K.R., and Whitman, G.D., "Pressurized-Thermal-Shock Test of 6-in.-Thick Pressure Vessels. PTSE-1: Investigation of Warm Prestressing and Upper-Shelf Arrest," *NUREG/CR-4106 (ORNL-6135)*, Oak Ridge National Laboratory, Oak Ridge, TN (April 1985).

(7) Rosenfield, A.R., Mincer, P.N., and Marschall, C.W., "High-Temperature Crack-Arrest Toughness Measurements Using Compact Specimens," *Fracture Mechanics: Eighteenth Symposium,*

ASTM STP 945, D.T. Read and R.P. Reed, eds., pp. 73–85 (1988).
(8) Crosley, P.B., and Ripling, E.J., "Plane Strain Crack Arrest Characterization of Steels," *Journal of Pressure Vessel Technology*, (November 1975).
(9) Chona, R., Fourney, W.L., Link, R.E., and Sanford, R.J., "The Recoverability of Plastic Zone Energy in Crack Arrest," *Proceedings, 1985 SEM Spring Conference on Experimental Mechanics*, Las Vegas, NV, pp. 13–19 (June 1985).
(10) Crosley, P.B., and Ripling, E.J., "Development of a Standard Test for Measuring K_{Ia} with a Modified Compact Specimen," *NUREG/CR-2294 (ORNL/Sub-81/7755/1)*, Materials Research Laboratory, Glenwood, IL (August 1981).
(11) Rosenfield, A.R., Mincer, P.N., Marschall, C.W., and Markworth, A.J., "Recent Advances in Crack-Arrest Technology," *Fracture Mechanics: Fifteenth Symposium, ASTM STP 833*, R.J. Sanford, editor, pp. 149–164 (1984).
(12) Freund, L.B., Hutchinson, J.W., and Lam, P.S., "Analysis of High-Strain-Rate Elastic-Plastic Crack Growth," *Engineering Fracture Mechanics*, Vol 23, No. 1, pp. 119–129 (1986).

APPENDIX II

DVM - MERKBLÄTTER September 1986

DEUTSCHER VERBAND FÜR MATERIALPRÜFUNG	Meßtechnische Anforderungen beim instrumentierten Kerbschlagbiegeversuch	DVM 001

German Association for Testing Materials.
Requirement for the measuring devices for instrumented Charpy tests.

Inhalt

1. Ziel

Das Ziel des Merkblattes ist die Schaffung der meßtechnischen Voraussetzungen für die Vereinheitlichung von Versuchsdurchführung und -auswertung. Dazu wurden meßtechnische Mindestanforderungen für die zum Einsatz kommenden Meßgeräte zur Erzielung der Gleichwertigkeit unterschiedlicher meßtechnischer Einrichtungen im Hinblick auf vergleichbare Meßergebnisse festgelegt.

Das vorliegende Merkblatt ist dem Stahl-Eisen-Prüfblatt 1315 "Kerbschlagbiegeversuch mit Ermittlung von Kraft und Weg; Empfehlungen zur Durchführung und Auswertung" zugeordnet.

2. Voraussetzungen

Die Instrumentierung muß an normgerechten Pendelschlagwerken nach DIN 51 222 anzuwenden sein. Die Durchführung des Kerbschlagbiegeversuches nach DIN 50 115 darf durch die Instrumentierung nicht beeinträchtigt werden.

3. Zu messende Größen

Über die im konventionellen Versuch ermittelte verbrauchte Schlagarbeit hinausgehende Informationen werden dem Schlagkraft-Durchbiegung-Diagramm entnommen. Zur Ermittlung des Schlagkraft-Durchbiegung-Verlaufes ist die Messung der Schlagkraft in Abhängigkeit von der Probendurchbiegung (Weg der Hammerfinne) oder der Zeit erforderlich. Siehe Punkt 4.

4. Generelle Vorgehensweise

Die unter 3. genannten Meßgrößen können durch

- direkte Messung der Kraft und der Probendurchbiegung
- Messung des Kraft-Zeit-Verlaufes und rechnerische Ermittlung der Probendurchbiegung

bestimmt werden.

Der an der Hammerfinne gemessene Kraft-Zeit-Verlauf F(t) ist proportional dem Beschleunigungsverlauf. Mit der Hammermasse m des als starrer Körper angenommenen Hammers, der Auftreffgeschwindigkeit v_o und der Zeit t ab Verformungsbeginn t_o läßt sich die Probendurchbiegung durch zweimaliges numerisches Integrieren errechnen:

$$v(t) = v_o - \frac{1}{m} \int_{t_o}^{t} F(t)dt;$$

$$s(t) = \int_{t_o}^{t} v(t)dt.$$

5. Kraftmessung

Die Kraftmessung erfolgt heute üblicherweise mit zwei aktiven Dehnungsmeßstreifen (DMS) an der als Dynamometer ausgebildeten normgerechten Schlagfinne, die durch zwei weitere kompensierende DMS oder Ergänzungswiderstände zu einer Vollbrücke geschaltet sind. Die Kompensations-DMS sollen nicht an einem bei Stoßeinwirkung zu Schwingungen neigenden Teil des Pendelschlagwerkes appliziert werden.

Ein für die Darstellung des Kraftsignals notwendiger Meßverstärker soll mindestens eine Bandbreite von 0 ... 100 kHz bei 3 dB Amplitudenabfall aufweisen (Gleichspannungsverstärker). Die für eine unverfälschte Wiedergabe erforderliche Grenzfrequenz f_g für die Darstellung eines Signales mit der kürzesten zeitlichen Änderung t_A (Anstiegszeit) zwischen 10 und 90 % des Abstandes von zwei aufeinanderfolgenden Extremwerten einer sinusförmig angenommenen Kraftschwingung ist durch die Beziehung $f_g = 0{,}35/t_A$ gegeben. Als die für die Signalauswertung maßgebliche kürzeste Anstiegszeit t_A wird für den Kerbschlagversuch die Zeit bis zum Erreichen des Scheitelwertes des Beschleunigungsstoßes genommen; sie liegt erfahrungsgemäß bei ca. 10 µsec, entsprechend einer oberen Grenzfrequenz von 35 kHz. Der noch kürzere Kraftabfall bei instabilem Rißfortschritt in der Probe bleibt dabei außer Betracht.

Es ist zu beachten, daß bei ausgeführten Systemen die Signalamplituden bei der angegebenen System-Grenzfrequenz üblicherweise um 3 dB (ca. 30 %) abgeschwächt dargestellt werden. Um auch bei der Grenzfrequenz $f_g = 0{,}35/t_A$ eine hinreichend genaue Amplitudendarstellung zu erhalten, wird empfohlen, die 3dB-Grenzfrequenz des verwendeten Meßverstärkers mindestens dreifach höher zu legen. Aus diesem Zusammenhang ergibt sich die o. a. Bandbreite 0 ... 100 kHz (-3 dB).

Die statischen Linearitäts- und Hysteresefehler der instrumentierten Finne im eingebauten Zustand einschließlich der betrieblichen Meßkette bis hin zum Registriergerät (Schreiber, Plotter) sollen von 100 - 50 % des Kraftnennwertes ± 2 % der jeweiligen Ablesung und von 50 bis 10 % des Nennbereiches ± 1 % des Kraftnennwertes nicht überschreiten. Siehe Anhang Seite 10, Bild 1 und 2.

Für den Aufnehmer allein in ausgebautem Zustand ist die Genauigkeitsanforderung herunter bis 10 % des Kraftnennbereiches ± 1 % der jeweiligen Ablesung. Unterhalb 20 % des Nennbereiches sind auch Abweichungen bis maximal ± 2 % zulässig, auf die dann aber mit einem entsprechenden Vermerk hingewiesen werden muß.

Die Instrumentierung der Finne ist auf den erforderlichen Kraftbereichs-Nennwert hin auszulegen. Die Erfahrung zeigt, daß bei ISO-V-Proben mit Nennwerten zwischen 10 und 40 kN fast alle bei Stahlwerkstoffen vorkommenden Schlagkräfte erfaßt werden können. Besondere Aufmerksamkeit ist auch der Unempfindlichkeit der instrumentierten Finne bezüglich außermittiger Krafteinleitung zu widmen.

Die dynamische Beurteilung der Kraftmeßkette erfolgt praxisgerecht vereinfacht durch die Messung des Scheitelwertes des Beschleunigungsstoßes (Inertial-Peak) bei ausreichend hoher Zeitauflösung des Kraftsignals. Die

Dynamik der Meßkette kann als ausreichend angesehen werden, wenn eine ISO-V-Stahlprobe bei einer Schlaggeschwindigkeit von rd. 5,5 m/sec einen Beschleunigungsstoß größer 8 kN als Scheitelwert erreicht. Dies gilt erfahrungsgemäß für eine Entfernung der aktiven DMS von der Spitze der Schlagfinne von 11 ... 15 mm.

6. **Wegmessung**

Wird der Weg nicht als abgeleitete Größe (s. Pkt.4) ermittelt, so wird die Probendurchbiegung durch berührungsloses Messen der Verschiebung der Hammerfinne oder des Hammerblattes relativ zum Auflager nach optischen, induktiven, kapazitiven o. a. Verfahren bestimmt. Die dafür verwendete Meßkette muß bezüglich des Laufzeitverhaltens dem der Kraftmeßkette entsprechen, damit eine zeitgleiche Registrierung möglich wird.

Die Wegmeßkette ist für Nennwerte bis zu 30 mm Probendurchbiegung auszulegen; der Linearitätsfehler einschließlich der betrieblichen Meßkette soll im Bereich 1 ... 30 mm, bezogen auf den jeweiligen Ablesewert, $< \pm 2$ % sein. Eine dynamische Kalibrierung kann über eine Zeitmessung erfolgen.

Für Wegmessungen im Bereich < 1 mm wird empfohlen, die Wegbestimmung aus der Zeitmessung und der Hammerauftreffgeschwindigkeit abzuleiten. Für diesen Wegbereich kann bis zu Schlagarbeiten < 4 % des jeweils eingestellten Arbeitsvermögens des Pendels die Schlagfinnengeschwindigkeit der Auftreffgeschwindigkeit v_o des Hammers mit einer Genauigkeit von < 2 % vereinfachend gleichgesetzt werden.

7. **Registriergeräte**

Die Registrierung der dynamischen Signale erfolgt vorzugsweise mit digitalen Speichereinrichtungen (Speicheroszilloskop, Transient-Recorder), die Ausgabe der Versuchsergebnisse auf X-Y-Schreiber oder Plotter.

Um die unter Punkt 5 und 6 angegebenen Genauigkeiten mit digitalen Meß- und Registriereinrichtungen zu erreichen, wird eine digitale Mindestauflösung von 0,4 % des jeweiligen Nennbereiches der Meßkette entsprechend einer Analog-Digital-Wandlung mit 8 Bit empfohlen. Zur unverfälschten Signaldarstellung ist eine Abtastrate erforderlich, die im wesentlichen von der höchsten zu erfassenden Frequenz des Meßsignals und der Art der Meßwert-Darstellung auf dem Registriergerät (Punktdarstellung, Datenpunkte linear oder sinusförmig verbunden) abhängig ist. Bei Punktdarstellung wird allgemein das 25-fache, bei linearer Interpolation das 10-fache und bei Sinusinterpolation mindestens das 2,5-fache der höchsten Signalfrequenz als Abtastfrequenz empfohlen. Ausgehend von den dem transienten Meßsignal überlagerten Schwingungen mit einer Frequenz von erfahrungsgemäß rd. 25 kHz ist somit bei der zumeist vorliegenden linearen Interpolation eine Abtastrate 250 kHz (4 µs Abtastintervall) notwendig. Für den erfahrungsgemäß bis max. 8 ms meßtechnisch zu erfassenden Zeitraum ist dann für eine uneingeschränkte Registrierung je Meßkette ein Signalspeicher für ca. 2000 Meßwerte erforderlich. Für Meßsignale kürzer 8 ms kann die erforderliche Speichergröße proportional verringert werden.

8. Kalibrierung

Die Aufnehmer- und Meßkettenkalibrierung erfolgt praxisgerecht statisch mit der unter 5 und 6 angegebenen Genauigkeit. Dabei wird empfohlen, die Kalibrierung mit der im Hammerblatt eingebauten Schlagfinne durchzuführen. Die Krafteinleitung in die Schlagfinne ist dabei in jedem Fall über ein der ISO-V-Probe entsprechendes Formstück (s. Anhang Seite 9) vorzunehmen.

Für die Bestimmung von Kräften und Wegen aus einer üblichen grafischen Darstellung (Diagramm, Plot) kann davon ausgegangen werden, daß die Koordinatenwerte mit einer Genauigkeit von ± 0,5 mm zu bestimmen sind. Unter dieser Voraussetzung wird für die Diagrammdarstellung auch in Anlehnung an DIN 51 221, Teil 1, eine Mindestkoordinatenlänge von 100 mm empfohlen. Dabei ist bei 50 % des Bereichsnennwertes die Entnahmegenauigkeit ± 1 % des aktuellen Wertes und damit der Ablesefehler kleiner als die unter Punkt 5 und 6 geforderte Gesamtgenauigkeit.

Zur Beurteilung der Güte der Instrumentierung kann ein Vergleich der aus der Instrumentierung berechneten und aus der standardmäßigen Maschinenanzeige (Schleppzeiger) ermittelten verbrauchten Schlagarbeit herangezogen werden. Erfahrungsgemäß lassen sich Abweichungen im Bereich ± 5 Joule erreichen.

Für die Kalibrierung einer Instrumentierung werden Abstände von einem Jahr empfohlen.

9. Anhang

9.1 Erprobte instrumentierte Schlagfinnen-Konstruktionen mit zugehörigem Kraft-Durchbiegung-Diagramm

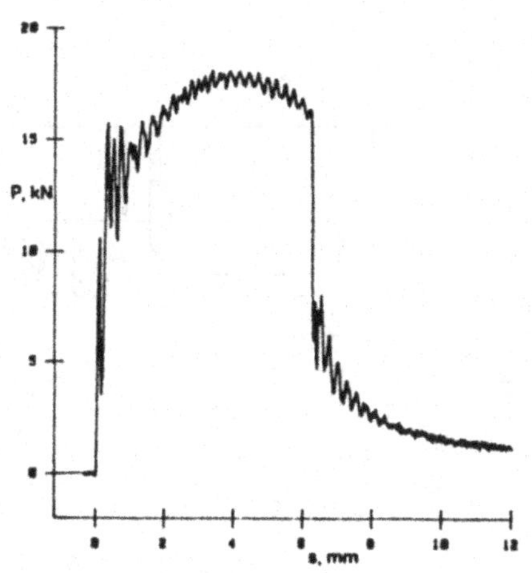

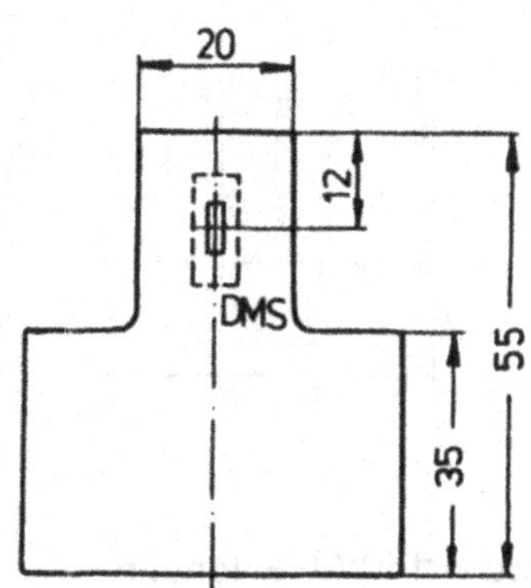

Bauart IWM-Freiburg

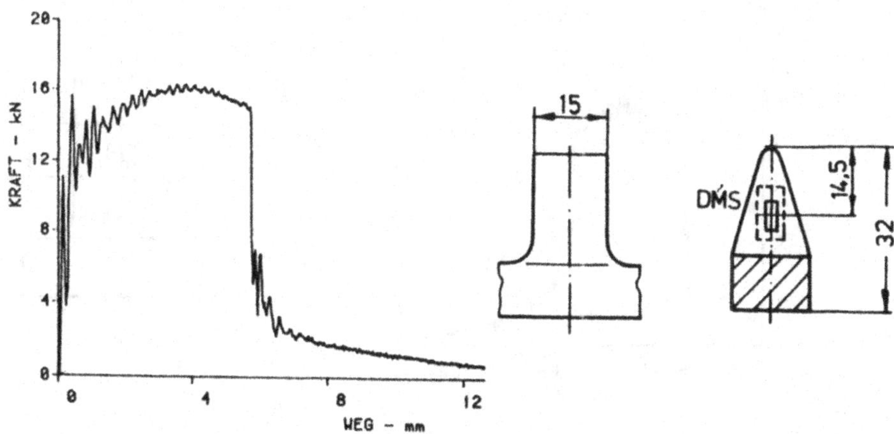

Bauart BAM - Berlin

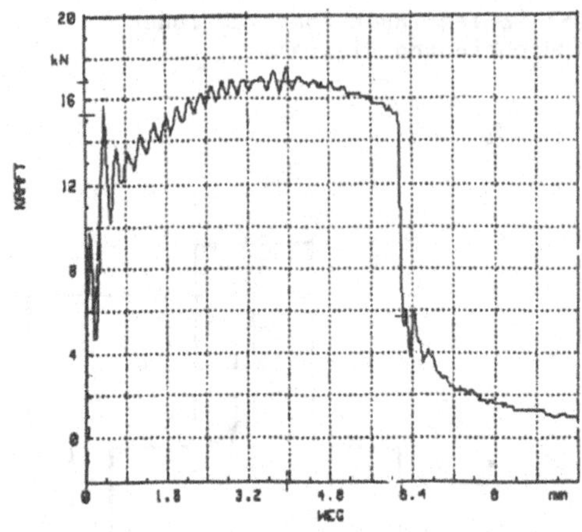

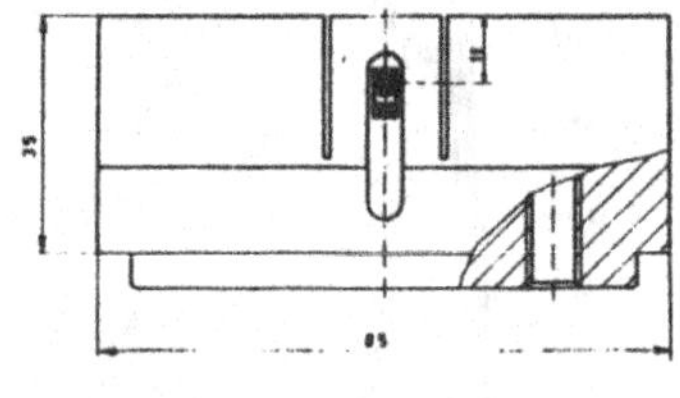

Bauart KWU-Erlangen

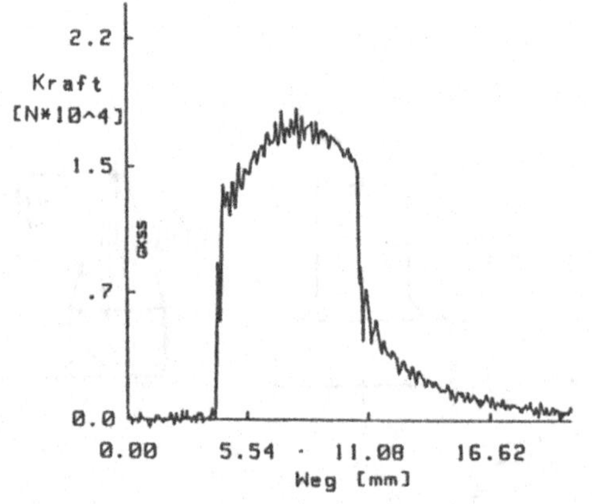

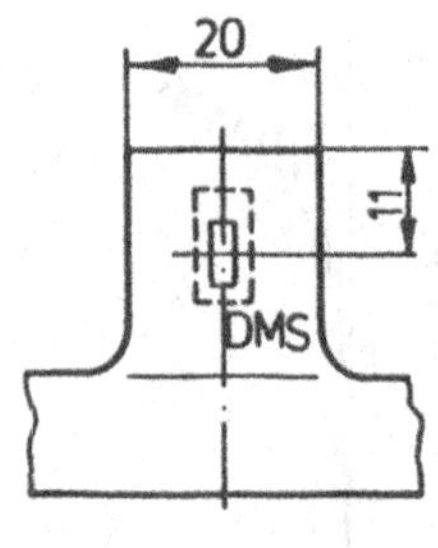

Bauart GKSS-Geesthacht

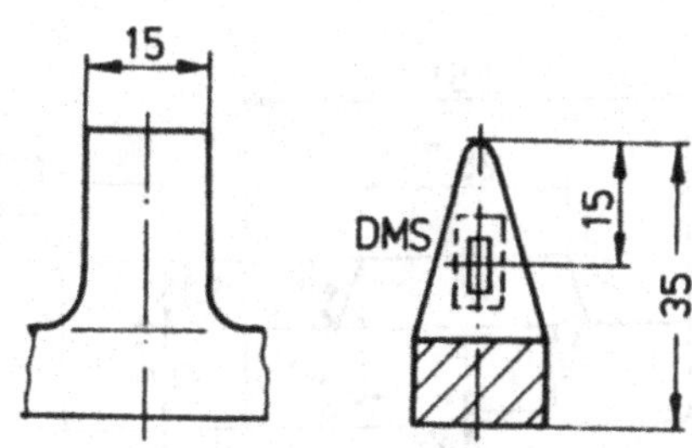

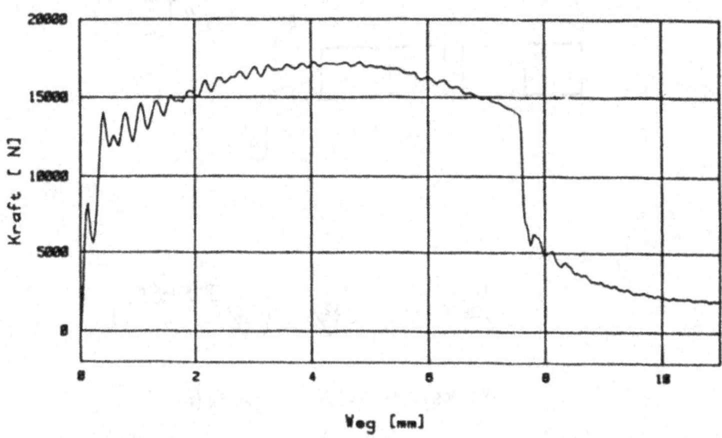

Bauart MPA - Stuttgart

9.2 Kalibrier-Formstück

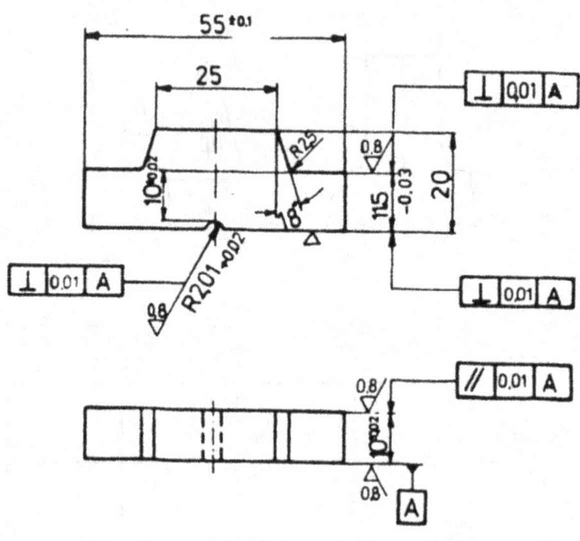

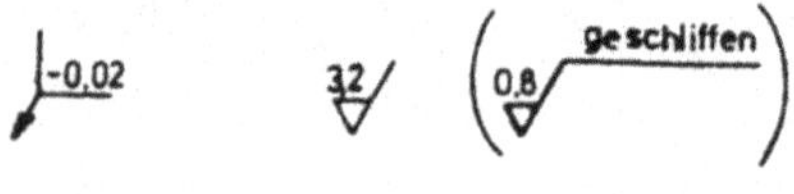

Werkstoff: 1.4122 gehärtet

Bauart Kraftwerk Union/Erlangen

9.3 Zulässige Fehler bei der Kraftmessung

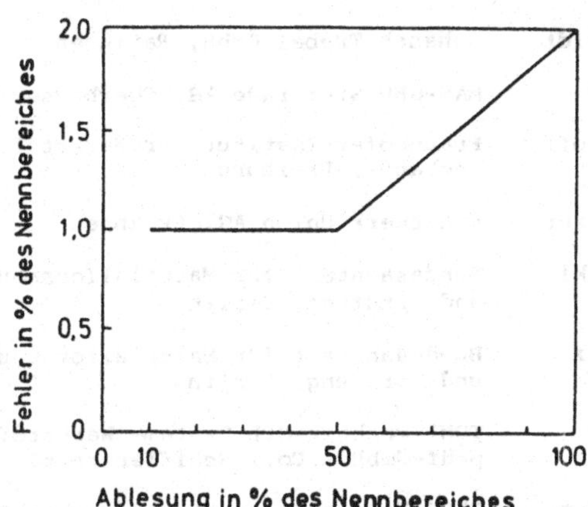

Bild 1 Auf den Nennbereich der Kraft bezogener zulässiger Fehler des Ablesewertes

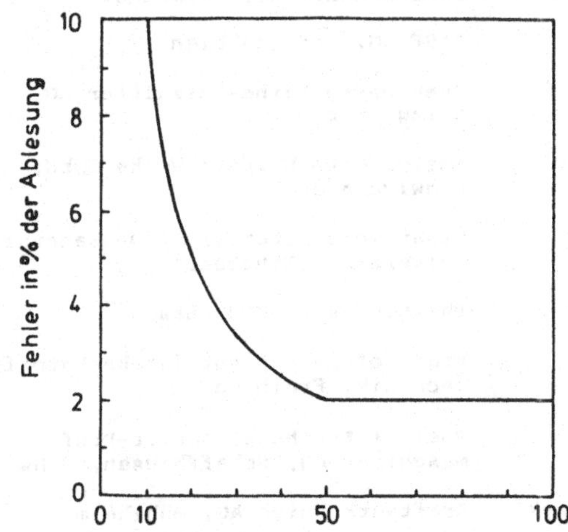

Bild 2 Auf die jeweilige Ablesung bezogener zulässiger Fehler der Kraftmessung

9.4 Mitgliederverzeichnis der für die Merkblatterstellung verantwortlichen DVM-Arbeitsgruppe "Instrumentierung von Pendelschlagwerken"

R. Helms, Obmann	Bundesanstalt für Materialforschung und -prüfung, Berlin
K. Achterfeldt	Schenck Trebel GmbH, Ratingen
H. Gathmann	MAN-GHH Sterkrade AG, Oberhausen
J. F. Kalthoff	Fraunhofer-Institut für Werkstoffmechanik, Freiburg
E. Klausnitzer	Kraftwerk Union AG, Erlangen
S. Ledworuski	Bundesanstalt für Materialforschung und -prüfung, Berlin
W. Markowski	Bundesanstalt für Materialforschung und -prüfung, Berlin
W. Mühl	TONI Technik-MFL Systeme Werkstoffprüf-GmbH & Co., Schifferstadt
J. Müller-Roos	GKSS-Forschungszentrum Geesthacht, Institut für Werkstofftechnologie, Geesthacht
A. Peuser	Carl Schenck AG, Darmstadt
R. Richter	BASF AG, Ludwigshafen
K. Seifert	Stahlwerke Peine-Salzgitter AG, Salzgitter
F. Spies	Amsler Otto Wolpert Werke GmbH, Ludwigshafen
W. Stoppler	Staatliche Materialprüfungsanstalt, Universität Stuttgart
G. Togler	Rhein.-Westf. TÜV, Essen
S. Winkler	Fraunhofer-Institut für Werkstoffmechanik, Freiburg
M. Witschi	Roell + Korthaus, Amsler-Prüfmaschinen AG, Schaffhausen, Schweiz
H. G. Wojaczyk	Kraftwerk Union AG, Mühlheim

STAHL-EISEN-Prüfblätter (SEP) des Vereins Deutscher Eisenhüttenleute Mai 1987

Kerbschlagbiegeversuch mit Ermittlung von Kraft und Weg

Empfehlungen zur Durchführung und Auswertung

SEP 1315
1. Ausgabe

1 Zweck und Wesen

Dieses Prüfblatt ergänzt die Festlegungen für den Kerbschlagbiegeversuch nach DIN 50 115 und ist nur in Verbindung mit DIN 50 115 anwendbar.

Der Kerbschlagbiegeversuch mit Ermittlung von Kraft und Weg unterscheidet sich von dem Kerbschlagbiegeversuch nach DIN 50 115 allein dadurch, daß mit einer Instrumentierung der Prüfeinrichtung nach DVM-Merkblatt 0001 die Kraft-Zeit-Kurven oder auch unmittelbar die Kraft-Weg-Kurven von im Kerbschlagbiegeversuch geprüften Kerbschlagproben registriert werden.

Anmerkung:
Das Vorgehen zur rechnerischen Ermittlung von Kraft-Weg-Kurven aus registrierten Kraft-Zeit-Kurven wird im DVM-Merkblatt 0001 beschrieben.

Das Flächenintegral unterhalb der Kraft-Weg-Kurve entspricht der von der Probe während des Versuches verbrauchten Schlagarbeit. Kraft-Weg-Kurven, die für unterschiedliche Werkstoffe oder bei unterschiedlichen Prüftemperaturen ermittelt werden, selbst wenn sie zu gleichen Werten des Flächenintegrals führen, können sich jedoch beträchtlich unterscheiden. Unterteilt man die Kraft-Weg-Kurven in mehrere kennzeichnende Abschnitte, lassen sich für verschiedene Phasen des Versuches kennzeichnende Merkmale ableiten, die über das Bruchverhalten der Probe weitergehende Aussagen liefern als der Wert der insgesamt verbrauchten Schlagarbeit allein. Die Kraft-Weg-Kurve liefert keinen Kennwert für die Festigkeitsberechnung. Auch kann aus ihr nicht unmittelbar auf die tiefste zulässige Beanspruchungstemperatur eines Werkstoffs in einem Bauteil geschlossen werden.

2 Anwendungsbereich

Der Kerbschlagbiegeversuch mit Ermittlung von Kraft und Weg hat den gleichen Anwendungsbereich wie der Kerbschlagbiegeversuch nach DIN 50 115. Er wird angewendet, wenn weitergehende Aussagen über das Bruchverhalten des zu prüfenden Werkstoffes erforderlich sind, als aus den Kennwerten abgeleitet werden können, die sich im Kerbschlagbiegeversuch nach DIN 50 115 ermitteln lassen.

3 Kurzbeschreibung des Verfahrens

Der Versuch wird in der gleichen Weise durchgeführt wie der Kerbschlagbiegeversuch nach DIN 50 115. Zusätzlich wird jedoch eine Kraft-Weg-Kurve ermittelt und im Hinblick auf kennzeichnende Phasen des Verformungs- und Bruchablaufes ausgewertet.

4 Begriffe

Hinweis:
Im folgenden werden nur diejenigen Begriffe definiert, die mit der Kraft-Weg-Kurve in Zusammenhang stehen. Sie sind im **Bild 1** *anschaulich dargestellt. Im übrigen gelten die Begriffe nach DIN 50 115.*

4.1 Kennzeichnende Werte der Kraft

1) Fließkraft F_{gy} (Grenzkraft bei Eintritt der Vollplastizierung des Ligaments, „general yield")

   Die Fließkraft F_{gy} ist die Kraft an der Übergangsstelle vom linear ansteigenden Teil zum gekrümmt ansteigenden Teil der Kraft-Weg-Kurve. Sie dient in erster Näherung zur Kennzeichnung des Eintritts der Vollplastizierung des Ligaments der Probe.

2) Höchstkraft F_m

   Die Höchstkraft F_m ist die größte Kraft im Verlauf der Kraft-Weg-Kurve.

3) Kraft bei Beginn der instabilen Rißausbreitung F_u

   Die Kraft F_u ist die Kraft beim Beginn des Steilabfalls der Kraft-Weg-Kurve. Sie kennzeichnet den Beginn der instabilen Rißausbreitung.

4) Kraft am Ende der instabilen Rißausbreitung F_a

   Die Kraft F_a ist die Kraft beim Übergang vom Steilabfall der Kraft-Weg-Kurve in den allmählich auslaufenden Teil der Kraft-Weg-Kurve. Sie kennzeichnet das Ende der instabilen Rißausbreitung.

   Anmerkung:
   Die Kraft, bei der stabiles Rißwachstum eintritt, prägt sich auf der Kraft-Weg- oder Kraft-Zeit-Kurve nicht aus und kann mit den Meßeinrichtungen nach DVM-Merkblatt 0001 nicht ermittelt werden. Diese Kraft ist deshalb nicht als kennzeichnender Wert in das Prüfblatt aufgenommen worden.

4.2 Kennzeichnende Werte des Weges

Den im Abschnitt 4.1 definierten Kräften sind entsprechende Weglängen s zugeordnet, die mit den gleichen Indizes wie die zugehörigen Kräfte bezeichnet werden. Zusätzlich wird die Weglänge s_t am Ende der Kraft-Weg-Kurve definiert. Es gilt:

1) Weglänge bei Eintritt der Vollplastizierung des Ligaments s_{gy}
2) Weglänge bei Erreichen der Höchstkraft s_m
3) Weglänge bei Beginn der instabilen Rißausbreitung s_u
4) Weglänge am Ende der instabilen Rißausbreitung s_a (wegen des Steilabfalles der Kraft-Weg-Kurve zwischen F_u und F_a gilt im allgemeinen $s_u \approx s_a$)
5) Weglänge am Ende der Kraft-Weg-Kurve s_t (dabei wird als Ende der Kraft-Weg-Kurve die Weglänge bei $F = 0{,}02 \cdot F_m$ definiert)

4.3 Kennzeichnende Werte der Schlagarbeit

4.3.1 Die als Flächenintegral unter der vollständigen Kraft-Weg-Kurve, d.h. vereinbarungsgemäß (siehe Abschnitt 4.2 (5)) bis $F = 0{,}02 \cdot F_m$, ermittelte Schlagarbeit wird mit dem Kurzzeichen W_t bezeichnet.

4.3.2 Anhand der in Abschnitt 4.1 definierten Kräfte läßt sich die Kraft-Weg-Kurve in Teilabschnitte unterteilen, für die auch die entsprechenden Flächenintegrale ermittelt

Seite 2 SEP 1315

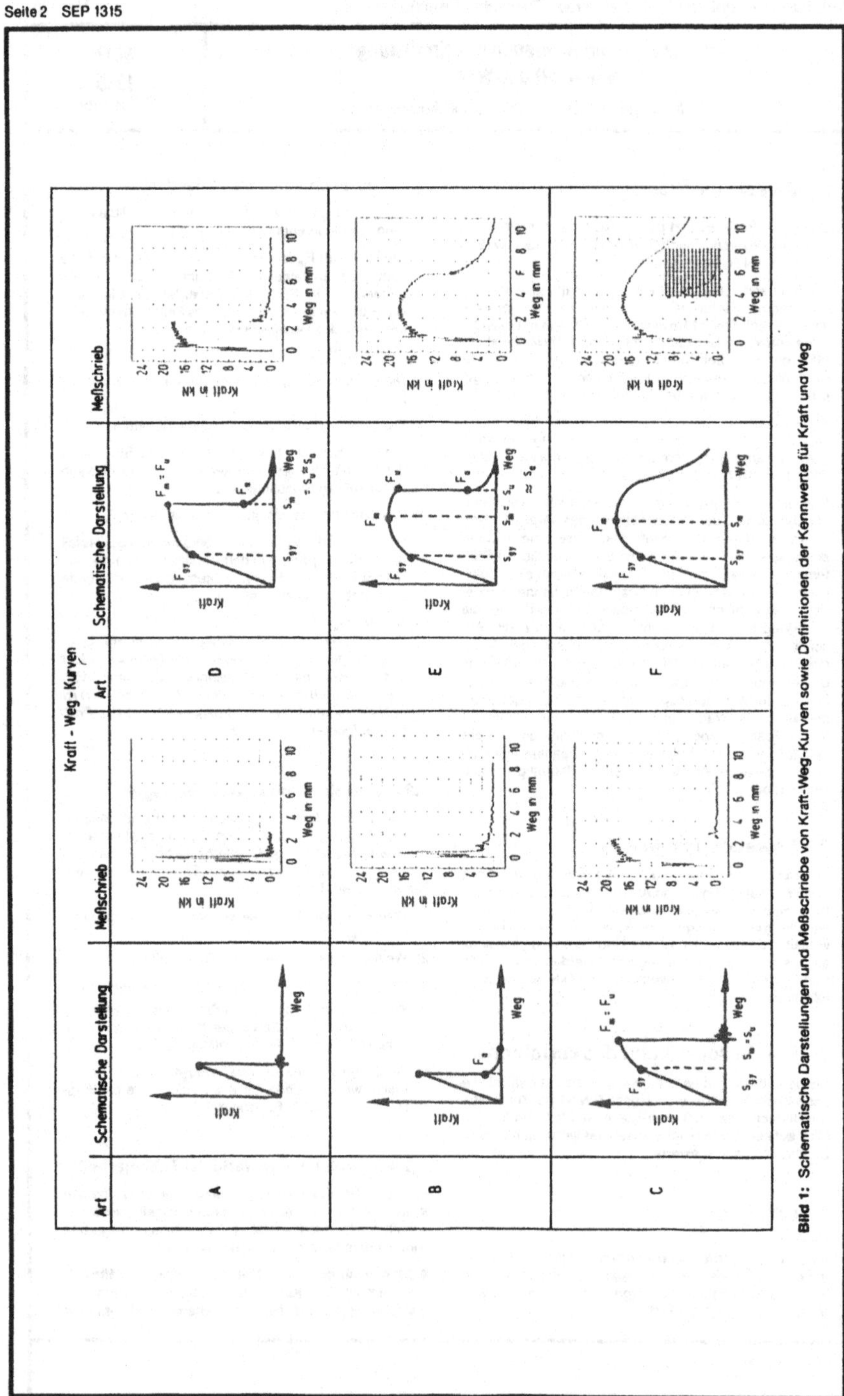

Bild 1: Schematische Darstellungen und Meßschriebe von Kraft-Weg-Kurven sowie Definitionen der Kennwerte für Kraft und Weg

werden können. Die Werte der Teilschlagarbeiten W werden mit den gleichen Indizes gekennzeichnet wie die Kräfte am Ende des jeweiligen Teilabschnittes der Kraft-Weg-Kurve. Es gilt:

1) Teilschlagarbeit W_m von $F = 0$ bis $F = F_m$
2) Teilschlagarbeit W_u von $F = 0$ bis $F = F_u$
3) Teilschlagarbeit W_a von $F = 0$ bis $F = F_a$ (wegen des Steilabfalles der Kraft-Weg-Kurve zwischen F_u und F_a gilt im allgemeinen $W_u \approx W_a$)

4.4 Arten von Kraft-Weg-Kurven

Zur Vereinfachung der Auswertung und Berichterstattung wird gemäß Bild 1 zwischen verschiedenen kennzeichnenden Arten von Kraft-Weg-Kurven unterschieden. Sie lassen sich der Kerbschlagarbeit-Temperatur-Kurve nach DIN 50 115 ungefähr wie folgt zuordnen:

Art A und B	Tieflage der Kerbschlagarbeit
Art C, D und E	Übergangsgebiet
Art F	Hochlage der Kerbschlagarbeit

Bei Kraft-Weg-Kurven der Arten D und E können stabiler und instabiler Rißfortschritt auch in mehrfacher Folge auftreten.

5 Prüfgerät

Die Prüfung wird mit Pendelschlagwerken nach DIN 51 222 durchgeführt, die zur Ermittlung von Kraft-Zeit- oder Kraft-Weg-Kurven mit einer Instrumentierung nach DVM-Merkblatt 0001 ausgerüstet sind.

6 Proben

Die Prüfung wird mit Proben nach DIN 50 115 durchgeführt.

7 Vorbereitung und Durchführung

Für die Vorbereitung und Durchführung der Prüfung gelten die Festlegungen in DIN 50 115.

8 Auswertung

Hinweis:
Im folgenden wird nur die Auswertung der Kraft-Weg-Kurve erläutert. Dabei ist zu beachten, daß das Kraft-Weg-Signal von Schwingungen überlagert ist, die zwangsweise aus der Wechselwirkung zwischen der instrumentierten Finne und der Probe herrühren. Die über Ausgleichskurven ermittelten kennzeichnenden Werte sind nach vorliegenden Erfahrungen reproduzierbar.
Im übrigen gelten die Festlegungen in DIN 50 115.

8.1 Ermittlung der Art der Kraft-Weg-Kurve

Durch Vergleich mit den schematischen Darstellungen in Bild 1 wird die Art der Kraft-Weg-Kurve ermittelt.

Voraussetzung für die weitergehende Auswertung der Kraft-Weg-Kurve ist das Auftreten einer ausgeprägten Fließkraft F_{gy}. Kraft-Weg-Kurven der Arten A und B sind nicht auswertbar.

8.2 Ermittlung der kennzeichnenden Werte der Kraft

8.2.1 Die Kraft F_{gy} wird ermittelt als Kraft im Schnittpunkt des Steilanstiegs der Kraft-Weg-Kurve und der Ausgleichskurve durch die Schwingungen der Kraft-Weg-Kurve nach Eintritt der Vollplastizierung des Ligaments (Bild 2).

8.2.2 Die Kraft F_m wird ermittelt als die Kraft im höchsten Punkt der Ausgleichskurve (Bild 2).

8.2.3 Die Kraft F_u wird ermittelt als Kraft im Schnittpunkt der Ausgleichskurve und des Steilabfalles der Kraft-Weg-Kurve (Bild 2). Tritt der Steilabfall bei Erreichen der Höchstkraft ein, so ist $F_u = F_m$ (Kraft-Weg-Kurven der Arten C oder D).

8.2.4 Die Kraft F_a wird ermittelt (Bild 2) als Kraft im Schnittpunkt des Steilabfalles der Kraft-Weg-Kurve und der Ausgleichskurve durch die Schwingungen des sich anschließenden Teiles der Kraft-Weg-Kurve (Kraft-Weg-Kurven der Arten D oder E).

Wird der sich instabil ausbreitende Riß nicht aufgefangen, so ist $F_a = 0$ (Kraft-Weg-Kurve der Art C).

8.3 Ermittlung der kennzeichnenden Werte des Weges

Die kennzeichnenden Werte des Weges nach Abschnitt 4.2 werden im Kraft-Weg-Schaubild als Abszissenwerte der nach Abschnitt 8.2 ermittelten kennzeichnenden Werte der Kraft ermittelt (siehe Bild 1).

Die Weglänge s_{gy} läßt sich bei Verwendung üblicher Meßeinrichtungen nur ungenau ermitteln. Infolgedessen wird im allgemeinen auf die Angabe von s_{gy} verzichtet.

Die Weglänge s_t wird nur ermittelt, wenn die Probe im Versuch vollständig gebrochen wurde und die Kraft-Weg-Kurve bis zum Bruch der Probe vorliegt. In diesem Falle nähert sich die Ausgleichskurve durch die Schwingungen der Kraft-Weg-Kurve asymptotisch dem Wert $F = 0$. Die Weglänge s_t wird angegeben als Abszissenwert der Ausgleichskurve im Punkt $F = 0{,}02 \cdot F_m$.

8.4 Ermittlung der kennzeichnenden Werte der Schlagarbeit

Die kennzeichnenden Werte der Schlagarbeit nach Abschnitt 4.3 werden ermittelt durch Bestimmen der Flächen unter den zugehörigen Teilabschnitten der Kraft-Weg-Kurve.

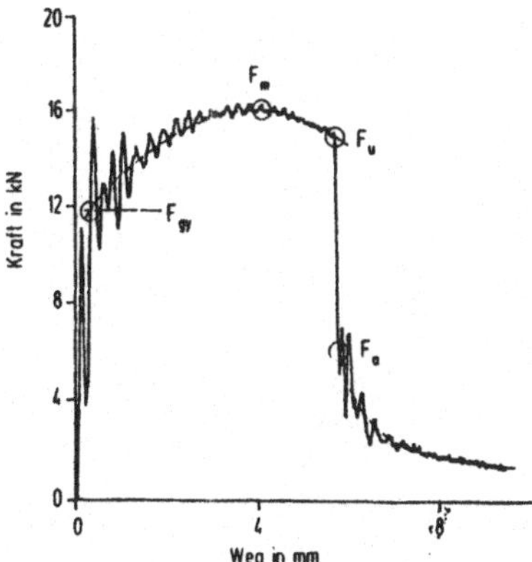

Bild 2: Ermittlung der kennzeichnenden Werte der Kraft nach Abschnitt 8.2

Seite 4 SEP 1315

8.5 Ermittlung des Anteils an matter Bruchfläche

Tritt im Verlauf der Kraft-Weg- oder Kraft-Zeit-Kurve kein Steilabfall der Kraft auf (Kurven der Art F nach Bild 1), darf dies als Hinweis dafür gewertet werden, daß der Anteil an matter Bruchfläche 100% der gesamten Bruchfläche beträgt. Tritt ein Steilabfall der Kraft auf, gibt der Betrag des Steilabfalles $F_u - F_a$, bezogen auf die größte Kraft F_m, einen Anhaltswert für den Anteil an matter Bruchfläche. Erfahrungsgemäß gilt

$$\text{Anteil an matter Bruchfläche} = \left(1 - \frac{F_u - F_a}{F_m}\right) \cdot 100\%.$$

9 Prüfbericht

Im Prüfbericht unter Hinweis auf dieses Prüfblatt sind anzugeben:

a) Werkstoff und Werkstoffzustand;

b) Probenlage;

c) Prüftemperatur in °C;

d) Arbeitsvermögen der Prüfeinrichtung;

e) Kerbschlagarbeit A_v nach DIN 50 115 mit Angabe des Kurzzeichens der Probenform in J;
bei einer Probe, die durch die Widerlager gezogen wurde, ohne zu brechen, wird das Prüfergebnis eingeklammert und die Probe somit als „nicht durchgebrochen" gekennzeichnet;

f) Art der Kraft-Weg-Kurve nach Abschnitt 4.4; falls die Kraft-Weg-Kurve den Arten A bis F nach Bild 1 nicht zugeordnet werden kann, ist sie dem Prüfbericht als Schaubild beizufügen;

g) die nach Abschnitt 8 ermittelten kennzeichnenden Werte der Kraft, des Weges und der Schlagarbeit, soweit die Kraft-Weg-Kurven auswertbar sind;

h) gegebenenfalls Schätzwert nach Abschnitt 8.5 für den Anteil an matter Bruchfläche.

Zitierte Normen und andere Unterlagen

DIN 50 115	Prüfung metallischer Werkstoffe; Kerbschlagbiegeversuch
DIN 51 222	Pendelschlagwerke
DVM-Merkblatt 0001	Meßtechnische Anforderungen beim instrumentierten Kerbschlagbiegeversuch (Bezug durch: Deutscher Verband für Materialprüfung e.V. (DVM), Unter den Eichen 87, 1000 Berlin 45)

APPENDIX III

ITEM — 1

ASTM E24.03.03
April 1980

Draft 2C

PROPOSED
STANDARD METHOD OF TEST FOR
INSTRUMENTED IMPACT TESTING OF PRECRACKED CHARPY SPECIMENS
OF METALLIC MATERIALS

TABLE OF CONTENTS

1. SCOPE

1.1 - This method describes fatigue precracked Charpy specimen impact testing of metallic materials by an instrumented (dynamic load monitoring) apparatus. It gives (a) a description of apparatus, (b) requirements for inspection and calibration, (c) dimensions and preparation of specimens, (d) testing procedures, and (e) calculations and interpretation of results.

1.1.1 - The nominal dimensions of the specimen are those given in E-23-72 (Figure 4, Charpy Type A). These are 10 mm (.394 in) thick by 10 mm (.394 in) deep by 55 mm (2.165 in) long. The machined notch is extended a minimum of 2.5 mm (.100 in) by fatigue cycling such that the total depth of the notch plus crack is between 4.5 mm (.177 in) and 5.5 mm (.217 in).

1.1.2 - the specimen is dynamically loaded in three point bending with a pendulum or falling weight machine.

1.1.3 - This method covers the determination of dynamic notch toughness $[K_{PCI}(t_f)]$ at nominal stress intensity rates, $\dot{K}$, of 5×10^4 to 5×10^5 MPa-$m^{1/2}$/s (5×10^4 to 5×10^5 ksi $\sqrt{in}$/s). Typical time to maximum load (t_f) values are 1.5×10^{-4} to 3.0×10^{-4}s.

2. APPLICABLE DOCUMENTS

2.1 ASTM Standards

E4	Verification of Testing Machines
E8	Tension Testing of Metallic Materials
E23	Notch Bar Impact Testing of Metallic Materials
E399	Plane-Strain Fracture Toughness of Metallic Materials
E602	Tentative Method for Sharp Notch Tension testing with Cylindrical Specimens
E604	Dynamic Tear Energy of Metallic Materials

3. SUMMARY OF METHOD

3.1 - The method involves three point bend testing of fatigue precracked specimens, impact loaded by a moving mass with a known kinetic energy great enough to break the specimen.

3.1.1 - Fatigue precracking of the specimen follows the general guidelines of Method E399.

3.1.2 - Dynamic notch toughness determinations require recording of the midspan load from the time at which impact occurs through fracture and subsequent unloading. If fracture occurs before general yielding, the maximum load is used to determine the dynamic notch toughness $[K_{PCI}(t_f)]$ from this value and the original dimensions of the specimen using the bend relationships of Method E399. The t_f is a time value that references the relative rate of loading. Acceptable $K_{PCI}(t_f)$ values can not be determined for fractures that occur after general yielding (i.e., the onset of elastic-plastic specimen deformation).

4. SIGNIFICANCE AND USE

4.1 - These toughness values can be of service for the following purposes but not necessarily limited to these examples.

4.1.1 - In research and development of materials, to study the effects of such variables as composition, heat treatment, mechanical processing, environmental degradation, etc., where a ranking of materials in terms of their dynamic notch toughness $[K_{PCI}(t_f)]$ may be useful.

4.1.2 - For specifications of acceptance and manufacturing quality control and for service evaluation to compare the resistance to dynamic fracture.

4.1.3 - To establish the variation in $K_{PCI}(t_f)$ with testing temperature and in particular to determine ranges of temperature in which substantial changes in the $K_{PCI}(t_f)$ occur (transition temperature).

2

5. TERMINOLOGY

Terminology E616 for Fracture Testing is applicable to this method.

5.1 - Definitions

5.1.1 - Stress-Intensity Factor, K_1, K_2, K_3, $(FL^{-3/2})$ - the magnitude of the ideal-crack-tip stress field for a particular mode in a homogeneous, linear-elastic body.

Note 1 - Values of K for the modes 1, 2, and 3 are given by:

$$K_1 = \lim_{r \to 0} [\sigma_y \sqrt{2\pi r}],$$

$$K_2 = \lim_{r \to 0} [\tau_{xy} \sqrt{2\pi r}], \text{ and}$$

$$K_3 = \lim_{r \to 0} [\tau_{yz} \sqrt{2\pi r}]$$

where r = a distance directly forward from the crack tip to a location where the significant stress is calculated.

Note 2 - In this method, mode 1 is assumed.

5.1.2 - Dynamic Notch Toughness, $K_{PCI}(t_f)$ $(FL^{-3/2}(s))$ - a measure of toughness of a material measured by the operational procedure specified in this test method.

Note 3 - K_{PCI} is computed in terms of a stress-intensity factor K. Dynamic notch toughness values by this method are reported as $K_{PCI}(t_f)$. In this method, determinations of $K_{PCI}(t_f)$ are restricted to fractures that occur at an applied load that is less than that required for general yielding and at a specified loading rate. For acceptable loading rates, the ratio of K_{PCI} to the time to maximum load (t_f) is between 5×10^4 and 5×10^5 MPa-m$^{1/2}$/s (5×10^4 to 5×10^5 ksi-in$^{1/2}$/s). Typical time to maximum load (t_f) values are 1.5×10^{-4} to 3.0×10^{-4}s.

Note 4 - The interpretation of $K_{PCI}(t_f)$ values determined by this method is that the fracture initiates and rapidly propagates before either general yielding or significant stable crack growth, either which normally can be identified by a definitive departure from linearity of the load-time record.

3

5.2 - Description of Terms Specific to This Method

5.2.1 - Time to Maximum load, t_f [s] — the nominal time to maximum load, computed as the difference between time at maximum load, and time at impact which is indicated by the initial rapid rise in load.

5.2.2 - Stress-Intensity Rate, $\dot{K}$ [$FL^{-3/2}/s$] — the nominal rate at which the stress-intensity factor (K) is increased up to maximum load.

Note 5- In this method $\dot{K}$, *is the ratio of* K_{PCI} *to the time to maximum load* (t_f).

5.3 - Symbols and Abbreviations

The following list of symbols used for terms found in this method are believed to be generally applicable to fracture testing. Symbols for other terms used in this method may be found where the term is first introduced.

a	crack size
B	specimen thickness
c_o	longitudinal bar-wave speed in the specimen material
C_{LL}	specimen load-line compliance
E	Young's modulus
U	energy
f	frequency
K	stress-intensity factor
$K_{PCI}(t_f)$	dynamic notch toughness from precracked charpy impact testing referenced to the time to maximum load
P_m	maximum load
σ	stress
σ_{YS}	0.2% offset tensile yield strength
σ_{YS1}	σ_{YS} at fatigue precracking temperature
σ_{YS2}	σ_{YS} at test temperature
t	time
T	temperature
v	velocity
W	specimen width, beam depth

4

6. APPARATUS*

6.1 - General Requirements - The procedure involves impact testing of notched specimens that have been fatigue precracked. Load versus time is recorded autographically. Total absorbed energy is determined from the difference between the kinetic energy of the hammer at the moment of impact and the final kinetic energy of the hammer upon completion of the fracture. Testing may be done in various machines (pendulum or drop weight) having suitable load sensing devices, recording instrumentation and sufficient energy to completely fracture the specimen in a single blow. The apparatus requirements can be separated into those for the machine or mechanical aspects and those for the instrumentation of the machine.

6.1.1 - Mechanical - The mechanical requirements are those for the actual loading to failure of the specimen. The impact test apparatus can be either a pendulum machine, as for the Charpy test (Method E23), or a drop tower, as for the 5/8-in (15.9 mm) DT tests (Method E604).

6.1.1.1 - Pendulum - Method E23 is the single best source of the pertinent information for proper operation of a pendulum impact test. For this method tests must be performed in strict accord with the requirements of Method E23 except as modified for acquisition of the load-time data and certain velocity reductions. The velocity requirements of the pendulum apparatus stated in paragraph 4.1.2 must be modified to accommodate the need for performance of instrumented impact tests of fatigue precracked specimens at various preselected velocities. The revision of Paragraph 4.1.2 of Method E23 includes the following additions:

6.1.1.1.1 - For $K_{PCI}(t_f)$ determination, tests may be made at various velocities (normally between 1.2 and 4 m/s (4 and 13.1 ft/s)), but the associated energies should be sufficient to completely break the specimen (see Section 6.1.3.1). Velocity shall always be stated as the maximum velocity of the striking member at the center of strike (see Figure 1, of Method E23).

*The apparatus for all dynamic tests should include proper safety devices and instructions for adequate protection of operators and observers.

5

6.1.1.2 - Drop Tower - For this type of testing, it is recommended that the crosshead (hammer) be guided on smooth columns. The maximum transverse slack should be less than 1 mm (.04 in) in any direction perpendicular to the center line of the vertical path.

The base plate, to which the specimen support fixture is attached, should be securely fastened to a rigid foundation having mass greater than or equal to ten times that of the falling crosshead. The requirements for maximum available kinetic energy of the hammer assembly are discussed in Section 6.1.3.1.

6.1.2 - Instrumentation

6.1.2.1 - General Requirements - The minimum instrumentation required for instrumented impact testing shall be a transducer to sense the specimen loading, the associated signal conditioning equipment, and autographic recording equipment suitable for providing a permanent record of the load sensing transducer output versus time or displacement.

The load-sensing transducer, associated signal conditioning equipment, and autographic recorder combination selected for the test shall allow determination of the maximum load to within ±3 percent as verified periodically in accordance with the calibration procedures of Section 6.1.3 of this method.

The autographic load recording apparatus shall be capable of recording the load-sensing transducer output during the time immediately preceding the test, in addition to the time period of specimen loading during the actual test. The purpose of this requirement is to define an actual zero load trace and to assure recognition of undesirable dynamic effects on the load-sensing transducer during the pendulum approach to the specimen. A means shall be provided for verification of signal amplification (e.g., a shunt calibration resistor). The equipment shall include a device capable of verification of the initial impact velocity of the hammer.

In addition to the above miminum requirements, the instrumented impact testing equipment may include a load-integrating device capable of superimposing upon the autographic load plot a record of the impulse, interpreted as energy absorbed during the test, as a function of time or displacement.

6.1.2.2 - Load Transducer - The load transducer for the impact machine is identified as an instrumented tup. The tup is the part of the hammer or crosshead which strikes the specimen and must conform to the general mechanical/geometrical requirements shown in Figure 1 of Method E23. It is recommended that semiconductor strain gages be placed on the tup in a protected recess to sense the compressive loading during impact with the test specimen. The gages should be positioned symmetrically on both sides of the tup, within 25 mm (1 in) of the point of contact with the specimen. It is recommended that a simple Wheatstone bridge circuit be employed for the strain gages. The inactive resistance portion of the strain gage circuit can be resistors.

6.1.2.3 - Dynamic Response System - The Dynamic Response System (DRS) is that portion of the instrumentation system which serves as the interface between the load transducer and the signal display system. The DRS includes the following basic functions for the strain gage bridge circuit of the instrumented tup:

1. Excitation power supply
2. Balance control
3. Output signal amplification

The power supply must be regulated constant voltage control. The circuit should also include a potentiometer for balancing the strain gage bridge. A means for verification of signal amplification shall be provided. It is also recommended that this amplifier be of the differential type and have electrical characteristics like that of a simple resistor capacitor circuit.

It is possible for the DRS to have improper frequency response, which results in the displayed signal being an unknown distortion of the actual signal transmitted from the tup. Reference (1) describes in detail the effect of limiting frequency response. This limitation shall be stated in terms of the response time T_R, (see Section 6.1.3.5) of the equipment and specific requirements for T_R are stated in Section 9.1.

6.1.3 - Inspection and Calibration - The mechanical and instrumentation portions of the apparatus must be carefully inspected and calibrated. The inspection and calibration of the instrumentation for instrumented impact testing is in the form of proof and verification tests. That is, the performance of each component of the instrumentation is evaluated by the general effect on a specific dynamic signal which is passed from the instrumented tup circuit through to the final display of that signal.

6.1.3.1 - Available Energy - The available energy requirements for $K_{PCI}(t_f)$ testing are based on minimizing the reduction in hammer velocity during the period required to reach maximum load. The energy absorbed by the specimen at maximum load is identified as U_I and to comply with this requirement for minimizing velocity reduction, the total available energy at impact (U_o) shall be

$$U_o \geq 3\, U_I$$

where

$$U_o = \frac{1}{2} m\, v_o^2 \quad ,$$

where v_o is the velocity at impact, m is the effective mass of the pendulum or hammer-tup assembly given by $m = w/g$, where g is the acceleration due to gravity, and w is the effective weight of the pendulum (determined by the procedure in paragraph 5.4.2.2 of Method E23) or hammer-tup assembly. In addition, U_o must be greater than the energy required to completely fracture the specimen. Requirements for low impact velocities (see Section 9.1) must be considered in light of the above available energy restriction

8

6.1.3.2 - Velocity - For instrumented impact testing, the impact velocity shall be known. The velocity, v_o, can be directly measured with an accuracy of ± 2 percent by use of photo-electric devices.

6.1.3.3 - Proof Tests - The proof testing shall be employed to check both the mechanical and instrumentation portions of the apparatus. The single best technique for checking impact machines for compliance with the requirements of Method E23 for proper mechanical operation is the proof testing described in paragraph 6.1 of that method. Standard calibration specimens must be obtained from the Army Materials and Mechanics Research Center (AMMRC).

6.1.3.4 - Load Transducer - The load transducer (instrumented tup) is a strain gage device which is probably the most vulnerable component in the instrumentation system to defective operation. Check of the operability of the transducer may be obtained by pressing against the edge of the tup, or lightly bumping the tup against a hardened bar resting against the anvils in the test position, and observing the displayed signal. This check demonstrates the ability of the strain gage circuit to generate a signal when subjected to strains in the tup.

The calibration of the load transducer requires dynamic tests to be performed and the results equated to the known reference values for those particular specimens. The reference variables are the known maximum load for strain rate insensitive materials and/or the known total absorbed energy for standard material. Suggested Maximum load material is aluminum alloy 6061-T651 plate. Standard energy material is the AMMRC calibration specimens.

The specimens employed for the above maximum load calibration procedure should be standard Charpy V-notch specimens from 6061-T651 aluminum plate stock. At least three specimens from this same stock and orientation should be tested by slow-bend testing in a conventional, universal-type static testing machine using a static load cell, which has been calibrated in accordance with the latest revision of Method E4. The maximum load for each test should be with ±3 percent of the average

9

value, which is then identified as P_m for this relatively strain rate insensitive material. The loading geometry and dimensions for these slow-bend test shall be carefully matched to those employed for the impact machine (see Method E23). When these specimens are tested in impact, the allowable tolerance on the average P_m of three tests is ± 3 percent of the average static value.

The samples employed for the total absorbed energy calibration procedure should be standard AMMRC Charpy specimens. The average energy value obtained from the instrumentation shall agree within 1.4J (1.0 ft-lb) or 5.0 percent, whichever is greater, of the nominal known values.

6.1.3.5 - Dynamic Response System - The inspection and calibration of the Dynamic Response System (DRS) is primarily concerned with the frequency respouse or response time, T_R, characteristics of the unit.

The response time, T_R, of a component has been defined [(1)] as the apparent rise time, t_r, of a signal which has been attenuated 10 percent. T_R must be determined experimentally by superimposing a sine wave on the output of the strain gage bridge circuit. The peak-to-peak amplitude of the signal must be equivalent to approximately one-half the full-scale capacity of the load transducer (e.g., 26.7 kN (6000 lb) for a typical Charpy tup). The frequency of the sine wave is then varied until the value for which the amplitude is attenuated 10 percent is found, and the response time is calculated from

$$T_R = \frac{0.35}{f_{.9db}} ,$$

where $f_{.9db}$ is the frequency at a .9db attenuation (10 percent voltage attenuation).

The response time (T_R) must be known for the specific DRS characteristics to be employed. Response time (T_R) must be determined for the entire system employed for data acquisition and display.

6.1.3.6 - Trigger - The trigger component is primarily used to supply an external signal to the display component just before the tup strikes the specimen. This component may also be designed to produce a signal which is useful for measurement of the maximum tangential velocity of the tup.

Systems which employ transient signal recording techniques may utilize internal triggering from the load signal and therefore, do not require a trigger component. However, care should be taken to ensure that the entire load-time record is recorded.

6.1.3.7 - Display - the display component of the instrumentation must also be subjected to the same inspection and calibration procedures for frequency response as was the Dynamic Response System. The single most important parameter to be inspected and calibrated for the display component is the unit controlling the sweep rate of this component. The unit shall be calibrated for an accuracy of 2 percent.

7. TEST SPECIMENS, DIMENSIONS AND PREPARATION

7.1 - Dimensions - The precracked Charpy type specimen is shown in Figure 1. The dimensions W, B, L and a are width, thickness, length and crack length, respectively. Width must be 10 mm (.394 in) to ensure proper alignment of tup contact surface at impact for pendulum machines. The length should also comply with the requirements of Method E23 and be 55 mm (2.165 in) maximum and 52.5 mm (2.067 in) minimum. The crack length is the sum of the machine starter notch and fatigue crack lengths. The crack length is normally equal to one-half of the width. However, for this method acceptable values of a are 0.45 to 0.55 W. Dimensional tolerances are as shown in the figure.

7.1.1 - The crack starter shall consist of a machined V-notch as for E23. The following variations are acceptable, see Figure 1. The depth of the machined notch should be no greater than 0.2 W. The radius at the bottom of the machined V-notch should be no greater than .25 mm (.010 in). A radius of .08 mm (.003 in) is obtainable by normal machine practice and is recommended.

7.2 - Fatigue Precracking - In order to ensure a sufficiently sharp fatigue crack and to facilitate the production of this crack, the following procedure shall be followed.

7.2.1 - The fatigue cracking shall be conducted with the specimen in the condition of heat treatment in which it is to be tested.

7.2.2 - The method used for fatigue cracking shall be such that the crack extension direction is perpendicular to the notched face of the specimen. Cantilever bending or three point bending may be used to produce fatigue cracks. Cantilever bending facilitates completely reversed loading which aids in starting and propagating the crack. For practical purposes, the K calibration for three point bending may be used to estimate the stress intensity factors.

Note 7 - The expression given in Method E399 for the K calibration in three point bending may be rewritten in terms of the bending moment, M, as follows and used to compute the stress intensity in cantilever bending: $K = (4M/BW^{3/2})\, f(a/W)$ *where the values of the function f(a/W) given in tabular form in Method E399 and listed below.*

a/W	f(a/W)	a/W	f(a/W)
0.450	2.29	0.500	2.66
0.455	2.32	0.505	2.70
0.460	2.35	0.510	2.75
0.465	2.39	0.515	2.79
0.470	2.43	0.520	2.84
0.475	2.46	0.525	2.89
0.480	2.50	0.530	2.94
0.485	2.54	0.535	2.99
0.490	2.58	0.540	3.04
0.495	2.62	0.545	3.09
		0.550	3.14

Note 8 - The stress distribution at the crack tip in cantilever bending can be influenced by excessive clamping forces to the extent that the fatigue crack will deviate from a plane normal to the notched face of the specimen.

12

7.2.3 - The fatigue crack shall extend at least 2.5 mm (.100 in) beyond the tip of the notch starter. During the last .5 mm (.020 in) of fatigue cracking, the ratio of the maximum stress intensity of the fatigue cycle to Young's modulus ($K_f(max)/E$) shall be in the range between 0.00013 and 0.00029 $m^{1/2}$ (0.0008 and 0.0018 $in^{1/2}$).

7.2.4 - The stress intensity range should not be less than 0.9 $K_f(max)$.

7.2.5 - When the fatigue cracking is conducted at a temperature, T_1, and then tested at a different temperature, T_2, where $T_1 < T_2$; the $K_f(max)$ of the fatigue cycle should be that value determined in 7.2.3 above multiplied by $\sigma_{YS1}/\sigma_{YS2}$, where σ_{YS1} is the tensile yield strength at temperature T_1 and σ_{YS2} is the tensile yield strength at temperature T_2. Where $T_1 > T_2$, proceed as in 7.2.3.

Note 9 - In all cases $K_f(max)$ *must be less than the subsequent measured value of* $K_{PCI}(t_f)$.

8. PROCEDURES

8.1 - Specimen Measurement - All specimen dimensions shall be within the tolerance shown in Figure 1.

8.1.1 - Measure the thickness, B, to the nearest .025 mm (.001 in) at not less than two positions between the fatigue crack tip and the unnotched edge of the specimen and record the average value. Measure the depth, W, to the nearest .025 mm (.001 in) at two positions, one on each side of the crack starter notch and record the average value.

8.1.2 - After fracture measure the initial fatigue crack length, a, to the nearest 0.5 percent at the following three positions: at the center of the crack front and midway between the center and the intersection of the crack front with the specimen surfaces. Use the average of these three measurements as the crack length in the calculation of $K_{PCI}(t_f)$ (see Section 9). If the difference between any two of the crack length measurements exceeds 5 percent of the average, or if part of the crack front is closer to the machined notch root than 5 percent of the average, the specimen should be discarded. Also, if the length of either surface

13

trace of the crack is less than 80 percent of the average crack length, as defined above, the specimen should be discarded. In addition, the crack plane shall be parallel to both the specimen width and thickness directions within $\pm$ 10 degrees.

8.2 - Specimen Support - specimen supports and tup (striker) shall conform to the requirements of Method E23.

8.2.1 - It is recommended that self-centering tongs similar to those suggested in Method E23 be used for placing the specimen on the anvil supports of the impact machine. The alignment of specimen on the anvil and with respect to the tup shall comply with the requirements of Method E23.

8.3 - Selection of Impact Velocity - The speed of testing for impact tests is defined as the maximum velocity of the center of strike immediately prior to impact with the specimen. There are four major reasons for selection of various speeds of testing, as follows:

1. Subject the material to a specific nominal stress-intensity rate, $\dot{K}$.
2. Provide sufficient velocity, such that, the attendant energy is compatible with the available energy requirements of the apparatus stated in Section 6.1.3.1.
3. Ensure the time of fracture (t_f) will be sufficiently greater than the limiting response time (T_R) of the instrumentation. The apparatus requirements for T_R are stated in Section 6.1.3.5. Section 9 discusses t_f and other interpretations of the load signal portion of the test record.
4. Minimize the relative effects of inertial loading[1] on the load signal through reduction of velocity. These effects are interrelated with T_R and t_f, and are regulated by the validity requirements of Section 9.

For this method the initial impact velocity for instrumented impact tests shall be in the range from approximately 1.2 to 4.0 m/s (4.0 to 13.1 ft/s). The specific speed selection shall depend on the above four reasons and the related apparatus requirements (Section 6.)

14

8.4 - Temperature of Testing - Requirements for achieving and controlling specimen temperatures shall conform to the procedures of Method E23. Requirements for temperature monitoring and recording apparatus for the instrumented impact test are those of Method E8.

8.5 - Test Record - $K_{PCI}(t_f)$ testing requires a record consisting of an autographic plot of the instrumented tup load output versus time. Acceptable test records can be obtained as photographs of oscilloscope CRT traces or from plotter type records produced by the secondary transmittal from transient signal recorders or similar computer aided devices. The entire apparatus employed for producing these test records must comply with the requirements of Section 6.1.3. The initial slope of the linear portion of the load-time record shall be between 1.0 and 3.0.

9. CALCULATION AND INTERPRETATION OF RESULTS

9.1 - Load Signal - A schematic diagram of an idealized display record of the load-time signal for a linear elastic fracture of a notched three-point bend specimen is shown in Figure 2. The load signal derived from an instrumented tup, in an impact test, oscillates around the actual load required to deform the specimen.[1,2] The amplitude of the oscillations decreases with time and the requirements for acceptable load data in this method are based upon assurance of sufficient dampening to provide confidence in the indicated fracture load values. These fluctuations of the tup signal are known as inertial oscillations.

9.1.1 - It has been shown that the period of the inertial oscillations can be predicted by the following empirical expressions:[2]

$$\tau = 1.68\ (SWEBC_{LL})^{1/2}/c_o$$

where S is the support span, c_o is the longitudinal bar-wave speed of sound in the specimen, E is Young's modulus, and C_{LL} is the specimen compliance or deflection per unit applied load. For tests of aluminum, titanium or steel alloys with W = 10 mm (.394 in) and S = 4W, it can be shown that τ is adequately predicted by

$$\tau = 7 \times 10^{-6}\ (EBC_{LL})^{1/2},$$

15

where EBC_{LL} is a non-dimensional specimen compliance determined by[3]

$$EBC_{LL} = 72[g(a/w)] + 20 ,$$

where

$$g(a/W) = 1.86(a/W)^2 - 3.95(a/W)^3 + 16.38(a/W)^4$$
$$- 37.23(a/W)^5 + 77.55(a/W)^6 - 126.87(a/W)^7$$
$$+ 175.53(a/W)^8 - 143.96(a/W)^9 + 66.56(a/W)^{10} .$$

9.1.2 - Acceptable load values for P_m are obtained when the instrumentation requirement for frequency response is satisfied and the inertial oscillations have been sufficiently dampened. These two conditions can be stated as follows in terms of the time value t_f obtained from the test records (see Figures 2 and 3):[1,2,4]

(INERTIA) $t_f \geq 3\tau ,$

(FREQUENCY RESPONCE) $t_f \geq 1.1T_R ,$

where T_R is the system response time defined in Section 6.1.3.5 and t_f is the time to the point on the test record for which the load value is to be used to determine $K_{PCI}(t_f)$.

Typical values of the minimum time (3τ) for which acceptable load values can be obtained from the test record of specimens of various crack depth ratios (a/W) are as follows:

a/W	3τ (10^{-3} s)
0.450 to 0.480	0.15
0.485 to 0.510	0.16
0.515 to 0.540	0.17
0.545 to 0.550	0.18

16

9.1.3 - To ensure compliance with the restrictions for acceptable load values, the single best method is selection of initial impact velocity (see Section 8.3). It is generally good practice to use the lowest applicable velocity.

9.2 - General Yielding Criterion - To determine if fracture occurred before general yielding, compare the ratios of load to deflection at maximum load and approximately one-half maximum load (see Figure 3). The two ratios must be within ten percent to be assured fracture occurred before general yielding.

When deflection data is not directly available, the deflection values can be determined as follows:

$$d = \bar{v}\, t ,$$

where t is time, $\bar{v}$ is average velocity and can be computed by

$$\bar{v} = v_o \left(1 - \frac{U_a}{4U_o}\right) ,$$

where U_a, v_o and U_o are defined in Section 9.5.

9.3 - Dynamic Notch Toughness - When fracture occurs before general yielding, see Figure 2, the dynamic notch toughness for elastic loading K_{PCI} can be calculated from the measured value of P_m and the initial specimen dimensions as is specified by the bend specimen relationship for S = 4W in Method E399. The resultant toughness values are reported as $K_{PCI}(t_f)$, where the time value t_f in parenthesis references the relative testing rate in units of seconds. t_f is defined in 5.2.1 and illustrated in Figures 2 and 3. The relationship for K_{PCI} in units of MPa-m$^{1/2}$ (psi-in$^{1/2}$) is as follows:

$$K_{PCI} = (P_m\, S/BW^{3/2})\, [2.9(a/W)^{1/2} - 4.6(a/W)^{3/2}$$
$$+ 21.8(a/W)^{5/2} - 37.6(a/W)^{7/2} + 38.7(a/W)^{9/2}] .$$

17

Note 10- This expression is specifically for a/W in the range 0.45 *to* 0.55.

where

P_m = maximum load, kN (lbf)
B = specimen thickness, cm (in)
S = support span, cm (in)
W = specimen depth, cm (in)
a = crack length as determined in 7.1, cm (in)

9.3.1 - For the Charpy specimen geometry, where B equals W and S equals 4W in units of inches, the above expression for K_{PCI} in units of psi-in$^{1/2}$ reduces to

$$K_{PCI} = P_m\, h(a/W)$$

where

$$h(a/W) = 46.9(a/W)^{1/2} - 74.4(a/W)^{3/2} + 352.6(a/W)^{5/2} - 608.2(a/W)^{7/2} + 625.0(a/W)^{9/2} .$$

18

To facilitate calculation of K_{PCI}, values of the power series given in the above expression are tabulated as follows for specific values of a/W:

a/W	h(a/W)	a/W	h(a/W)
0.450	36.86	0.520	45.92
0.455	37.41	0.525	46.70
0.460	39.98	0.530	47.49
0.465	38.55	0.530	48.31
0.470	39.15	0.535	49.14
0.475	39.75	0.540	50.00
0.480	40.37	0.550	50.88
0.485	41.01	0.555	51.78
0.490	41.66		
0.495	42.33		
0.500	4 301		
0.505	43.71		
0.510	44.43		
0.515	45.17		

9.4 - Loading Rate - The relative rate of loading and test temperature shall be used to reference the resultant value of $K_{PCI}(t_f)$. Calculate relative stress intensity rate, $\dot{K}$, as follows:

$$\dot{K} = K_{PCI} / t_f ,$$

where t_f is the time between impact and maximum load (P_m) which is used to compute $K_{PCI}(t_f)$. For this method, $\dot{K}$ values in the range of approximately 5×10^4 to 5×10^5 MPa-m$^{1/2}$/s ($5 \times 10^4 \times 5 \times 10^5$ ksi- in$^{1/2}$/s) can be expected. Typical time to maximum load (t_f) values are 1.5×10^{-4} to 3.0×10^{-4} s.

9.5 - Absorbed Energy - The energy absorbed (U) at any time (t) during the impact event can be computed from the instrumented tup load-time record as follows[(1)]

$$U = U_a \left(1 - \frac{U_a}{4U_o}\right) ,$$

19

where U_o is the total available energy at impact (see Section 6.1.3.1) and U_a is determined as follows from the area under the load-time record:

$$U_a = v_o \int_o^t Pdt \quad .$$

where v_o is the initial impact velocity and Pdt is the incremental impulse or area under the load-time record. The total absorbed energy U_T is determined by the above expression for U when $t = t_n$, (see Figures 2 and 3). The energy required to reach maximum load (U_I) is estimated by the above U computation for $t = t_f$ (see Figures 2 and 3).

9.6 - Crack Plane Orientation - The toughness of a material often depends on the orientation and direction of propagation of the crack in relation to the anisotropy of the material, which in turn depends on the principal directions of mechanical working or grain flow. Reference should be made to Method E399 for a description of the system used to designate crack plane orientation and crack propagation directions for precracked bend specimens.

9.7 - Fracture Appearance - The appearance of the fracture may furnish valuable supplementary information. A means for describing the fracture appearance of cracked bend specimens is given in Method E399 and may be used to characterize the fracture appearance of the precracked Charpy specimens.

10. REPORT

10.1 The report shall include the following for each specimen tested.

10.1.1 All specimen dimensions listed in 7.1.

10.1.2 The maximum load (P_m)

10.1.3 The critical time value (t_f).

10.1.4 The absorbed energy to total fracture (U_T)

10.1.5 The initial impact velocity (v_o).

10.1.6 The total energy available at impact (U_o).

10.1.7 Test temperature

10.1.8 The maximum stress intensity [(K_f(max)] during the final 0.5 mm (0.02 in) of fatigue cracking.

10.1.9 Crack plane orientation.

10.1.10 Fracture appearance.

10.1.11 Dynamic notch toughness [$K_{PCI}(t_f)$] where applicable.

10.1.12 Loading rate ($\dot{K}$) as determined by the method of Section 9.4.

10.1.13 Response time (T_R).

10.1.14 Load time and/or load displacement record.

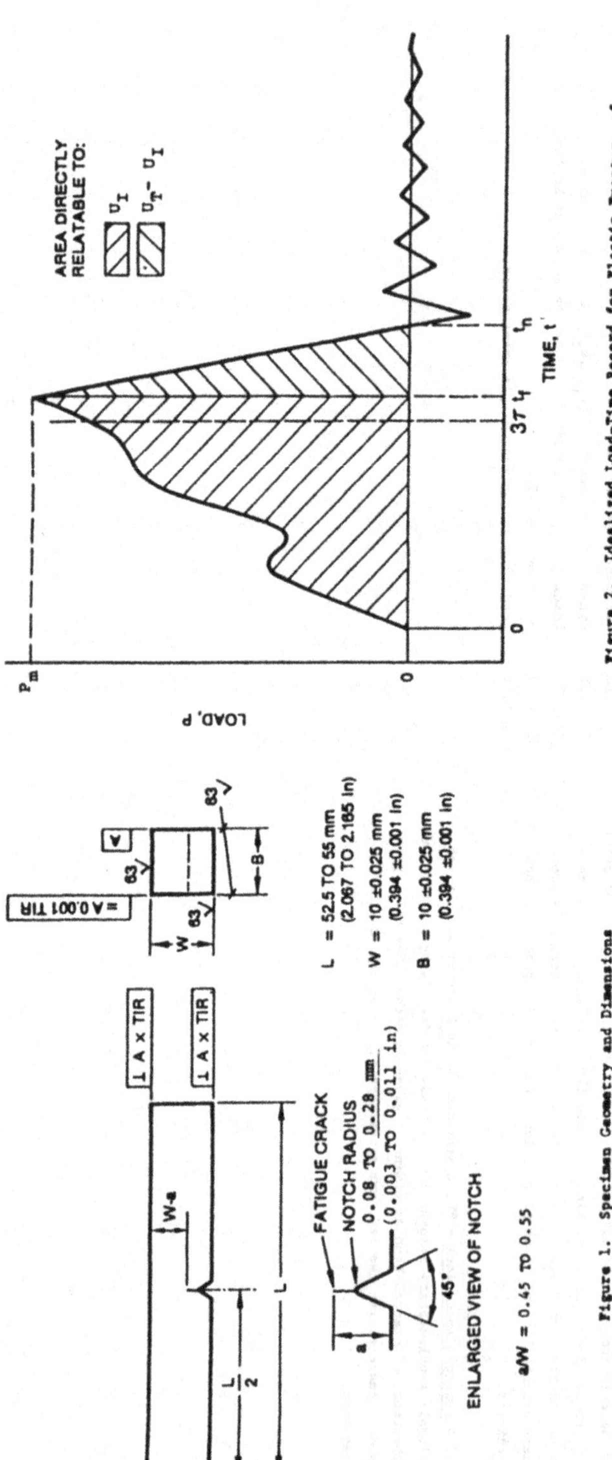

Figure 1. Specimen Geometry and Dimensions

Figure 2. Idealized Load-Time Record for Elastic Fracture of Three-Point Bend Specimen

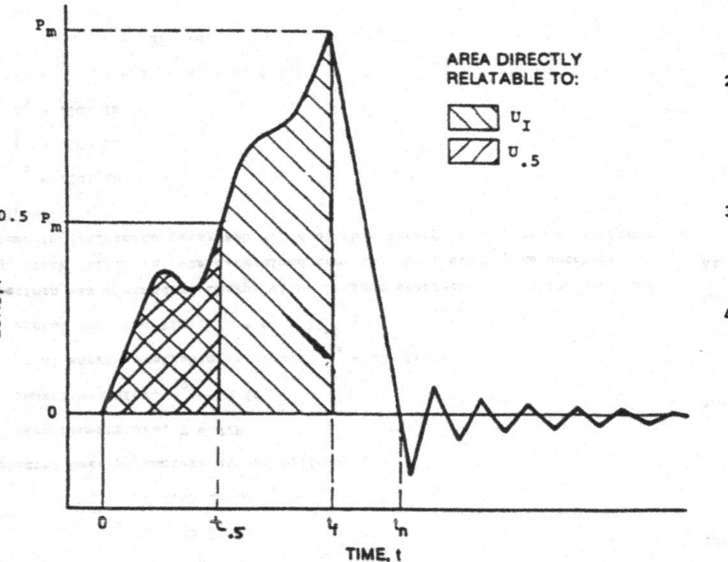

Figure 3 Idealized Load-Time Record Showing Load, Time and Area Values Used For Checking General Yielding Criterion.

24

REFERENCES

1. Ireland, D. R., "Procedures and Problems Associated with Reliable Control of the Instrumented Impact Test," ASTM STP 563, Instrumented Impact Testing, 1974.

2. Ireland, D. R., "Critical Review of Instrumented Impact Testing," Dynamic Fracture Toughness International Conference, London 1976, The Welding Institute of American Society for Metals, Effects Technology, Inc., Technical Report No. 79-55.

3. Bucci, R. J., P. C. Park, J. D. Landes, and J. R. Rice, "J-Integral Estimation Procedures," Scientific Paper 71-1E7-FMPWR-P1, Westinghouse Research Laboratories, December 1971.

4. Server, W. L., "Impact Three Point Bend Testing for Notched and Precracked Specimens," Journal of Testing and Evaluation, Vol. 6, No. 1., January 1978, pp. 29-34.

ANNEX

A 1. TYPICAL CALCULATION AND INTERPRETATION OF RESULTS

Let the idealized load-time record shown in Figures 2 and 3 represent an actual test where the specific values shown on the figures are as follows:

P_m = 1150 lb

$t_f = 0.2 \times 10^{-3}$ s

$t_{.5} = 0.11 \times 10^{-3}$ s

U_a at the following time values

$t_{.5}$, 0.32 ft-lb

t_f , 1.15 ft-lb

t_n , 1.45 ft-lb

The selected test parameters are as follows:

test temperature, T = 72F

impact velocity, v_o = 10 ft/s

total energy available at impact, U_o = 100 ft-lb

system response time, $T_R = 60 \times 10^{-6}$ s

The specimen was a standard Charpy with fatigue precrack. After the test the average crack length (a) was determined from the three crack plus notch depth measurements (reference paragraph 8.1.2 of this method) across the crack front as follows:

a_1 = .201 in.

a_2 = .206 in.

a_3 = .202 in.

$a = 1/3\ (a_1 + a_2 + a_3)$

a = .203 in.

The W dimension for the Charpy is 0.394 in. and the a/W value for this test is then

a/W = 0.203/0.394

a/W = 0.515

To verify that acceptable load values have been obtained compare the measured t_f value with the requirements for frequency response and inertial oscillations as stated in paragraph 9.1.2 of this method.

$t_f \geq 3\tau$

$3\tau = 1.7 \times 10^{-4}$ s, for a/W = 0.515

$t_f = 2 \times 10^{-4}$ s

therefore, the inertial oscillations have been sufficiently dampened.

$t_f \geq 1.1\ T_R$

$1.1\ T_R = 0.66 \times 10^{-4}$ s

$t_f = 2 \times 10^{-4}$ s

therefore, there was adequate frequency response.

To insure that fracture occurred before general yielding compare the ratios of load to deflection at P_m and 0.5 P_m (reference paragraph 9.2 of this method).

At P_m

U_a = 1.15 ft-lb

U_o = 100 ft-lb

v_o = 10 ft/s

$t_f = 2 \times 10^{-4}$ s

$\bar{v} = v_o\left(1 - \frac{U_a}{4U_o}\right)$, $\bar{v}$ = 10 ft/s

$d = \bar{v} t_f$, $d = 2 \times 10^{-3}$ ft

Ratio = P_m/d , Ratio = 5.75×10^5 lb/ft

At .5 P_m $\quad U_a = 0.32$ ft-lb

$t_{.5} = 1.1 \times 10^{-4}$ s

$\bar{v} = 10$ ft/s

$d = 1.1 \times 10^{-3}$ ft

Ratio $= 5.23 \times 10^5$ lb/ft

Fracture occurred before general yielding because the load deflection ratios are within 10 percent of each other. That is,

$5.23 \times 10^5 / 5.75 \times 10^5 = 0.91$

Dynamic notch toughness, K_{PCI}, is computed as follows, (reference paragraph 9.3.1 of this method).

$P_m = 1150$ lb

$a/W = 0.515$, $\quad h(a/W) = 45.17$

$K_{PCI} = P_m\, h(a/W)$, $\quad K_{PCI} = 52{,}000$ psi-in$^{1/2}$

$t_f = 2 \times 10^{-4}$ s

This toughness value is reported as

$K_{PCI}(t_f) = 52{,}000$ psi-in$^{1/2}(2 \times 10^{-4}$ s).

The relative rate of loading (reference paragraph 9.4) is reported as a relative stress intensity rate as follows:

$\dot{K} = K_{PCI}/t_f$, $\quad \dot{K} = 2.6 \times 10^8$ psi-in$^{1/2}$ s

The total absorbed energy for fracture (U_T) of this specimen is computed as follows (reference paragraph 9.5 of this method).

$$U = U_a \left(1 - \frac{U_a}{4U_o}\right) ,$$

U_a at t_n is 1.45 ft-lb

$U_o = 100$ ft-lb

$U_T = 1.44$ ft-lb

The available energy was 100 ft-lb and is sufficiently large to assure there has not been adverse velocity reduction during the test (reference paragraph 6.1.3.1 of this method).

8. REFERENCES

1. P. Manogg, "Anwendung der Schattenoptik zur Untersuchung des Zerreissvorgangs von Platten", Dissertation, Freiburg, Germany, 1964.

2. P. Manogg, "Schattenoptische Messung der spezifischen Bruchenergie wahrend des Bruchvorgangs bei Plexiglas", Proc. Int. Conf. Phys. Non-Crystalline Solids, Delft, The Netherlands, 1964, 481-490.

3. P.S. Theocaris and N. Joakimides, "Some Properties of Generalized Epicycloids Applied to Fracture Mechanics", J. Appl. Mech., 22, 1971, 876-890.

4. P.S. Theocaris, "The Reflected Caustic Method for the Evaluation of Mode III Stress Intensity Factor", Intern. J. Mech. Sci., 23, 1981, 105-117. 5. P.S. Theocaris, "Stress Concentrations at Concentrated Loads", Exp. Mech., 13, 1973, 511-528.

6. A.J. Rosakis, and L.B. Freund, "Optical Measurement of the Plastic Strain Concentration at a Tip in a Ductile Plate", J. Eng. Mater. Technol., 104, 1982, 115-125

7. A.J. Rosakis, C.C. Ma, and L.B. Freund, "Analysis of the Optical Shadow Spot Method for a Tensile Crack in a Power-Law Hardening Material", J. Appl. Mech., 50, 1983, 777-782.

8. J.F. Kalthoff, J. Beinert, and S. Winkler, "Analysis of Fast Running Arresting Cracks by the Shadow-Optical Method of Caustics", IUTAM Symp. Opt. Methods Mech. Solids (A. Lagarde, Ed.), University of Poitiers, France, Sept. 10-14, 1979, Sijthoff-Noordhoff, Alphen aan den Rijn, The Netherlands, 1980, 497-508.

9. J. Beinert, and J.F. Kalthoff, "Experimental Determination of Dynamic Stress Intensity Factors by Shadow Patterns", in Mechanics of Fracture Vol. 7, Experimental Fracture Mechanics (G.C. Sih, Ed.), Martinus Nijhoff Publishers, Hingham, Mass. 1981, 280-330.

10. J.F. Kalthoff, "Stress Intensity Factor Determination by Caustics", Proc. Int. Conf. on Experimental Stress Analysis, organized by Japan Society of Mechanical Engineers (JSME) and American Society for Experimental Stress Analysis (SESA), Honolulu-Maui, Hawaii, May 23-29, 1982, 1119-1126.

11. J.F. Kalthoff, W. Boehme, and S. Winkler, "Analysis of Impact Fracture Phenomenon by Means of the Shadow Optical Method of Caustics", Proc. 7th Int. Conf. Exp. Stress Anal., organized by SESA, Haifa, Israel, Aug. 23-27, 1982, 148-160

12. J.F. Kalthoff, "Shadow Optical Method of Caustics", in "Handbook on Experimental Mechanics", Ed. A.S. Kobayashi, Prentice Hall, Englewood Cliffs, N.Y., 1987, 430-500.

13. J.F. Kalthoff, "D 2.9 Schattenoptisches Kaustikenverfahren", in "Handbuch der experimentellen Spannungsanalyse", Hrsg. Ch. Rohrbach, N. Czaika, VDI-Verlag, Duesseldorf, 1988.

14. M. Born, and E. Wolf, "Principles of Optics", Pergamon Press, New York, 1970.

15. P.S. Theocaris, "Complex Stress Intensity Factors of Bifurcation Cracks", J. Mech. Phys. Solids, 210, 1972, 265-279.

16. U. Seidelmann, "Anwendung des schattenoptischen Kaustikenverfahrens zur Bestimmung bruchmechanischer Kennwerte bei ueberlagerter Normal- und Scherbeanspruchung", Bericht 2/76 des Fraunhofer-Instituts fuer Festkoerpermechanik, Freiburg, 1976.

17. G.C. Sih, "Handbook of Stress Intensity Factors", Institute of Fracture and Solid Mechanics, Lehigh University, Bethlehem, Pa., 1973.

18. L. B. Freund, "Crack Propagation in an Elastic Solid Subjected to General Loading - I. Constant Rate of Extension", J. Mech. Phys. Solids, 20, 1972, 129-140.

19. J.F. Kalthoff, "Zur Ausbreitung und Arretierung schnell laufender Risse", Fortschritt-Berichte der VDI-Zeitschriften, Reihe 18, 4, VDI-lag, Duesseldorf, 1978, 1-95.

20. J.W. Hutchinson, "Singular Behavior at the End of a Tensile Crack in a Hardening Material", J. Mech. Phys. Solids, 16, 1986, 13-31.

21. J.R. Rice, and G.F. Rosengreen, "Plane Strain Deformation Near a Crack Tip in a Power-Law Hardening Material", J. Mech. Phys. Solids, 16, 1968, 1-12.

22. J.W. Hutchinson, "Plastic Stress and Strain Fields at a Crack Tip", J. Mech. Phys. Solids, 165, 1968, 337-347.

23. J.F. Kalthoff, Fortschritt-Berichte der VDI-Zeitschriften, Reihe 18, 4, VDI-Verlag, Duesseldorf, 1978, 1-95.

24. J.F. Kalthoff, "Bruchdynamik laufender und arretierender Risse", in "Grundlagen der Bruchmechanik", Hrsg. H.-P. Rossmanith, Springer-Verlag, Wien, New York, 1982, 191-219.

25. M. Yoshiki, T. Kanazawa, and S. Machida, "Some Basic Considerations on Crack Arresting Mechanisms in Welded Steel Structures", Fac. of Engineering, Dept. of Naval Architecture, University of Tokyo, Japan, 1965.

26. J.I. Bluhm, "Fracture Arrest", Chapter 1 in Vol. V of Fracture, An Advanced Treatise, ed. by H. Liebowitz, Academic Press, New York, London, 1969.

27. H. Kihara, and K. Ikeda, "Technische Anwendung der Bruchmechanik auf duennwandige Bauteile", Archiv fuer das Eisenhuettenwesen, Bd. 45, 1974, 413.

28. ASTM, "Fast Fracture and Crack Arrest", ASTM STP 627, American Society for Testing and Materials, Philadelphia, 1977.

29. ASTM, "Crack Arrest Methodology and Applications", ASTM STP 711, American Society for Testing and Materials, Philadelphia, 1980.

30. Robertson T., Journal of the Iron and Steel Institute, 1953, 306-312.

31. ASTM E 1221, "Standard Test Method for Determining Plain-Strain Crack-Arrest Fracture Toughness, K_{Ia}, of Ferritic Steels", Annual Book of ASTM Standards, Vol. 03.01, American Society for Testing and Materials, Philadelphia, 1988 (see Appendix I)

32. P.B. Crosley, and E.J. Ripling, "Crack Arrest Toughness of Pressure Vessel Steels", Nuclear Engineering and Design, 17, 1971, 32-45.

33. P.B. Crosley, and E.J. Ripling, "Towards Development of a Standard Test for Measuring K_{Ia}", in "Fast Fracture and Crack Arrest", ASTM STP 627, American Society for Testing and Materials, Philadelphia, 1977, 372-391.

34. J.F. Kalthoff, J. Beinert, and S. Winkler, "Measurements of Dynamic Stress Intensity Factors for Fast Running and Arresting Cracks in Double-Cantilever-Beam Specimens", in "Fast Fracture and Crack Arrest", ASTM STÜP 627, (G.T. Hahn and M.F. Kanninen, Eds.), American Society for Testing and Materials, Philadelphia, 1977, 161-176.

35. J.F. Kalthoff, J. Beinert, S. Winkler, and W. Klemm, "Experimental Analysis of Dynamic Effects in Different Crack Arrest Test Specimens", in "Crack Arrest Methodology and Applications", ASTM STP 711 (G.T. Hahn and M.F. Kanninen, Eds.), American Society for Testing and Materials, Philadelphia, 1980, 109-127.

36. G.T. Hahn, R.G. Hoagland, M.F. Kanninen, and A.R. Rosenfield, "A Preliminary Study of Fast Fracture and Arrest in the DCB Test Specimen", Proc. Int. Conf. Dynamic Crack Propagation, Ed. G.C. Sih, Lehigh University, Bethlehem, Pa., U.S.A., July 10-12, 1972.

37. G.T. Hahn, P.C. Gehlen, R.G. Hoagland, M.F. Kanninen, C. Popelar, A.R. Rosenfield, et al., "Critical Experiments, Measurements and Analyses to Establish a Crack Arrest Methodology for Nuclear Pressure Vessel Steels", Reports BMI-1937, 1959, 1965, Battelle Columbus Laboratories, Columbus, Ohio, 1975, 1976, 1978.

38. J.F. Kalthoff, "Crack Arrest Toughness Measurememt", Proc. CSNI-Workshop on Application of Crack Arrest Concepts, U.S. Nuclear Regulatory Commission Report, Fraunhofer-Institut fur Werkstoffmechanik, Freiburg F.R. of Germany, June 4-5, 1984, 1-14.

39. ASME Boiler and Pressure Vessel Code, Section XI, A 4000, A 5000. The American Society of Mechanical Engineers, New York, 1974.

40. J.F. Kalthoff, and J. Beinert, "Bericht uber die Ergebnisse des ASTM-Round-Robin-Tests zur Messung der Rissarrestzaehigkeit", 18. Sitzung des Arbeitskreises Bruchvorgaenge im DVM, Aachen, 18/19. 2. 1986, 183-190.

41. J. Beinert, and J.F. Kalthoff, " The Development of a Crack Arrest Test Specimen with Reduced Dynamic Effects", Proc. Int. Conf. on Application of Fracture Mechanics to Materials and Structures, Eds. G.C. Sih, E. Sommer, W. Dahl, Freiburg, F.R.G., June 20-24, 1983, 493-507.

42. ASTM, Round Robin Test Program of Crack Arrest, Ed. W. Corwin, American Society for Testing and Materials, Philadelphia 1988.

43. J.F. Kalthoff, J. Beinert, and S. Winkler, "Einfluss dynamischer Effekte auf die Bestimmung von Rissarrestzaehigkeiten und auf die Anwendung von Rissarrestsicherheitsanalysen", 8. Sitzung des Arbeitskreises Bruchvorgaenge im DVM, Koln, 1976, 138.

44. J.F. Kalthoff, and A.P. Reisch, "Instrumentierter Kerbschlagbiegeversuch zur Ermittlunhg bruchmechanischer Kennwerte", Vortragsband der Tagung Werkstoffpruefung 1984, Deutscher Verband fuer Materialpruefung e.V., Bad Nauheim, 6./7. Dezember 1984, 1-10.

45. DVM, "Messtechnische Anforderungen beim instrumentierten Kerbschlagbiegeversuch", DVM-Merkblatt 0001, Deutscher Verband fuer Materialpruefung, Berlin, 1986, 1-11, (see Appendix II).

46. VDEh, "Kerbschlagbiegeversuch mit Ermittlung von Kraft und Weg", STAHL-EISEN-Pruefblaetter (SEP-1315) des Vereins Deutscher Eisenhuettenleute, Verlag Stahleisen, Duesseldorf, 1987, (see Appendix II)

47. ASTM E 24.03.03, "Proposed Standard Method of Tests for Instrumented Impact Testing of Precracked Charpy Specimens of Metallic Materials", Draft 2d, American Society for Testing and Materials, Philadelphia, U.S.A., 1981.

48. D.R. Ireland, "Critical Review of Instrumented Impact Testing", Proc. Int. Conf. Dynamic Fracture Toughness, London, July 5-7, 1976.

49. J.F. Kalthoff, W. Boehme, S. Winkler, and W. Klemm, "Measurements of Dynamic Stress Intensity Factors in Impacted Bend Specimens", C.S.N.I. Specialist Meeting on Instrumented Precracked Charpy Testing, Palo Alto, Calif., Dec. 1-3, 1980, EPRI NP-2102-LD, Electric Power Research Institute, Nov. 1981, 1-17.

50. A. Krisch, "Anforderungen an Kraft und Durchbiegungsmessungen beim Kerbschlagbiegeversuch", Archiv Eisenhuettenwesen, Vol. 43, 1972, 901-905.

51. C.E. Turner, "Dynamic Fracture Toughness Measurements by Instrumented Impact Testing", Advanced Seminar on Fracture Mechanics, Oct. 20-24, 1975, Commission of European Communities, Joint Research Centre, ISPRA, Italy.

52. W. Boehme, and J.F. Kalthoff, "The Behaviour of Notched Bend Specimens in Impact Testing", Int. Journal of Fracture, Vol. 20, 1982, 139-193.

53. S. Venzi, A.H. Priest, and J.J. May, "Impact Testing of Metals", ASTM STP 466, The American Society for Testing and Materials, Philadelphia, 1970, 165-180.

54. W. Goldsmith, "Impact", Edward Arnold Ltd., London , 1960.

55. D.M. Norris, "Engineering Fracture Mechanics", 11, 1979, 261-274.

56. J.F. Kalthoff, S. Winkler, W. Boehme, and W. Klemm, "Determination of the Dynamic Fracture Toughness K_{Id} in Impact Tests by Means of Response Curves", Proc. 5th Int. Conf. on Fracture, Cannes, March 29 April 3, 1981, in "Advances in Fracture Research", Pergamon Press, Oxford, New York, 1980, 363-373.

57. J.F. Kalthoff, "The Concept of Impact Response Curves", in "Metals Handbook - Vol. 8, Mechanical Testing", American Society for Metals, Metals Park, Ohio, 1985, 272.

58. J.F. Kalthoff, S. Winkler, and W. Boehme, "A Novel Procedure for Measuring the Impact Fracture Toughness K_{Id} with Precracked Charpy Specimens", Proc. Int. Conf. on Mechanical and Physical Behaviour of Materials under Dynamic Loading, Paris, Sept. 2-5, 1985, Journal de Physique, Tome 46, Colloque C5, supplement au n° 8, 1985, 179-186.

59. F.J. Loss, "Dynamic Toughness Analysis of Pressure Vessel Steels", Third Water Reactor Safety Research Information Meeting, organized by U.S. Nuclear Regulatory Commission at National Bureau of Standards, Gaithersburg, Md., Oct. 1, 1975.

60. W. Boehme, "Eine einfache Methode zur Bestimmung der dynamischen Rissspitzenbeanspruchung bei schlagbelasteten Dreipunktbiegeprpoben", IWM-Report Z 2/84, Fraunhofer-Institut fuer Werkstoffmechanik, Freiburg, F.R. of Germany, March, 1984.

61. D.R. Ireland, "Procedures and Problems Associated with Reliable Control of the Instrumented Impact Test", in "Instrumented Impact Testing", ASTM STP 563, American Society for Testing and Materials, Philadelphia, 1974, 3-29.

62. J.F. Kalthoff, and S. Winkler, "Vorrichtung zur Erfassung des Rissstarts bei einer Bruchmechanikprobe", Patentanmeldung P 33 34 570.8, Deutsches Patentamt, Muenchen, Sept. 24, 1983, and "Instrument for Detecting the Instant of which a Crack Begins in a Mechanical Strength Test of a Ferromagnetic Metal", U.S. Patent and Trademark Office, Washington, Patent application, Serial No. 06/652 320, Filed Sept. 19, 1984.

63. J.F. Kalthoff, S. Winkler, W. Boehme, and D.A. Shockey, "Mechanical Response of Cracks to Impact Loading", Proc. Int. Conf. on Dynamical Mechanical Properties and Fracture Dynamics of Engineering Materials, Ed. Z. Bilek, Czechoslovak Academy of Sciences, Brno-Valtice, June 16-18, 1983, 1-16.

64. J. Eftis, and J.M. Krafft, "A Comparison of the Initiation with the Rapid Propagation of a Crack in a Mild Steel Plate", J. Basic Engineering, Trans. ASME, March, 1965, 257-263.

65. G.C. Sih "Handbook of Stress Intensity Factors", Institute of Fracture and Solid Mechanics, Lehigh University, Bethlehem, Pa., 1973.

66. D.A. Shockey, and D.R. Curran, "A Method for Measuring K_{Ic} at Very High Strain Rates", in "Progress in Flow Growth and Fracture Toughness Testing", ASTM STP 536, 1973, 297-311.

67. D.A. Shockey, J.F. Kalthoff, and D.C. Ehrlich," Evaluation of Dynamic Instability Criteria", Int. J. Fracture 22, 1983, 217-279.

68. L.S. Costin, J.Duffy, and L.B. Freund, "Fracture Initiation in Metals Under Stress Wave Loading Conditions", in "Fast Fracture and Crack Arrest", (G.T. Hahn and M.F. Kanninen, Eds.), ASTM STP 627, 1977, 301-308.

69. J.R. Klepaczko, "Applications of the Split-Hopkinson Pressure Bar to Fracture Dynamics", in Proc. 2nd Conf. Mech. Prop. High Rates of Strain (J. Harding, ed.), Oxford, 201-204, The Institute of Physics, Conf. Ser. No. 45, Bristol, London, 1979.

70. J.R. Klepaczko, "Loading Rate Spectra for Fracture Initiation in Metals", Theoretical and Applied Fracture Mechanics, 1, 1984, 181-191.

71. K. Ravi-Chandar, and W.G. Knauss, "An Experimental Investigation Into Dynamic Fracture: 1. Crack Initiation and Arrest," Int. J. Fracture, 25, 1984, 247-262.

72. J.F. Kalthoff, "Fracture Behaviour Under High Rates of Loading", Proc. Int. Conf. on Dynamic Fracture Mechanics, San Antonio, Texas, Nov. 7-9, 1984, Eng. Frac. Mech., Vol. 23, 1986, 289-298.

73. J.F. Kalthoff, and S. Winkler, "Fracture Behavior Under Impact", First and Second Anual Reports (W8/83 and W10/83) and Final Report (W10/86) on the Project DAJA 37-81-C-0013 prepared for ERO of ARO, London, Fraunhofer-Institut fuer Werkstoffmechanik, Freiburg 1982, 1983, 1985.

74. W. Boehme, and J.F. Kalthoff, "Der Einfluss der Probengroesse auf dynamische Effekte bei der K_{Id}-Bestimmung im Kerbschlagbiegetest", Report W3/83 on the Project Ka 443-7 prepared for Deutsche Forschungsgemeinschaft, Fraunhofer-Institut fuer Werkstoffmechanik, Freiburg, 1983.

75. J.F. Kalthoff, "Shadow Optical Analysis of Dynamic Shear Fracture", Proc. Int. Conf. on Photomechanics and Speckle Metrology, (SPIE), San Diego, Calif., Aug. 16-21, 1987, 531-538, and Optical Engineering, 27, 1988, 835-840.

76. J.F. Kalthoff, "Failure Mode Transition at High Rates of Shear Loading", Int. Conf. on Impact Loading and Dynamic Behaviour of Materials", Bremen, May 18-22, 1987, Deutsche Gesellschaft fuer Metallkunde (DGM), 185-196.

77. R. Dormeval, "Adiabatic Shear Phenomenon", Proc. Int. Conf. on Impact Loading and Dynamic Behaviour of Materials, Bremen, May 18-22, 1987, Deutsche Gesellschaft fuer Metallkunde (DGM).

78. D.R. Curran, L. Seaman, and D.A. Shockey, "Dynamic Failure of Solids", Physics Reports (Review Section of Physics Letters), 147, Nos. 5,6, 1987, 253-388, North Holland, Amsterdam.

79. H. Neuber, "Kerbspannungslehre", Springer-Verlalg, 1958.

80. D.A. Shockey, J.F. Kalthoff, and D.C. Ehrlich, "Evaluation of Dynamic Crack Instability Criteria", Int. Journ. of Fracture 22, 1983, 217-229.

81. H. Homma, D.A. Shockey, and Y. Murayama, "Response of Cracks in Structural Materials to Short Pulse Loads", J. Mech. Phys. Solids, 1983.

82. D.A. Shockey, J.F. Kalthoff, H. Homma, and D.C. Ehrlich, "Criterion for Crack Instability under Short Pulse Loads", Proc. Int. Conf. on Fracture, Cannes, March 29 - April 3, 1981, in "Advances in Fracture Research", Pergamon Press, Oxford, New York, 1980, 415-423.

83. J.D. Achenbach, "Dynamic Effects in Brittle Fracture", in Mechanics Today, I. ed. S. Nemat-Nasser, Oxford: Pergamon-Press, 1972, 1-57.

84. L.B. Freund, "Crack Propagation in an Elastic Solid Subjected to General Loading - III. Stress Wave Loading", J. Mech. Phys. Solids, 21, 1973, 47-61.

85. G. Sih, "Handbook of Stress Intensity Factors" Inst. of Fracture and Solid Mechanics, Lehigh University, Bethlehem, Pa., 1973.

86. J.F. Kalthoff, and D.A. Shockey, "On the Instabillity of Cracks Loaded by Tensile Stress Pulses of Short Duration", Stanford Research Institute, Poulter Laboratory Technical Report 001-75, 1975.

87. J.F. Kalthoff, and D.A. Shockey, "Instability of Cracks under Impulse Loads", J. Appl. Phys., Vol. 48, No. 3, 1977, 986-993.

DYNAMIC CRACK INITIATION, SOME EXPERIMENTAL METHODS AND MODELLING

J.R. Klepaczko
University of Metz, Metz, France

ABSTRACT

The main purpose of this part of the book is to review new experimental methods which are useful and effective in determination of resistance to fracture under fast and impact conditions of loading. In recent decade a substantial progress has been made in this domain.

The first part concentrates on loading rate effects in fracture initiation and its theoretical basis. In general, the small scale yielding is considered, however, some cases of the large scale yielding are also discussed.

The loading rate spectrum is thoroughly analyzed. Experimental techniques and some results obtained within the low and medium loading rate are both considered. Over the region of higher loading rates application of elastic waves for testing fracture resistance is discussed. An emphasis is placed on application of Split Hopkinson Bar (SHB) to fracture dynamics (Modes I and III are discussed).

In the final part attention is being given to experimental results in crack initiation over a wide range of loading rates and temperatures. A modelling of the loading rate spectra is attempted. A generalized model for quasi-static, fast and impact loading of a stationary crack has been developed and discussed.

The experimental results are presented in such a way to be useful in practical applications. It is believed that a better understanding of the nature of the loading rate effects in metallic materials can be useful in preventing catastrophic failures under impact.

1. RATE EFFECTS IN FRACTURE, GENERALITIES

The definition of "Dynamic Fracture Mechanics" is very general, it applies not only to those problems where inertia forces are important but also to the problems where loading rates are fast enough to develop a time-dependent reaction of a crack. It follows that the class of problems associated with "Dynamic Fracture Mechanics" is very broad, from fast loading, impact, to shock waves interacting with cracks.

An isolated crack, oriented in the spatial coordinates (x_1, x_2, x_3) or (x,y,z) and with zeros at the crack tip, can be loaded in three basic modes :

Mode I - the opening mode,

a. plane strain ; $\varepsilon_{33} = 0$ or $\varepsilon_{zz} = 0$
b. plane stress ; $\sigma_{33} = 0$ or $\sigma_{zz} = 0$

Mode II - the sliding mode,

a. plane strain ; $\varepsilon_{33} = 0$ or $\varepsilon_{zz} = 0$
b. plane stress ; $\sigma_{33} = 0$ or $\sigma_{zz} = 0$

Mode III - the tearing mode, (no dilatation).

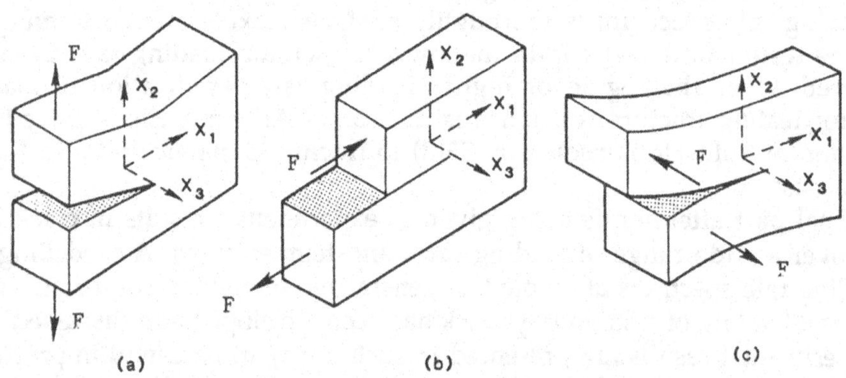

Fig. 1.1, Three basic loading modes : (a) Mode I-opening mode ; (b) Mode II-sliding mode ; Mode III-tearing mode ; F_i are external forces which must be applied to obtain specific mode.

The external forces (Fig. 1.1) $\pm F_i$ per unit thickness, develop stress fields $\sigma_{ij}(r,\theta)$ around cracks, and consequently a stress singularity at $x_i = 0$ is created. A cracked body can also be loaded in the Mixt Mode, i.e. by any combination of Mode I, II or III.

It is important to recognize that external forces $F_i(t)$ can be applied at different rates $\dot{F}_i$. Consequently, different loading rates will develop different reactions around crack tip. This will be caused by inertia or by intrinsic rate sensitivity of a body (visco-elasticity or visco-plasticity).

If external forces are high, they will develop high intensity stress fields around cracks, and it is obvious that it may lead to catastrophic fracture. Thus, an isolated crack may be characterized by the following states :

1. stationary crack of length a, $\dot{a} \neq 0$;

2. moving crack

   a. stable movements , $\dot{a} \neq 0$, (for example, creep fracture) ;
   b. fast propagation , $a \neq 0$.

A fast crack propagation results in a rapid separation of a body into two or more pieces, i.e. a catastrophic fracture occurs.

If a metal shows ability to plastic deformation a plastic zone will form around crack tip. A much smaller zone, called the process zone, will also form very close to the crack tip. Within the process zone material separation will occur. The external part of a body outside the plastic zone is under an elastic stress field. In the case of brittle materials the plastic zone is practically reduced to zero, however, the process zone with microcracks may exists.

In every mode of crack tip loading stress and strain singularities are observed. Within the framework of Linear Elastic Fracture Mechanic (LEFM) the stress and displacement distributions around elastic crack in an infinite body are, (Mode I)

$$\sigma_{ij} = \frac{K_I}{\sqrt{a\pi r}} f_{ij}(\theta) \tag{1.1}$$

$$u_i = \frac{K_I}{\sqrt{2\pi r}} g_{ij}(\theta) \tag{1.2}$$

where K_I is the Stress Intensity Factor (SIF) in Mode I. The stress intensity factor is an indicator of the strength of stress and displacement fields around cracks. Similar solutions for Mode II and Mode III can be found in literature Schematic illustration of stress and strain distribution near crack tip is shown in Fig. 1.2.

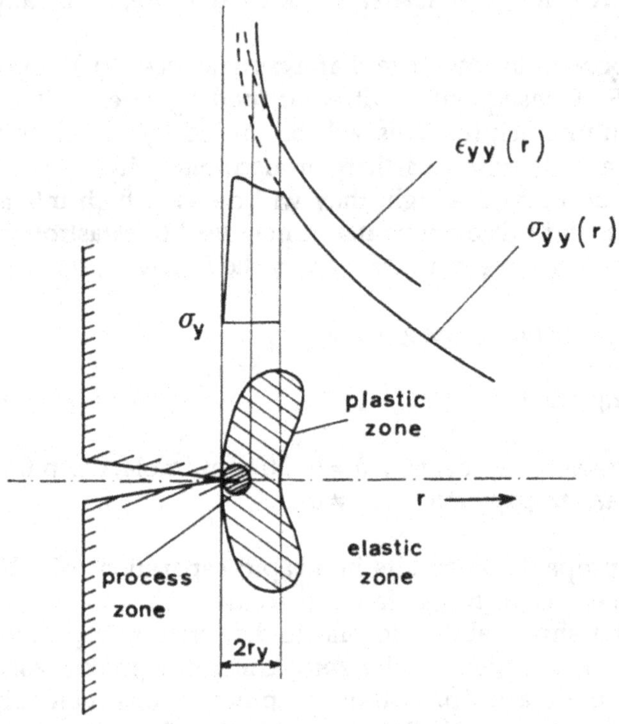

Fig. 1.2, Schematic illustration of stress and strain distribution near a crack tip.

Griffith (1920) postulated that during increment of crack extension da the sum of the potential energy of deformation and the surface energy is constant. If γ denotes the surface energy density per unit thickness of the body and per unit area, then the increment of the surface energy S is $dS = 2\gamma da$ and

$$G = 2\gamma \quad \text{or} \quad G = -\frac{dE_p}{da} \tag{1.3}$$

where E_p is the potential energy of deformation per unit thickness, G is called the energy release rate, and it can be also derived from a potential function, so it is often referred to as a crack driving force. When all three modes of deformation are present the energy release rate can be related to the stress intensity factors as follows

$$G = \frac{1}{E^*}\left(K_I^2 + K_{II}^2\right) + \frac{1}{2\mu}K_{III}^2 \tag{1.4}$$

$E^* = E$ for plane stress

$E^* = E/1\text{-}\nu^2$ for plane strain.

According to the elastic stress field solution a stress singularity exists at the tip of an elastic crack, i.e. $\lim_{r\to 0} \sigma_{22} = \infty$. The infinite σ_{22} component of the stress tensor is not acceptable in reality since real materials tend to exhibit some

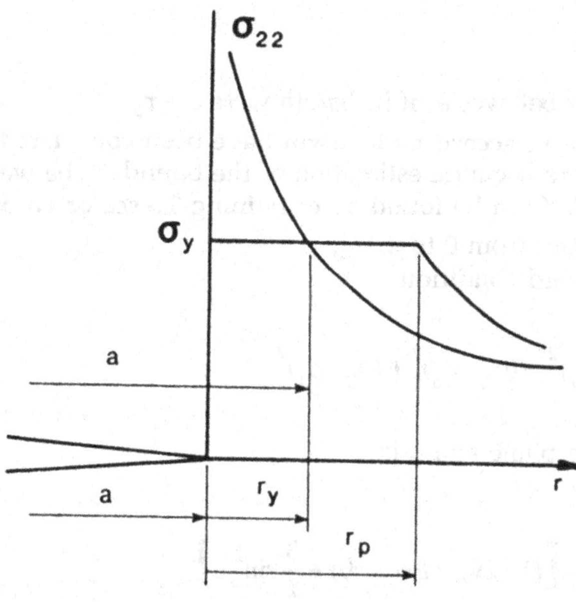

Fig. 1.3, Elastic plane stress distribution and plastic stress correction.

plasticity. Irwin [1], proposed a simple procedure to account for a plastic zone.

This approach in Linear Elastic Fracture Mechanics (LEFM) is called the framework of Small Scale Yielding (SSY), and is shown schematically in Fig. 1.3.

The σ_{22} component of the stress tensor is cut-off at the level of the yield stress σ_y and resulting plastic zones are :

$$r_p = 2v = \frac{1}{\pi}\left(\frac{K_I}{\sigma_y}\right)^2 \quad ; \quad \theta = 0 \tag{1.5}$$

for plane stress, and

$$r_p = 2v = \frac{1}{3\pi}\left(\frac{K_I}{\sigma_y}\right)^2 \quad ; \quad \theta = 0 \tag{1.6}$$

for plane strain. Thus, after correction the effective crack length is

$$a_e = a + r_y$$

It follows that crack behaves as if its length were $a + r_y$.

Estimations of r_p according to Irwin have been considered for $\theta = 0$ (i.e. x_1-direction). A more occurate estimation of the boundary between elastic and plastic zones for SYY can be found by examining Tresca or Huber-Mises yield condition for θ-angles from 0 to π.
For Huber-Mises yield condition

$$2\sigma_y^2 = (\sigma_1 - \sigma_2)^2 + (\sigma_2 - \sigma_3)^2 + (\sigma_3 - \sigma_1)^2 \tag{1.7}$$

the solution for the plane strain is

$$r_y(\theta) = \frac{K_I^2}{4\pi\sigma_y^2}\left[(1 - 2v)^2 (1 + \cos\theta) + \frac{3}{2}\sin^2\theta\right] \tag{1.8}$$

and for the plane stress

$$r_y(\theta) = \frac{K_I^2}{4\pi\sigma_y^2}\left[1 + \cos\theta + \frac{3}{2}\sin^2\theta\right] \tag{1.9}$$

If the Tresca yield condition is used, i.e. $\tau_{max} = 1/2\ \sigma_y$, one obtains for the plane strain

$$\sup \begin{cases} r_y = \dfrac{K_I^2}{2\pi\sigma^2}\cos^2\dfrac{\theta}{2}\left[1 - 2\nu + \sin\dfrac{\theta}{2}\right]^2 & (1.10a) \\ r_y = \dfrac{K_I^{2y}}{2\pi\sigma_y^2}\cos^2\dfrac{\theta}{2} & (1.10b) \end{cases}$$

and for the plane stress

$$r_y(\theta) = \frac{K_I^2}{2\pi\sigma_y^2}\left[\cos\frac{\theta}{2}\left(1 + \sin\frac{\theta}{2}\right)\right]^2 \tag{1.11}$$

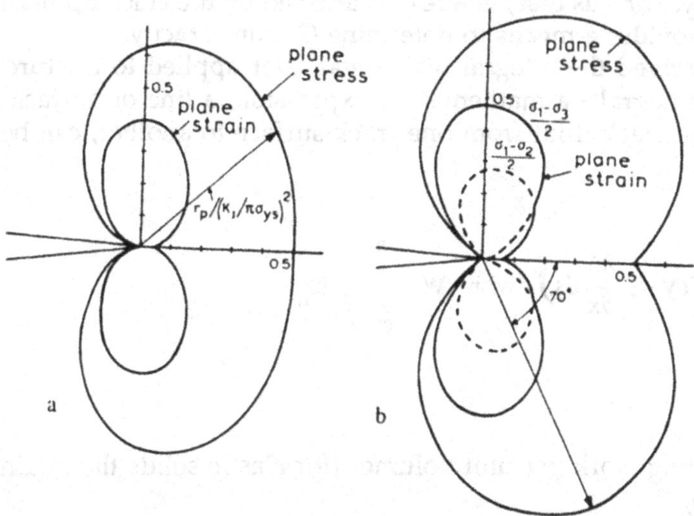

Fig. 1.4, Plastic zone shapes according to Huber-Mises and Tresca yield criteria ; a. Huber-Mises criterion ; b. Tresca criterion ; reproduced with permission from M. Nijhoff Pub, after [151].

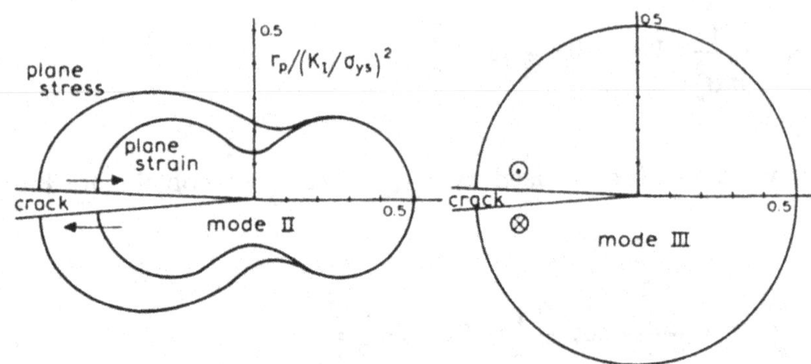

Fig. 1.5, Plastic zone shapes for Modes II and III, after [3].

From a comparison of these solutions it is clear that the plane strain boundary is significantly smaller. Moreover, if the Tresca yield condition is applied the plastic zones are slightly different. It is interesting to note that for the case of Mode III the plastic zone is a circle, [2]
Of course, more exact estimations of development of plastic zones in metals are possible by applications of numerical methods.

The SSY approach is limited to relatively small plastic zones. If there is appreciable plasticity around crack tip the energy release rate G can not be determined exactly. For this case, where G is affected by the crack tip plastic zone, the J-integral provides a means to determine G more exactly.

Eshelby derived the integral which was later applied to fracture mechanics by Rice. J-integral - a mathematical expression, a line or surface integral that encloses the crack front from one crack surface to another, can be formulated as follows

$$J = \int_{\Gamma} \left(W dy - T \frac{\partial \bar{u}}{\partial x} ds \right) \quad \text{with} \quad W = \int_{0}^{\varepsilon} \sigma_{ij} \, d\varepsilon_{ij} \tag{1.12}$$

where :

W is the loading work per unit volume (for elastic solids the strain energy density G),

Γ is the path of the integral which contains the crack tip,

ds is the increment of the contour path,

T is the outward fraction vector on ds,

u is displacement vector at ds.

The second term, i.e. T ∂u/∂x ds, is the rate of work input from the stress field into the area enclosed by Γ.

Begley and Landes [4,5] developed an experimental procedure to determine J-values using the following formula

$$J = -\frac{1}{B}\left(\frac{\partial U}{\partial a}\right)_a \tag{1.13}$$

where J is interpreted as the energy release rate, U is the strain energy and B is the specimen thickness.

Another formula, for specific geometries of specimen uses the area under load vs. load-point displacement record (energy), up to critical force F_{cr}

$$J = \eta \frac{A}{Bb} \quad ; \quad A = \int_0^{F_{cr}} F \, d\delta \tag{1.14}$$

where b is the initial uncracked ligament of specimen, η is a dimensionless coefficient that corrects for the tensile component of loading, $\eta \approx 2$.

It may be mentioned that the J-integral is not only invariant integral in fracture mechanics. Another possibility is to include inertia forces in a path-independent integral, this is the case of propagating crack. Some recent attempts in this direction has been done by Bui [6] and by Nakamura, Shih and Freund [7]. Initially, a crack tip contour integral for dynamic energy release rate was proposed by Atkinson and Eshelby [8].

The recent derivation of the contour integral which include inertia forces is by Nakamura et al., this leads to the result

$$F = \int_\Gamma \left[\sigma_{ij} n_j \frac{\partial u_i}{\partial t} + (U + T) v n_1\right] d\Gamma \tag{1.15}$$

for a crack in (x_1,x_2) plane with the tip at $x_1 = x_2 = 0$ $x_2 = 0$ and growth in the x_1-direction with the speed v. The contour Γ is fixed in size and orientation, and it translates with the crack tip ; F is the instantaneous rate of energy flow out of the body through Γ, where T is the kinetic energy density and U is the stress work density for any material response

$$U = \int_0^t \sigma_{ij} \frac{\partial^2 u_i}{\partial t^* \partial x_j} dt^* \quad ; \quad T = \frac{1}{2} \rho \frac{\partial u_i}{\partial t} \frac{\partial u_i}{\partial t} \tag{1.16}$$

ρ is the mass density of the material.

The first term in F is the rate of work of the material outside of Γ on the material inside Γ, and the second term represents the energy flux through Γ due to mass transport associated with crack motion.

The integral F underlies virtually all crack tip energy integrals that have been defined and applied in fracture mechanics. In general, $F(\Gamma)$ cannot be a path-independent integral. In the limiting case the dynamic energy release rate G_o can be found for the near tip elastodynamic field, Freund [9], in the form

$$G_o = \frac{1-\nu^2}{E} f(v) K_I^2 \tag{1.17}$$

where f(v) is a universal function of the instantaneous crack tip speed v, f(0) = 1, f'(0) = 0 and $f(v) \to \infty$ as $v \to C_R$, where C_R is the Raleigh wave speed. Thus, $G_o = G$ for v = 0.

Another quantity which characterize intensity of the stress field around crack tip or the energy release rate is the crack tip opening displacement (CTOD). Wells [10] advanced this concept in the purpose of application to cracks with a substantial plastic zone. For the case of SSY the CTOD = δ_t is

$$\delta_t = \alpha \frac{K^2}{E\sigma_y} \quad ; \quad 1 \leq \alpha \leq \frac{4}{\pi} \tag{1.18}$$

Note that δ_t is related to r_y via strain, i.e. by the yield strain $\varepsilon_y = \sigma_y/E$, $\delta_t = 2\pi\, r_y\, \varepsilon_y$.

The crack tip opening displacement δ_t is related to the Crack Opening Displacement δ determined outside of the crak tip (COD = δ). The CTOD concept has been rationalized by Dügdale solution with small plastic zones, [11]. Dugdale obtained the closed-form solution which permits for estimation of δ_t

$$\delta_t = \frac{8}{\pi}\frac{\sigma_y}{E} a \log\left[\sec\left(\frac{\pi}{2}\frac{\sigma}{\sigma_y}\right)\right] \tag{1.19}$$

For $\sigma << \sigma_y \delta_t$ is compatible with SSY approach, for this case

$$G = \delta_t \sigma_y \tag{1.20}$$

A more general expression relating G and δ is

$$G = \lambda \delta \sigma_y \; ; \tag{1.21}$$

where λ is the proportionality constant. Application of the J-integral to the Dugdale model yields the following result

$$J = \delta_t \sigma_y \tag{1.22}$$

Thus, for the case of SSY the equivalence can be shown for all introduced fracture mechanics parameters (at least for plane strain Mode I)

$$G = J = \delta_t \sigma_t = \frac{1-v^2}{E} K_I^2 \tag{1.23}$$

Many experimental procedures have been designed to determine those parameters experimentally and to use them in engineering practice.

IT MUST BE, HOWEVER, REMEMBERED THAT FOR REAL MATERIALS THEY ARE TEMPERATURE AND RATE DEPENDENT.

Rate and temperature sensitivity of fracture parameters stem from the fact that mechanical properties of materials, including atom separation and plasticity, are rate and temperature sensitive. Temperature and rate sensitivity of plastic properties of metals are related via Arrhenius relation

$$\frac{\dot{\varepsilon}}{\dot{\varepsilon}_o} \ln\left(\frac{\Delta H}{kT}\right) = \text{const.} \tag{1.24}$$

where ΔH is the activation enthalpy, k is Boltzmann constant, ε_o is the characteristic strain rate, ε and T are respectively plastic strain rate and absolute temperature.

Rate sensitivity of yield stress or flow stress can be defined in several ways. The following parameters show constant values in different regions of strain rate and temperature, [12]

i. logarithmic rate sensitivity m at constant microstructure and constant homologous temperature θ ($\theta = T/T_m$, where T_m is the melting point)

$$m = \left(\frac{\partial \log\sigma}{\partial \log\dot{\varepsilon}}\right)_{\theta,STR} \tag{1.25}$$

ii. rate sensitivity of flow stress β at constant microstructure and at constant homologous temperature θ

$$\beta = \left(\frac{\partial \sigma}{\partial \log\dot{\varepsilon}}\right)_{\theta,STR} \tag{1.26}$$

iii. viscosity η at constant microstructure and at constant homologous temperature θ

$$\eta = \left(\frac{\partial \sigma}{\partial \dot{\varepsilon}}\right)_{\theta,STR} \tag{1.27}$$

Very frequently state of a metal is characterized by plastic strain which is not correct, since microstructure depends also on history of plastic deformation. Those definitions of rate sensitivity are related to each other, and if one is known the others can be found.

Also, temperature sensitivity of flow stress (or yield stress) can be defined as follows

$$\zeta = \left(\frac{\partial \sigma}{\partial T}\right)_{\theta,STR} \tag{1.28}$$

usually ζ shows negative values. Three rate sensitivities defined above may show constant values at different ranges of strain rates and temperatures. Usually m is constant at higher temperatures and lower strain rates, β is constant at medium strain rates and medium and low temperatures, finally η (apparent viscosity) is constant at high and very high strain rates. Figure below shows characteristic trends in changes of rate sensitivities for low carbon steel, Campbell and Fergusson, [13].

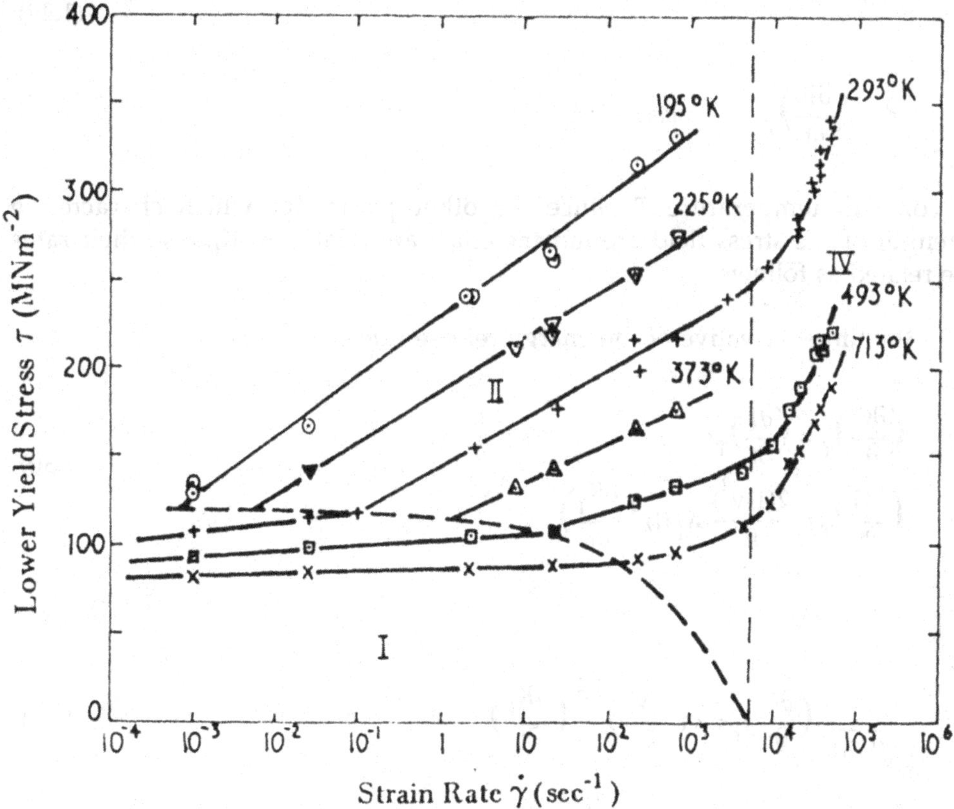

Fig. 1.6, Variation of lower yield stress with strain rate, at constant temperature, [13].

The strain rate spectrum in material testing can be very broad and can be extended over several decimal orders in $\dot{\varepsilon}[s^{-1}]$, from quasi-static values $\dot{\varepsilon} \approx 1 \times 10^{-4}$ s^{-1} to very fast plastic deformation $\dot{\varepsilon} \approx 1 \times 10^6\ s^{-1}$.

There is great interest in relating mechanical properties of structural metals determined from unnotched specimens loaded in tension/compression/torsion, to properties determined from notched (prefatigued) specimens. One of fundamental parameters to be related with strain rate $\dot{\varepsilon}$ is the loading rate of crack tip $\dot{K}_I$, defined as

$$\dot{K}_I = \left(\frac{\partial K_I}{\partial t}\right)_T \quad , \quad \text{plane strain}$$

or (1.29)

$$\dot{K} = \left(\frac{\partial K}{\partial t}\right)_T \quad , \quad \text{plane stress}$$

at constant temperature T. Since the other parameter which characterize strength of the stress field around the crack are related to K_I, also their rates are related as follows

i. the time-derivative of the energy release rate G

$$\left(\frac{\partial G}{\partial t}\right)_T = \left(\frac{\partial J}{\partial t}\right)_T$$
$$\left(\frac{\partial G}{\partial t}\right)_T = \frac{2(1-v^2)}{E} K_I\,(t) \left(\frac{\partial K_I}{\partial t}\right)_T \tag{1.30}$$

or

$$\frac{1}{\sqrt{G(t)}} \left(\frac{\partial G}{\partial t}\right)_T = 2\left(\frac{1-v^2}{E}\right)^{1/2} \left(\frac{\partial K_I}{\partial t}\right)_T \tag{1.31}$$

ii. time-relations between G(t) and δ(t)

$$\left(\frac{\partial G}{\partial t}\right)_T = \lambda \sigma_y(t) \left(\frac{\partial \delta}{\partial t}\right)_T + \lambda\, \delta(t) \left(\frac{\partial \sigma_y}{\partial t}\right)_T \tag{1.32}$$

Since σ_y is rate- or time-dependent the rate sensitivity can be introduced into this relation. After introduction $(\partial\sigma_y/\partial t)_T = \theta_T(t)\,\varepsilon$, where $\theta_T = (\partial\sigma/\partial\varepsilon)_T$, into the equation above one obtains

$$\left(\frac{\partial G}{\partial t}\right)_T = \lambda \sigma_y\,(t) \left(\frac{\partial \delta}{\partial t}\right)_T + \lambda \delta(t)\, \theta_T(t)\, \dot{\varepsilon} \tag{1.33}$$

If the crack tip is loaded by a constant rate of opening then $\delta(t) = (\partial\delta/\partial t)_T = \dot{\delta}t$, and $(\partial G/\partial t)_T$ can be determined as

$$\left(\frac{\partial G}{\partial t}\right)_T = \lambda\dot{\delta}\,[\sigma_y(\dot{\varepsilon}) + \theta_T(\dot{\varepsilon})\,\dot{\varepsilon}] \tag{1.34}$$

Evidently, the term in the brackets is related to the rate-dependent plasticity parameters like rate-sensitive yield stress σ_y and rate-sensitive modulus of strain hardening θ_T. This equation can be rewritten in the following form, with $\dot{\delta}$ as a parameter

$$\left(\frac{\partial G}{\partial t}\right) = \lambda\dot{\delta}\,[\sigma_y(\dot{\varepsilon}) + \theta_T(\dot{\varepsilon})\,\dot{\varepsilon}] \tag{1.35}$$

As a rule, specimens used for fracture toughness testing are loaded at constant rate of force, $\dot{P}$ = const. Since K is proportional to P

$$K = P\,f(a,w) \tag{1.36}$$

where f (a,w) is a function of specimen geometry, a is the crack length and W is the specimen width, all specimens loaded with $\dot{P}$ = const. are in fact loaded with $\dot{K}_I$ = const. Thus, the constant rate of loading is the most popular in fracture toughness testing, the units of $\dot{K}_I$ are [MPa√m/s]. The standard quasi-static value of $\dot{K}_I$ is 1MPa√m/s. If specimen is loaded at $\dot{K}_I$ = const. then the rate of G is

$$\left(\frac{\partial G}{\partial t}\right)_T = \frac{2(1-\nu^2)}{E}\,t\,\dot{K}_I^2 \tag{1.37}$$

and the rate of G increases in proportion to time of loading.

It is obvious that the loading rate parameter $\dot{K}_I$ may be changed over several orders of magnitude (an analogy to strain rate in testing of the regular-geometry specimens), simply by changing the loading rate $\dot{P}$. It is assumed that the correction function f (a,w) is rate independent. Assumed value of $\dot{K}_I$ = 1MPa√m/s as quasi-static reference level provides an estimation of time period for a quasi-static test for fracture toughnes, for example if K_{Ic} = 50MPa√m the test will take time t_c = 50 s.

The whole possible spectrum of loading rates is shown in the logarithmic scale in Fig. 1.7.

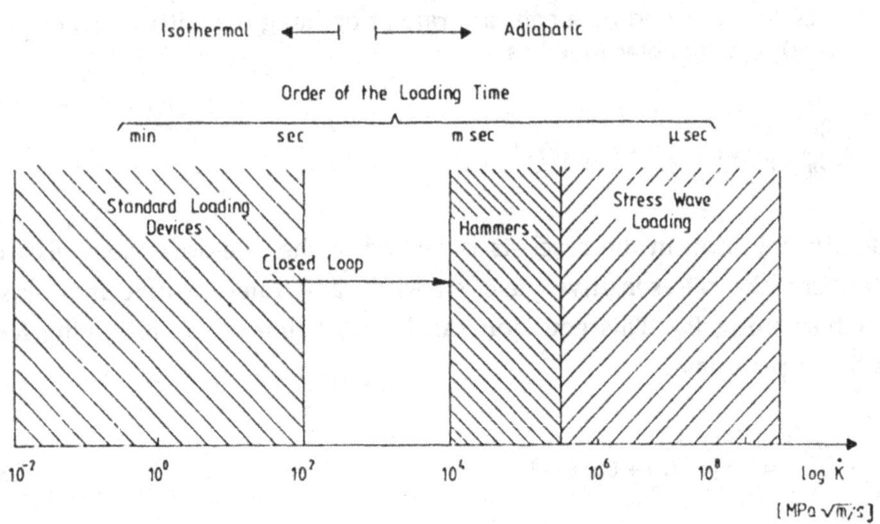

Fig. 1.7, The loading rate spectrum for stress intensity factor K_I.

Such a spectrum can be constructed for any initial temperature T_o. As shown in Fig. 1.7, higher loading rates than 1 MPa√m/s can be achieved with standard testing machines and standard specimen geometries only up to $\dot{K}_I \approx 1 \times 10^4$ MPa√m/s. To achieve still higher values of $\dot{K}_I$ other experimental techniques have been developed to cover the region 1×10^4 MPa√m/s $\leq \dot{K}_I \leq 10^6$ MPa√m/s. Besides the instrumented Charpy test with $\dot{K}_I \approx 1 \times 10^5$ MPa√m/s, a promising direction in fracture toughness testing at high loading rates is the use of the Hopkinson bar concept.

Other specimen configurations (different from standard geometries) and methods of stress wave loading have been proposed recently. For example, stress pulse technique, together with the shadow optical analyses, can indicate the entire time history of crack tip loading. Still higher loading rates, up to $\dot{K}_I \approx 1 \times< 10^9$ MPa√m/s, can be attained by the plane wave loading technique applied to an isolated crack.

The main purpose of this text is to provide fundamental information on the techniques which can be applied within the whole spectrum of the loading rate $\dot{K}_I$.

Since it is desirable to search for a correlation between temperature and strain rate sensitive unnotched properties of metals and fracture parameters, the relation between strain rate $\dot{\varepsilon}$ at the crack tip and the loading rate expressed as $\dot{K}_I$ is important. In the literature a variety of approaches to this problem can be spotted. Here, the solution for elastic singularity will be adopted to find correlation in between $\dot{\varepsilon}$ and $\dot{K}_I$. Using plane strain solution with Irwin's correction

$$r_y = \frac{(1-2\nu^2)^2}{2\pi}\left(\frac{K_I}{\sigma_y}\right)^2 \qquad (1.38)$$

together with time differentiation

$$\dot{\varepsilon} = \frac{\dot{K}_I}{E\sqrt{2\pi r}}\,(1-2\nu)^2 \qquad (1.39)$$

and eliminating singularity $(2\pi r_p)^{-1/2}$ one obtains

$$\dot{\varepsilon} = \frac{\sigma_y}{EK_I}\,\dot{K}_I \qquad (1.40)$$

In both cases, $\dot{\varepsilon}$ and $\dot{K}_I$, the logarithmic scales are used to demonstrate changes in flow stress and fracture parameters as a function of rate. Taking logarithm of both sides

$$\log\dot{\varepsilon} = \log\left(\frac{\sigma_y}{EK_I}\right) + \log\dot{K}_I \qquad (1.41)$$

the Logarithmic Conversion Factor is defined as

$$LCF = \log\left(\frac{\sigma_y}{EK_I}\right) \qquad (1.42)$$

It indicates the scale shift between $\log\dot{\varepsilon}$ and $\log\dot{K}_I$. For a more brittle metals the logarithmic conversion factor is LCF = - 4. It must be remembered that the LCF has been determined for the specific distance from the crack tip $r = r_y$, i.e. for the limit of idealized plastic zone defined in the SSY. In the vicinity of the crack tip, i.e. in the process zone, the local strain rates can be much higher. The mean value of $\dot{\varepsilon}$ in between $r_p \le r \le r_y$ will be then higher in comparison to the estimation based on the elastic singularity.

Such estimations are useful in modelling of loading rate sensitivity of fracture toughness.

A very useful expression relating $\dot{\varepsilon}$ and $\dot{K}_I$ is following, [139]

$$\dot{\varepsilon} = k \frac{\sigma_y}{E K_I} \dot{K}_I \tag{1.43}$$

where k is the coefficient of intensification for strain singularity, $1 \leq k \leq 10$, k is larger for more ductile metals.

2. FRACTURE PARAMETERS DETERMINED AT LOW AND MEDIUM LOADING RATES

It is obvious that an increase of the stress field around an isolated crack cannot be increased to infinity. Every atom bonding has its capacity and resistance to separation. Also capacity of plastic shearing is limited. Those limitations cause INSTABILITY in crack response to increasing external load. Before conditions for instability are reached, crack will be stable. For more plastic materials crack will blunt and will move slightly in a stable manner. When conditions for instability are reached (the critical state of a crack) crack will start to propagate with a high velocity which depends on driving force, and this volocity may be a high fraction of Ragleigh wave speed C_R.

In both cases, slow and fast loading, it is crucial to formulate conditions for crack instability or FRACTURE CRITERION. At present there is no one universal fracture criterion. The formulation of fracture criterion may be as :

i. global
ii. local.

The global formulation is related to the whole conditions around a crack, defined for example by one parameter like G, K and J. For this case the following fracture criteria can be formulated :

i.	$G = G_c$	the critical energy release rate ;
ii.	$K = K_c$	the critical stress intensity factor in plane stress ;
	$K_I = K_{Ic}$	the critical stress intensity factor in plane stress ;
iii.	$J = J_c$	the critical value of J-integral.

The kinematic formulation of fracture criterion is the critical crack tip opening displacement

i.	$\delta_t = \delta_{yc}$	the CTOD (determined at the tip) ;
ii.	$\delta = \delta_c$	the COD (determined ouside the tip).

More recent formulations of fracture criteria are based on the local approach. For example, the critical conditions are reached when certain characteristic parameters exceed some characteristic quantities in the process zone (stress or strain). Two fracture criteria of this kind are used more frequently :

i. $\sigma_{22} = \sigma_F$ at $x_1 = l_F$ i.e. the normal stress σ_{22} must be equal to the fracture stress σ_F at the characteristic distance l_F from the crack tip ;

ii. $\varepsilon_{22} = \varepsilon_c$ at $x_1 = l_F$ i.e. the normal strain ε_{22} must be equal to the critical strain ε_c at the characteristic distance from the crack tip.

Anoter criterion derived from i. is the following :

iii. $(\sigma_{22})_{max} = \sigma_F$ the maximum normal stress $(\sigma_{22})_{max}$ must be equal to the fracture stress σ_F.

This criterion stems from the fact that blunting "unloads" the region of process zone and the σ_{22} component has a maximum.

All global criteria do not specify in which way a material is separated, i.e. micromechanics of fracture is not specified. Whereas, the local fracture criteria do specify to some extend the conditions within the process zone. It may be mentioned that besides J_c-fracture criterion the local criteria are very useful in computational fracture mechanics. They permit for prediction of fracture toughness from uniaxial properties.

Generally, fracture mechanics lacks of universal fracture criteria. It is difficult to establish sound physical basis related to ONE universal criterion. It is clear that fracture criteria should be based upon physical models. Investigations of physical fracture mechanisms bear largely on fractography, i.e. optical microscopy and electron microscopy. With the introduction of observation techniques based on the scanning electron microscopy it has become possible to observe the fracture surface directly.

Fractography has revealed a variety micromechanism of surface separation. The most common is classification as brittle or ductile separation. Those terms are not so precise, the classical fracture morphologies are as follows :

i. surface separation by cleavage (brittle) :

a. quasi-cleavage ;
b. intergranular cracking ;
c. transgranular cracking.

ii. fibrous separation (ductile) :

a. microvoid coalescence.

It is commonly assumed that mechanisms of separation associated with cleavage are controlled by the critical cleavage stress σ_F, whereas all fibrous mechanisms of separation are controlled by the critical strain ε_c.

An ideal cleavage fracture of metals occurs by direct separation along crystallographic planes oriented the most favorably to the normal stress σ_{22}.

The mechanism of cleavage is typical for low temperatures and also for a short time of loading. Cleavage dominates in BCC crystallographic structures. Very frequently cleavage is combined with twining.

Fracture associated with plastic deformation within process zone is essentially ductile. Extremes in fracture behavior are usually observed in tension tests by neck formation with almost 100% striction at higher temperatures and by entirely brittle fracture without any traces of plasticity at low temperatures.

Such dramatic change in fracture behavior and ductility is observed with decreasing temperatures and is called ductile to brittle transition. This transition is also rate-dependent.

To some extend the situation is similar within the process zone in a notched specimen or in specimen with a sharp crack. However, the geometric scale is different and some triaxiality of stress is present (i.e. triaxial tension). It is the reason why the critical cleavage stress σ_F is not exactly the same in the process zone and in an unnotched bulk specimen.

If plastic deformation is easier within the process zone, the process of material separation occurs by the mechanism of void growth and coalescence. The fracture surface after such separation shows characteristic dimples. Since engineering materials always contain second phase particles which are more brittle than matrix, the more brittle spots initiate microvoids. Consequently, size of dimples is related to the distribution of second phase particles.

Transition from brittle to ductile behavior is observed in certain structural materials, primarily structural steels, independently if bulk specimens are tested or the notch-toughness is investigated. It should be noted that certain structural materials, for example aluminum, aluminum alloys, titanium and titanium alloys (generally other structures than BCC) do not exhibit a transition-temperature behavior.

The transition-temperature behavior of 0.16%C steel with grain diameter D = 34 μm is shown in Fig. 2.1, after Hahn et al. [14]. The results have been obtained with cylindrical-tensile specimens.

At temperature T ≈ 70K a dramatic reduction of plasticity is observed. The following regions are present :

A - necking and cup-and-cone fracture ;
B - necking and mixed fibrous / cleavage fracture ;
C - begining of fracture (no necking) ; D and F - cleavage.

Boundary B/C defines the ductility transition temperature, T_d. Value of the critical cleavage stress σ_F is $\sigma_F \approx 965$ MPa. Below the ductility transition temperature T_d the fracture is completely cleavage and is initiated by cleavage microcracks. The frequency of microcracks rises with decreasing temperature in

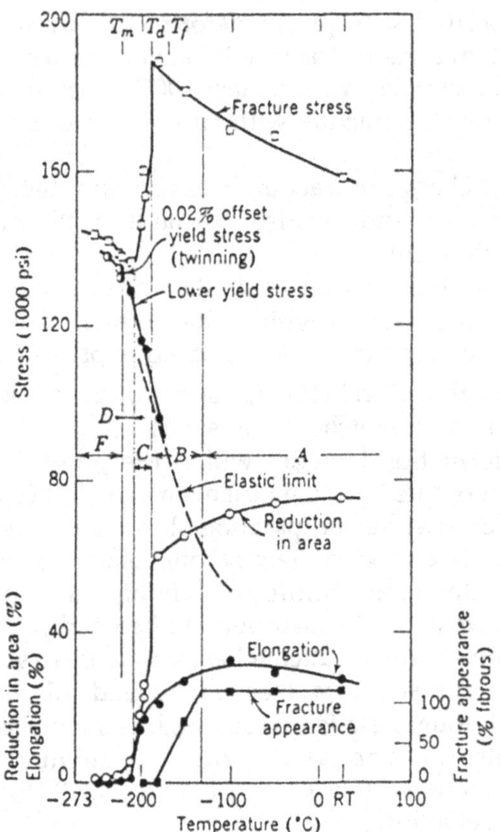

Fig. 2.1, Tensile properties and fracture appearance of steel with grain diameter D = 34 μm, redrawn after [14].

region C. Nevertheless, the samples were not entirely brittle in this range and elongations before fracture were within the range $0.1 \leq \varepsilon_F \leq 0.2$.

In region D, the σ_y and σ_F are almost identical, but yielding was always observed prior tof fracture. The number of microcracks decreased as the temperature is lowered. In region E, cleavage fracture occurs abruptly. Microcracks are not found in region E, presumally because the first one propagates to failure. In region F, specimens break abruptly with a slightly lower σ_F, the cleavage range in this region of temperatures is associated with mechanical twinning. These findings indicate that twins initiate cleavage in region F.

In conclusion, plastic deformation in the temperature ranges B, C and D appears to be a basic requirement for cleavage initiation. Since atom separation (cleavage) and dislocation movements (plasticity) are assisted by thermal vibration of atomic lattice, all features discussed above will be also RATE DEPENDENT. For example, the relation of microcrack formation to strain rate is shown in Fig. 2.2 (after Hahn et al.).

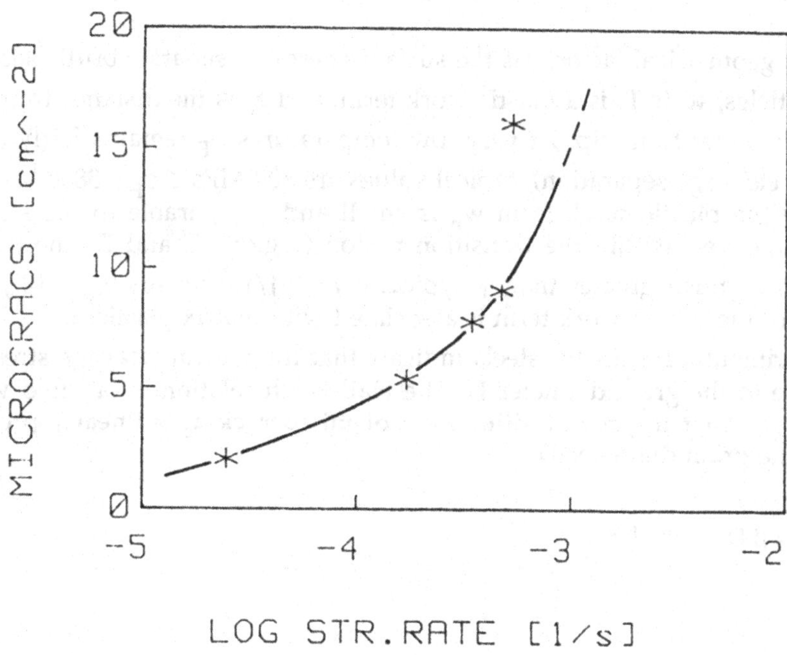

Fig. 2.2, Surface observation of microcrack density as a function of strain rate ; steel 0.22%C, D = 41 μm, test temperature T = 113K ; $\dot{\varepsilon} = 1 \times 10^{-3}\ 1/\text{min} = 1.66 \times 10^{-6} s^{-1}$; $\varepsilon = 0.06$, data after [14].

It appears that changes in σ_y ($\dot{\varepsilon}$) that are achieved by varying the strain rate had about the same effect on the number of microcracks as changes produced by varying the test temperature ; regions C and D.

Thus, microcrack dynamics is the key factor for understanding cleavage separation at different temperatures and strain rates. Many physical models for cleavage mechanism in steel have been proposed, the most representative is the Griffith-Orowan formulation for σ_F

$$\sigma_F = \left[\frac{\xi E(\gamma + w_p(\dot{\varepsilon},T))}{(1-v^2)\, a_c}\right] \tag{2.1}$$

$\xi = \pi/2$ for penny-shape microcrack

$\xi = 2/\pi$ for through-thickness microcrack

where ξ is geometrical factor, γ is the surface energy to separate brittle second-phase particles, $w_p(\varepsilon,T)$ is a plastic work term, and a_c is the distance from the microcrack center to its tip. At very low temperatures σ_F remains fairly constant (pure cleavage separation), typical values are 800 MPa $\leq \sigma_F \leq$ 3800 MPa. In this range the plastic work term w_p is small and comparable to the surface energy. However, within the transition region (regions C and D) the plastic work term is much greater than γ, typically $\gamma \approx 2\ \mathrm{J/m^2}$ versus $w_p \approx 10\ \mathrm{J/m^2}$. Existence of the plastic work term is associated with matrix plasticity.

Experimental results for steels indicate that the critical cleavage stress σ_F is sensitive to the grain diameter D. The Hall-Petch relation is satisfied when it is assumed that the critical dimension of microcrack a_c is linearly proportional to the grain diameter D

$$a_c = \beta D \quad \text{with} \quad \beta \leq 1$$

then

$$\sigma_F = \left\{\frac{\xi E[\gamma + w_p(\dot{\varepsilon},T)]}{(1-v^2)\, \beta D}\right\}^{1/2} \tag{2.2}$$

At higher temperatures, higher than for the regions C and D, the controlling factor of fracture is plastic strain (a critical value) and not σ_F. As it is shown in Fig. 2.1 the band of transition temperatures is narrow, $\Delta T \approx 50\mathrm{K}$. A very convenient plot to identify transition temperature is $\sigma = f(1/T)$, Louat and Wain, [15], as shown schematically in Fig. 2.3.

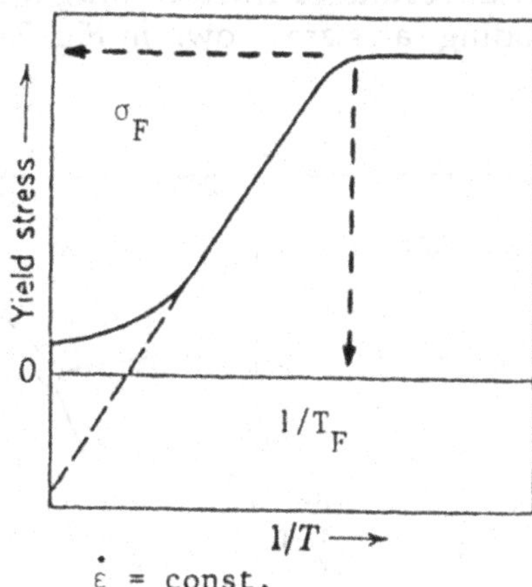

Fig. 2.3, Schematic relation between yield stress and inversion of temperature showing ductile/brittle transition.

The highest plateau identifies the critical cleavage stress σ_F. For example, for 4340 steel quenched and tempered 2 hrs at ~ 873K the cleavage stress is σ_F = 1655 MPa. It is known that transition temperature T_F of unnotched specimens (the ductile-brittle transition) depends on grain size D (at constant strain rate). For smaller D temperature T_F is lower.

According to thermodynamic principles, and Arrhenius relation based on statistical thermodynamics [16], values of T_F and $\dot{\varepsilon}_F$ should be related as follows

$$\frac{\dot{\varepsilon}_F}{\dot{\varepsilon}_{FO}} \ln\left(\frac{\Delta H_F}{kT_F}\right) = \text{const.} \tag{2.3}$$

where $\dot{\varepsilon}_{FO}$ is the reference characteristic strain rate and ΔH_F is the activation enthalpy for the ductile-brittle process. Since ΔH_F depends on hydrostatic pressure the transition temperature T_F and the cleavage stress σ_F are also pressure-dependent or depend on stress triaxiality.

There is a great interest to relate the unnotched tensile properties described above to the crack tip behavior. The main parameters to be investigated from the point of view of fracture mechanics is fracture toughness K_{Ic} and brittle-ductile transition temperature T_F. Because of spatial distribution of σ_{ij}, ε_{ij} and $\dot{\varepsilon}_{ij}$ around the crack tip the brittle-ductile transition is not so sharp as for bulk specimens. Typical results of fracture toughness testing at different temperatures and two loading rates are shown in Fig. 2.4.

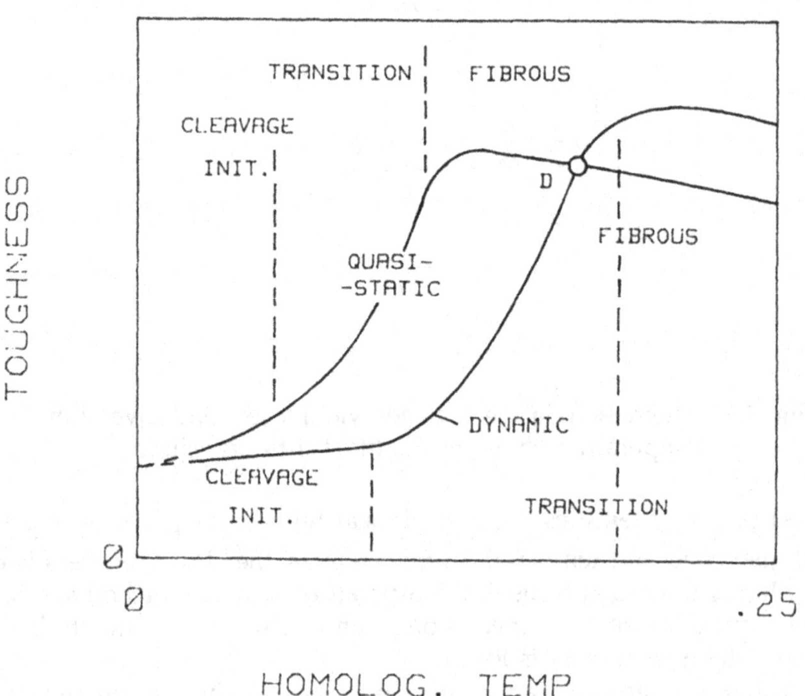

Fig. 2.4, Typical results of fracture toughness testing as a function of homologous temperature and two loading rates.

This schematic figure is not very exact, it shows however the main trends. According to the loading rate spectrum, Fig. 1.7, there is no such definition as "static" and "dynamic". In every case the loading rate $\dot{K}_I$ MUST be given for each curve. ASTM standards recommend that at each value of fracture toughness the critical time t_c [s] should be given in brackets $K_{Ic}(t_c)$. Concerning modelling the system with $\dot{K}_I$ is more flexible and rational. Of course, if t_c is known value of $\dot{K}_I$ can be calculated from the formula

$$\dot{K}_I = \frac{K_{Ic}}{t_c}\ \text{, the case of proportional loading} \tag{2.4}$$

In reference to this schematic figure the "static" curve is usually referred to the loading rate $\dot{K}_I = 1$ MPa$\sqrt{m}$/s whereas "dynamic" one is referred to $\dot{K}_I \approx 1 \times 10^5$ MPa$\sqrt{m}$/s.

A good example of experimental results for two steels on temperature transition is given in Fig. 2.5, after Wilson et al. [17]. In the case of 1018 cold rolled steel (1018 CRS) and 1020 hot rolled steel (1020 HRS) temperature transition from cleavage to fibrous separation is clearly indicated. For the high loading rate the transition occurs at higher temperature and is usually better defined.

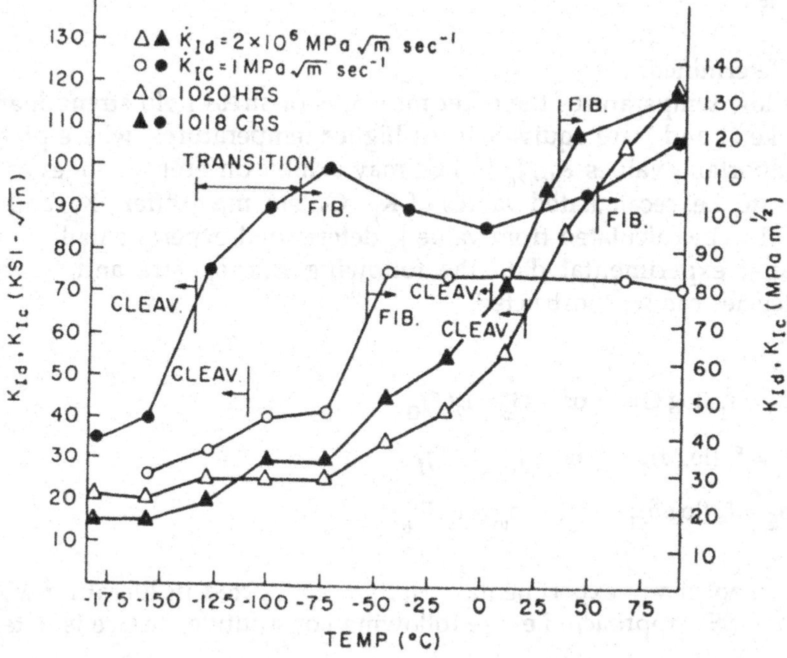

Fig. 2.5, Fracture toughness for 1020 hot-rolled steel ; quasi-static $\dot{K}_I = 1$ MPa$\sqrt{m}$/s, and fast loading $\dot{K}_I = 2\times10^6$ MPa$\sqrt{m}$/s, after [17] ; courtesy of Pergamon Press.

Within the region of very low temperatures fracture toughness exhibits a plateau, so called lower shelve of fracture toughness, whereas in the region of higher temperatures (usually room temperature) another plateau or maximum is observed, called the upper shelve of fracture toughness. It is important to note that up to certain temperature, close to point D in the schematic figure 2.4, the loading rate sensitivity of fracture toughness is NEGATIVE. On the other hand, the loading rate sensitivity of the upper shelve is POSITIVE.

The largest horizontal difference in temperature between low and high loading rate is the temperature shift which defines translation of brittle to ductile transition as a function of loading rate. The translation obeys usually the Arrhenius law. The highest negative rate sensitivity is observed in the region of transition temperature.

The most accepted approach to determine transition temperature is based on SSY and fracture toughness. Also, loading rate sensitivity of fracture resistance is usually determined via SSY. For this case the loading rate spectrum in the form

$$K_{Ic} = f\,(\log \dot{K}_I)_T \tag{2.5}$$

can be determined.

At low temperatures the other measures of stress field strength around a crack, like G and J are equivalent. At higher temperatures, where plasticity is more important, values K_{Ic}, J_c and G_c may exhibit different changes as a function of rate, i.e. recalculated values of K_{Jc}, G_{Jc} etc. may differ ; K_{Jc} and G_{Jc} are supposed to be calculated from value J_c determined experimentally.

Using experimental data the following rate spectra and temperature dependencies can be constructed

$$\begin{aligned} G_c &= f_1\,(\log \dot{G})_T \quad \text{or} \quad G_c = f_2(T)_{\dot{G}} \\ J_c &= f_3\,(\log \dot{J})_T \quad \text{or} \quad J_c = f_4(T)_{\dot{J}} \\ \delta_{tc} &= f_5\,(\log \dot{\delta})_T \quad \text{or} \quad \delta_{tc} = f_6(T)_{\dot{\delta}} \end{aligned} \tag{2.6}$$

However, not many experimental results are reported in literature which use other than SSY approach, i.e. the following constitutive surface is determined

$$K_{Ic} = f_1\,(\log \dot{K}_I) \quad \text{and} \quad K_{Ic} = f_2\,(T)_{\dot{K}}$$

For example, Krasovsky et al. [18] reported experimental results on stereographic measurements of the total stretched zone height 2h, $\delta_{tc} \approx 2h$, as a function of temperature for two loading rates. Determined changes of 2h are similar to changes of K_{Ic} in similar conditions of T and $\dot{K}_I$. The stretched zone height is defined as shown in Fig. 2.6.

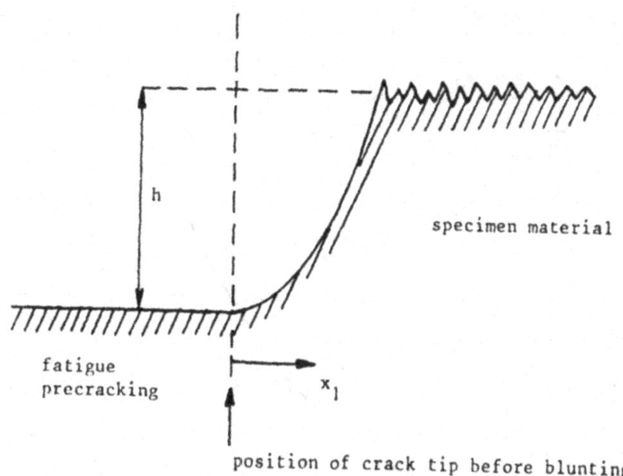

Fig. 2.6, Definition of the stretch zone h.

It is shown in Fig. 2.7 that changes of the stretch show some similarities as a function of temperature to the analogous changes of K_{Ic} $(T,\dot{K}_I)$ from Fig. 2.5. Reduction of the stretch zone at the higher COD rate indicates for a more brittle behavior when the rate is increased.

All informations discussed in this chapter are very useful in constructing fundamental relations between unnotched tensile properties (dynamic plasticity) from one side and fracture resistance observed for a stationary crack loaded at different rates and temperatures.

General relation in viscoplasticity is defined in the form of the constitutive surface

$$\sigma = f_1\,(\varepsilon,\dot{\varepsilon},T) \tag{2.7}$$

On the other hand, general relation for stationary cracks in SSY is

$$\begin{aligned} K_{Ic} &= f_2\,(\dot{K}_I,T)_{v=0} \quad \text{plane strain} \\ K_c &= f_2^q\,(\dot{K},T)_{v=0} \quad \text{plane stress} \end{aligned} \tag{2.8}$$

v = 0 means the stationary crack. Some concepts how to correlate rate and temperature dependent uniaxial properties in the form (2.7) with rate and temperature dependence of fracture toughness in the form (2.8) will be discussed in the next parts of this book.

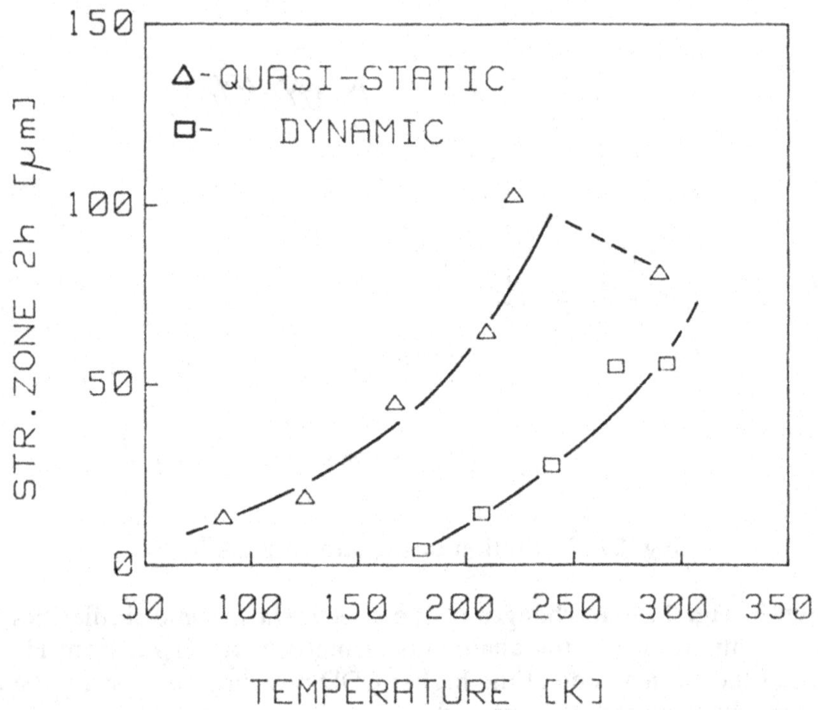

Fig. 2.7, Temperature dependence of the total stretch zone 2h for a low-alloy steel, quasi-static -1, and dynamic loading -2 ; data after [18].

3. SOME EXPERIMENTAL METHODS TO DETERMINE FRACTURE TOUGHNESS AT LOW AND MEDIUM LOADING RATES, TYPICAL RESULTS

During last decade a growing interest has been observed in studies on rate effects in fracture mechanics. The main reason of those studies is that the techniques of fracture mechanics are proving to be of great value when used to asses the safety of engineering structures. This has led to the adoption of the plan strain fracture toughness K_{Ic}, the critical value of the CTOD = δ_{tc}, and the critical value of J-integral J_c, as meaningful design parameters. While there is a tendency to treat those parameters by engineers as material constants it must be remembered, as it was shown in precedent parts of this text, that those parameters are temperature and rate dependent.

Many available experimental results have indicated that under certain conditions of temperature and loading rate a substantial decrease of fracture toughness may occur. This is a very important engineering problem. Recent results show that the resistance to fracture of some structural alloys, including steels, may vary in a highly nonlinear manner when the loading rate is changed over several orders of magnitude.

Thus, it is of great interest to determine fracture resistance of structural materials over the widest possible range of loading rates at different temperatures. Such methodology would permit to specify the constitutive surface, for example within the framework of SSY, as

$$f(K_{Ic}, \dot{K}_I, T) = 0 \quad , \quad v = 0$$

It may be mentioned that the loading rate $\dot{K}_I$ for the stationary crack, even during relatively fast loading, may substantially differ from that determined during crack propagation, i.e. K_{ID}, where K_{ID} is dynamic fracture toughness determined during unstable crack propagation with $v \neq 0$, thus, $\dot{K}_{ID} = (\partial K_{ID}/\partial t)_v$. In the case of crack propagation the loading rate $\dot{K}_{ID}$ is usually very high, typically $\dot{K}_{ID} \approx 1\text{x}10^9$ MPa$\sqrt{\text{m}}$/s in comparison to the typical very high loading rate $\dot{K}_I \approx 1\text{x}10^6$ MPa$\sqrt{\text{m}}$/s for a stationary crack.

In order to cover a large part of the loading rate spectrum several experimental techniques must be employed. Higher loading rates than 1 MPa$\sqrt{\text{m}}$/s can be achieved with standard testing machines and standard specimen geometries only up to $\dot{K}_I \approx 1\text{x}10^3$ MPa$\sqrt{\text{m}}$/s. The fast closed-loop, servo-controlled testing machines enable to reach the loading rate up to ~ $1\text{x}10^4$ MPa$\sqrt{/\text{m}}$/s.

The most popular specimen geometries used in the systems for fast loading are the Compact Tension (CT), Three Point Bend specimen (3PB) and Notched Round specimen loaded in Tension (NRT). Those geometries are shown in Fig. 3.1.

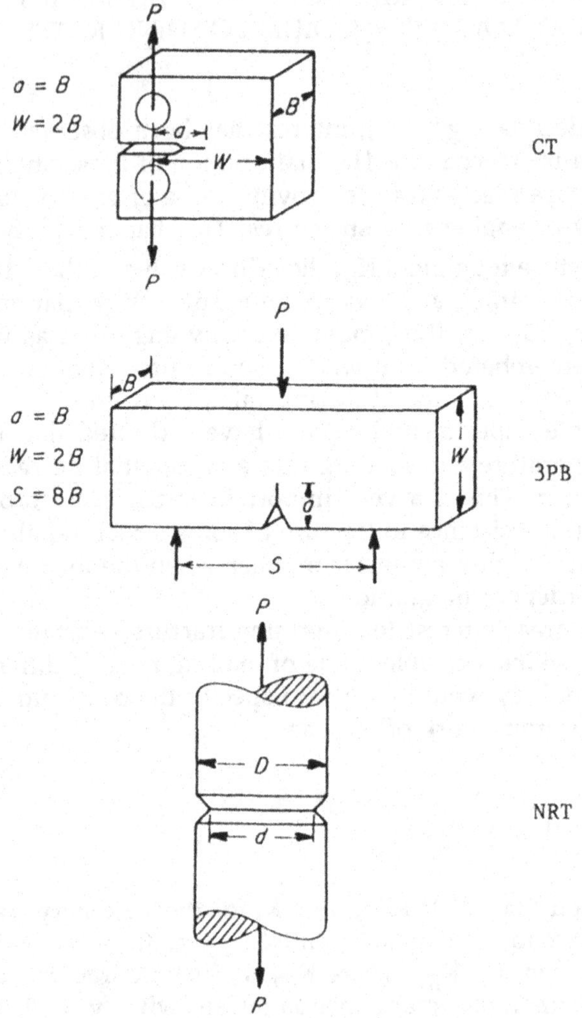

Fig. 3.1, Specimen geometries used most frequently for impact tests ; a - Compact Tension (CT), b - Three Point Bent (3PB), c - Notched Round Tensile (NRT).

Relations between stress intensity factor K_I, force P and specimen geometry can be found, for example, from national standards. For CT and 3-point bent specimens

$$K_I = P f_i \left(\frac{a}{W}\right) \tag{3.1}$$

where f_i is a compliance function,

$i = 1$, CT specimen

$f_1 = 1.107 - 2.120\,(a/W) + 7.71\,(a/W)^2 - 13.55\,(a/W)^3$

$i = 2$, 3PB specimen , $S/W = 8$ (S/W - the span ratio)

$f_2 = 29.6\,(a/W)^{1/2} - 185.5\,(a/W)^{3/2} + 655.7\,(a/W)^{5/2} - 1017\,(a/W)^{7/2} + 638.9\,(a/W)^{9/2}$

For NRT specimen with d_f, where d_f is the diameter of prefatigued ligament, $d_f < d$, the compliance function is

$$K_I = P f_3 \left(\frac{d_f}{D}\right) \tag{3.2}$$

$$f_3 = \frac{1}{2}\left[1 + \frac{1}{2}\left(\frac{d_f}{D}\right) + \frac{3}{5}\left(\frac{d_f}{D}\right)^2 + 0.363\left(\frac{d_f}{D}\right)^3 + 0.731\left(\frac{d_f}{D}\right)^4\right]$$

Those geometric corrections are derived for quasi-static loadings. They must be used in caution in the case of impact loadings. Even at moderate loading rates some technical difficulties are experienced. For example, vibrations are usually superimposed on the main signal measuring force P. This causes that it is difficult to analyse load-displacement curves. The idealized load-displacement characteristics, as shown in Fig. 3.2 are important to determine proper values of the critical force P_c.

In addition to vibrations excited in the loading system, displacement gage (clip gage) usually recommended by standards, may also vibrate. One possibility to reduce vibrations is to reduce masses of moving parts, i.e. specimen size, etc. Reduction of specimen geometry is not always possible if the SSY approach should be retained. Thus, some extra precautions are important to deal with vibrations.

The most difficult problems are created by the 3-point bend specimen. The reason is that this geometry has the lowest frequency of free vibrations of all three geometries discussed.

Two configurations of specimen loading (CT) which are relatively efficient at elevated loading rates are shown in Fig. 3.3 and Fig. 3.4. The first one,

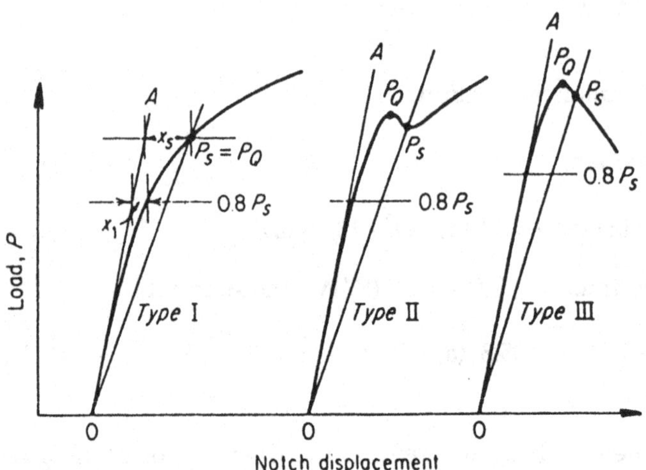

Fig. 3.2, Idealized load-displacement curves as defined by many standards ; A - elastic slope, P_c - critical force.

reported by Marandet et al. [19], can be used with closed loop actuator. As shown in Fig. 3.3, CT-25 specimen 1 is loaded with predetermined rates by the actuator 7 through the electrically isolated specimen holder 2. The impact produced at higher loading rates is damped by the shock absorber 5. The force in the system of loading is converted into an electrical signal by the load cell 6. Tests can be performed at different temperatures since the specimen is submerged into the cooling mixture, temperature is controlled by the cooling system which operates with liquid N_2. The measuring system consists, besides the load cell 6, the potential drop circuit and the clip gage 8 which measures the crack opening displacement. All measured quantities like actuator stroke, the electric potential, the load and the COD are simultaneously recorded as a function of time. An analysis of the signals enables for determination of fracture toughness K_{Ic} or critical value of J-integral, J_c at different loading rates. The maximum loading rate reached with this system is $\dot{K}_I \approx 2 \times 10^4$ MPa$\sqrt{m}$/s.

A similar system, as shown in Fig. 3.4, was applied by Krabiell and Dahl [20]. A high speed hydraulic testing machine was used and variable masses 7 were placed in series with the actuator 8 to control the kinetic energy of the system. In addition to the mass a shock absorber with the rubber washers 6 was used to reduce vibrations. The force in the system is monitored by the piezo-quartz load cell 4 with the natural frequency ~ 20 kHz. In addition to the load cell 4, the force acting on the CT specimen can be monitored by two strain gages 9 cemented to its outer surface. The crack opening displacement is

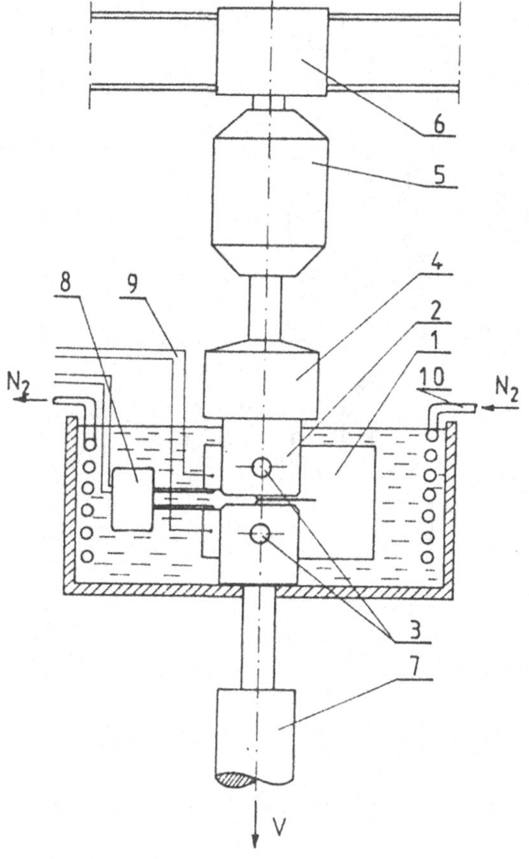

Fig. 3.3, The scheme for fast loading of CT-25 specimen, [19].

monitored by the LVDT displacement gage 10. All tests can be performed over the range of temperatures between 77K and room temperature. The highest loading rate is about $\dot{K}_I \approx 4\text{x}10^5$ MPa√m/s. However, interpretation of the records obtained at such loading rate is questionable, mainly due to the oscillations caused by the vibrating masses. The signals from strain gages 9 placed on the specimen sides appeared to be advantageous in measurements of the load-time records.

Both systems indicate certain similarities in design and operation. They can cover relatively wide range of loading rates, approximately from quasi-static to $\dot{K}_I \approx 5\text{x}10^4$ MPa√m/s, and simultaneously a good resolution in fractu-

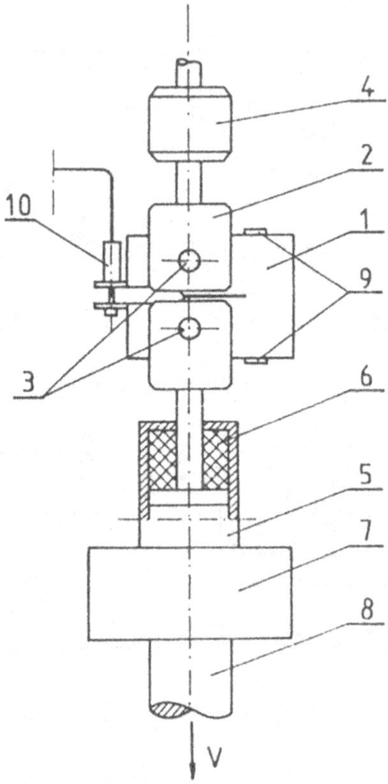

Fig. 3.4, The scheme for fast loading of CT specimen, [20].

re toughness determination can be achieved. Such experimental techniques are suitable to cover the lower part of the loading rate spectrum.

Very efficient methods for measurement of kinematics in impact testing and fracture dynamics are based on optics and optoelectronics. One of those applications, reported by Kussmaul et al. [21], consists of two optical fiber leads inserted into the CT specimen by the drilled holes of 0.9 mm diameter. The axis of holes are perpendicular to the crack plane and they are close to the crack tip. Such configuration permits for exact measurements of CTOD. The opposite fiber ends are fixed to the crack vicinities, when crack opens the distance between the two fibers increases accordingly. In the result the beam of light transmitted through the system is attenuated. The intensity of attenuation is related to CTOD. The system is shown schematically in Fig. 3.5.

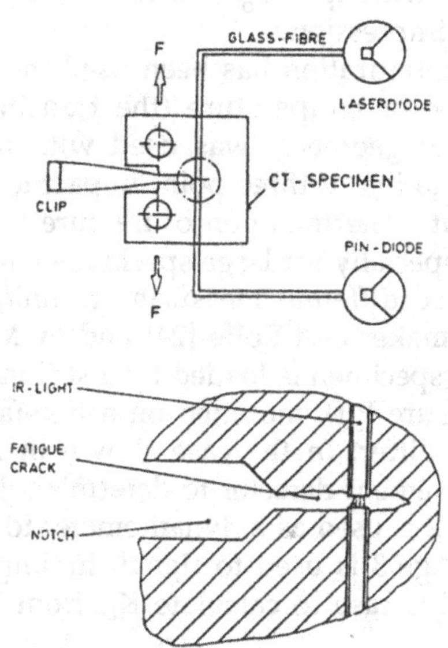

Fig. 3.5, Application of optoelectronics to CTOD measurement, [21], reproduced with permission from Les Editions de Physique.

The frequency response of the system depends on characteristics of the photodiode and in the discussed case was ~ 10 MHz.

Measuring systems based on the principle of Moiré fringes are also popular in inertia-free displacement measurements for example [22] and [23]. Some of such concepts will be discussed later.

The test geometry most widely used for determining K_{Ic} at high loading rate (impact) is a prefatigued three-point bend specimen. The three point bend configuration is not optimal concerning dynamics, vibrating masses with a low frequency, slow bending waves, sometime too large dimensions cause a lot of difficulties in interpretation of test recorde according to Fig. 3.2. This geometry was automatically transferred from the quasi-statics. Specimen dimensions in bending tests may be quite different, starting from the smallest - the Charpy specimen with supports 40 mm apart and the cross section 10 mm x 10 mm, with V-notch 2 mm deep, to very large with support distance 410 mm, the cross section 25 mm x 80 mm with notch depth ~ 30 mm. Even larger specimens have been reported as 600 mm, cross section 18 mm x 150 mm and notch depth 75 mm.

The dynamicf tear test with three point bend specimen loaded by impact is standarized in many countries. The specimen is a sharply notched at the center with increased height W = 120 mm and thickness B = 25 mm, the support distance L = 410 mm, notch depth α_0 = 45 mm. The notch in this specimen is produced by the tool impression.

The three point bent configuration has been used initially to determine fracture energy as a function of temperature (the transition curve). In the recent decade the three point geometry was used with instrumentation to fracture dynamics. Implementation of three point impact loaded specimens to fracture mechanics is difficult. Determination of fracture toughness from P(t) records is not always exact (specially for large specimens). Many researches in different laboratories proposed different measuring techniques. For example, the early approach by Schoemaker and Rolfe [24] and by Madison and Irwin [25] is shown in Fig. 3.6. The specimen is loaded by a striking tup, usually instrumented with gages to mesure P(t), mounted on a free-falling weight. Specimen instrumentation may differ. In the case shown in Fig. 3.6 the strain gage 1 is used as a crack movement detector to determine the instant of crack propagation. The strain gage 2 is used as a dynamometer to determine the critical bending moment and gage 3 is used to signal the impact. The standard static relations are used in this case to calculate K_{Ic} from the strain gage signals.

A more modern and rational approach to the three point bend configuration has been introduced by Kalthoff et al. [26]. Application of the so-called impact response curve and the method of caustics have allowed for dynamic analyses of the test and its better interpretation. In addition, a magnetic gage applied to ferromagnetic materials (steels) permits for determination the onset of crack propagation, [27].

The experimental methods which apply closed-loop testing machines and drop weight hammers are useful to load cracks with loading rate up to $\dot{K}_I \approx 1\text{x}10^5\ \text{MPa}\sqrt{\text{m}}/\text{s}$. The advantage of the closed-loop testing machines (they can be very fast now) lies in a possibility of displacement regulation and programing of actuator. Devices with a falling weight develop a lot of vibrations caused by an abrupt impact.

These experimental techniques can be applied at different temperatures T to find the constitutive surface for fracture initiation, i.e.

$$K_{Ic} = f\,(\dot{K}_I, T)_{v=0} \quad ; \quad B = \text{const.} \tag{3.3}$$

Such a three-dimensional surface characterizes completely the resistance of a material to fracture initiation within the framework of the SSY. The constitutive surface (2.8) can be shown in two dimensions by one of two ways. The first one is the cross-section

$$K_{Ic} = f_1\,(T)_{\dot{K}} \quad , \quad v = 0 \tag{3.4}$$

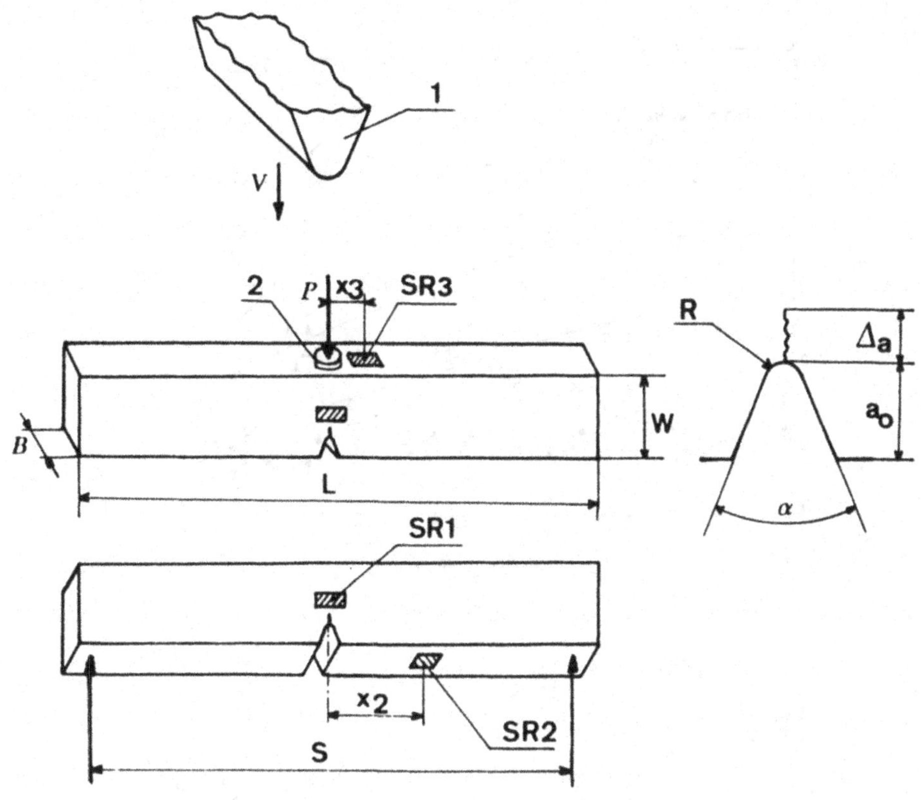

Fig. 3.6, Dynamic test configuration, 3PB specimen, after [24] and [25].

the second is to construct the loading rate spectra (3.5)

$$K_{Ic} = f_2 (\log \dot{K}_I)_T \quad , \quad V = 0 \tag{3.5}$$

Currently, few complete experimental studies exist that shows the large part of the surface. The most well known are experimental results of Westinghouse, Shabbits, [30], for A533-B steel. Those results together with additional cross-section $K_{Ic}(T)_K$ are shown in Fig. 3.7.

To characterize rate sensitivity of fracture toughness the following definition is introduced

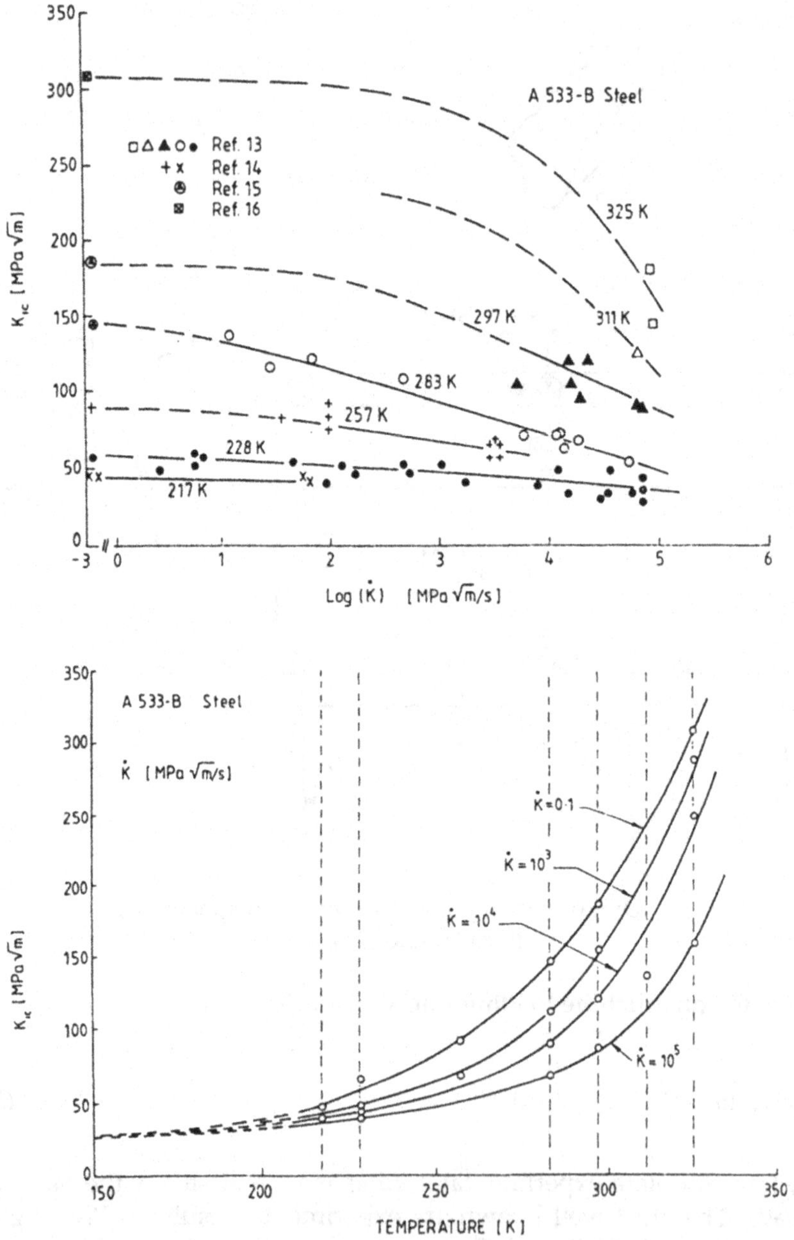

Fig. 3.7, Fracture toughness as a function of loading rate and temperature for A533-B steel ; after [28] and [29] ; (a) - $K_{Ic}(\log \dot{K}_I)$ for seven temperatures, the references as indicated are given in [28] ; (b) - $K_{Ic}(T)$ for four loading rates.

$$\eta = \left(\frac{\partial K_{Ic}}{\partial \log \dot{K}_I}\right)_T \quad [\mathrm{MPa}\sqrt{\mathrm{m}}] \tag{3.6}$$

The rate sensitivity η has NEGATIVE value in the range of loading rates studied for A533-B steel and the absolute value of η_{min} increases with increasing temperature, as shown in Fig. 3.8.

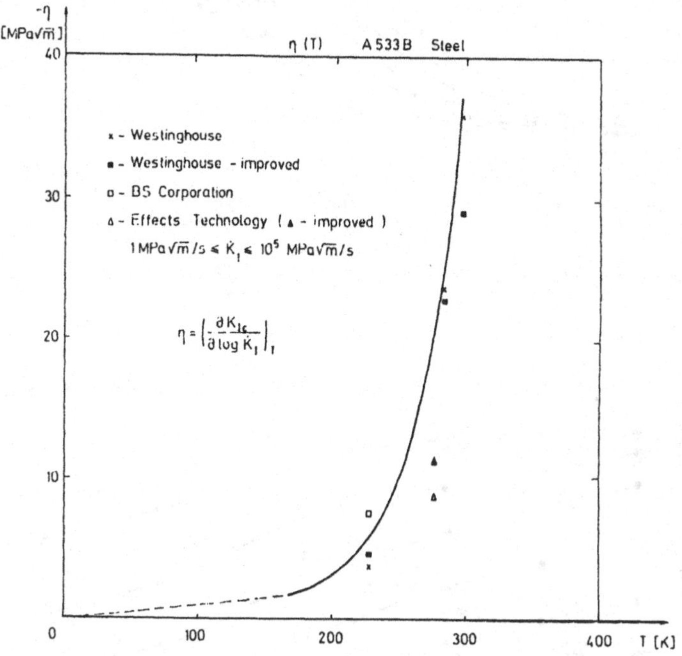

Fig. 3.8, Rate sensitivity of fracture toughness $-\eta$ for A508 Cl.3 steel determined for two extreme loading rates : $\dot{K}_{min} = 2.0$ MPa$\sqrt{\mathrm{m}}$/s and $\dot{K}_{max} = 1\times10^6$ MPa$\sqrt{\mathrm{m}}$/s ; after [28].

A very systematic experimental study of fracture toughness at different temperatures and loading rates has been performed for three structural steels by Krabiell and Dahl [20]. The most complete data for Fe E 460 steel (German Standards) with additional cross-sections $K_{Ic} = f_2(\log \dot{K})$ published in [28] are shown in Fig. 3.9.

The conclusion which can be drawn from the above figures is the similarity as a whole of the loading rate spectra, $K_{Ic}(\log \dot{K}_I)_T$, and also the entire surfaces $K_{Ic}(T,\dot{K}_I)$ do show substantial similarities. Thus, such behavior as

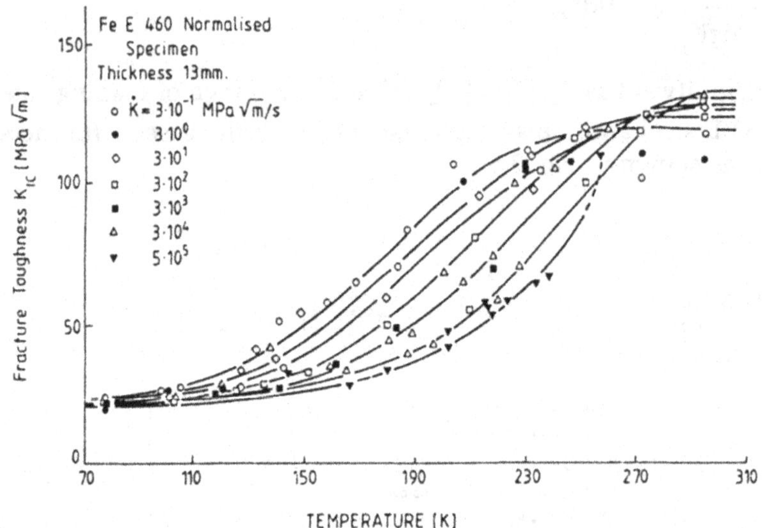

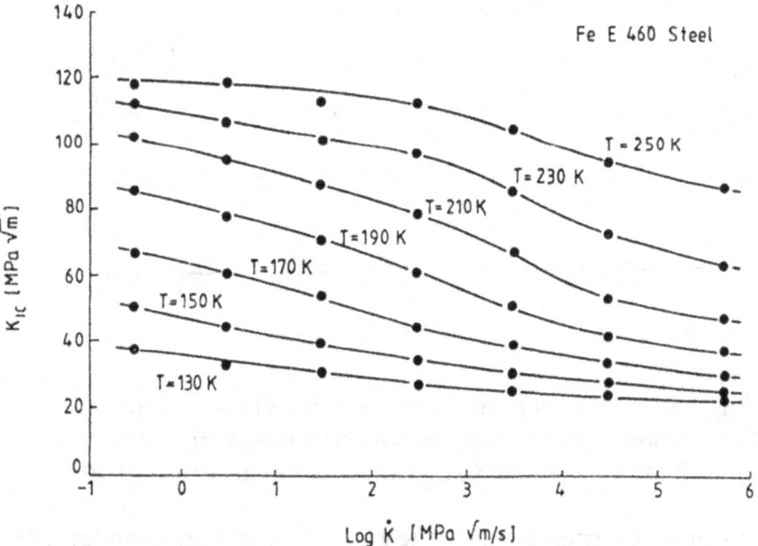

Fig. 3.9, Fracture toughness as a function of loading rate and temperature for Fe E 460 steel (German Standards) ; after [20] and [28] ; (a) - $K_{Ic}(T)$ for seven loading rates ; (b) - $K_{Ic}(\log\dot{K}_I)$ for seven temperatures.

demonstrated in these figures is assumed to be typical for a variety of steels. The constitutive surface $K_{Ic}(T,\dot{K}_I)$ is usually characterized by two regions. Low and medium temperature domain with NEGATIVE loading rate sensitivity, and a higher temperature domain with mild, POSITIVE loading rate sensitivity.

The experimental results shown in Fig. 3.9 served for the construction of the three-dimensional surface as shown in Fig. 3.10. The surface $K_{Ic}(T,\dot{K}_I)$ is constructed only for the lower shelf response where cleavage separation dominates. However, the loading rate $\dot{K}_I$ and temperature T in the constitutive relation $K_{Ic}(T,\dot{K}_I)$ are understood as parameters and not as independent variables, consequently the pulse loading will not be here considered.

Numerical calculations and theoretical analyses of the experimental results indicate that there exists an absolute minimum of fracture toughness $(K_{Ic})_{min}$ which is characteristic for zero absolute temperature. This is rate insensitive material constant which exclusively depends on microstructure. For many structural steels $(K_{Ic})_{min}$ appears to be around 20 MPa$\sqrt{m}$. For example, the estimated value of $(K_{Ic})_{min}$ for Fe E 460 steel is 22.7 MPa$\sqrt{m}$. This value is indicated in the 3D figure as the solid line for T = 0, (Fig. 3.10).

Interpretation of experimental results in the region of the upper shelf is difficult and frequently impossible within the framework of the SSY. The main problem is that the degree of triaxial constraint at the crack tip is substantially reduced by enhanced plasticity and by reduction of the specimen thickness B. This effect is different for different loading rates and is well illustrated by the data of Krasovsky et al. [18] that are shown in Fig. 3.11. It can be seen that the apparent values of K_Q for the upper shelf occur at different temperatures and different loading rates depending upon the specimen thickness B. The dashed lines in Fig. 3.11 are obtained from the equation

$$K_Q^2 = 0.4\,B\,\sigma_{02}^2 \quad \text{(ASTM condition)} \tag{3.7}$$

where σ_{02} is the yield stress at 0.2% offset strain. Of course, values of K_{Jc} calculated from determined experimentally J_c-values are different than K_Q values for the upper shelf (they are much higher).

It is preasumed that values of K_Q for the upper shelf are very close to the plane stress fracture toughness K_c. This is illustrated by the plot $K_c(B)$ in the case of quasi-static loading for 4340 steel, Jones and Brown [31], which is shown in Fig. 3.12.

Since values of K_Q (i.e. non-valid K_{Ic}) depend on specimen thickness B, the thickness should be given if a whole rate and temperature spectrum is covered. In another words, for each constitutive surface determined the specimen thickness is an essential information.

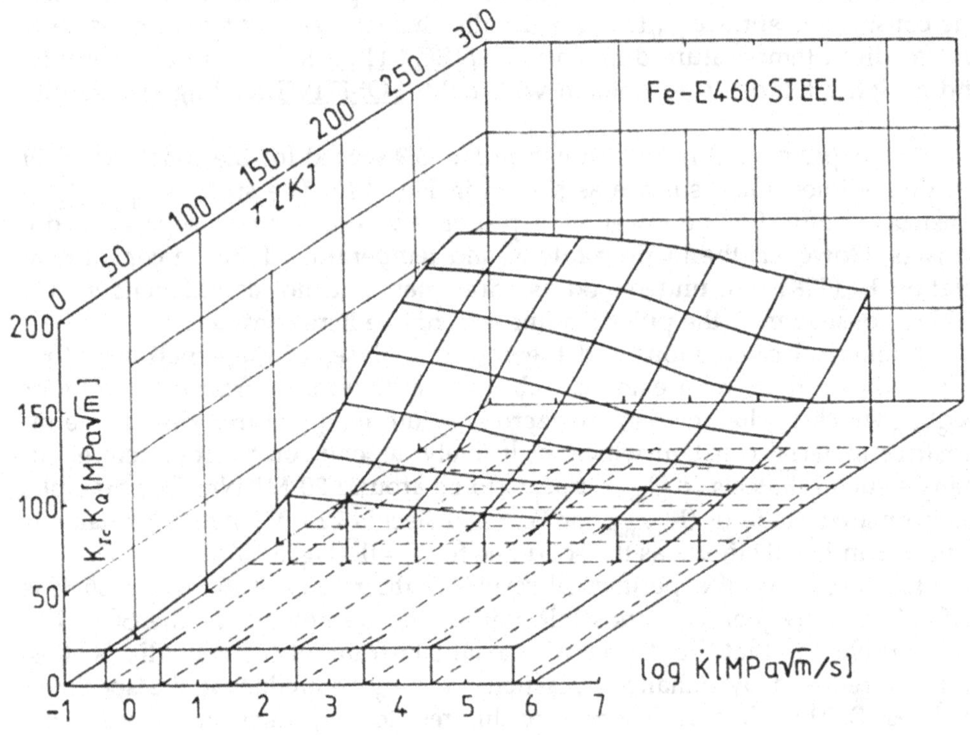

Fig. 3.10, Three-dimensional constitutive surface for crack initiation constructed using data of Fig. 3.9 for the lower shelf. Broken lines indicate test temperatures and loading rates, after [30].

If at higher temperatures a positive loading rate sensitivity is found, it means that the mechanism of fibrous separation probably occurs and the upper shelf is tested.

In the case of alloy steels or steels with a higher carbon content which are heat-treated to high values of hardness the effect of loading rate on fracture toughness is negligible. For example, it has been reported by Nunomura et al. [31] that in the case of ball bearing steel values of K_{Ic} determined up to the loading rate $\dot{K}_I = 2.2 \times 10^5$ MPa$\sqrt{m}$/s retain the constant value $K_{Ic} \approx 21.1$ MPa$\sqrt{m}$.

In conclusion, it is clear that plasticity present within the crack tip vicinity plays a decisive role in fracture toughness performance. If the yield limit is

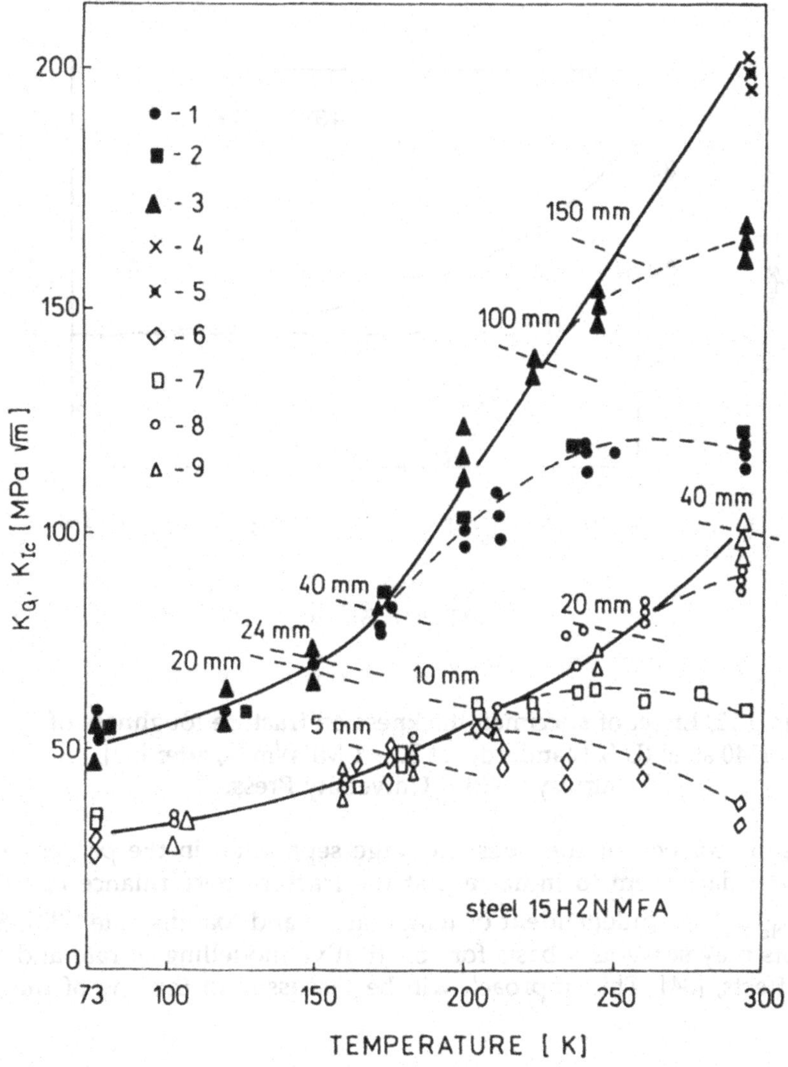

Fig. 3.11, Fracture toughness for a low alloy steel (Soviet Standards) determined as a function of temperature at two loading rates quasi-static $\dot{K}_I \approx 1$ MPa√m/s and impact $\dot{K}_I \approx 1\text{x}10^5$ MPa√m/s ; specimen thickness is shown in milimeters, after [18].

higher (due to decrease of temperature or increase of strain rate) fracture toughness will change accordingly as illustrated in Fig. 3.13 - it will decrease.

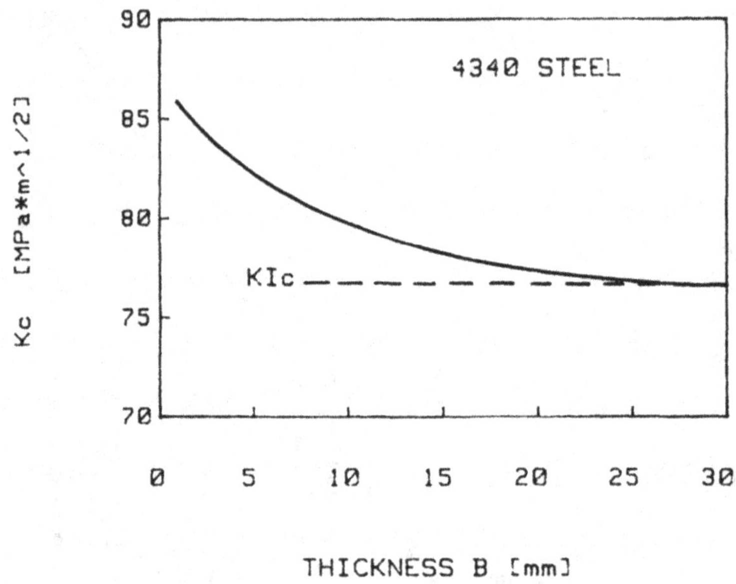

Fig. 3.12, Effect of specimen thickness on fracture toughness of 4340 steel (U.S. Standards) at $\dot{K}_I \approx 1$ MPa$\sqrt{m}$/s, after [32], courtesy Oxford University Press.

Such a rule is correct for the quasi-cleavage separation in the process zone. Experimental data seem to indicate that the fracture performance curves in the form $K_{Ic}(\sigma_y)$ are independent of temperature and loading rate, [33]. Such master plots may serve as a basis for constitutive modelling of rate and temperature effects, [34]. This approach will be discussed in the one of the next chapters.

4. APPLICATION OF ELASTIC WAVES TO FRACTURE DYNAMICS - THE HOPKINSON BAR AND KOLSKY APPARATUS

Hopkinson in 1914, [35,36], was the first to devise a satisfactory method of measuring short pressure pulses. Hopkinson's original apparatus, which has become known as the "Hopkinson pressure bar", consisted of a steel bar suspended balistically, so it could swing after impact in a vertical plane, and a short cylindrical pellet, called the "time-piece". The pressure to be measured was applied to one end of the bar whilst at the other end a cylindrical "time piece", was attached. These "time piece" (short bars) were of the same diameter as the bar and of the same steel. The short bar could be attached to the end of the bar with very little grease.

The principle is that when a pulse of compression travels down the bar with the velocity C_o

$$C_o = \sqrt{\frac{E}{\rho}} \tag{4.1}$$

where E is Young's modulus and ρ is the bar density, it is transmitted through the joint of a short bar without change of form. When it reaches the free end of the short bar it is reflected as a pulse of tension. Since the joint is very weak the short bar flies off at the instant when any tension appears at the interface. The momentum trapped in the short bar corresponds to such part of the length of the incident pulse which is twice that of the short bar, i.e. the length of the trapped pulse is λ

$$\lambda = C_o t \quad ; \quad \lambda = 2L_t \tag{4.2}$$

where L_t is the length of the short bar. If the experiment is repeated with short bars of different length, and the momentum is measured each time by ballistic pendulum, the nature of the pressure-time relation for the incident pulse can be evaluated. More precisely, the maximum pressure and the time for which any value of the pressure was exceeded could be deduced.

An electrical version of the Hopkinson bar was devised by Davies [37]. The bar has been devised with the condenser microphone measuring displacement as a function of time of the free end, (without the "time piece"). This has enabled the pulse to be exactly measured, provided that its length is comparable with the diameter of the bar. One of disadvantages in measuring displacement U(t) is necessity to differentiate the measured pulse to determine longitudinal stress (or pressure) $\sigma(t)$

$$\sigma(t) = \rho C_o \frac{\partial u}{\partial t} \quad , \quad v = \frac{\partial u}{\partial t} \tag{4.3}$$

where u is bar displacement and v is the mass velocity.

In 1945 Kolsky [38] proposed a new configuration of TWO Hopkinson bars to determine the stress-strain relation of materials when deformation time is short, typically from 20 μs to 100 μs. The specimen in the form of a short disc was inserted in between two bars, which were fitted with condenser microphones. The incident pulse was generated by a detonator. The general arrangement of the Kolsky apparatus is shown in Fig. 4.1 (version of 1949).

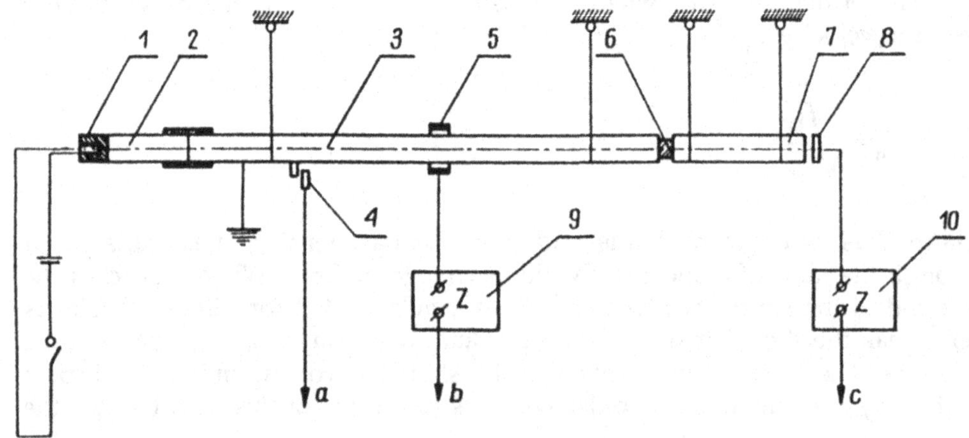

Fig. 4.1, General arrangement of the Kolsky apparatus [38] ; Split Hopkinson Pressure Bar (SHPB), from [39] ; a, b, c outputs to the oscilloscope.

The pressure pulse is produced by a detonator 1 which is hold against a replacable steel anvil 2. After explosion the compression elastic wave travels down the incident bar 3 and compresses cylindrical specimen 6 which is placed between the bars. The shape of the incident wave is measured with a cylindrical condenser gage 5 the output of which is amplified and fed by the unit 9. The electric signal from the unit 9 is recorded by a double beam oscilloscope. The time base of the oscilloscope is triggered by the switch 4. Part of the incident wave is reflected from the specimen and part is transmitted into the transmitter bar 7. The transmitted wave travels down the transmitter bar and is measured by the second condenser gage 8. The shape of the transmitted wave is recorded with the oscilloscope following amplification in the unit 10.

The oscilloscope traces are the incident and reflected waves in the form of pressure-time relations (gage 5) called $\sigma_I(t)$ and $\sigma_R(t)$, and the displacement-time relation of the free end (gage 8) called $\xi(t)$. From $\sigma_I(t)$, and $\sigma_R(t)$ and $\xi(t)$ the stress-strain curve of the specimen may be determined.

The transmitted stress-time relation $\sigma_T(t)$ may be obtained from the relation

$$\sigma_T(t) = \frac{1}{2} \rho\, C_o \frac{d\xi_B}{dt} \tag{4.4}$$

In order to determine the strain-time relation for the specimen $\varepsilon_s(t)$ the displacements of the two ends of the specimen must be considered separately. The displacement of the incident bar due to incident wave is given by

$$\xi_1 = \frac{1}{\rho C_o} \int_o^t \sigma_I(\theta)\, d\theta \tag{4.5}$$

whilst the displacement due to reflected wave is

$$\xi_2 = \frac{1}{\rho C_o} \int_o^t [-\sigma_R(\theta)]\, d\theta \tag{4.6}$$

Thus the total displacement of the specimen face is

$$\xi_A = \frac{1}{\rho C_o} \int_o^t [\sigma_I(\theta) - \sigma_R(\theta)]\, d\theta \tag{4.7}$$

The displacement due to transmitted wave is

$$\xi_B = \frac{1}{\rho C_o} \int_o^t \sigma_T(\theta)\, d\theta \tag{4.8}$$

The change in length of the specimen is $\Delta l = \xi_A - \xi_B$

$$\Delta l = \frac{1}{\rho C_o} \int_o^t [\sigma_I(\theta) - \sigma_R(\theta) - \sigma_T(\theta)]\, d\theta \tag{4.9}$$

Since $-\sigma_R \approx \sigma_I$ the total displacement ξ_A is

$$\Delta l = \frac{1}{\rho C_o} \int_o^t [2\sigma_I(\theta) - \sigma_T(\theta)]\, d\theta \tag{4.10}$$

Finally, Δl and the specimen deformation $\varepsilon_s = \Delta l / l_o$ can be determined as a function of measured quantities $\sigma_I(t)$ and $\xi_B(t)$

$$\Delta l(t) = \frac{2}{\rho C_o} \int_o^t \sigma_I(\theta)\, d\theta - \frac{1}{2} \rho C_o \xi_B(t) \tag{4.11}$$

As the first approach the stress which is supported by the specimen $\sigma_s(t)$ is proportional to the transmitted stress $\sigma_T(t)$, using (4.4) one finds

$$\sigma_s(t) = \frac{1}{2} \rho C_o \left(\frac{d_E}{d_s}\right)^2 \frac{d\xi_B}{dt} \tag{4.12}$$

The main disadvantage of the configuration with the condenser displacement gages is necessity to differentiate the transmitted oscillogram $\xi_B(t)$ to find the specimen stress σ_s.

The main assumptions made in this technique are that both bars remain elastic and the simplest wave mechanics in elastic bars is acceptable. Under those assumptions it is possible to analyze the incident, reflected and transmitted waves generated in the bars, and a soft specimen, to determine tension/compression stress-strain curve obtained at high rate of strain ($\dot{\varepsilon} \approx 1 \text{x} 10^3$ s^{-1}).

Further developments in the Kolsky apparatus occured during last decades. The use of strain gages to measure longitudinal elastic waves on the pressure bars has been introduced by Simmons, Hauser and Dorn, [40], (1960). The explosives have been replaced by the mechanical impact ; the bar projectile was used to develop regular incident wave, Lindholm, [41], (1964). Some configurations have been introduced for testing plastic properties of materials in tension. Also, torsional configuration has been introduced by Duffy, Cambell and Hawley, [42,43], (1970). The modern configuration of the Kolsky apparatus is shown in Fig. 4.2.

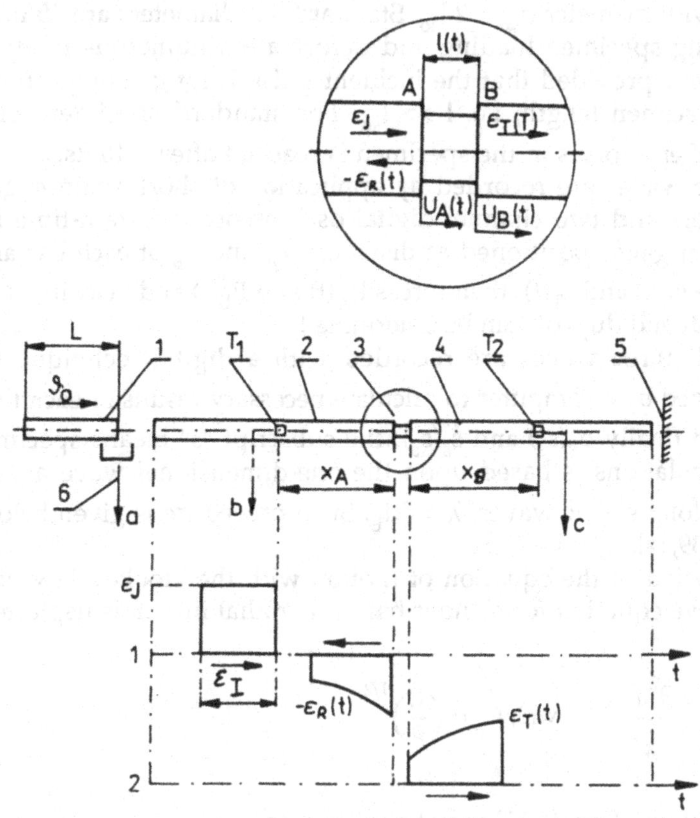

Fig. 4.2, The Split Hopkinson Bar arrangement with three bars : striker-1, incident bar-2 and transmitter bar-4.

As shown in Fig. 4.2 the incident wave $\varepsilon_I(t)$ is initiated by the impact of a striker bar 1 of length L with velocity v_o against the incident bar 2 of length $L_I \approx$ 1000 mm. The striker bar may be accelerated to predetermined impact velocity v_o (usually $(v_o)_{max} \approx 30$ m/s) by a compressed gas gun. The impact of the striker bar produces a compression elastic wave of constant amplitude $\varepsilon_I = v_o/2C_o$ and a finite duration λ depending upon the length of the striker bar, $\lambda = 2L$. The resulting incident wave is propagated at the elastic wave speed C_o into specimen. Since the specimen is capable for plastic deformation, the incident wave is partially reflected from the interface, and partially transmitted through specimen into the transmitter 4 bar of length $L_T \approx 1000$ mm. The situation is shown in Fig. 4.2. The relative magnitudes of the incident $\varepsilon_I(t)$, reflected - $\varepsilon_R(t)$ and transmitted $\varepsilon_T(t)$ waves depend upon the plastic properties of the specimen of the initial length l_{so}. Specimens are usually short 5 mm $\leq l_{so} \leq$ 20 mm with diameter $d_{so} \approx 2\,l_{so}$. Standard bar diameters are 15 mm $\leq d_E \leq$ 25 mm. During specimen loading and deformation numerous internal wave reflections occur provided that the incident pulse is long enough in comparison to the specimen length, i.e. $l >> l_{so}$. For standard specimen length the equilibrium of end forces in the specimen is reached after ~ 10 µs.

All three waves are recorded by application of short strain gages, wide band amplifiers and two channel digital oscilloscope. If strain-time histories from the strain gages positioned at distances x_A and x_B of each bar are recorded, i.e. $\varepsilon_I(t)$, - $\varepsilon_R(t)$ and $\varepsilon_T(t)$, then forces $P_A(t)$ and $P_B(t)$ and velocity at each interface, du_A/dt and du_B/dt, can be determined.

If the all three waves are recorded with a digital technique they are usually analyzed by a computer to calculate necessary results of each test : $\varepsilon_S(t)$, $\sigma_s(t)$, $\dot{\varepsilon}_s(t)$ and finally $\sigma_s(\varepsilon_s)$ and $\dot{\varepsilon}_s(\varepsilon_s)$, the subscript "s" means specimen. The theory of calculations is based upon the one-dimensional wave analysis appropriate for long elastic waves, $\lambda > 5d_E$. Such procedure is given below following papers [39,44].

Combination of the equation of motion with the Hooke's law yields the simplified wave equation for slender bars (the radial inertia is neglected)

$$\frac{\partial^2 u}{\partial t^2} = C_o^2 \frac{\partial^2 u}{\partial x^2} \quad ; \quad \text{with } C_o = \left(\frac{E}{\rho}\right)^{1/2} \tag{4.13}$$

A general solution of eq. (4.13) may be written as

$$u = f\,(C_o t - x) + g\,(C_o t + x) \tag{4.14}$$

where u is displacement, x coordinate of propagation and C_o is the elastic wave speed as a function of Young's modulus E and bar density ρ. The relation between strain $\varepsilon = \partial u/\partial x$ and the particle velocity $v = \partial u/\partial t$ is the result of the solution (4.14)

$$\frac{\partial u}{\partial t} = C_o \frac{\partial u}{\partial x} \quad \text{or} \quad v = C_o\,\varepsilon \tag{4.15}$$

Introducing Hooke's law into eq. (4.15), relation between particule velocity v and longitudinal stress σ in the bar is obtained as follows

$$\sigma = E\frac{\partial u}{\partial x} \quad \text{and} \quad \sigma = \frac{E}{C_o}\frac{\partial u}{\partial t} \tag{4.16}$$

finally

$$\sigma = \rho\, C_o\, v \tag{4.17}$$

In addition, relations between displacement u, particle velocity v and deformation are

$$u(t) = \int_o^t v(\theta)\, d\theta \quad , \quad \varepsilon(t) = \frac{v(t)}{C_o} \tag{4.18}$$

Since the resistance strain gages, which measure strain e(t), are commonly used for measurements of elastic waves in bars the following relations are the most important in application of the Hopkinson bar concept. After eq. (4.15)

$$v(t) = C_o\, \varepsilon(t) \tag{4.19}$$

$$\sigma(t) = \rho\, C_o^2\, \varepsilon(t) \tag{4.20}$$

$$U(t) = C_o \int_o^t \varepsilon(\theta)\, d\theta \tag{4.21}$$

In addition, to relations (4.19) to (4.21) a very important quantity in the experimental methods based on Hopkinson bar is the energy of elastic wave. The increment of energy of the elastic wave as a function of the increment of time is

$$d\,W = F_E\, \rho\, C_o^3\, \varepsilon(t)\; dt \tag{4.22}$$

and after integration replacing $F_e = \pi\, d^2/4$ one yields

$$W = \frac{1}{4}\pi \rho d^2 C_o^3 \int_o^t \varepsilon(\theta)\, d\theta \tag{4.23}$$

where d is bar diameter.
For the case of harmonic waves of the form

$$u = u_o \exp[i(\omega t \pm kx)] \tag{4.24}$$

where u_o is the displacement amplitude, w the circular frequency and k the wave number. The phase velocity C_p of harmonic waves is

$$C_p = \frac{\omega}{k} \quad \text{and} \quad C_p = \frac{\omega \Lambda}{2\pi} \tag{4.25}$$

i.e. the points u (x,t) of and Λ is the wavelength, $\Lambda = 2\pi/k$. The group velocity C_g of harmonic

$$C_g = C_p - \Lambda \frac{dC_p}{d\Lambda} \tag{4.26}$$

Substituting eq. (4.24) into the wave equation (4.13) one obtains

$$C_p = C_o \tag{4.27}$$

Since Λ is constant eq. (4.26) yields

$$C_g = C_o \tag{4.27}$$

Equations show that the phase velocities of longitudinal elastic waves are independent of the wavelength. This implies that very short waves will propagate with the same phase velocity as long waves. This is a very crude approximation for the case of short wavelengths $\Lambda_{cr} \leq 5$ d, [37]. Thus, if the elastic waves in cylindrical bars are shorter than Λ_{cr} a more exact approximation of the wave mechanics is needed which takes into account the radial motion of a bar material, for example Love approximation or even exact formulation,

[45,46]. Recent analyses of the SHPB technique include to some extent the so called Pochhamer-Chree vibrations, [47], caused by the radial inertia. Fig. 4.3 shows differences between elementary approximation (broken line) and exact calculations of u $(t)_x$.

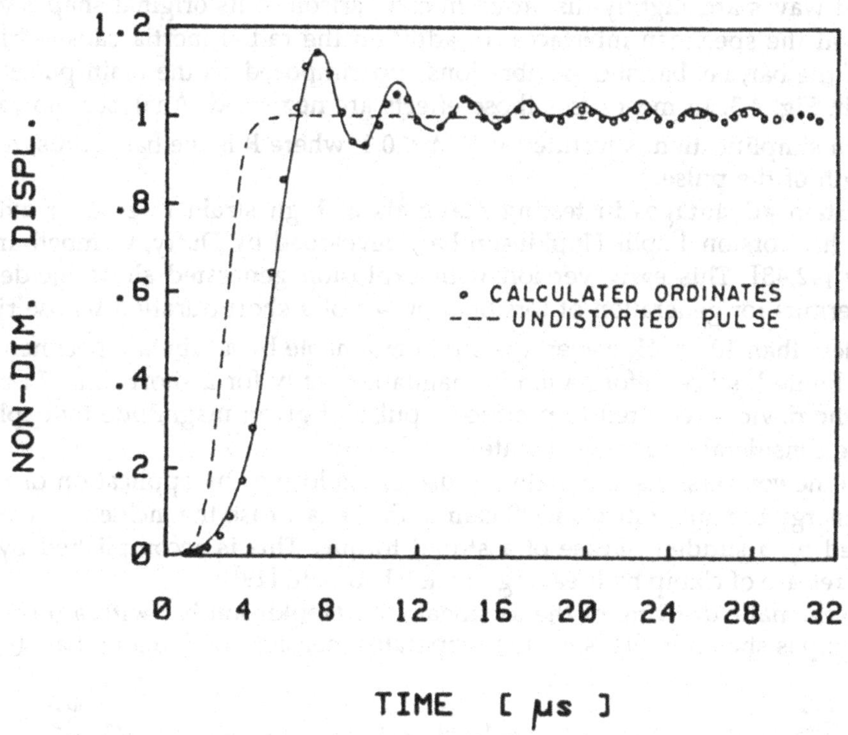

Fig. 4.3, Comparison between elementary approximation of wave displacement $u(t)_x$-broken line and exact calculation - points approximated by the solid line ; redrawn after [48].

The wave equation (4.13) has been clearly formulated for the non-dissipative medium where the phase velocity C_p does not depend on the wavelength Λ.

All real materials do display dissipation behavior and the phase velocity C_p depends in this case on the wavelength and the group velocity C_g determined by eq. (4.26) will slightly differ from C_0. Dispersion of elastic waves is an important phenomenon because it governs the changes of shape of a wave traveling along a bar. In conclusion, the Hopkinson bar will show in every case some dispersion and changes of a propagating wave may be sometimes important in interpretation of an experiment. It is recommended to evaluate dispersion effects and to find its importance (or non importance) in each particular case of experiment.

More exact formulation of elastic wave propagation in bars and more exact solutions clearly indicate the dispersive nature of wave propagation. This is specially true for short waves comparable with bar diameter.

Since strain gages are usually cemented far apart from the specimen the recorded waves are slightly distorded in comparison to its original shapes generated at the specimen interfaces. In addition the radial inertia causes "ringing" of the bar, i.e. bar shows vibrations superimposed on the main pulse as shown in Fig. 4.3. In most cases those effects are neglected. Analyses indicate that such simplification is justified if $R/\Lambda \leq 0.1$, where R is the bar radius, Λ is the length of the pulse.

Certain advantages in testing materials at high strain rates over axial loading has torsional split Hopkinson bar, developed by Duffy, Cambell and Hawley [42,43]. This early version with explosion generated shear incident wave permits for generation of torsional pulses of a short duration whose rise time is less than 10 µs. However, the strain attainable by a tubular specimen is severly limited, since deformation is maintained only for a short time. In addition, the devices required to produce a pulse of given magnitude by explosives are considerably more elaborate.

The newer versions eliminate the use of explosives by application of the stored energy bar and a mechanical clamp. In the last case the incident wave is produced by a sudden release of a stored torque. This is accomplished by a sudden release of clamp by breaking the notched bolt, [49].

A schematic drawing of the torsional split Hopkinson bar with a mechanical clamp is shown in Fig. 4.4. The apparatus consists of two long bars (~ 2

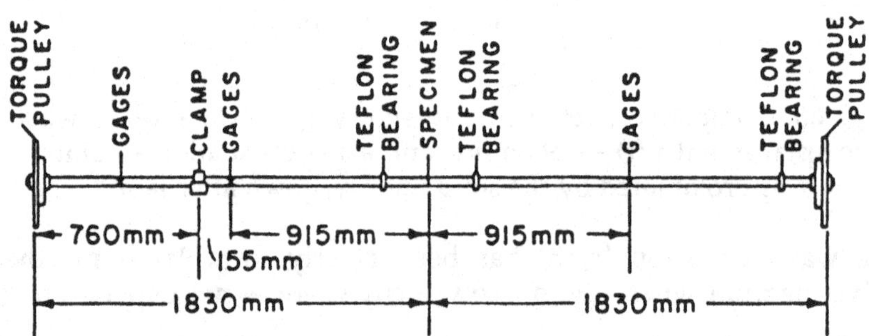

Fig. 4.4, Schematic diagram of the torsional split Hopkinson bar with a mechanical clamp, [51], courtesy The Institute of Physics.

m each) of diameter ~ 25 mm. To create a short torsional wave, one end of the loading bar is rotated slowly by means of a hydraulic jack and a torque pulley. The applied torque is stored in the part of the bar in between the pulley and the clamp. Fracture of the notched bolt in the clamp provides the sudden release of the stored elastic energy and the torsional pulse propagates to the tubular specimen cemented between bars. Similarly as in compression system of the gages provide a measure of stress and strain in the specimen as a function of time. The gage on the right-hand side provides a direct measure of rotation rate. The right hand pulley is used in the quasi-static tests to load the specimen against clamp. The concept of the mechanical clamp which is now commonly used in laboratory practice is shown in Fig. 4.5. Another solution of mechanical clamp was introduced by Campbell and Lewis, [43]. The release mechanism consists of a tapered flange, machined as an integral part of the loading bar and located near its center, cemented with epoxy adhesive to a fixed external holding ring. To create a steep-fronted torsional wave, one end of the loading bar is rotated slowly by means of a motor and gear box. When the torque is high enough a torsional wave is released from the clamp by fracturing of the epoxy joint.

The mechanical clamp as shown in Fig. 4.5 has many advantages over the other constructions, like small inertia, rigidity and an ease of use.

The technique of the split Hopkinson pressure bar as used for compression or tension tests has several inherent difficulties, which limits its range of application also to fracture dynamics. These difficulties stem from wave dispersion effects in the Hopkinson bars and specimen behavior, for example the effect of interfacial friction between bars and specimen, [50]. These problems are eliminated by the use of a system which operates in torsion. The most important advantage of the torsion system is that torsional pulses propagate along bars without geometric dispersion. Thus, a short-time loading is free of distortions and analysis of specimen behavior is more exact. The torsional split Hopkinson bar can also be applied in fracture mechanics as it is demonstrated in one of the next parts of this review.

In cylindrical coordinates (r,θ,z) the equation governing the motion of particles in the a homogeneous, isotropic elastic bar is given by [46],

$$\frac{\partial^2 \delta}{\partial r^2} + \frac{1}{r}\frac{\partial \delta}{\partial r} - \frac{\delta}{r^2} + \frac{\partial^2 \delta}{\partial z^2} = \frac{1}{C_2^2}\frac{\partial^2 \delta}{\partial t^2} \tag{4.29}$$

where δ is the circumferential displacement which is independent of θ and $C_2 = \sqrt{\mu/\rho}$ is the velocity of elastic shear waves. For the case of harmonic torsional vibrations the displacement δ may be written as

$$\delta\,(r,z,t) = \frac{1}{q} B_2 J_1\,(qr) \exp\,[i\,(kz - wt)] \tag{4.30}$$

where

$$q^2 = \left(\frac{\omega}{C_2}\right)^2 - k^2 \tag{4.31}$$

and w and k are respectively the angular frequency and wave number.

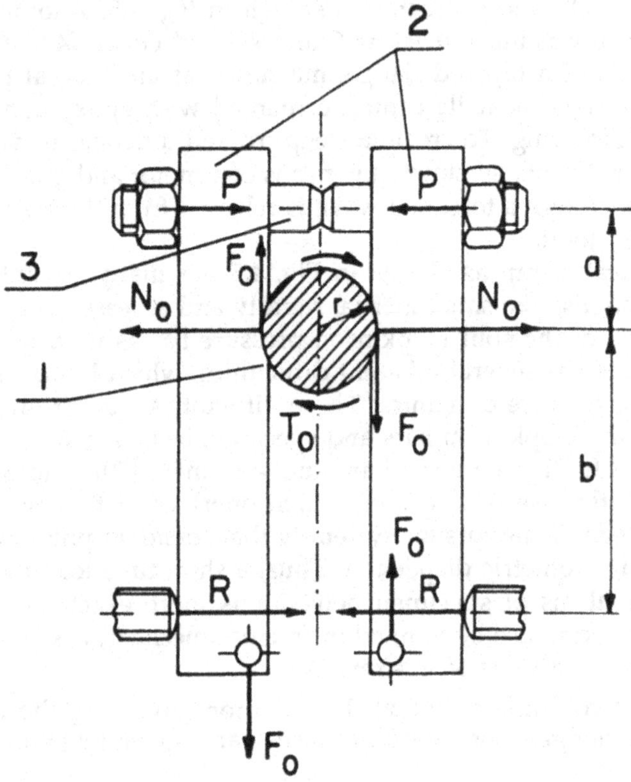

Fig. 4.5, Quick-release frictional clamp which is commonly used with torsional split Hopkinson bar, after [49].

Since for an infinity long bar with a solid circular cross section of diameter $d = 2r_o$ the external surface is free of stresses the condition at $r = r_o$ are

$$\sigma_r = 0 \;\; ; \;\; \tau_{r\theta} = 0 \;\text{ and }\; \tau_{rz} = 0 \tag{4.32}$$

These conditions yield the frequency for displacement δ

$$(qr_o)\, J_o\, (qr_o) - 2\, J_1\, (qr_o) = 0 \tag{4.33}$$

where J_o and J_1 are the Bessel functions of the zero and first order. Equation (4.33) has several roots, but it may be noted that q = 0 is also a solution of the frequency equation. The limit $q \rightarrow 0$ in eq. (4.30) yields

$$\delta = \frac{1}{2} B_2\, r \exp\, [i\, (kz - \omega t)] \tag{4.34}$$

This solution is associated with propagation of displacement δ at the lowest torsional mode. In the lowest mode δ is proportional to the radius r and the rigid rotation of each cross section occurs about its center. In the case of the lowest torsional mode (q = 0) the phase velocity is equal to C_2, thus

$$C_p = C_2$$

and propagation of this mode is non-dispersive. This is a very important conclusion concerning applications of torsional Hopkinson bar. All wave records from torsional split Hopkinson bar are free from Pochhamer-Chree oscillations characteristic of the longitudinal arrangement. Thus, the wave equation in displacements (4.29) can be transformed for the zero-mode propagation into wave equations with the angle of rotation θ

$$\frac{\partial^2 \theta}{\partial t^2} = C_2^2 \frac{\partial^2 \theta}{\partial z^2} \tag{4.35}$$

The torque transmitted by the torsional wave is related to the angle of rotation θ as follows

$$T = J\, \rho\, C_2^2 \frac{\partial \theta}{\partial z} \tag{4.36}$$

where J is the polar moment of intertia $J = 0.5\, \pi\, r_o^4$. The torque T can be also related to the angular velocity $\omega = \partial\theta/\partial t$. Since

$$\frac{\partial\theta}{\partial z} = \frac{\omega}{C_2} \tag{4.37}$$

the equation for torque is

$$T = J\,C_2\,\rho\,\omega \tag{4.38}$$

Usually, the shear strain γ is measured at the surface of the torsional bars and the following relations are important

$$\gamma = r_o \frac{\partial\theta}{\partial z} \tag{4.39}$$

Substituting for $\partial\theta/\partial z$ from (4.39) in (4.36) and (4.37) the following relations are obtained as a function of shear strain γ

$$T = J\,\rho\,C_2^2 \frac{\gamma}{r_o} \quad \text{or} \quad T = J\,\mu \frac{\gamma}{r_o} \tag{4.40}$$

and

$$\omega = \frac{C_2}{r_o}\gamma \tag{4.41}$$

In the case of tubular specimen (thin-walled) of mean radius r_s, length l_s and wall thickness h_s, subjected to torque T_s, the mean stress and strain rate are given by

$$\tau_s = \frac{T_s}{2\pi r_s h_s} \tag{4.42}$$

$$\dot{\gamma}_s = \frac{r_s}{l_s}\dot{\omega}_s \tag{4.43}$$

where $\omega_s = d\theta_s/dt$ is the relative rotation between the end planes of the specimen. The specimen torque T_s is assumed to be equal to the transmitter bar (the right-hand bar in Fig. 4.4) torque T_T, $T_s = T_T$ neglecting the time-shift between the transmitter gage station and the specimen. The twist rate of the specimen ω_s is given by the difference between the angular velocities of the incident bar (the left-hand bar in Fig. 4.4) and transmitter bar

$$\omega_s = \omega_I + |\omega_R| - \omega_T \tag{4.44}$$

If the incident, reflected and transmitted shear waves are recorded as a function of time, i.e. $\gamma_I(t)$, $\gamma_R(t)$ and $\gamma_T(t)$, the behavior of specimen can be determined from the relations

$$T_s = J\rho C_2^2 \frac{\gamma_T(t)}{r_o} \quad \text{and} \quad \tau_s = \frac{J\rho C_2^2}{2\pi r_s h_s r_o} \gamma_T(t) \tag{4.45}$$

and

$$\omega_s = \frac{C_2}{r_o}[\gamma_I(t) + |\gamma_R(t)| - \gamma_T(t)] \tag{4.46}$$

finally

$$\dot{\gamma}_s = \frac{C_2}{l_s}\frac{r_s}{r_o}[\gamma_I(t) + |\gamma_R(t)| - \gamma_T(t)] \tag{4.47}$$

The torsional split Hopkinson bar is a powerful tool for testing of materials at high strain rates, the level of strain rate $\dot{\gamma} \approx 10^4\ s^{-1}$ can be easily attained. The technique was also successfully applied to test the resistance to adiabatic shearing and fracturing, for example [52]. In recent years the split Hopkinson bar with different configurations is more frequently applied to determine fracture resistance of stationary cracks in Mode I. A configuration for Mode III is presented in one of the next parts of this review.

5. EXPERIMENTAL METHODS FOR DETERMINATION OF FRACTURE RESISTANCE PARAMETERS AT HIGH RATES - APPLICATION OF HOPKINSON BAR

5.1 Quasi-static

High fundamentals rate fracture testing at loading rates greater than $\dot{K}_I \approx 1x10^4$ MPa√m/s is of interest for many applications. It is obvious that many machine parts and structural components are subjected in service to impact and high stress rates. In such situations all mechanical components must be designed against crack initiation under high loading rates. On the other hand ability to crack arrest is of great importance.

Since the fracture resistance of many metals and alloys is lower, at least up to the loading rate $1x10^4$ MPa√m/s, in the case of a fast loading as compared to quasi-static rates, designing must take this fact into account.

The following question arises : is fracture resistance at loading rates higher than $\dot{K}_I = 1x10^5$ MPa√m/s is greater or smaller than observed at lower rates ? This question can be answered ony if new and reliable experimental techniques of fracture toughness testing would be available. One possibility is to use the Hopkinson bar concept in fracture testing at high loading rates, typically at $\dot{K}_I = 1x10^6$ MPa√m/s. Obviously, measurements and analyses of fracture behavior within the part of the rate spectrum at high loading rates are more complex than under quasi-static conditions. On the other hand experimental techniques are not so established as to warrant very exact evaluation of fracture mechanics criteria. More recently developed test techniques do provide, however, more reliable fracture parameters.

The following methodologies are the most advanced at the moment :

i. Split Hopkinson pressure bar with Wedge Loaded Compact Tension specimen (WLCT technique),

ii. Dynamic testing of Notched Round Bar (NRT technique).

Other techniques based on the Hopkinson bar concept are not so advanced and more difficult in application. The WLCT method has been proposed by Klepaczko in 1979, [53]. The aim was to develop a relatively inexpensive and universal experimental technique to study the fracture toughness of structural materials over a wide range of loading rates. The method would be suited for application in all sub-ranges of the $\dot{K}_I$ spectrum, such as quasi-static and stress-wave loadings, and is characteristic in application of universal, easy-to-make specimen suitable for slow and fast loading, a device for quasi-static experiments, and a special arrangement of the split Hopkinson pressure bar, in its popular version, to cover loading rates up to $\dot{K} \approx 1x10^6$ MPa√m.

The system enables fracture toughness data to be obtained over six orders of magnitude in $\dot{K}_I$.

To evaluate the fracture toughness at low and moderately high loading rates, up to $\dot{K}_I \approx 1x10^4$ MPa√m/s, the ASTM standard was used in the specimen design (specifications for plane strain fracture toughness ASTM E-399-83). More exactly geometry of the compact tension (CT) specimen was adopted. The main reason for the choice of the CT specimen is its high rigidity and consequently a high frequency of free vibrations. Since the compression is desirable in a fast loading experiment, the CT specimen has been modified in such a way that the part above the axis of pinholes was cut off and an angular incision was milled to acomodate a wedge, by which the modified specimen was loaded. Geometries of both specimens (CT and WLCT) are shown in Fig. 5.1. Application of a wedge made of a hard steel with angle α = 45° enabled for application of compressive force, whereas the crack tip remainded loaded in tension. A more exact details of the wedge-specimen contact is shown in Fig. 5.2.

In addition, the K-calibration of the WLCT specimen remains almost the same as for CT one, so that the standard formulate could be used to calculate the plain strain fracture toughness K_{Ic} using the critical tensile force F_c

$$K_{Ic} = \frac{F_c}{B\sqrt{W}} f\left(\frac{a}{W}\right) \tag{5.1}$$

in the present case F_c is the critical tensile force acting on the WLCT specimen at the angle incision. The compliance function f(a/W) is assumed, as a first approach, in the same form as for CT specimen, eq. (3.1). However, the finite element calculations by Gambin and Lipinski have shown that coefficient of friction slightly changes the compliance function f(a/W), [54]. In a more general case the intensity factor in Mode I, K_I, is given by the formula

$$K_I = \frac{F}{B\sqrt{W}} \left(\frac{BWE}{2(1-\nu^2)} \frac{\partial C}{\partial a}\right)^{1/2} \tag{5.2}$$

where E and ν are respectively the Young's modulus and Poisson's ratio, C is the compliance of a specimen. The compliance is defined as C = F/δ where δ is the displacement developed by the force F, δ is measured along the force axis. The compliance function is then defined by (5.3)

COMPACT WEDGE LOADED SPECIMEN

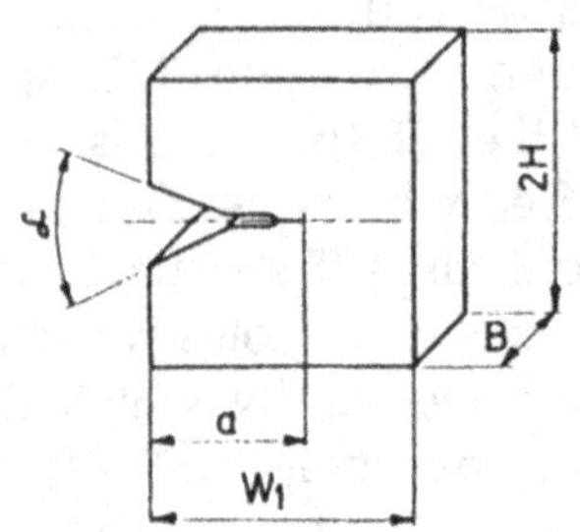

	B	W_1	a	H	α
PA6	8--20	20	12,0	20	43°
K63∥	10	25	125	12	43°
K63⊥	10	24	12	125	43°

CT SPECIMEN

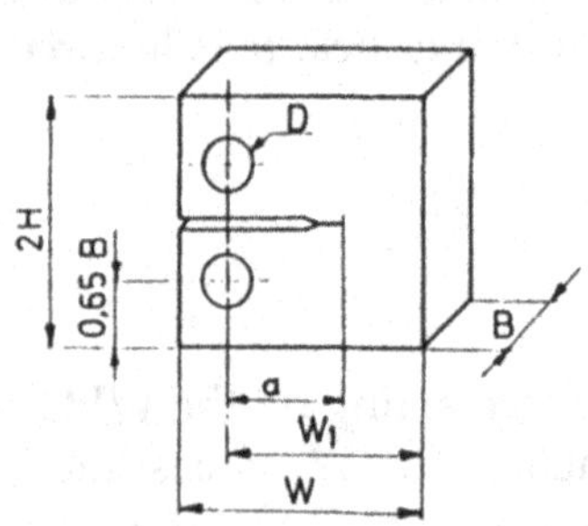

$a = B = \frac{W_1}{2}$

$H = 1{,}2\,B$

$W_1 = 2\,B$

$W = 2{,}5B$

$D = 1{,}5B$

Fig. 5.1, Geometry of the wedge-loaded specimen versus standard CT geometry ; the table shows typical range of dimensions applied for two Al alloys PA6 and K63, (all dimensions in milimeters), [53].

$$f\left(\frac{a}{W}\right) = \left(\frac{BWE}{2(1-\nu^2)}\,\frac{\partial C}{\partial a}\right)^{1/2} \tag{5.3}$$

Since sometime the WLCT geometry is slightly different than in the standard CT specimen, i.e. dimension is 2H = W in the WLCT specimen versus 2H = 1.2 W in the case of CT, then such modification changes the compliance of the WLCT specimen. Reduction to 2H is motivated by reduction of inertia forces of rotation. Numerical calculations by FEM have shown that for the modified

WLCT specimen with 2H = W the following compliance function is adequate

$$f\left(\frac{a}{W}\right) = A_o\left(\frac{a}{W}\right)^{1/2} + A_1\left(\frac{a}{W}\right)^{3/2} + A_2\left(\frac{a}{W}\right)^{5/2} + A_3\left(\frac{a}{W}\right)^{7/2} + \\ + A_4\left(\frac{a}{W}\right)^{9/2} \tag{5.4}$$

with the coefficient A_i given in Table 5.1 for three values of friction coefficient μ. Because the modified geometry with 2H = W is slightly less rigid than the

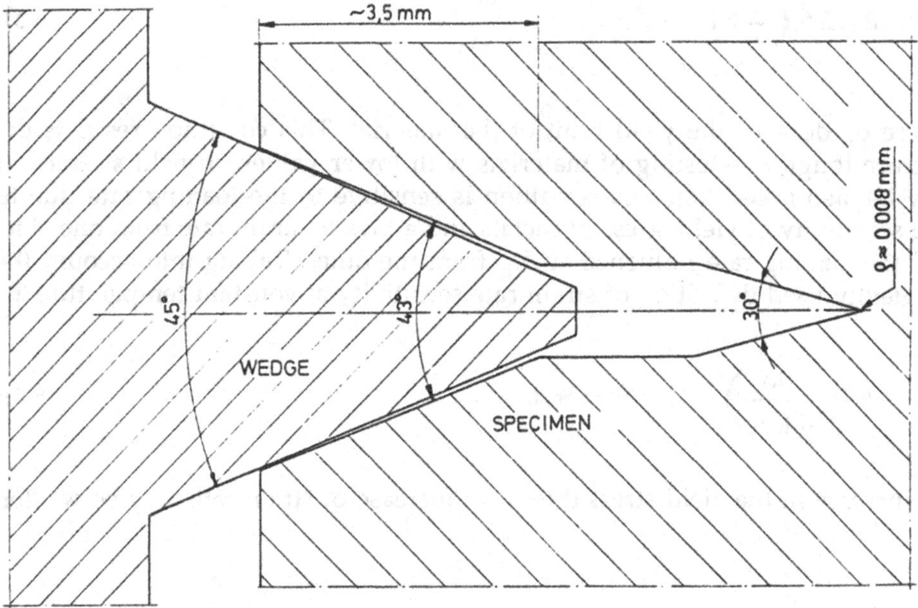

Fig. 5.2, Details of the contact between WLCT specimen and loading wedge.

standard one the compliance function f (a/W) for the WLCT specimen is slightly lower than for CT. For real values of the coefficient of friction μ (μ ≈ 0.1) changes of the compliance function are within ~ 3% for a/W ≈ 0.5 as compared to the case with μ = 0. Those changes of f (a/W) due to friction are within the range of experimental errors.

It may be remembered that there is a limitation on a minimum specimen thickness B imposed by the ASTM Standard E 399 to satisfy the plan strain conditions.

Table 5.1

μ	A_0	A_1	A_2	A_3	A_4
0.0	-12.7459	138.474	-357.673	446.958	-175.262
0.1	-14.0418	147.627	-384.913	479.136	-189.571
0.2	-17.0004	168.741	-447.494	553.184	-222.422

$$B \geq 2.5 \left(\frac{K_{Ic}}{\sigma_y}\right)^2 \tag{5.5}$$

where σ_y denotes the yield limit of the material. This condition restricts the fracture toughness testing of materials with lower values of yield stresses. It may be also noted that this condition is sensitive to the loading rate due to rate sensitivity of yield stress of metals and alloys to strain rate from one side, and the loading rate influence on K_{Ic} from the other. Taking into account the frequently used definition of strain rate sensitivity at constant temperature T

$$\beta = \left(\frac{\partial \sigma}{\partial \log \dot{\varepsilon}}\right)_T \quad ; \quad T = \text{const.} \tag{5.6}$$

the increase in the yield stress due to an increase of strain rate may be written as

$$\sigma_y = \sigma_{ys} + \beta \log (\dot{\varepsilon}/\dot{\varepsilon}_s) \tag{5.7}$$

where σ_{ys} is the quasi-static yield stress obtained at $\dot{\varepsilon}_s$; typically $\varepsilon_s = 1 \mathrm{x} 10^{-4} \mathrm{s}^{-1}$. Assuming further that

$$\frac{\dot{\varepsilon}}{\dot{\varepsilon}_s} = \frac{\dot{K}_I}{K_{Is}} \tag{5.8}$$

where K_{Is} is the quasi-static loading rate, typically $\dot{K}_{Is}$ = 1 MPa$\sqrt{m}$/s (ASTM standard recommendation 0.55 to 2.75 MPa√m/s), one obtains

$$\sigma_y = \sigma_{ys} + \beta \left(\frac{\dot{K}_I}{\dot{K}_{Is}}\right) \quad ; \quad \dot{K}_I \geq \dot{K}_{Is} \tag{5.9}$$

In addition, it may be assumed, as the first approximation, that the fracture toughness K_{Ic} diminishes proportionallay to $\dot{K}_I$. Thus

$$K_{Ic} = K_{Ic}^s - \eta \log \left(\frac{\dot{K}_I}{\dot{K}_{Is}}\right) \quad ; \quad \dot{K}_I \geq \dot{K}_{Is} \tag{5.10}$$

where $K_{Ic}{}^s$ is the quasi-static value of the fracture toughness obtained at $\dot{K}_{Is}$. Taking into account σ_y (K_I) and $K_{Ic}(K_I)$ in the condition for B one obtains

$$B \geq 2.5 \left[\frac{K_{Ic}^s - \eta \log (\dot{K}_I/\dot{K}_{Is})}{\sigma_{ys} + \beta \log (\dot{K}_I/\dot{K}_{Is})}\right]^2 \tag{5.11}$$

The important conclusion to be drawn from this relation is that the specimen used at higher loading rates $\dot{K}_I$ can be substantially miniaturised. A similar observation, on a purely experimental basis, was mentioned by Shoemaker and Rolfe in [116]. In order to visualize this condition the result of calculations for the aluminum alloy PA6 in the form B(log $\dot{K}_I$) is shown in Fig. 5.3 with β = 17.4 MPa and η = - 2.5 MPa√m. At the loading rate $\dot{K}_I \approx 1$ MPa√m/s the thickness B ≈ 24 mm satisfies the ASTM condition, and at $\dot{K}_I \approx 1\text{x}10^6$ MPa√m/s the thickness B ≈ 4 mm is sufficient.

When the ASTM condition for B is satisfied, the fracture toughness may be calculated using the relation, [53]

$$F_c = \frac{P_c}{2\text{tg}\left(\frac{\alpha}{2} + \text{tg}^{-1}\mu\right)} \tag{5.12}$$

where P_c is the compressive force acting on the wedge, α is the angle of the wedge, and μ denotes the coefficient of friction between the steel wedge and the specimen. If the formula (5.1) for K_{Ic} is taken into consideration one obtains

$$K_{Ic} = \frac{P_c f(a/W)}{2B\sqrt{W}\, tg\left[\frac{\alpha}{2} + tg^{-1}\mu\right]} \tag{5.13}$$

Thus, when the coefficient of friction μ is known, the fracture toughness K_{Ic} may be calculated from the record P(t) or $P(\delta_x)$, where δ_x is the wedge displacement.

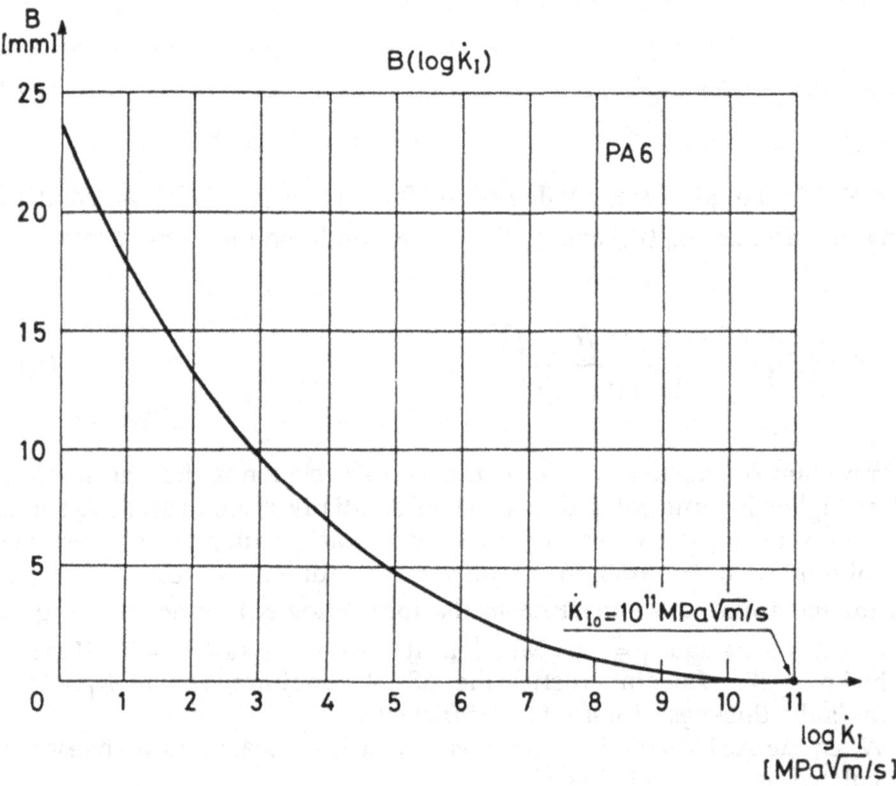

5.3, Theoretical condition for thickness of CT specimen B calculated for PA6 aluminum alloy (~ 6% Cu) as a function of loading rate $\dot{K}_I$.

5.2 Fast and stress-wave loading

The self-consistent loading device suitable for slow and fast loadings is shown schematically in Fig. 5.4.

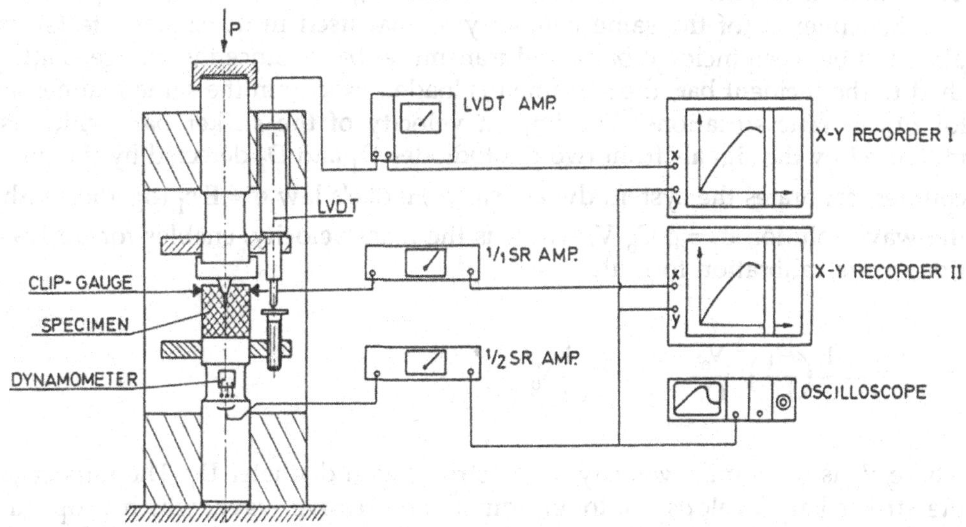

Fig. 5.4, Scheme of the device for slow and fast loadings of the WLCT specimen ; the force P acting on the wedge is applied by a testing machine, [53].

Any universal testing machine may be used to accommodate this device. The compressive force P is applied to the WLCT specimen through the upper plunger, which is fitted with lubricant in the body of the device. The lower support bar is tightly pressed into the lower part of the device body and plays the role of an independent dynamometer with SR gages. The vertical displacement of the wedge is measured precisely by a LVDT displacement gage. In addition, the transverse displacement of the specimen opening can be measured by a clip gage. All signals, i.e. the force P, the wedge displacement δ_x and the transverse displacement δ_t, may be recorded by an analog or digital system.

Thus, accurate measurements of force and crack opening displacement as a function of time can be obtained for slow or fast loadings. Such data are sufficient to be used in any theoretical analysis of fracture.

The device described above permits for fracture toughness testing with loading rates up to $\dot{K}_I \approx 1x10^4$ MPa$\sqrt{m}$/s or even up to higher rates, $\dot{K}_I \approx 1x10^5$ MPa$\sqrt{m}$/s if the LVDT displacement gage is replaced by an optical one. Still higher rates of loading, within the region of the rate spectrum 1x10^5 MPa$\sqrt{m}$/s $\leq \dot{K}_I \leq$ 1x10^6 MPa$\sqrt{m}$/s are possible with experimental procedure has been described in [53] and [55]. A schematic diagram of the experimental setup is shown in Fig. 5.5.

The experimental setup consists of a gas gun, the system of two properly instrumented Hopkinson bars and the measuring and recording equipment.

Specimen 7 (of the same geometry as that used in quasi-static tests), is placed in between incident bar 4 and transmitter bar 5. Since the wedge is attached to the incident bar, the specimen is loaded exactly in the same manner as for quasi-static situations. The impact velocity of the striker bar, which is measured by the signals from two photodiodes D_1 and D_2 detected by the time counter, calibrates the system dynamically. Hooke's law $\sigma = E\,\varepsilon_I$ together with the wave solution $\sigma_I = \rho\, C_o\, V$, where is the mass velocity, enables for derivation of the calibration formula

$$\varepsilon_I^o = \frac{1}{2}\left(\frac{D_1}{D_E}\right)^2 \frac{V_o}{C_o} \quad ; \quad v = \frac{1}{2} V_o$$

where V_o is the impact velocity of the striker with diameter D_1. The impact of the striker bar develops the longitudinal compressive wave, which propagates along the incident bar 4. The incident wave is measured by the SR gage station T_1, and an analogue signal after amplification in the SR amplifier is recorded in the memory of the dual-trace digital oscilloscope. Since the striker bar is at least ten times longer than the total length of the wedge 6 and specimen 7, the process of the specimen loading by the incident wave through the wedge may be assumed to be as quasi-static. This is the basic assumption of the method.

The assumption holds only for the long incident waves. Application of short waves with lengths comparable to the specimen dimensions (~ 2 mm) leads to the short pulse fracture mechanics discussed in the next part of this manuscript.

Next, part of the compressive incident wave is reflected in the incident bar as a tensile pulse, and part (which depends on the specimen behavior) is transmitted into the transmitter bar. The transmitted compressive wave is measured by the SR gage station T_2, and an analogue signal after amplification in the SR amplifier is recorded in the memory of the second trace of the digital oscilloscope. At the same time, the reflected tensile wave is recorded on trace one by the SR station T_1. All signals stored in the memory of digital oscilloscope can be visualized in an analog form or transmitted to a computer for further analyses. Thus, registration of the incident, reflected and transmitted pulses provides all the information concerning specimen loading and fracturing.

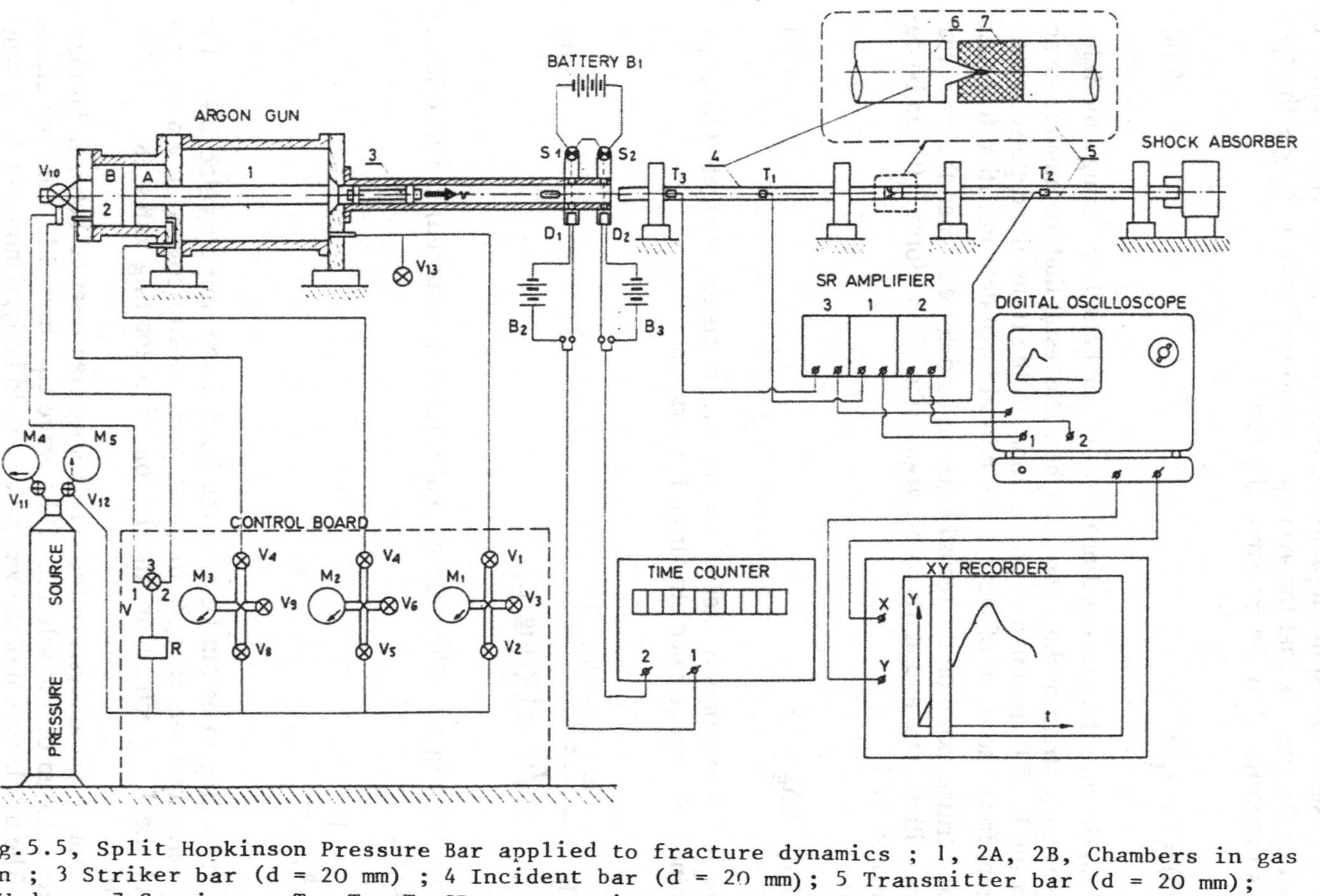

Fig.5.5, Split Hopkinson Pressure Bar applied to fracture dynamics ; 1, 2A, 2B, Chambers in gas gun ; 3 Striker bar (d = 20 mm) ; 4 Incident bar (d = 20 mm); 5 Transmitter bar (d = 20 mm); 6 Wedge ; 7 Specimen ; T_1, T_2, T_3 SR gage stations ; V_1-V_{13} values ; M_1-M_5 Manometers ; B_1, B_2, B_3 Supply voltage units ; D_1, D_2 Photodiodes ; S_1, S_2 Light sources, /53/

To describe quantitatively the dynamics of specimen fracturing the simplified analysis of the split Hopkinson pressure bar will be employed. It is assumed in the analysis that the average axial force P which charges the specimen is proportional to the transmitted pulse

$$P(t) = EA_E\, \varepsilon_T(t) \tag{5.14}$$

where A_E denotes the cross sectional area of the bars, E is Young's modulus, and $\varepsilon_T(t)$ elastic strain of the transmitter bar. The result that the mean force acting on the WLCT specimen is proportional to the transmitted pulse is very important from the point of view of the critical force determination to obtain K_{Ic}. If the critical point on the record $\varepsilon_T(t)$ can be detected, when the crack reaches the critical state and starts to propagate, the critical force $P_c(t)$ can be determined as well

$$P_c(t) = EA_E\, \varepsilon_T^c(t_c) \tag{5.15}$$

where t_c is the time period after which the crack starts to propagate. Once P_c is determined, K_{Ic} can be calculated from the formula

$$K_{Ic} = \frac{EA_E\, \varepsilon_T^c(t_c)\, f(a/W)}{2B\sqrt{W}\ \mathrm{tg}\left(\frac{\alpha}{2} + \mathrm{tg}^{-1}\mu\right)} \tag{5.16}$$

The mean loading rate (the proportional loading is assumed) can be calculated using t_c

$$\dot{K}_I = \frac{K_{Ic}}{t_c} \tag{5.17}$$

Different loading rates can be achieved by different impact velocities of the striker bar. However, changes in the impact velocity covers only a narrow band of the $\dot{K}_I$ spectrum in the region high loading rates ($K_I \approx 1\text{x}10^6$ MPa$\sqrt{m}$/s).

A more exact analysis of the WLCT specimen inserted between Hopkinson bars is based on the solution of the wave equation (4.13) for long elastic waves. The displacement of the wedge, as a rigid body, denoted by U_A is written as

$$U_A(t) = C_o \int_o^t [\varepsilon_I(\vartheta) - \varepsilon_R(\vartheta)]\, d\vartheta \tag{5.18}$$

and the displacement of the face B of the transmitter bar U_B is written by analogy

$$U_B(t) = C_o \int_o^t \varepsilon_T(\vartheta)\, d\vartheta \tag{5.19}$$

This situation is shown in Fig. 5.6. The net displacement of the wedge can be written as a function of time

$$\delta_x(t) = U_A(t) - U_B(t) \tag{5.20}$$

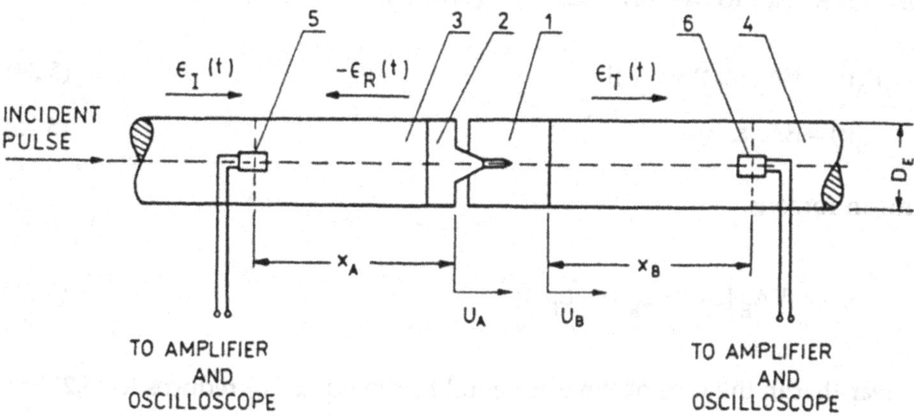

Fig. 5.6, Scheme of the split Hopkinson pressure bar as applied to fracture at high loading rates, [55].

Introducing (5.16) and (5.17) into (5.18) the displacement $\delta_x(t)$ can be directly related to the measured pulses $\varepsilon_I(t)$ and $\varepsilon_T(t)$

$$\delta_x(t) = C_o \int_o^t [\varepsilon_I(\vartheta) - \varepsilon_R(\vartheta) - \varepsilon_T(\vartheta)]\, d\vartheta \tag{5.21}$$

If equilibrium of the forces $P_A(t)$ and $P_B(t)$ is attained on both sides of the WLCT specimen, i.e. $P_A = P_B$, it leads to the following condition

$$\varepsilon_I(t) = \varepsilon_T(t) - \varepsilon_R(t) \tag{5.22}$$

when the condition (5.22) for the equilibrium of forces is introduced into eq. (5.21) the simplified version for $\delta_x(t)$ is obtained

$$\delta_x(t) = -2\, C_o \int_o^t \varepsilon_R(\vartheta)\, d\vartheta \tag{5.23}$$

Analysis of forces based on the solution of the wave equation and the Hooke's law lead to the formulas for P_A and P_B

$$P_A(t) = EA_E\, [\varepsilon_I(t) + \varepsilon_R(t)] \tag{5.24}$$

$$P_B(t) = EA_E\, \varepsilon_T(t) \tag{5.25}$$

the mean force is

$$\bar{P}(t) = \frac{1}{2} EA_E\, [\varepsilon_I(t) + \varepsilon_R(t) + \varepsilon_T(t)] \tag{5.26}$$

It is clear that in the case of the force equilibrium eq. (5.26) reduces to (5.27)

$$\bar{P}(t) = EA_E\, \varepsilon_T(t) \quad ; \quad \bar{P}(t) = P_B(t) \tag{5.27}$$

Equation (2.27) has been used in the fundamental relation (5.16) to determine K_{Ic}.

On the basis of equations (5.21) and (5.26) or (5.23) and (5.27) the dynamic load-displacement curve can be constructed after elimination of time.

Since WLCT specimens may be loaded with variety of initial conditions, like different impact velocities of strikers of different lengths, it is recommended to verify the condition of force equilibrium for every performed test.

5.3 Analysis of friction effects

Because WLCT specimen is loaded by the wedge and the coefficient of Coulomb friction μ enters as a parameter into the formula for K_{Ic} the important problem appears how to determine m. It is obvious that in spite of ample lubrication in both experimental situations, static and dynamic, friction between the wedge and the specimen is always present. An experimental procedure how to estimate μ will be outlined below. This procedure was established by Klepaczko and Andrzejewski [56] and discussed further by Klepaczko [57]. The main assumption made to solve this problem is that the specimen has the same compliance when its sides are loaded in tension or compression. This assumption holds in the elastic range with maximum load $\Delta K_I \approx 1/2\ K_{Ic}$. Specimen on which the coefficient of friction has to be tested is loaded twice as shown in Fig. 5.7.

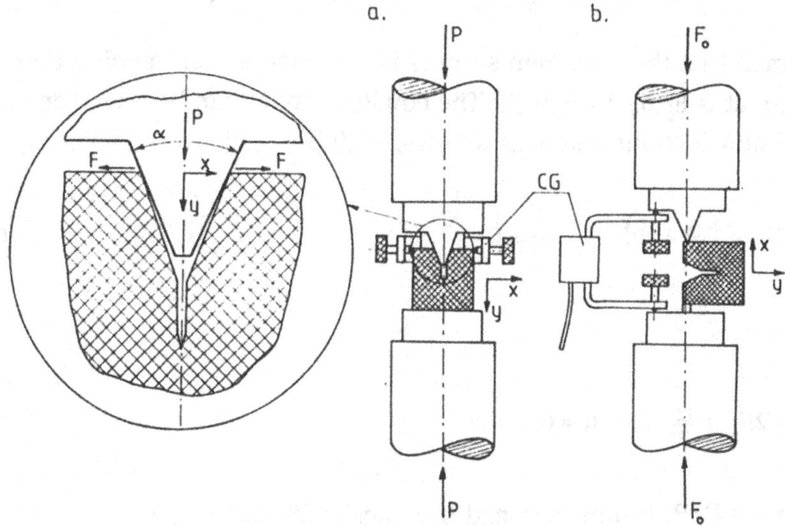

Fig. 5.7, Two loading configurations of the WLCT specimen (without fatigue crack) for the quasi-static determination of μ ; CG - clip gage for δ_x measurements, [57].

The same geometry and conditions of lubrication should be applied in the case of determination of μ as in the case of K_{Ic} testing. However, the WLCT specimen designed for measurements of μ is usually without fatigue crack.

The first step is loading of the WLCT specimen in the normal configuration, that is with the normal wedge lubrication and the normal support, Fig. 5.7a. Two records are commended with this configuration, namely the axial force P(t) as a function of time (direction y) and the transversal displacement of specimen $\delta_x^+(t)$ along the line of transversal force F (direction x). For this case the opening mode in specimen loading is employed. Elimination of time provides the $P(\delta_x^+)$ record. The second type of loading is shown in Fig. 5.7b. The WLCT specimen is turned 90° and the closure mode of loading is applied. This time the WLCT specimen is loaded without friction effects, and the force $F_o(t)$ is recorded as a function of time along with displacement $\delta_x^-(t)$ from the clip gage (CG in Fig. 5.7). Elimination of time provides $F_o(\delta_x^-)$ record. Having measured $P(\delta_x^+)$ and $F_o(\delta_x^-)$ the coefficient of friction μ can be calculated. For this purpose a general form of equation (5.12) is applied

$$F(t) = \frac{P(t)}{2\,\mathrm{tg}\left(\frac{\alpha}{2} + \mathrm{tg}_{\mu}^{-1}\right)} \tag{5.27}$$

It is assumed that the specimen stifness is the same under opening and closure modes, that is $F_o(\delta_x^+) = F_o(\delta_x^-)$. The equation above can be rewritten for frictional and non frictional cases as follows, with $\beta = \mathrm{tg}_{\mu}^{-1}$

$$P = 2F\,\mathrm{tg}\left(\frac{\alpha}{2} + \beta\right) \quad ; \quad \mu \neq 0 \tag{5.28}$$

and

$$P_o = 2F_o\,\mathrm{tg}\,\frac{\alpha}{2} \quad ; \quad \mu = 0 \tag{5.29}$$

Thus the ratio P/P_o can be obtained from eqs (5.28) and (5.29)

$$\frac{P}{P_o} = \frac{F\,\mathrm{tg}\,(\alpha/2 + \beta)}{F_o\,\mathrm{tg}\,(\alpha/2)} \tag{5.30}$$

For the same displacement $\delta_x = \delta_x^+ = \delta_x^-$ the equality of forces is obvious (i.e. $F = F_o$), and the coefficient of friction μ can be calculated from eq (5.30) after rearrangement

$$\mu = \mathrm{tg}\left\{\mathrm{tg}^{-1}\left[\frac{P}{P_o}\mathrm{tg}\frac{\alpha}{2}\right] - \frac{\alpha}{2}\right\} \tag{5.31}$$

Determination of μ is possible provided the ratio P/P_o is calculated in advance using the record $F_o(\delta_x^-)$ and equation (5.29).

Since the coefficient of friction μ is not a constant but usually decreases as a function of the surface pressure, it is desirable to determine μ as a function of P. This should be performed for particular conditions of fracture toughness experiments, the same wedge and the same lubricant must be used To illustrate the range of parameters which enter into the equation for μ the Fig. 5.8 shows μ as a function of P/P_o and μ is shadowed, that is for $40° \leq \alpha \leq 60°$ changes in $1.30 \leq P/P_o \leq 1.50$ provide the following range of μ : $0.12 \leq \mu \leq 0.15$.

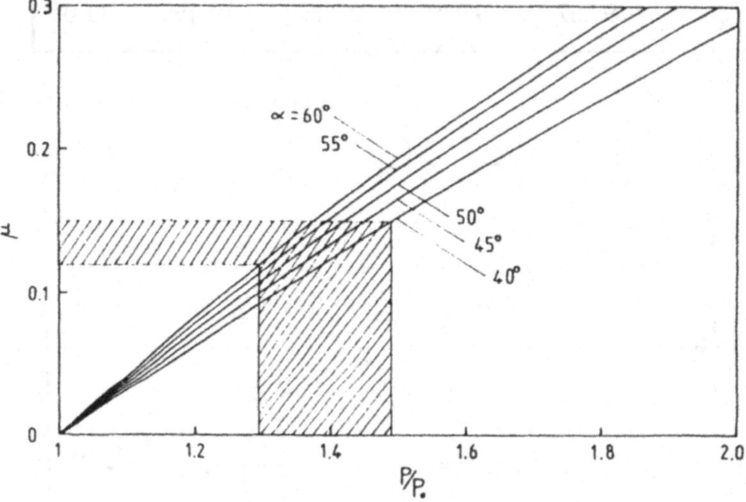

Fig. 5.8, Visualization of eq. (5.31) for five values of the wedge angle α ; shadowed region indicates expected range of m and P/P_o, [57].

The Fig. 5.8 demonstrates that the linear approximations to $\mu = f(P/P_o)$ is sufficient enough for expected ranges of μ and P/P_o. Consequently, the following linear approximate formula is plausible

$$\mu = C(\alpha)\frac{P}{P_o} \tag{5.32}$$

where C is the nondimensional coefficient which depends on the wedge angle α. Values of $C(\alpha)$ are provided in Table 5.2 for five values of α.
Application of such procedure for the determination of the coefficient of friction μ leads to the result as shown in Fig. 5.9 where calculated values of μ versus axial force P are plotted. This is the result of experiment with a titanium alloy wedge with $\alpha = 60°$ and a steel specimen lubricated with molybdenum disulfide ($Mo\ S_2$) grease.

Table 5.2 Values of the coefficient C(a) in eq. (5.32)

α deg	40	45	50	55	60
$C \times 10^2$	9.167	9.239	9.885	10.45	11.01

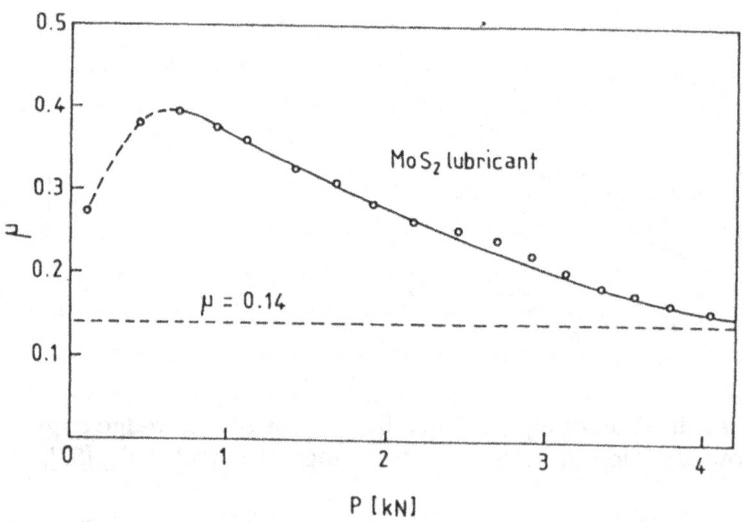

Fig. 5.9, Result of determination of the coefficient of friction μ as a function of the axial force P ; value μ = 0.14 was taken as adequate for the critical force P_c, [57].

The coefficient of friction decreases, as it is expected, when pressure or force increases, the asymptotic value $\mu = 0.14$ was used in the calculations of fracture toughness parameters.

The formula which relates compressive force P and K_I depends on the coefficient of friction μ. The coefficient of friction, when neglected or false, may introduce some errors in estimation of fracture toughness. In order to analyse this case the relative error can be defined as follows

$$W_\mu = \frac{F - F_{\mu=0}}{F} \quad \text{or} \quad W_\mu = \frac{\Delta F}{F} \tag{5.33}$$

Using the fundamental formula for $F(\mu,\alpha)$ one obtains

$$W_\mu = - \left(\frac{\mathrm{tg}\,(\alpha/2+\beta)}{\mathrm{tg}\,(\alpha/2)} - 1 \right) \ , \quad P = \text{const.} \tag{5.34}$$

where β is the angle of friction, $\beta = \mathrm{tg}^{-1}\mu$.

Thus, in the case when the coefficient of friction is completely neglected, $\mu = 0$, or taken smaller than real, it causes an overestimation of the tensile force F, and consequently an overestimation of fracture toughness. This effect is shown in Fig. 5.10 where - W[%] and β is shown as a function of μ. Note that for the range of angles α which are of interest ; i.e. $\alpha \approx 45°$, values of the angle of friction β is proportional to a good approximation to μ, in such case

$$\beta = A\,\mu \quad \text{with } A = 57.06° \text{ for } \alpha = 45°$$

The range of - 10% error is marked by the dashed line. Since in the range of interest the error is almost proportional to μ and $\alpha = 45°$ the approximate relation is

$$- W = A\,\mu \quad \text{with } A \approx 3.1 \tag{5.35}$$

In the expected range of μ, $0 \le \mu \le 0.15$, values of W are $0 \le W \le 0.45$.

A more exact analysis of the effect of friction is possible by taking the following partial derivative

$$\left(\frac{\partial K_I}{\partial \beta}\right)_{\alpha,\mu} = - \frac{A_o}{\sin^2\left(\frac{\alpha}{2} + \beta\right)} \quad ; \quad A_o = \frac{P\, f\,(a/W)}{2B\sqrt{W}} \tag{5.36}$$

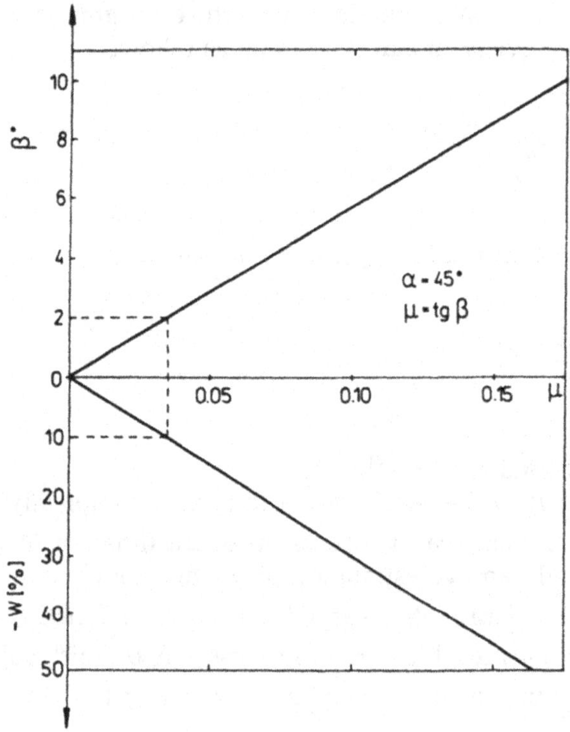

Fig. 5.10, Angle of friction $\beta = tg^{-1}\mu$ and the error in K_I determination W as a function of the coefficient of friction μ.

For small angles of friction, β, it may be assumed $\mu \approx \beta$ and finally

$$\left(\frac{\Delta K_I}{K_I}\right)_{\alpha,\mu} \approx - \frac{\Delta\mu}{\sin\left(\frac{\alpha}{2} + \mu\right)\cos\left(\frac{\alpha}{2} + \mu\right)} \tag{5.37}$$

where $\Delta\mu$ is an error in the estimation of the coefficient of friction. As it was discussed above an overestimation of μ resulting in + Δμ underestimates the measured fracture toughness, whereas underestimation of μ leads to an overestimation of fracture toughness. For real values, i.e. $\alpha/2 = 0.39$ rd (22.5°) and $\mu = 0.1$ rd (5°71'), changes $\pm\,\Delta\mu = 0.01$ develop an error in $(\Delta K_{Ic}/K_{Ic}) = \pm\,0.03$ or 3%, but for $\pm\,\Delta\mu = 0.04$ the error rises to ± 0.12 or 12%.

It is concluted from above considerations that the change of μ by 0.034 results in 10% error in fracture toughness estimation. This example demonstrates that underestimation or overestimation of μ may introduce errors in determination of fracture toughness. To minimize those errors the attention must be paid to obtain a very exact values of μ.

The effect of negligence of μ during calculations of K_{Ic} is demonstrated in Fig. 5.11 for aluminum alloy PA6 (4% Cu). Each point is the mean of five quasi-static tests.

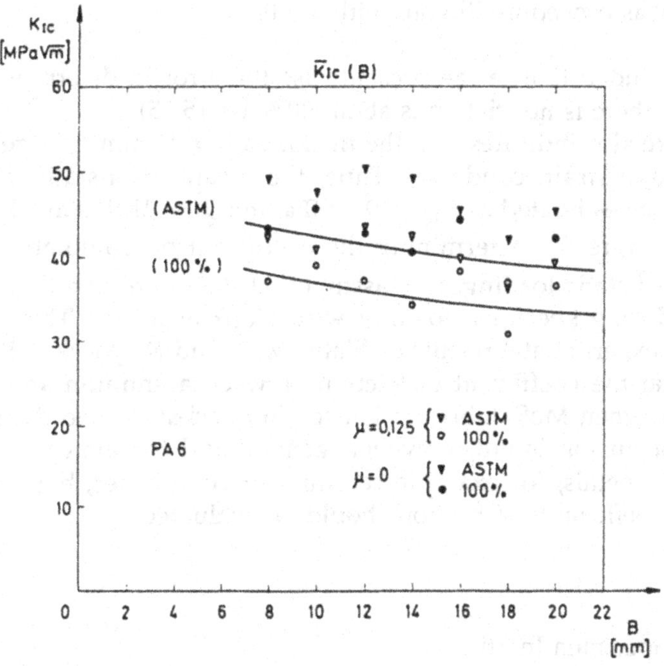

Fig. 5.11, Determined values of K_{Ic} as a function of thickness B by four procedures ; aluminum alloy PA6, each point is the mean of five tests, [56].

The results of K_{Ic} or K_Q calculation for seven thickness B demonstrate differences developed by different calculation procedures. Since there is no pop-in point observed on P(t) or P(δ) records, the 5% offset as recommended by the ASTM procedure has been applied to find the critical force P_c, and next K_{Ic} values (or K_θ).
Each point is an average value of five tests.

Four types of the following calculations were performed :

i. the strict ASTM procedure with μ determined by the method described previously, μ = 0.125 ;

ii. the same procedure ASTM but no frictional effects were assumed, μ = 0 ;

iii. the critical force P_c was determined as the first departure from the linearity of the $P_c(\delta)$ record with μ = 0.125 ;

iv. the same as procedure iii., but with μ = 0.

It may be concluded that in the present case the error in determination of K_{Ic} assuming that there is no friction is about 30%, eq. (5.35).

The figure also indicates that the thickness B = 20 mm is large enough to satisfy the plane strain condition. Thus, the mean quasi-static value of for this alloy may be estimated as K_{Ic} = 39.0 MPa√m ; ($\dot{K}_I \approx$ 1MPa√m/s).

The procedure for determining the coefficient of Coulomb friction μ is limited to quasi-static loading. It is assumed at the same time that μ does not differ much during specimen loading with Hopkinson bar. This assumption is justified by experimental results of Malinowski and Klepaczko [58], where it was found that the coefficient of friction between aluminum and steel is almost the same when MoS_2 lubricant is used in quasi-static and dynamic situations. This conclusion is not, however, general and for the case of different materials than metals, for example ceramics or composites, both quasi-static and dynamic coefficients of friction should be evaluated.

5.4 Effects of Specimen Inertia

Since during dynamic loading the fracture time is relatively short and specimen is subjected to inertia forces, the fracture toughness calculations neglecting these forces may introduce some errors. Thus, the estimation of specimen inertia leads to an estimation of errors introduced in K_{Ic}. Because of small scale yielding the wedge displacement $\delta_x(t)$ and the transverse displacement $\delta_y(t)$ are relatively small and they do not cause high inertia effects. Nevertheless, due to the short time to fracture initiation of the order of ~ 20 μs, even for small wedge displacement these effects have to be estimated.

Generally there are two sources of inertia : the longitudinal and transverse accelerations of specimen material developed by the wedge movement. The axial displacement of the specimen as a rigid body remains small in comparison to the axial displacement of the wedge. Thus the main source of inertia effects lies in the transverse displacement δ_y and accelerations in this direction. Of course, this is only true at larger wedge angles.

The transverse displacement $\delta_y/2$ of one half of the specimen opens it up causing the fracturing process, and the upper part of the specimen starts to rotate around the axis localized at the crack tip. The center of rotation moves along the crack tip displacement during the process of crack propagation. To estimate the transverse inertia effects it has been assumed that both halves of specimen rotate as rigid bodies around the axis localized at the specimen end backed by the transmitter bar (the face B of the Hopkinson bars). It is assumed then that the specimen consists of two disconnected and rotating pieces. This is a conservative assumption which overestimates the inertia forces. The differential of the transverse reaction dR_{ym} acting on the wedge in the y direction is

$$dR_{ym} = B\rho_s \frac{a_{ym}}{r_m^2} r^3 \, dr \, d\theta \tag{5.36}$$

where ρ_s is the density of specimen material, a_{ym} is the transverse acceleration of the maximum radius r_m, and r is a current radius. The formula for R_{ym} is

$$R_{ym} = B\rho_s \frac{a_{ym}}{r_m^2} \int_0^{\pi:2} d\theta \int_0^{r(\theta)} r^3 \, dr \tag{5.37}$$

where $r(\theta)$ is the half specimen contour in cylindrical coordinates. The maximum radius r_m for the WLCT specimen is

$$r_m = \sqrt{W^2 + H_0^2} \tag{5.38}$$

and the limits of integration are

$$r(\theta) = \frac{H_0}{\cos\theta} \quad \text{for } 0 \le \theta \le \theta_m$$

$$r(\theta) = \frac{W}{\cos\left(\frac{\pi}{2} - \theta\right)} \quad \text{for } \theta_m \le \theta \le \frac{\pi}{2}$$

where

$$\theta_m = \cos^{-1} \frac{H_0}{\sqrt{W^2 + H_0^2}}$$

To simplify further calculations it may be assumed that $r(\theta) = W$ along with $r_m \approx W$, which is still an overestimation of the inertia forces, thus

$$R_{ym} = B\rho_s \frac{a_{ym}}{W^2} \int_0^{\pi/2} d\theta \int_0^{W} r^3\, dr \tag{5.39}$$

after integration

$$R_{ym} = \frac{\pi}{8} B\rho_s W^2 a_{ym} \tag{5.40}$$

where

$$a_{ym} = \frac{dv_{ym}}{dt}$$

But accelerations are related by the kinematic relation

$$a_{ym} = a_x \operatorname{tg}\frac{\alpha}{2} \text{ or } \frac{dv_{ym}}{dt} = \frac{dv_x}{dt} \operatorname{tg}\frac{\alpha}{2} \tag{5.41}$$

And finally the axial acceleration is related to the reflected pulse as follows

$$\frac{dv_x}{dt} = -2C_o \frac{d\varepsilon_R}{dt} \tag{5.42}$$

Thus introducing (5.42) into (5.41) and next into (5.40) R_{ym} is derived

$$R_{ym} = -\frac{\pi}{4} B\rho_s C_o W^2 \, tg\left(\frac{\alpha}{2}\right) \frac{d\varepsilon_R}{dt} \tag{5.43}$$

The axial reaction due to the transverse inertia acting on the wedge, and decreasing reflected pulse, is

$$R_x = 2R_{ym} \, tg \frac{\alpha}{2} \tag{5.44}$$

Introducing (5.43) into (5.44) the final expression is obtained

$$R_x^T = -\frac{\pi}{2} B\rho_s C_o W^2 \left(\frac{\alpha}{2}\right) \frac{d\varepsilon_R}{dt} \tag{5.45}$$

The effect of the longitudinal inertia, which diminishes the transmitted pulse $\varepsilon_T(t)$, is

$$R_x^L = 2\rho_s C_o M_o BW \frac{d\varepsilon_T}{dt} \tag{5.46}$$

Again, it has been assumed that the whole specimen moves as a rigid body in the longitudinal direction, and the axial acceleration a_x is measured by the transmitted pulse $\varepsilon_T(t)$.

The second, and a simpler way to estimate transverse inertia, is to assume that the specimen opens up in an elastic manner and the velocity and acceleration fields are in the form

$$v_y = \frac{x}{W} v_{ym} \text{ and } a_y = \frac{x}{W} a_{ym} \tag{5.47}$$

Thus, the increment of transverse reaction acting on the wedge in the y direction is

$$dR_{ym} = \rho_s H_o Ba_{ym} \left(\frac{x}{W}\right)^2 dx \tag{5.48}$$

and it produces the transverse reaction

$$R_{ym} = \frac{1}{3} B\rho_s H_o Wa_{ym} \tag{5.49}$$

Introducing (5.41), (5.42), and (5.48) into (5.44), the axial reaction acting on the wedge due to lateral inertia is obtained

$$R_x^T = -\frac{4}{3} \rho_s C_o H_o BWtg^2\left(\frac{\alpha}{2}\right) \frac{d\varepsilon_R}{dt} \tag{5.50}$$

The comparison of equations (5.45) and (5.50) gives the result that the formula (5.45) leads to higher values of the estimated lateral inertia effects. The ratio of these two estimations is

$$\frac{(R_x^T)_{ROT}}{(R_x^T)_{EL}} = \frac{3\pi}{8} \frac{W}{H_o} \tag{5.51}$$

for $W = 2H_0$ this ratio takes the value $3\pi/4$. The higher value obtained from equation (5.45) is caused by the assumption that the specimen shape was taken as one quarter of the rotating disk with the maximum radius $r_m = W$. The formula (5.50) seems to provide a more realistic estimation. The difference between numerical values of $R_x{}^T$ for $W = 2H_0$ in equation (5.51) reaches 135 percent. In conclusion, the crude integration of equation (5.37) is not recommended for practical purposes and the remainingf analysis will be performed using equation (5.50).

The total axial reaction acting on the wedge due to both lateral and axial inertia is

$$R_x = R_x^L + R_x^T \tag{5.52}$$

Introducing (5.46) and (5.50) into (5.52) the total axial reaction is

$$R_x = 2\rho_s C_o H_o BW \left(\frac{d\varepsilon_T}{dt} - \frac{2}{3} tg^2 \left(\frac{\alpha}{2}\right) \frac{d\varepsilon_R}{dr}\right) \tag{5.53}$$

It must be remembered that ρ_s is the density of specimen material, whereas C_o is the longitudinal elastic wave speed in the Hopkinson bars, and $\rho_s C_o$ does not represent an impedance.

Since it has been assumed that the axial forces acting on the WLCT specimen are in equilibrium, the following relation holds (with the exclusion of the first few microseconds of the loading time)

$$P_A - P_B = R_x^L + R_x^T \tag{5.54}$$

Defining forces without inertia effects acting on the specimen as F^o, $P_A{}^o$, and $P_B{}^o$, the following relations are obtained

$$\begin{aligned} P_A &= P_A^o + R_x^L + R_x^T \\ P_B &= P_B^o \text{ or } P_B^o = P_A^o \end{aligned} \tag{5.55}$$

Introducing (5.55) into (5.26) an average force P acting on the specimen can be calculated

$$\bar{P} = \bar{P}^o + \frac{1}{2}(R_x^L + R_x^T) \tag{5.56}$$

where an average force without inertia is (see equation (5.26)).

$$\bar{P}^o = \frac{1}{2}(P_A^o + P_B^o) \tag{5.57}$$

In addition, remembering (5.12) and (5.25), a new relation between F and P_B can be written which includes inertia forces,

$$F = \frac{P_B(t)}{2tg\left(\frac{\alpha}{2} + \beta\right)} \tag{5.58}$$

where P_B is measured by the transmitted pulse $\varepsilon_T(t)$. Introducing the relation $F = F^o + R_y$ and equation (5.55) into (5.58), the real force F^o which opens the crack can be obtained as follows

$$F^o = \frac{P_A(t) - 2\,(R_x^T + R_x^L)}{2\mathrm{tg}\,\frac{\alpha}{2}}\ , \qquad \mu = 0 \tag{5.59}$$

where $P_A(t)$ can be obtained from equation (5.24). The second possibility is

$$F^o = \frac{P_B(t) - R_x^T}{2\mathrm{tg}\,\frac{\alpha}{2}}\ , \quad \mu = 0 \tag{5.60}$$

Equation (5.60) is more useful for practical purposes since it is related directly to the measurable quantity taken from experiment, i.e., $\varepsilon_T(t)$. Due to the fact that the force opening the specimen is measured directly by the transmitted pulse $\varepsilon_T(t)$, which provides value of $P_B(t)$, the influence of the longitudinal inertia is automatically eliminated from equation (5.60). The final form of the relation (5.60) to calculate F^o, after introducing (5.25) into (5.50), is

$$F^o = \frac{EF_E\varepsilon_T(t)}{2\mathrm{tg}\left(\frac{\alpha}{2} + \mathrm{tg}^{-1}\mu\right)} - \frac{2}{3}\rho_s C_o BW\mathrm{tg}\left(\frac{\alpha}{2}\right)\frac{d\varepsilon_R}{dt} \tag{5.61}$$

The second term in equation (5.61) has a relatively high value only during the rise time, usually within ~ 20 μs then $d\varepsilon_R/dt \approx 0$ and the inertia effects are negligible. The conditions of a WLCT dynamic experiment should be chosen such that the specimen would fracture at a time t_c longer than the rise time t_r of the reflected pulse. In this case the inertia effects are negligible. If this is not the case the error introduced in the calculation of F, and consequently in K_{Ic}, due to specimen inertia is

$$W_i = \frac{F^o - F}{F^o} \tag{5.62}$$

Introducing of (5.58) and (5.60) into (5.62) yields

$$W_i = -\left(\frac{P_B}{R_x^T} - 1\right)^{-1} \tag{5.63}$$

The expression for the error estimation takes the form following introducing (5.25) and (5.50) into (5.63)

$$W_i = -\left(\frac{EF_E\,\varepsilon_T(t)}{\frac{3}{4}\rho_s C_o H_o BW tg^2\left(\frac{\alpha}{2}\right)\frac{d\varepsilon_R}{dt}}\right)^{-1} \tag{5.64}$$

Equation (5.64) estimates an error due to inertia effects and takes into account both measured signals, the transmitted pulse $\varepsilon_T(t)$, and the time derivative of the reflected pulse $d\varepsilon_R/dt$. But this equation contains only the transverse inertia. The longitudinal inertia is automatically eliminated because the force is measured directly at the face B. Equation (5.56) for an average axial force P acting in the loading system, bars and specimen, contains both inertia components R_x^L and R_x^T.

The relations (5.53), (5.59), (5.62), and (5.63) make it possible to estimate quantitatively the inertia effects. A general conclusion can be reached that neglecting the inertia effects in equation (5.60) causes an underestimation of the calculated fracture toughness. This happens only when at the critical time t_c the acceleration of the wedge is not equal to zero.

The calculations of R_x^L and R_x^T were performed to determine the order of the inertia effects in the WLCT method. All input data were provided from the oscillogram for 0.45% C steel which is shown in Fig. 5.15. This oscillogram shows that the rise time t_c of the incident pulse $\varepsilon_R(t)$ is shorter, $t_r \approx 30$ µs, than the critical time t_c shown by the transmitted pulse $\varepsilon_T(t)$, $t_c \approx 34$ µs. Thus the maximum values of R_x^L and R_x^T do not exactly coincide. However, for estimation purposes both maximum values are taken as occurring at the same time. As the first step the axial inertia force R_x^L was calculated according to equation (5.46), and the following values of the parameters were assumed : $\rho_s = 7.8$ g/cm^3 ; $C_o = 5\text{x}10^5$ cm/s ; $H_o = 1$ cm ; $B = 2$ cm ; $W = 2$ cm ; $\varepsilon_T^c = 3.037 \times 10^{-4}$; $t_c =$ 34 µs. The result of calculation is

$$R_x^L = 2.79 \text{ kN}$$

To calculate the inertia force R_x^T developed by the transverse inertia, equation (5.50) was used, and the following values of the entering parameters were assumed : ρ_s, C_o, H_o, B and W - the same as above, in addition $\alpha/2 = 22.5°$; $-\varepsilon_R = 9.22 \times 10^{-4}$; $t_r = 30$ µs, the result is

$$\frac{d\varepsilon_R}{dt} = -30.73 \text{ s}^{-1} \text{ and } R_x^T = 1.097 \text{ kN}$$

These calculations indicate that the axial component of the transverse inertia R_x^T is smaller than the longitudinal component. The ratio is

$$\frac{R_x^T}{R_x^L} = 0.393 \quad \text{and} \quad R_x^L + R_x^T = 3.183 \text{ kN}$$

To estimate a maximum possible error developed by the lateral inertia the value of P_B at its peak for $t_c = 34$ µs was calculated from equation (5.27). In this calculation the following values are taken : $E = 2.06 \times 10^5$ MPa ; $F_E = 3.1416$ cm^2 (diameter of the Hopkinson bar $D_E = 2.0$ cm) ; $\varepsilon_T^c = 3.037 \times 10^{-4}$. The resulting maximum axial force acting on the face B is

$$P_B^c = 19.65 \text{ kN}$$

Introducing values of R_x^T and P_B^c into equation (5.63) the error W_i for the steel specimen can be calculated. Its value is

$$W_i = -0.0591 \text{ or } W_i = -5.91 \text{ percent}$$

These calculations indicate that the fracture toughness of steel calculated without taking into account the inertia effects is underestimated by ~ 6 percent, this is true $\dot{K}_I \approx 1\text{x}10^6$ MPa$\sqrt{m}$/s. This value of error falls within the error of the experimental method. The calculated values of the inertia forces R_x^L and R_x^T as well as P_B^c for a hard aluminum alloy, which was also tested, are far lower than for steel. For example, the value of the inertia error does not exceed - 3.3 percent, more exactly $W_i = -0.03286$.

It then may be concluded that the inertia effects do not exceed the usual experimental scatter observed in fracture toughness experiments. In addition the effects of inertia and friction cancel each other, as shown by equations

(5.34) and (5.63). To obtain zero error due to this interaction the following relation holds

$$W + W_i = 0 \tag{5.65}$$

where W is the error introduced by neglecting the effect of friction (e.g. (5.34)), equations (5.34) and (5.64) can be introduced into (5.65), and the result is

$$\mu_o = \text{tg}\left\{ -\frac{\alpha}{2} + \text{tg}^{-1}\left[-\frac{P_B}{R_x^T}\left(\frac{P_B}{R_x^T} - 1\right)^{-1} \text{tg}\frac{\alpha}{2} \right\} \tag{5.66}$$

where μ_o is the coefficient of friction at which the inertia effect is completely compensated by friction. For example for steel and for $\alpha/2 = \pi/8$ the value of μ_o is 0.0207, for aluminum alloy $\mu_o = 0.0116$. These values of μ_o are still lower than the measured coefficient of friction μ, which in the real experiment is $\mu \approx 0.1$. Thus, even for the dynamic conditions of loading the frictional effects generally prevail.

It has been demonstrated that the lateral inertia effects are more important only when the fracture initiation takes place during the rise time of the incident pulse. This corresponds to the period of constant acceleration $d\varepsilon_R/dt$. For the critical times t_c longer than the rise time of the incident pulse the wedge moves approximately at a constant velocity and the transverse inertia is zero. The axial acceleration of the specimen, which is usually fairly constant up to fracture initiation (see Fig.), is far lower than the lateral acceleration and, if the force is measured at the face B, does not contribute significantly to the value of the critical force.

Friction and inertia can be controlled fairly well in the experimental technique described here. However, the loading rate $\dot{K}_I \approx 10^6$ MPa m s^{-1} constitutes the upper limit for this technique. A higher loading rates can still be achieved within this technique, but in this case an exact wave analysis of the specimen behavior is required.

5.5 Some Results for Aluminum Alloys and Steels

Aluminum alloys (PA6) with 4 percent Cu was extensively tested over a wide spectrum of loading rates $\dot{K}_I$. This alloy is both ductile and rate-sensitive and shows a rate sensitivity $\beta = 17.4$ MPa, where $\beta = \partial\sigma/\partial\log\dot{\varepsilon}$. In the case of this alloy definite pop-in points were observed in a majority of oscillograms

$\varepsilon_T(t)$, and coordinates of these points (ε_T^c, t_c) were used to calculate K_{Ic} from eq. (5.16). A typical oscillogram for PA6 alloy, as recorded by the digital oscilloscope, is shown in Fig. 5.12. There is an important remark that the sensitivity setting for the transmitted wave ε_T (channel 2) is ten times higher than the sensitivity for incident and reflected waves (channel 1). Duration of the incident wave is about 130 μs with the risetime ~ 20 μs. It means that the wedge displacement, being proportional to the sum of the integrals of the incident and reflected waves, is much more larger than the face B displacement of the transmitter bar, i.e. $U_A(t) >> U_B(t)$.

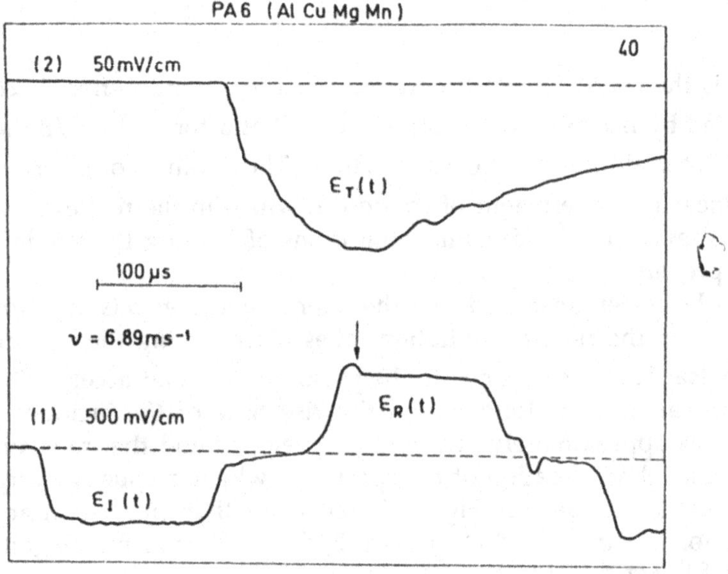

Fig. 5.12, Incident ε_I, reflected ε_R, and transmitted ε_T waves for WLCT specimen made of aluminum alloy PA6 ; lower trace : incident and reflected pulses ; upper trace : transmitted pulse ; v is the impact velocity of the striker bar, [53].

Moreover, the wedge velocity, which is proportional to the sum of incident and reflected waves, is almost constant during specimen loading and fracturing. This indicates for proportionality of the wedge displacement δ_x to t^2 at the rising portion of the wave, and a direct proportionality to t during the later time. The critical time t_c for the PA6 series was an average ~ 17 μs. The oscillogram shown in Fig. 5.12 also indicates tat there is some shift of time between the beginning of the incident wave reflection and the initiation of the specimen loading. This short delay, as it is marked by the vertical arrow, is due to a small distance reduction existing initially in between WLCT specimen and the wedge. The results obtained with WLCT specimens over the range of loading rates 0.5 MPa√m/s^{-1} ≤ $\dot{K}_I$ ≤ 1.2 x 10^6 MPa√m/s^{-1} are shown in Fig. 5.13, [55]. A substantial drop in fracture toughness is observed beginning from $\dot{K}_I \approx 10^2$ MPa√m s^{-1} ≤ $\dot{K}_I$ ≤ 10^6 MPa√m s^{-1}, it is difficult to presume a minimum value for K_{Ic} within this range. The results for PA6 aluminum alloy have been compared with results obtained for aluminum alloy DTD 502A (British Standards) in [59], and this comparison is also shown in Fig. 5.13. The DTD 502A aluminum alloy is slightly more brittle, but the effect of loading rate is similar for both materials. The fracture toughness shows a decrease at increased loading rates.

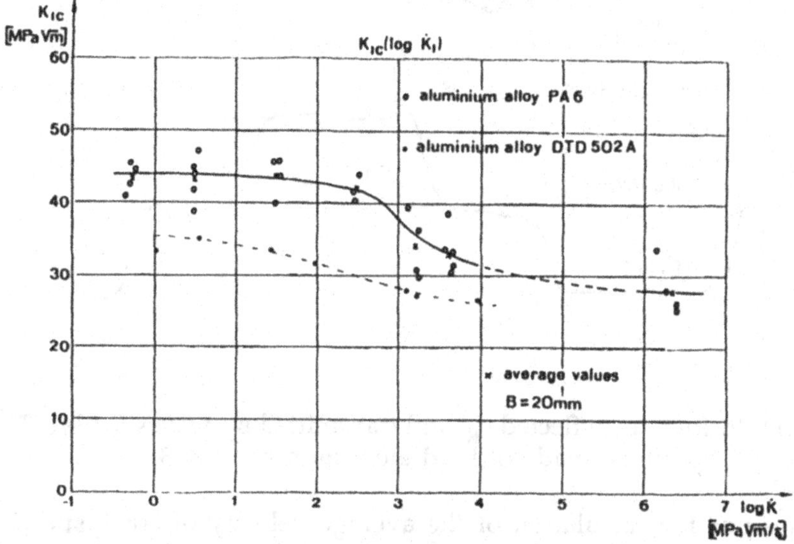

Fig. 5.13, Spectra of fracture toughness for two aluminum alloys : PA6 - WLCT technique, and DTD502A (British Standard), after [55].

The results demonstrates a substantial drop in fracture toughness beginning from $\dot{K}_I \approx 10$ MPa√m/s. Values of K_{Ic} determined at the stress wave loading are almost twice lower as those obtained with quasi-static rates, $\dot{K}_I \approx 1$ MPa√m/s. Moreover, the fracture toughness reduction is not proportional to log K_I, but shows a more complicated behavior. Because of the gap in the experimental data over the region 10^4 MPa√m/s ≤ $\dot{K}_I$ ≤ 10^6 MPa√m/s, it is difficult to presume a minimum of K_{Ic} in this range. It may be also noted that the scatter of experimental points, including stress wave loading, is relatively small for such kind of experiment.

A more brittle aluminum alloy designated by K63 (Al Zn Mg Cu) and tested after rolling and thermal treatment have shown different response to quasi-static and dynamic loadings. One of oscillograms for K63 alloy is shown in Fig. 5.14, the fracture plane was choosen along direction of rolling. A different shape of the transmitted wave $\varepsilon_T(t)$ can be noted as compared to a softer alloy PA6. After the initial rising portion of the wave the maximum is reached, and then during fast fracture the signal drops steadily to zero. This fea-

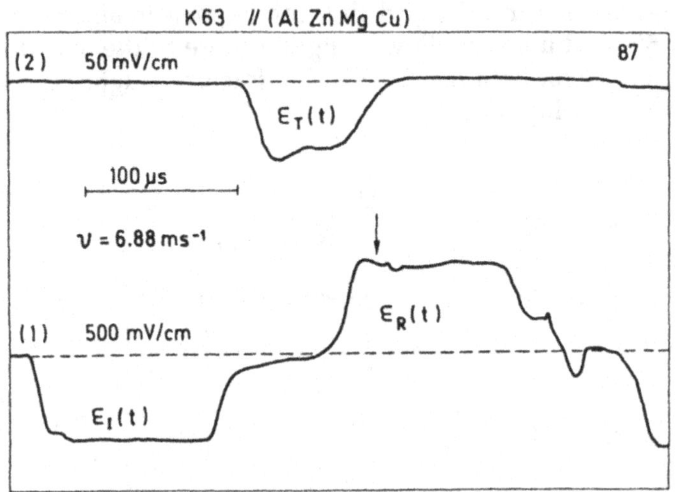

Fig. 5.14, Incident ε_I, reflected ε_R, and transmitted ε_T, waves for WLCT specimen made of hard aluminum alloy K63.

ture enables one for calculation of the average velocity of crack speed. For example, K63 alloy fractures in an essentially brittle manner, and an estimated mean crack speed ~ 0.4 mm/μs. Some results for the alloy K63 are given in Table 5.3. Since this alloy shows anisotropy after rolling the WLCT method has been applied in two directions, parallel and perpendicular to the direction of rolling. In both cases, when the loading rate $\dot{K}_I$ has been changed over six orders, the fracture toughness diminishes from about 20% to 36% depending upon direction of testing.

Another series of tests have been performed on 1045 steel (0.45%C) in the as-received state (after rolling). An representative oscillogram for this series of tests is shown in Fig. 5.15. The overall response of the steel specimen is similar to the brittle behavior of K63 alloy. A brittle fracture occurs after ~ 30 μs, the crack propagation follows and finally two parts of specimen separate. The whole event takes about 100 μs.

Table 5.3 Results for static and stress wave loading experiments for hard aluminum alloy K63.

Type of loading	Stating loading $\dot{K}_I = 1.0$ $MPa\sqrt{m}\,s^{-1}$	Wave loading $\dot{K}_I = 1.15x10^6$ $MPa\sqrt{m}\,s^{-1}$	Percentage change in K_{Ic}
Crack direction	$K_{Ic}(MPa\sqrt{m})$	$K_{Ic}(MPa\sqrt{m})$	$\frac{\Delta K_{Ic}}{K_{Ic\,stat}}$ x 100
Parallel ∥	21.75	17.12	- 21.29
Perpendicular ⊥	29.65	19.05	- 35.75

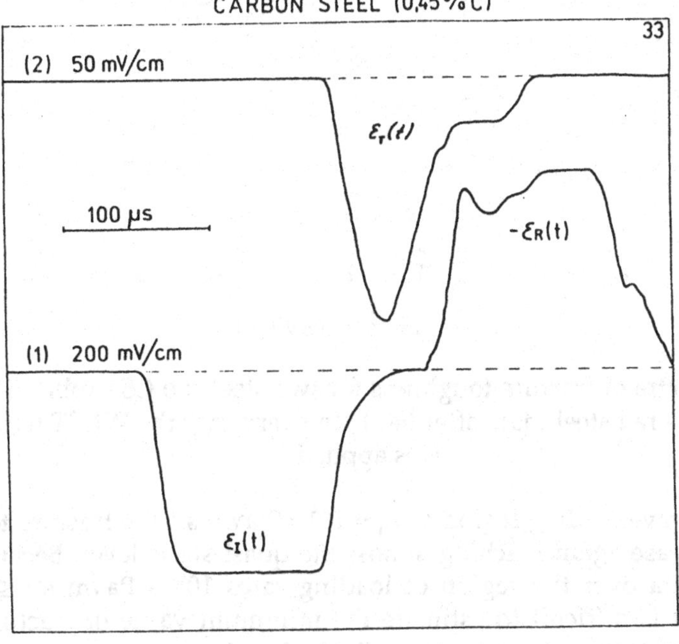

Fig. 5.15, Incident ε_I, reflected ε_R, and transmitted ε_T, waves for WLCT specimen made of carbon steel 1045 (0.45%C).

Tests result on carbon steel (0.45 percent C) are shown in Fig. 5.16. These results cover the loading rate range 0.02 MPa√m $s^{-1} \le \dot{K}_I \le 10^6$ MPa√m s^{-1}. In addition to experimental results for 1045 carbon steel (denoted by CS in Fig. 5.16) experimental results for the rail steel denoted by RS have also been included in Fig. 5.16. The distribution of experimental points demonstrate a very consistent behavior. It was found that the behavior of steel differs substantially from that observed for aluminum alloys. Between 10^2 MPa√m $s^{-1} \le \dot{K}_I \le 10^3$ MPa√m s^{-1} a considerable decrease of fracture toughness is to be noted.

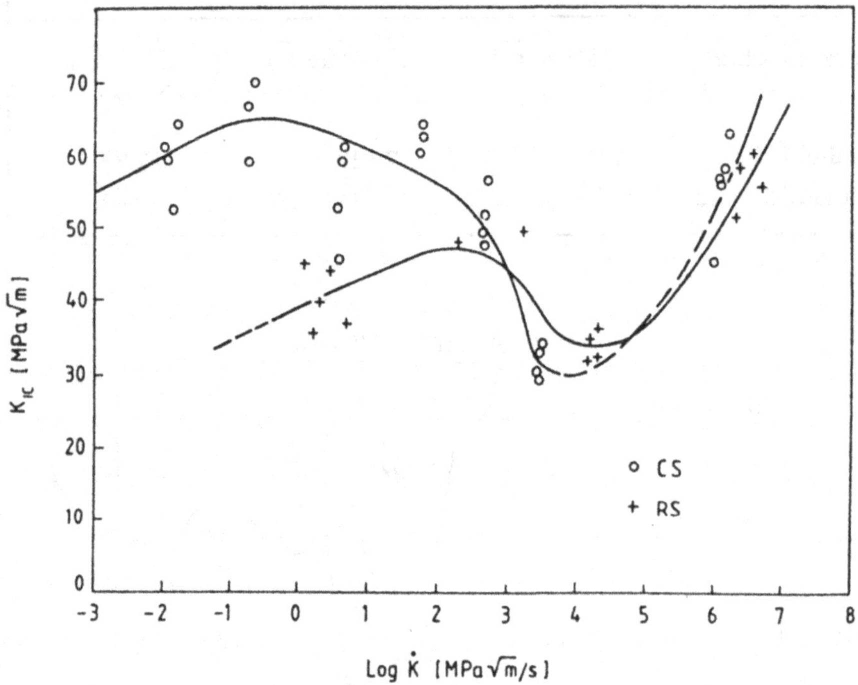

Fig. 5.16, Spectra of fracture toughness for two steels : o CS - rolled 1045 steel, [55] ; + RS - rail steel (data after [60]) ; in every case the WLCT technique was applied.

In the stress wave loading region at $\dot{K}_I \approx 10^6$ MPa√m s^{-1} the fracture toughness seems to increase again reaching almost the quasi-static level. Because of the gap in the data over the region of loading rates 10^4 MPa√m $s^{-1} \le \dot{K}_I \le 10^6$ MPa√m s^{-1}, it is difficult to estimate the minimum value of fracture toughness. Since thus far there is no other reliable data for a carbon steel concerning the whole loading spectrum, further experimental work is needed to clarify this point. Previous studies have provided the loading rate spectra up to $\dot{K}_I \approx 10^4$ MPa√m s^{-1}, for example [61]. More recently experimental fracture toughness data were reported for SAE 4340 and AISI 1020 hot-rolled steel [62]. Only quasi-static, $\dot{K}_I \approx 1$ MPa√m s^{-1}, and dynamic, $\dot{K}_I \approx 10^6$ MPa√m s^{-1}, experiments were performed. In contrast to the results from 4340 steel, there was a considerable difference between the dynamic and quasi-static K_{Ic} values for 1020 steel, respectively, quasi-static K_{Ic} = 126 MPa√m, and dynamic K_{Ic} = 70 MPa√m, in comparison to K_{Ic} = 63 MPa√m (quasi-static) and 55 MPa√m (dynamic) for 0.45 percent C steel. Almost most all existing experimental results show a decrease of fracture toughness up to a loading rate $\dot{K}_I \approx 10^4$ MPa√m s^{-1}, this is true for materials exhibiting strain rate effects in their plastic properties. However, at higher loading rates the situation is unclear. It is known that for running cracks K_{ID} values are far higher than values determined for the process of cracks initiation [63].

The experimental technique described in this part was designed to determine the parameters used in the description of fracture initiation in mode I over a wide spectrum of loading rates. Using of this methodology it enables to achieve high loading rates with only a moderate cost of specimen preparation. This makes it possible to perform extensive series of screening experiments to design and inspect structural materials against fracture under impact loading.

5.6 Procedure to determine J-integral with WLCT specimen at high loading rates

Since its adaptation to fracture mechanics by Rice, [64], the critical value of J-integral has proved to be a successfull fracture criterion for behaviour ranging from linear to fully plastic response of test materials. Within the framework of a total strain formulation of elastic-plastic deformation, the J-integral may be defined for two-dimensional problems

$$J = \int_{\Gamma} (W dy - \bar{T} \frac{\partial \bar{u}}{\partial x} ds) \tag{5.67}$$

with

$$\bar{T} = \sigma_{ij} n_j \quad \text{and} \quad W = \int_0^{\varepsilon_{mn}} \sigma_{ij} d\varepsilon_{ij}$$

where (x,y) are rectangular coordinates normal to the crack front, y being perpendicular to the crack surface, ds is an increment of arc length along any contour Γ, beginning along the bottom surface of the crack and ending along the top surface, T is the stress vector exerted on the material within the contour, u is the displacement vector and W is the integral called strain energy density, ε_{ij} is an infinitesimal strain tensor, and σ_{ij} is the stress tensor. Properties of the J-integral are widely reviewed in the literature, for example [65,66], and it is not intended to attempt such a review. Analytical estimation procedures have been developed for its use, Begley and Landes [4], Landes and Begley [5] and independently Broberg [67] by application of an alternate and equivalent definition of J for linear or non-linear materials as a potential energy difference for identically loaded configurations having neighbouring crack sizes a and a + da. In particular for the crack length a

$$J = -\frac{1}{B}\frac{\partial U}{\partial a} \quad \text{or} \quad J = -\frac{1}{B}\int_0^{\delta}\left(\frac{\partial F}{\partial a}\right)_{\delta} d\delta \tag{5.68}$$

wherer U is the potential energy, F is the load, and d is the load-point displacement. It can be also shown that U is given by the area under the load-displacement curve F(δ), Rice et al. [68]. Further analyses by Merkle and Corten [69] and Kanazawa et al. [70] have lead to more specimen expressions correlating J with area under F(δ) curve. Finally, when specimen ligament is both under bending and tensile forces, the formula recommended by the ASTM Standard E813-81, [71] is as follows

$$J = \frac{1}{B(W - a_0)} f\left(\frac{a_0}{W}\right) \int_0^{\delta_0} F(\delta)\, d\delta \tag{5.69}$$

where W is the specimen width, $W - a_0 = b$ is the initial uncracked ligament, a_0 is the original crack size, including fatigue precrack. Values of dimensionless coefficient $f(a_0/W)$ that corrects for the tensile component of loading, Clarke and Landers [72], is given in the tabulated form in the Standard E813-81.

Since it has been demonstrated that quasi-brittle and elastic-plastic fracture is a thermally activated event, for example Krabiell and Dahl [20], Klepaczko [73,74], the fracture toughness must be not only temperature but also rate dependent. The early studies within the framework of small scale yielding indeed do show under certain conditions a substantial rate effects on fracture toughness, Eftis and Krafft [63].

For structural metallic materials with strong temperature and strain rate dependence of their plastic properties, the fracture toughness within the region of lower shelf usually decreases with decreasing temperature and increased loading rate; For the upper shelf where plastic response dominates the positive rate sensitivity is observed, Marandet et al. [19], Krabiell and Dahl [20], Klepaczko [73]. Thus, it is desirable from the practical point of view to evaluate both shelves experimentally. For the upper shelf response the critical value of J-integral is especially convenient as the fracture criterion because plasticity effects are included in its analysis.
Application and analysis of J-integral for dynamic situation has been studied by Bui [75] and also by Ehrlacher [76]. The potential application of dynamic J-integral in the case of instrumented impact test was studied by Mall [77]. Those analyses of J-integral take into account inertia effects. It may be mentioned here that integral formulated as eq. (5.67) is not truly a path-independent in the dynamic case. The dynamic J-integral formulated by Bui and reformulated by Mall can be written as follows

$$J_d = J + \int_A \rho\, \ddot{u}_i \frac{\partial \bar{u}}{\partial x}\, dA \tag{5.70}$$

where J is defined as (5.67) and the second term is the integral over the area A enclosed by an arbitrary contour Γ. Thus, the second term is the modification of the static J-integral due to the presence of inertia forces. It can be shown that integral J_d is path-independent. Basically, the integral (5.70) has been devised for a running crack and its role in the dynamic crack initiation remains to be analysed more exactly.

In order to determine J-integral experimentally the force versus displacement record F(δ) must be obtained. Moreover, in order to determine the critical value of J_{Ic}, under quasi-static or dynamic loading conditions, the first crack advancement following blunting must be detected. Many experimental procedures have been developed in the past to accomplish those goals. It is not, however, the purpose of this text to review all experimental possibilities. Application of the multiple specimen J-resistance curve technique, as recommended by the E813-81 Standard, is not possible for the case of fast loading. Thus, a single specimen technique remains as an alternative, but the compliance method with partial several unloadings cannot be applied as well. One of attractive methods to determine crack growth as a function of time without test interruptions is the electrical potential method, for example Marandet et al. [19]. Another solution to this problem was proposed in [57,78], namely the use of properly placed small strain gage on the specimen surface

near crack tip is sufficient to detect a crack advancement. This technique can be used for both, quasi-static and impact loading.
Again, the Wedge Loaded Compact Tension specimen (WLCT) was used in this purpose. This technique has been modified to determine J-integral at different loading rates, from quasi-static to stress-wave loading. Three types of specimen support, and consequently reaction forces, are shown in Fig. 5.17. This is an important point since during determination of force-displacement, support conditions change due to rotation of both halves of WLCT specimen approaching case C. For K_{Ic} testing, support of the type B is adequate. Compressive force P is developed by the wedge and specimen is open-up by force F which depends on geometry and coefficient of friction between wedge and specimen edges at contact points. The angle of the wedge must be larger than specimen incision a to assure proper kinematics of loading. The relation bet-

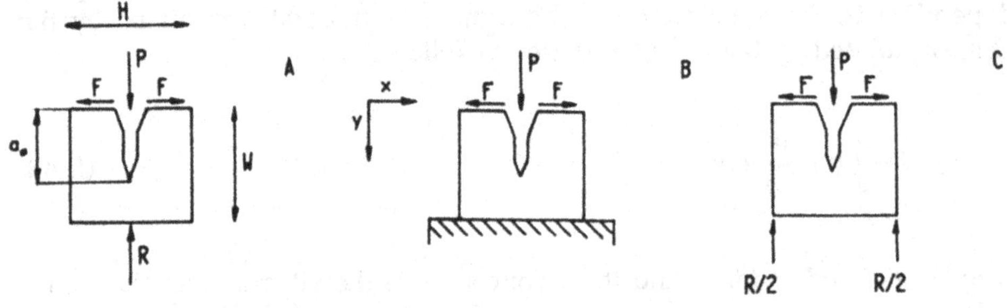

Fig. 5.17, Three possible supports for the WLCT specimen.

ween P and F is given by eq. (5.27). If the wedge displacement $\delta_y(t)$ is measured as a function of time the force-displacement record $F(\delta_x)$ can be obtained after elimination of time in (5.27) and (5.71)

$$\delta_x = \delta_y \, tg \frac{\alpha}{2} \tag{5.71}$$

Thus, the opening force F as a function of displacement provides $F(\delta_x)$ record needed to calculate work dissipated during crack blunting and advancement. When the WLCT specimen is put on the flat surface the support of type C will prevail, and a larger force P will be needed to initiate yielding and continue process of fracturing. The condition of force and moment equilibrium provides the formula for force F opening the crack for the case C.

$$F = \frac{P}{2}\left(\frac{1}{\mathrm{tg}\left(\frac{\alpha}{2} + \mathrm{tg}^{-1}\mu\right)} - \frac{H}{W + a_o}\right) \tag{5.72}$$

Such a support substantially increases the tensile component of stress acting over the ligament cross section. It has been concluded that the best way to determine J-integral is the central support placed at the specimen axis (case A in Fig. 5.17). For this case the potential energy which must be introduced into eq. (5.69) is

$$\int_o^{\delta_{xc}} F(\delta_x)\, d\delta_x$$

Since $P(d_X)$ is recorded during both quasi-static and stress-wave loading it is useful to express the above integral as follows

$$\int_o^{\delta_{xc}} F(\delta_x)\, d\delta_x = \frac{\mathrm{tg}\,\alpha/2}{\mathrm{tg}\left(\frac{\alpha}{2} + \beta\right)} \int_o^{\delta_{yc}} P(\delta_y)\, d\delta_y \tag{5.73}$$

with $\beta = \mathrm{tg}^{-1}\mu$ and $\delta_{xc} = \delta_{yc}\,\alpha/2$. Introducing coefficient of friction m in the explicit form and taking into account eq. (5.69) into eq. (5.73) the final expression for critical value of J-integral is obtained

$$J_{Ic} = \frac{1 - \mu}{B(W - a_o)\left(1 + \frac{\mu}{\mathrm{tg}\,\alpha/2}\right)}\, f\left(\frac{a_o}{W}\right) \int_o^{\delta_{yc}} p(\delta_y)\, dy \tag{5.74}$$

where μ is the coefficient of Coulomb friction and δ_{yc} is the critical displacement detected by strain gage during recording of $P(\delta_y)$ diagram. Elastic-plastic finite element calculations for WLCT specimen by Lipinski (Klepaczko et al. [79]) have indicated that the plastic zone domain for the critical conditions extends not far but about ~ 0.3 mm on the side of fatigue crack. Calculations were performed for both quasi-static and dynamic loadings taking into account quasi-static and dynamic strain hardening curves of 1045 steel. In order to detect crack advancement after blunting the gage should be cemented in the second or third quadrant with orientation along x axis and slightly behing crack tip, as it is shown in Fig. 5.18. During crack advancement this region is partially unloaded which can be detected on $\varepsilon_y(t)$ or $\varepsilon_y(\delta_y)$ record. Typical records of $P(\delta_y)$ and $\varepsilon_y(\delta_y)$ where δ_y is the wedge displacement (wedge angle, α = 50°) is shown in Fig. 5.19. The first unloading jump on $\varepsilon_y(\delta_y)$ curve indicates the critical force P_c and critical displacement δ_{yc}. Material tested was 1045 steel in the annealed state σ_y = 300 MPa and σ_u = 600 MPa. Prefatigued specimens had the following geometry : H = 30 mm, W = 25 mm, a_0 = 12.5 mm (with fatigue crack), α^* = 45° (angle of specimen incision), B = 12.5 mm. The wedge planes were properly lubricated with MoS_2 grease, and the wedge volocity was kept constant $d\delta_y/dt = 4\text{x}10^{-3}$ mm. As expected the critical force P_c occurs well before the maximum load takes place, this was observed for all three quasi-static tests performed. Since the central support was applied in the form of a small beam at the specimen bottom (Fig. 5.20), relation (5.74) could be directly applied to determine critical value of J-integral. Quasi-static experiments with WLCT specimens were mainly performed in the purpose to check reliability of this method by comparison of the results with earlier result obtained for the same steel at transition temperature region (213K < T < 203K) but by R-curve method with three point bend specimens, Bayoumi and Bassim [80]. The mean value of J_{Ic} for five tests at room temperature (T) was 72.0 kJ/m^2,

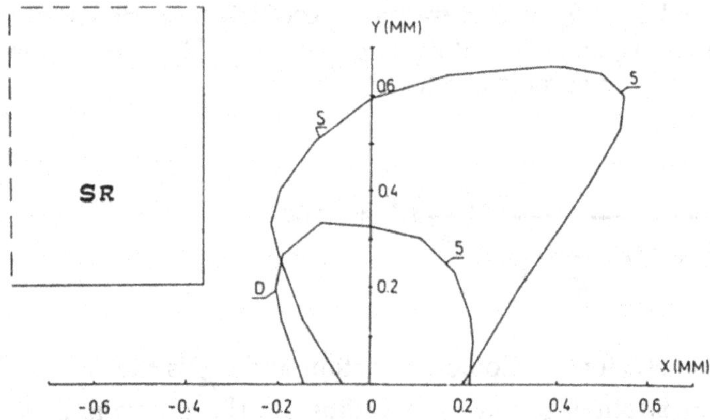

Fig. 5.18, Plastic zones, quasi-static S and dynamic D ; estimated loading rates $\dot{K}_{IS} \approx 1$ MPa$\sqrt{m}$/s and $\dot{K}_{ID} \approx 10^6$ MPa$\sqrt{m}$/s ; calculations for 1045 steel ; SR shows position of strain gage in the same scale as (x,y) axes.

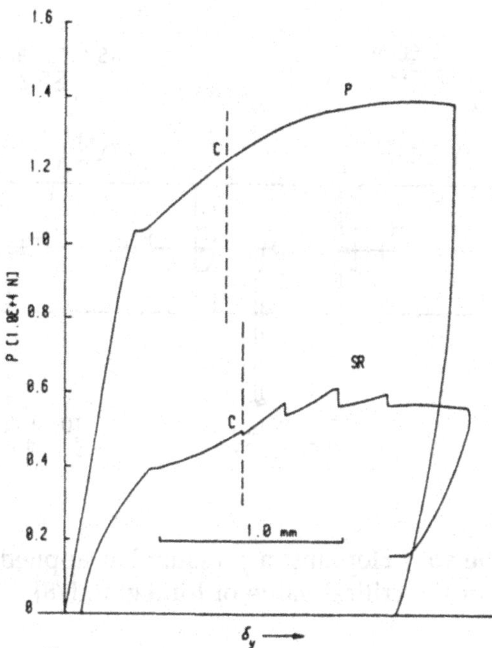

Fig. 5.19, Records $P(\delta_y)$ and $\varepsilon(\delta_y)$ for quasi-static loading by wedge and with the central support of WLCT specimen ; 1045 annealed steel, [78].

whereas three WLCT tests gave the mean value J_{Ic} = 73.5 kJ/m^2 with μ = 0.14 and $f(a_o/W) \approx 2.25$. Overall the mean value of J_{Ic} calculated from eq. (5.74) for three specimens tested is only 2.1% larger than that obtained from the R-curve method. It is probable that the contact lines between wedge and specimen move slightly inward specimen when gross yielding occurs, and thus increasing specimen stiffness. Also underestimation of coefficient of friction may lead sometime to larger values of J_{Ic} can be still evaluated.

Experimental procedure employed in determination of critical values of J-integral under stress wave loading is similar to the methodology developed for determination of the critical stress intensity factor K_{Ic} within the framework of small scale yielding. The fundamentals of this procedure, as applied to determination of critical value of J-integral, is shown in Fig. 5.20. The experimental setup employed for J_{Ic} determination is shown in Fig. 5.21. A simple system has been designed and constructed, (Klepaczko [81]). The experimental setup consists of a gas gun ; the system of two instrumented Hopkinson bars made of hard aluminum alloy (ALCAN-2024-T3, diameter 38 mm), and the recording equipment. WLCT specimen of the same geometry as that used in

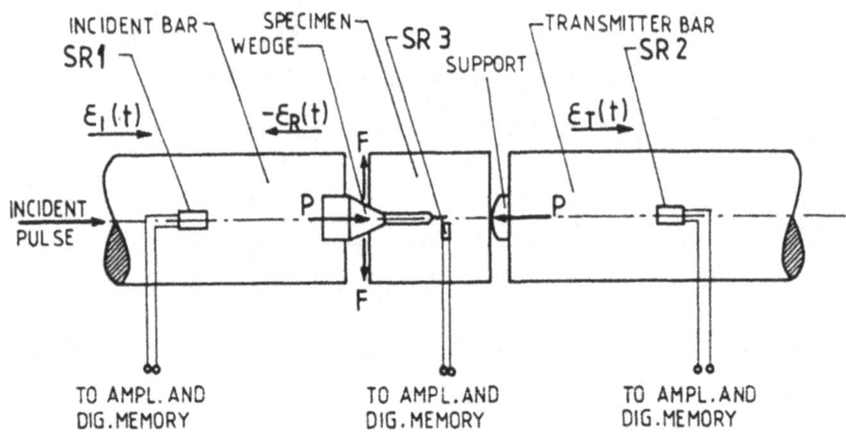

Fig. 5.20, Scheme of the split Hopkinson pressure bar applied for determination of the critical value of J-integral, [78].

quasi-static tests is placed between incident and transmitter bars. Since the wedge is attached to the incident bar, the specimen is loaded in exactly the same manner as for the quasi-static configuration. On the face of transmitter bar (face B) a small beam of the same length as bar diameter and thickness about 8 mm is attached. This is specimen support (case A in Fig. 5.17) allowing for specimen opening without force reactions from specimen corners (case C in Fig. 5.17). The same support was applied during the course of quasi-static tests. In addition a small strain gage is cemented to the specimen to detect crack advancement during stress wave loading. Striker bars of different lengths and constant diameter D_E = 38 mm are launched from the gas gun at a desired velocity v_o. The impact of the striker bar onto the face of the incident bar develops the longitudinal compressive wave which propagates along this bar. The incident wave $\varepsilon_T(t)$ is measured by the strain gage station on the incident bar. Since the striker bar is at least ten times longer than the length W of the WLCT specimen, the process of its loading and opening by the incident wave through the wedge attached to the face A of the incident bar may be assumed to be like quasi-static. This is the basic assumption of the method to determine J-integral for a short time loadings. When the incident compressive wave reaches WLCT specimen, part of this pulse in reflected back from the face A as a tensile pulse $-\varepsilon_R(t)$, and part is transmitted through the specimen into the transmitter bar as a compressive wave $\varepsilon_T(t)$. The reflected pulse is also measured by the strain gage station at the transmitter bar. Electrical signals from all three strain gage stations are recorded with the aid of digital transient

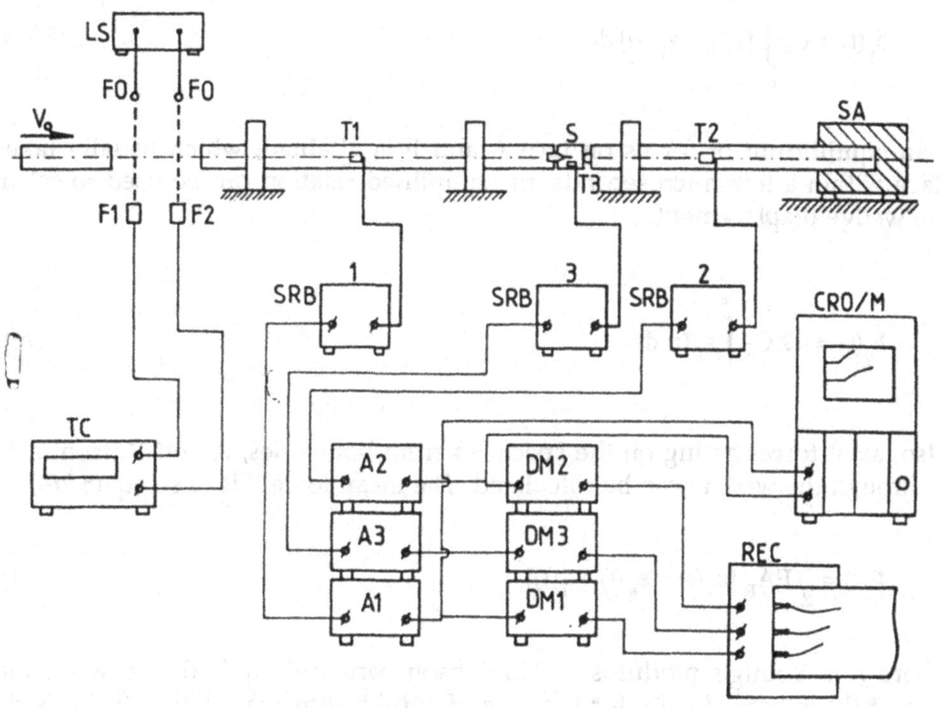

Fig. 5.21, Application of large diameter split Hopkinson pressure bar to determination of J_{Ic} ; LS - light source ; FO - optic fibers ; F1,F2 - photodiodes ; Tc - time counter ; T1,T2,T3 - SR gages ; S - WLCT centrally supported specimen ; SRB - SR conditioners ; A1,A2,A3 - amplifiers ; DM1,DM2,DM3 - digital memories ; CRO - oscilloscope ; REC - analog recorder.

recorders. Thus, recordings of incident $\varepsilon_I(t)$, reflected - $\varepsilon_R(t)$ and transmitted $\varepsilon_T(t)$ pulses plus a signal $\varepsilon_y(t)$ from strain gage attached to WLCT specimen provide all the information concerning specimen loading and fracturing. The proper distances x_A and x_B between gages make it possible to separate in time $\varepsilon_I(t)$ and - $\varepsilon_R(t)$ pulses. The following analysis of three waves to describe the dynamics of fracturing is similar to that introduced in Chapter 4 and at the beginning of this Chapter. The solution for long elastic bars without any dispersion provides the suitable expressions as discussed above. The net displacement of the wedge $\delta_y(t) = U_A(t) - U_B(t)$ is

$$\delta_y(t) = C_o \int_0^{t_c} [\varepsilon_I(t) - \varepsilon_R(t)]\, dt \qquad (5.75)$$

If an equilibrium of forces on face A and B is attained, which usually takes place within a few microseconds, the simplified relation can be used to calculate wedge displacement

$$\delta_y(t) = -\,2\, C_o \int_0^{t_c} \varepsilon_R(t)\, dt \qquad (5.76)$$

Also, axial forces acting on the specimen from both sides, A and B (from side A through the wedge) can be calculated. The mean force P is (also eq. (5.26))

$$\bar{P}(t) = \frac{1}{2} EA_E\, [\varepsilon_T(t) + \varepsilon_R(t) + {}_T(t)] \qquad (5.77)$$

where E is Young's modulus of Hopkinson bars and A_E is the cross section area of those bars. Again, for the case of equilibrium the relation (5.26) is reduced to, (also eq. (5.14))

$$P(t) = EA_E\, \varepsilon_T(t) \qquad (5.78)$$

It is concluded from above relations that equations (5.75) or (5.76) enable one to calculate the wedge displacement as a function of time, equations (5.77) or (5.78) make it possible to calculate of the mean force as a function of time. Consequently, the dynamic force-displacement diagram can be constructed by elimination of time from $\delta_y(t)$ and P(t). If the critical time t_c is detected by the strain gage cemented on the specimen then the force-displacement curve can be integrated within the limits $0 < \delta_y(t) < \delta_y(t_c)$ and the critical value of J-integral can be determined. A number of prefatigued WLCT specimens of 1045 annealed steel have been tested under stress wave loading conditions. A representative record of dynamic test is shown in Fig. 5.22. Duration of the incident pulse $\varepsilon_T(t)$ is about 180 μs and this is the record on channel 1, also on channel 1 the reflected pulse - $\varepsilon_R(t)$ is recorded. The transmitted pulse $\varepsilon_T(t)$ is recorded on channel 2, and the signal from the strain gage $\varepsilon_y(t)$ is recorded on channel 3. A time shift of about 50 μs is observed between $\varepsilon_T(t)$ and $\varepsilon_y(t)$, this is caused by the fact that the strain gage station on the transmitter bar is positioned at the distance $x_A = 5D_E$ where D_E is the bar diameter. It is quite clear from the $\varepsilon_y(t)$ record that the initiation of fracturing can be easily detected from such a test. In Fig. 5.23 the force-displacement diagram $P(\delta_y)$ is shown. This diagram has been obtained using the simplified relations (5.76) and (5.78). The use of these relations does not introduce much error in the $\delta_y(t)$ calculations since $\varepsilon_R >> \varepsilon_T$ (sensitivity of the channel 2 is higher than that for channel 1).

Since three quasi-static tests with WLCT specimens gave satisfactory mean value of J_{Ic} ; $J_{Ic} = 73.5$ kJ/m^2, with standard deviation s = 7.67, larger series of tests were performed with stress wave loading. Several specimens were tested under different support conditions (type A or B of Fig. 5.17). It was learned that the central support gives good results, and those will be shown and discussed here. Five specimens have been analyzed more throughly. The first noticeable fact drawn immediately after comparison of quasi-static and dynamic force-displacement diagrams is that under impact loading annealed 1045 steel behaves in a much more brittle manner. Practically the ductile-brittle transition has been observed for this steel when loading rate was increased

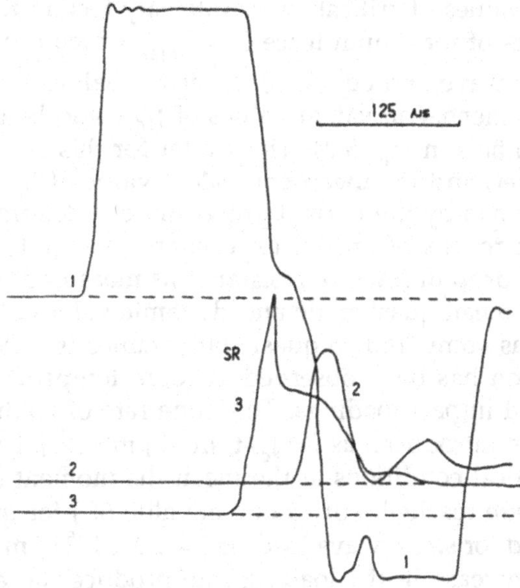

Fig. 5.22, Analog records obtained from the stress wave loading of WLCT centrally supported specimen ; 1 - records of $\varepsilon_I(t)$ and - $\varepsilon_R(t)$; 2 - record of $\varepsilon_T(t)$; 3 - strain gage signal $\varepsilon_y(t)$.

from slow to impact loading. During quasi-static loading gross-yielding of WLCT specimen was always observed, whereas during stress wave loading only yielding was present. All these observations have been confirmed after determination of critical values of J-integral. Values of J_{Ic} determined from the stress wave loading experiments are substantially lower than those obtained from quasi-static tests. Results of both quasi-static and dynamic experiments and impact experiments are shown in Fig. 5.24. Determined values of J_{Ic} in both quasi-static and impact experiments at room temperature are compared with earlier results obtained for the same 1045 steel at different temperatures using three-point bend specimens and the R-curve technique.

Quasi-static results are indicated by the open triangles and the mean value of J_{Ic} is $J_{Ic} = 73.5$ kJ/m^2 with s = 7.67, where s is the standard deviation. It is clear from above that this value of J_{Ic} agrees well with $J_{Ic} \approx 72$ kJ/m^2 obtained at room temperature using the R-curve method. Values of J_{Ic} obtained from stress-wave loadings of WLCT specimens are indicated in Fig. 5.24 by dark triangles. The mean value for five tests is $J_{Ic} = 30.7$ kJ/m^2 with s = 6.48. The eq. (5.69) was again used to calculate J_{Ic} with $f(a_o/W)$ recommended by E813-81 ASTM standard. Since under stress-wave loading specimens behaved in the quasi-brittle manner, values of critical stress intensity factors K_{Ic} could be also calculated using values of maximum force $P_c = P_{max}$ which can be determined directly from the $\varepsilon_T(t)$ wave and eq. (5.16). After K_{Ic} values were determined for the same five specimens, equivalent values of J_{Ic} could be calculated, they are indicated by x symbols in Fig. 5.24. The scatter for this case, as may be expected, is much smaller, and the mean equivalent value of $J_{Ic} = 31.8$ kJ/m^2 is very close to that obtained by the normal procedure of J determination. It may be concluded that the results of impact experiments are quite consistent and do show a substantial drop in fracture resistance as measured by the J-integral criterion. The ratio of mean quasi-static and dynamic values of J_{Ic} is 0.42, and the percentage drop as compared to quasi-static values is - 58%. Indeed, the ductile-brittle transition has been observed at room temperature by application of quasi-static and impact loadings. To define rate of loading the loading rate parameter J can be introduced as $\dot{J} = J_c/t_c$ in [kJ m^{-2} s^{-1}]. Parameter J indicates how fast the critical conditions at the mean the moment of crack propagation after blunting are reached. For example, value of J for quasi-static tests is $\dot{J} = 0.33$ kJ m^{-2} s^{-1} and for stress wave loading $\dot{J} = 1.0 \times 10^6$ kJ m^{-2} s^{-1}. As shown in Fig. 5.24 such an increase in the loading rate produces an equivalent shift in testing temperature of about $\Delta T = -32°C$.

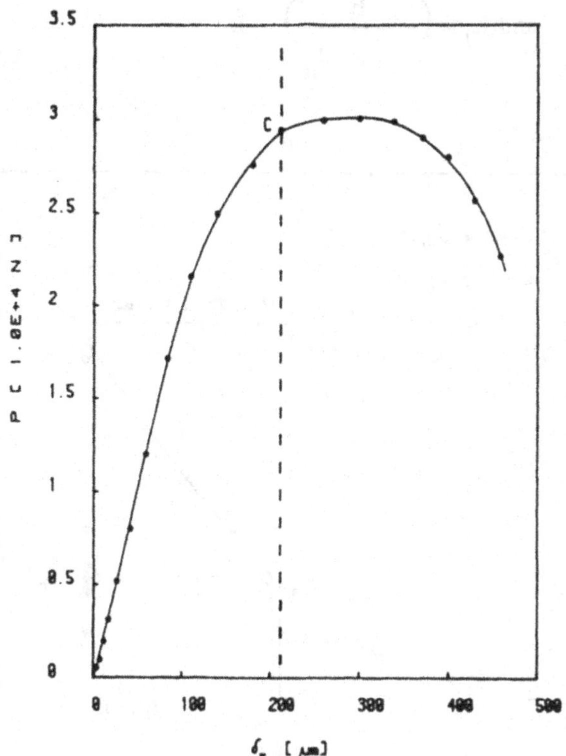

Fig. 5.23, Force-displacement plot $P(\delta_y)$ determined from the wave records of Fig. 5.22 for 1045 annealed 1045 steel ; vertical line denotes the critical wedge displacement at $t_c = 30$ μs.

Results for steel shown in Fig. 5.16 are quite different as compared to behavior aluminum alloys. Test results on carbon steel 1045 (0.45%C) after rolling which were obtained within a very wide range of loading rates $2\text{x}10^{-2}$ MPa$\sqrt{}$m/s $\leq \dot{K}_I \leq 1\text{x}10^6$ MPa$\sqrt{}$m/s, i.e. almost eight decimal orders and shown in Fig. 5.16 are compared in Fig. 5.25 the with results for annealed 1045 steel. The most important is the existence of minimum of K_{Ic} at the loading rate $\dot{K}_I \approx 1\text{x}10^4$ MPa$\sqrt{}$m/s. Since rolled 1045 steel was tested for K_{Ic}, values of J_{Ic} for annealled 1045 steel were converted into the equivalent K_{Jc} values using the standard relations

$$K_{Jc} = \left(\frac{EJ_{Ic}}{1 - \nu^2}\right)^{1/2} \text{ and } \dot{K}_I = \left(\frac{E}{(1 - \nu^2) J_{Ic}}\right)^{1/2} \dot{J}_I \tag{5.79}$$

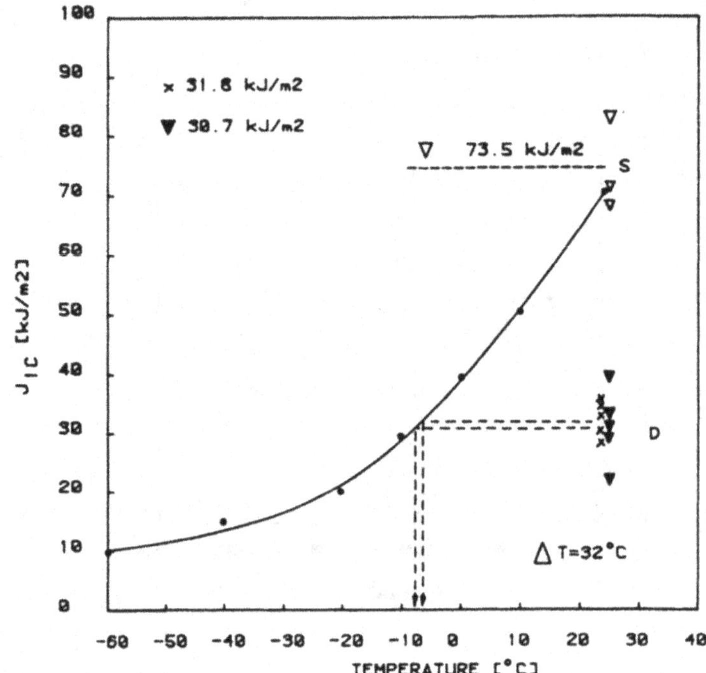

Fig. 5.24, Quasi-static and dynamic values of J_{IC} for annealed 1045 steel ; ∇ - quasi-static ; ▼ - values obtained with stress wave loading by the normal procedure ; x equivalent J_{Ic} determined from K_{Ic} ; o - mean values for the same steel by R-curve method.

The solid line in Fig. 5.25 represents the arithmetic means of K_{Ic} for the rolled 1045 steel as obtained via K_{Ic} procedure. The squares represent equivalent values of K_{Jc} determined for the annealed 1045 steel via the J-procedure.

It is important to note that for the annealed 1045 steel the ductile-brittle transition has been observed when the loading rate was increased from the quasi-static to the stress-wave loading.

It has been demonstrated in this part of the review that the experimental technique with WLCT specimen is flexible enough to provide data based on the linear as well as nonlinear fracture mechanics. The split Hopkinson pressure bar configuration shows many advantages in comparison to the instrumented Charpy test.

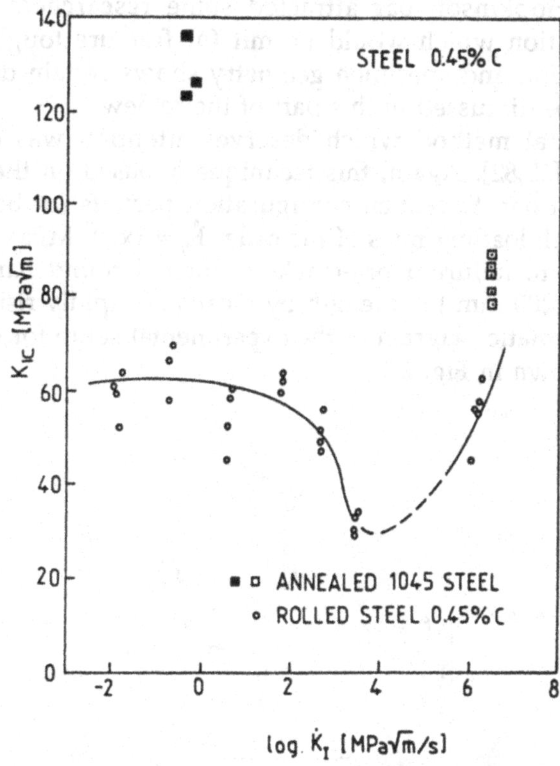

Fig. 5.25, Experimental results for the annealed 1045 steel (squares) after transforming into equivalent K_{Jc} values, and results for the rolled 1045 steel (circles) as a function of loading rate $\dot{K}_I$, [57].

6. DIFFERENT EXPERIMENTAL METHODS FOR DETERMINATION OF FRACTURE PARAMETERS USING THE HOPKINSON BAR CONCEPT

6.1 Tensile schemes in Mode I

The idea of Hopkinson bar attracted some researchers to find bar-specimen configuration which would permit for fracture toughness testing. Each bar configuration and specimen geometry shows certain disadvantages, some of them will be discussed in this part of the review.

An experimental method which deserves attention was developed at Brown University, [62,82]. Again, this technique is based on the principle of the split Hopkinson bar. Specimen configuration permits for both K_{Ic} and J_c determination at high loading rates of the order $\dot{K}_I \approx 1 \times 10^6$ MPa$\sqrt{m}$/s. The test consists in loading to failure a pre-cracked notched round bar of 25.4 mm diameter and of ~ 1200 mm total length by means of rapidly rising tensile incident wave. A schematic diagram of the experimental setup for dynamic fracture initiation is shown in Fig. 6.1.

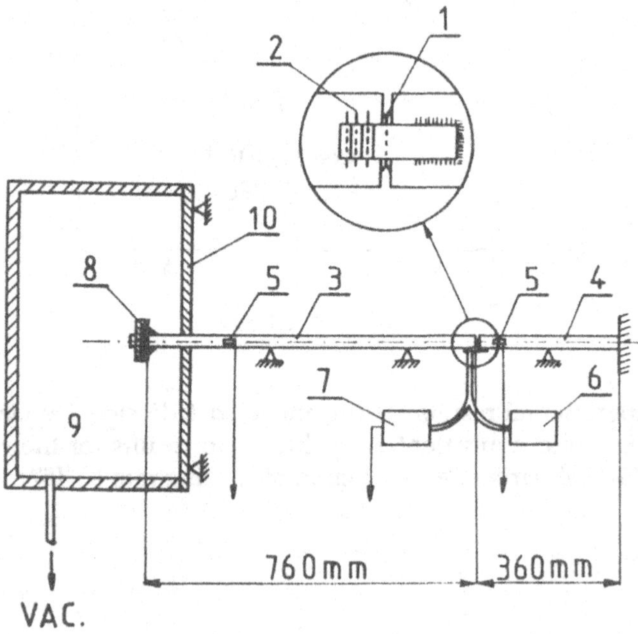

Fig. 6.1, Schematic diagram of the experimental setup developed at Brown University, [62].

The test setup is rather elaborate, and the long bar with the circumferential notch constitutes Notched Round Tensile (NRT) specimen. Thus, after each test the whole bar with strain gage instrumentation must be replaced. As a result, the technique is not suitable for screening or routine testing. On the other hand the technique has certain advantages when a precise evaluation of the fracture initiation properties of a particular material is required. Since the prefatigued notch is loaded symmetrically by an incident tensile wave the loading rates achieved with the Brown University technique can be generally higher than in the case of the WLCT specimen loaded by standard split Hopkinson pressure bar. In the case of NRT specimen loaded by longitudinal elastic wave maximum loading rates are of the order $\dot{K}_I \approx 10^7$ MPa$\sqrt{m}$/s.

During the test, a tensile incident wave is initiated at one end of the bar (NRT specimen) by the detonation of an explosive charge placed against the specially shaped loading head. The incident wave of sufficiently high magnitude and the rise time ~ 40 μs propagates down the bar to the prefatigued notch. The incident tensile pulse produces a fracture of the prefatigued circumferential crack. During the process of fracturing part of the incident wave is reflected as compression wave and the second part is transmitted through the notch until the specimen breaks completely. Thus, the complete analogy to the split Hopkinson bar exists. The electric resistance strain gages are used to monitor all three waves. The transmitted wave is proportional to the force carried by the notch and provides a measure of average stress as a function of time in the notch cross-section. In addition to wave measurements an optical device is used to measure crack opening displacement as a function of time (COD) on the bar surface. Only a brief description of this measuring technique will be given here. The principle of displacement coding by frequency modulation produced by Moiré fringes is applied. Moiré fringes are produced when light passes through two transparent gratings mounted face to face with their grids parallel or not. Using photoresist, a grid of 33 lines per milimeter is printed on the small glass slide, and the glass slide is cemented to the transmitter side of the NRT specimen to span both sides of the notch. A matching grid is deposited on the metal surface of the NRT specimen, covering a small, polished area, located just close to the notch on the incident side and beneath the glass slide. As the NRT specimen is opened by the tensile pulse, the crack opens and the two grids slide past each other. Their relative displacement as a function of time is monitored from changes of light intensity reflected from the grids by a photodiode. Light is supplied to the grids and returned to the photodiode through bifurcated fiber-optics. To eliminate the effect of a small misaligument a pair of optical displacement gages is used, one on either side of the notch. Altogether four recording channels are needed to monitor the experiment. Resolution of COD measurement is about $\Delta\,\delta \approx 4$ μm.

Since explosives are used in this technique and similarly the use of the optical displacement gages is difficult, Dormeval et al. [84], introduced several modifications in the test technique. The NRT specimen was made shorter, the total length is equal to 700 mm with the prefatigued notch positioned at

The projectile which impacts the cross bar is guided by two parallel rails, and NRT specimen is supported by three bearings. Impact velocities which are applied in this experiment are $3\ m/s \le v_o \le 40\ m/s$.

In order to determine various fracture parameters from NRT specimen, like K_{Ic} or J_{Ic} the following records are of importance : $P_T(t)$ - the transmitted force, $P_T(t) = A_E E \varepsilon_T(t)$,$\varepsilon_T(t)$ is the transmitted wave measured by strain gage, $\delta(t)$ is the crack opening displacement on the bar surface, and finally the critical time is t_c.

When the SSY crack response is expected and the plain strain condition is satisfied

$$r_o \ge 2.5\left(\frac{K_I}{\sigma_y}\right)^2 \tag{6.1}$$

where b is the radius of the remaining circular ligament after fatiguing, σ_y is the yield stress of the material determined at strain rate estimated to that achieved near the crack tip during the fracture test, value of K_{Ic} can be calculated from the standard formula

$$K_{Ic} = \frac{P_T(t_c)}{b\sqrt{\pi a}}\, f\left(\frac{r_o}{R}\right) \quad ; \quad a = R - r_o \tag{6.2}$$

where f (r_o/R) is the compliance function, and $R = 1/2\ D$ is the outer radius of the NRT specimen, $P_T(t)$ is the critical force (usually maximum) determined from $P_T(t_c) = A_E E \varepsilon_T(t_c)$.

Elimination of time from $P_T(t)$ and $\delta(t)$ permits for constructing $P(\delta)$ plot, and next, application of the ASTM procedure using 5% slope offset to determine K_{Ic}. Calculation of the loading rate is done as usual

$$\dot{K}_I = \frac{K_{Ic}}{t_c} \tag{6.3}$$

In the case of more ductile materials the condition (6.1) may not be satisfied and one must use the J-integral approach. According to Rice et al. [68] the J-integral for a notched round bar can be written as

$$J_{Ic} = \frac{1}{2\pi r_o^2}\left[3\int_o^{\delta_c} P d\delta - P\delta_c\right] \tag{6.4}$$

where δ_c is the critical COD. This expression for J_{Ic} is valid under certain assumptions which will not be here discussed. The mean rate of J is determined from the formula

$$\dot{J}_I = \frac{J_{Ic}}{t_c} \tag{6.5}$$

Once J_{Ic} is determined, an equivalent K_{IJc} can be calculated from

$$K_{IJc} = \left[\frac{EJ_{Ic}}{(1 - \nu^2)}\right]^{1/2} \tag{6.6}$$

It may be mentioned that the COD can be determined independently from the recorded incident, reflected and transmitted pulses using standard relations, (5.20), (5.21)

$$\delta(t) = U_A(t) - U_B(t) \tag{6.7}$$

Values of $\delta(t)$ calculated in this way should be compared with those determined directly from the electro-optical gage.

A critical analysis of the NRT specimen used for determination of K_{Ic} and J_{Ic} under quasi-static and wave loading conditions has been performed by Nakamura et al. [7]. It has been shown that the incident and reflected waves and the transmitted wave through the ligament can be accurately inferred from surface strain measurements taken at the distance larger than one bar diameter from the crack plane. Those numerical results confirm that the location of the strain gages on the NRT specimen are appropriate for determination of K_{Ic} from the transmitted wave and J_{Ic} from the deep crack formula (6.4). In addition, a closer inspection of the crack tip region reveals that the near-tip stress fields are SYMMETRICAL with respect to the notch plane, i.e. they can be characterized as MODE I.

The following conclusions concerning the deep crack formula for J_{Ic} can be drawn. In the elastic and contained plastic regime, the J_{Ic} formula underestimates the actual value of J by ~ 20% (or K_{Ic} by ~ 10%) if the relative crack depth r_0/R is about 0.7. As yielding progresses, the accuracy of the deep crack formula improves and the formula is very accurate at fully yielded conditions. The numerical analysis has indicated that the relative crack depth be greater than $r_0/R = 0.7$.

Because of symmetry and an optimal position of the crack plane to the tensile waves the maximum loading rates attainable with this configuration are ~ 1×10^7 MPa$\sqrt{m}$/s. Since the method is expensive and relatively complicated it is practically not possible to perform screening tests or to cover the whole loading rate spectrum. So far only quasi-static and stress wave loading results have been compared using the NRT specimens. The Brown University method has also been applied to determine brittle-ductile transition for several steels.

Quasi-static tests with the NRT specimens are similar to techniques employed for different geometries. The geometry of the NRT quasi-static specimen is the same as for the dynamic loading, except the overall length is usually reduced to ~ 700 mm. An environmental chamber is employed for tests at other than room temperature. The quasi-static fracture test is accomplished by using an universal testing machine. Fig. 6.4 shows schematically the apparatus for quasi-static testing in use at Brown University.

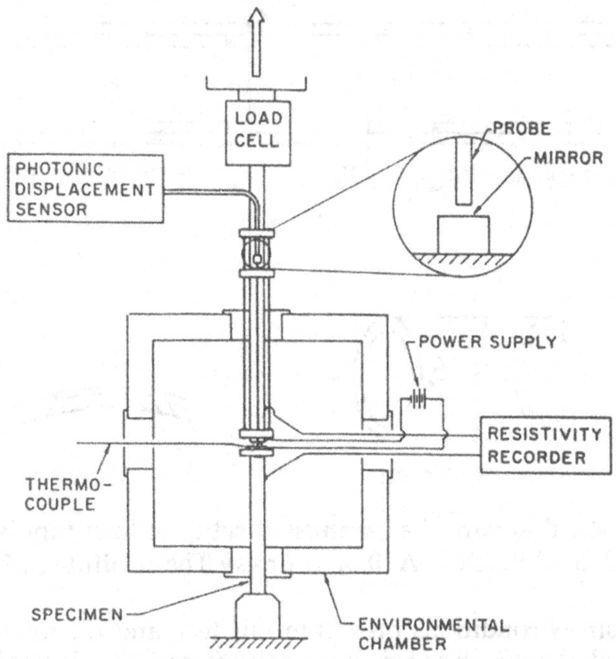

Fig. 6.4, Apparatus for quasi-static fracture testing which can be mounted in an universal testing machine, after [84].

The system of measurements consists of the load cell of testing machine, an photonic sensor for measurement of the notch opening displacement and electric resistivity gage which determines the instant of fracture initiation. Of course, the apparatus for quasi-static loading can be used in the range of small rates.

The Brown University technique is one of the best concerning symmetry of the crack loading and thus for determination of dynamic K_{Ic} or J_{Ic}. The main drawback is its cost which limits number of tests. Since only two rates are usually employed, ~ 1 MPa√m/s and ~ 5x10^6 MPa√m/s, the minimum of fracture toughness which is observed for some steels at ~ 1x10^5 MPa√m/s is not covered in existing so far publications.

One of the less conventional specimen design is that proposed by Ohlson [84]. A tubular specimen of length l_o = 25 mm and the wall thickness g = 0.5 mm, filled with high-viscosity oil, was inserted in between sealed Hopkinson bars. A pressure pulse developed by the incident wave builts up a hydrostatic pressure in the oil, producing more or less uniform radial load in the specimen. The scheme of specimen loading is shown in Fig. 6.5. A machined notch is made on the outer surface of the specimen at an arbitrary angle with the axis.

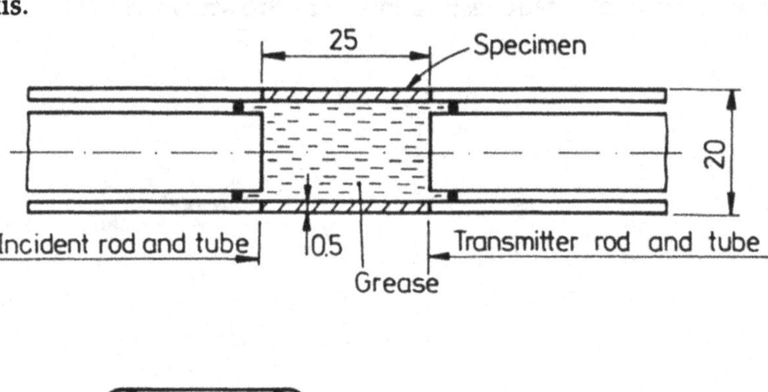

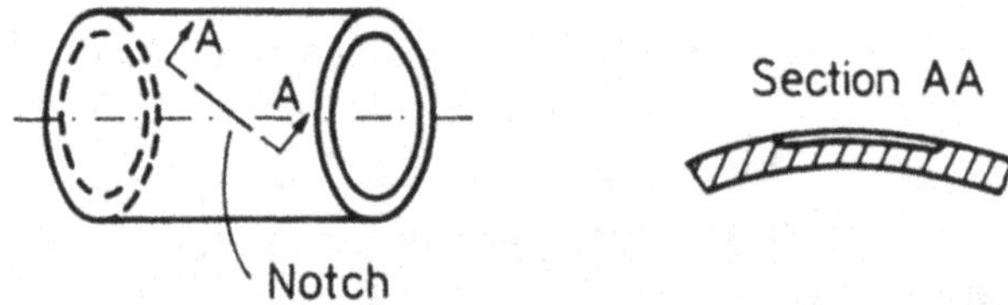

Fig. 6.5, Schematic diagram of experimental setup to load tubular specimen with a machined notch A-A, [85], courtesy The Institute of Physics.

The plane stress conditions prevail in this test, and the mixed mode I an II can be achieved. An effective stress intensity factor, K_{eff}, is used for the evaluation of the combined loading tests

$$K_{eff} = (K_I^2 + K_{II}^2)^{1/2} \quad , \text{ plane stress} \tag{6.8}$$

Two materials were investigated and it was found that the effective critical stress intensity factor (K_{eff}) increases in both cases when the loading rate was increased from $\dot{K} = 1$ MPa$\sqrt{m}$/s to $\dot{K} \approx 1\text{x}10^5$ MPa$\sqrt{m}$/s. For Al-alloy similar to PA6 (4.4% Cu, 0.6% Mg and 0.8% Mn) the values for the pure Mode I are K_c = 8.8 MPa$\sqrt{m}$ for quasi-static rate and K_c = 12.2 MPa$\sqrt{m}$ for the fast loading (Hopkinson bar). Similar values were obtained for the mixed mode.

In addition, the same aluminum alloy was tested in the plane strain conditions with central cracked flat specimens loaded in tension in the range of loading rates 1 MPa$\sqrt{m}$/s $\leq \dot{K}_I \leq$ 50 MPa$\sqrt{m}$/s. It was found that plane strain fracture toughness K_{Ic} diminishes from $K_{Ic} \approx$ 30 MPa$\sqrt{m}$ to $K_{Ic} \approx$ 20 MPa$\sqrt{m}$ at $\dot{K}_I$ = 50 MPa$\sqrt{m}$/s, as shown in Fig. 6.6. This result is in agreement with the PA6 plane strain results, as it is shown in Fig. 5.13. It is interesting to note that the same aluminum alloy shows positive loading rate sensitivity in plane stress and at the same time negative rate sensitivity is observed for plane strain.

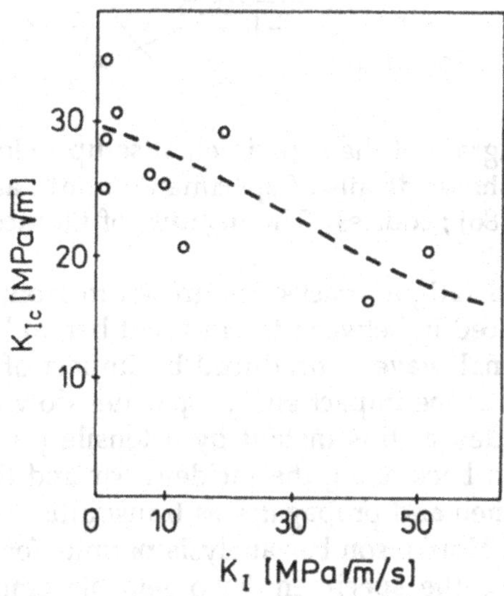

Fig. 6.6, Plane strain fracture toughness K_{Ic} versus $\dot{K}_I$ for aluminum alloy 4332, [85].

Another experimental configuration, based on the split Hopkinson pressure bar concept, has been proposed by Kishida et al. [86]. In this case a specimen geometry proposed earlier for tension tests by Lindholm and Yeakley,

[87], has been adapted for fracture testing in plane stress conditions (Mode I).

A schematic diagram of the apparatus for dynamic loading along with the specimen geometry is shown in Fig. 6.7. A hat-shaped specimen has two incisions as shown in Fig. 6.8. Those slits are made symmetrically by electric

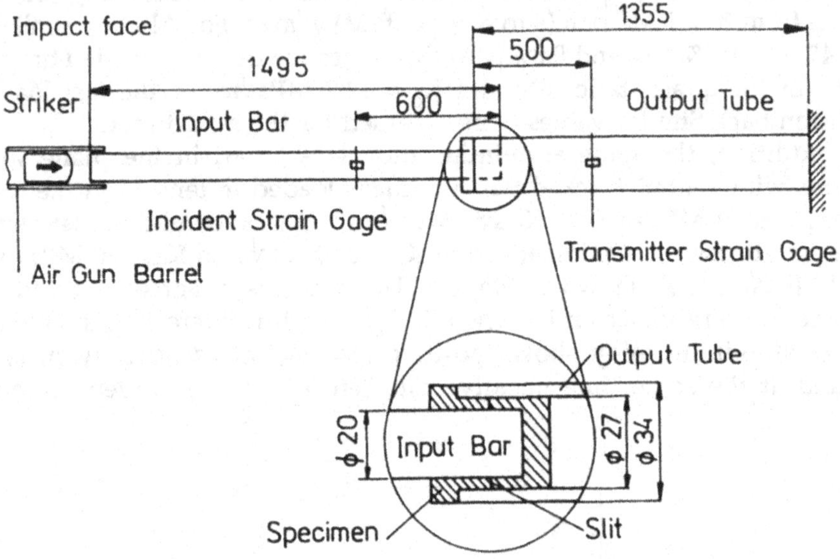

Fig. 6.7, Schematic diagram of the experimental setup to load hat-shaped specimen ; the inset shows details of specimen mount ; all dimensions in mm ; [86] ; courtesy The Institute of Physics.

discharge machining and fatigue cracks are grown in from the roots of the slits. The specimen is placed in between the incident bar and transmitter tube. The incident compressional wave is produced by impact of a projectile. The incident wave generated at the impact end propagates down the incident bar and the specimen is loaded at this instant by a tensile pulse. A part of the incident pulse is reflected back along the incident bar and the second part is transmitted by the specimen and propagates as transmitted compression wave down the tube. Standard Hopkinson bar analysis permits for determination of the loading conditions of the specimen. Two sets of semiconductor strain gages are used to monitor all three waves. In addition two strain gages are cemented on each specimen to detect fracture initiation, as shown in the lower part of Fig. 6.8. The strain gages have 0.3 mm gage length.

The formula for calculations of K as a function of geometry for the hat-shaped specimen is as follows, [86],

$$K = \beta\, \sigma \sqrt{\pi a}\, \left(\frac{W}{\pi a} \tan \frac{\pi a}{W}\right)^{1/2} \tag{6.9}$$

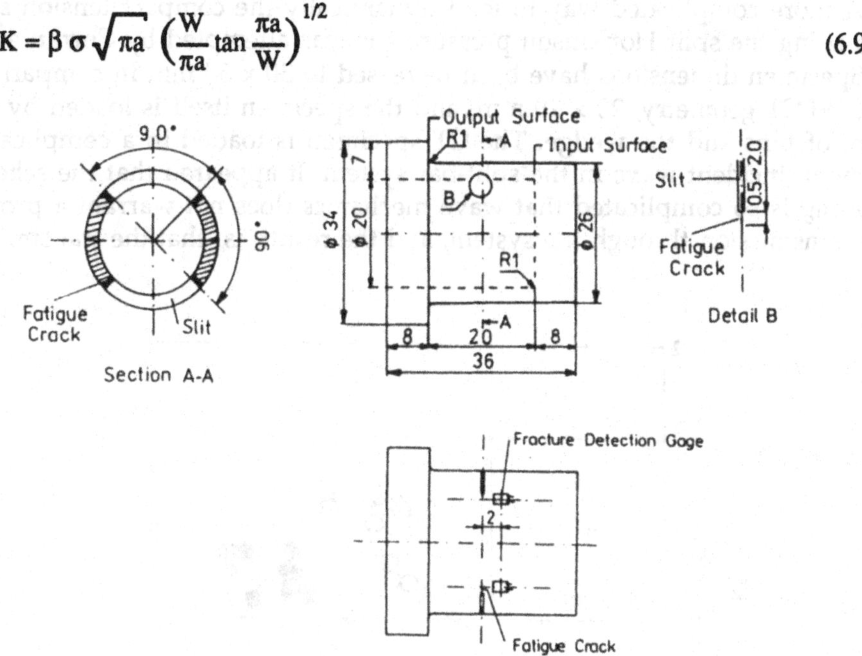

Fig. 6.8, Details of hat-shaped specimen design, [86] ; courtesy The Institute of Physics.

where β is the correction factor, $\beta \approx 1$ and σ is the mean stress in the specimen. At high loading rate value of K is determined using "the time function" $\alpha(t)$, defined as

$$K(t) = \alpha(t)\, K^{o} \tag{6.10}$$

where K^{o} is the value of K at $\dot{K} = 1$ MPa$\sqrt{m}$/s. The so called "time function" $\alpha(t)$ was found experimentally comparing quasi-static and high rate loading outputs from detection gages.

Two materials were tested with this technique, aluminum alloy 7075-T651 and Ti alloy 6A1-4V. The loading rates were as follows : 0.1 ~ 0.2 MPa$\sqrt{m}$/s in quasi-static range, and $\dot{K}$ at high loading rate 1.5 ~ 2.5 x 10^6 MPa$\sqrt{m}$/s. The results in the form of the ratio K_c ($1x10^6$) / K_c (0.15) are shown in Fig. 6.9.

It is interesting to note that those authors have obtained a negative loading rate sensitivity of fracture toughness η even for the case of the plane stress.

A more complicated way to load dynamically the compact tension specimen using the split Hopkinson pressure bar was attempted by Corran et al. [88]. Specimen dimensions have been increased to 30 x 30 mm in comparison to the WLCT geometry, 20 x 20 mm, and the specimen itself is loaded by the system of pins and the wedge. The CT specimen is loaded in a complicated way by an incident wave in the split-bar system. It appeared that the scheme of loading is so complicated that wave mechanics does not warrant a proper wave transmission through the system, and the result is that the transmitted

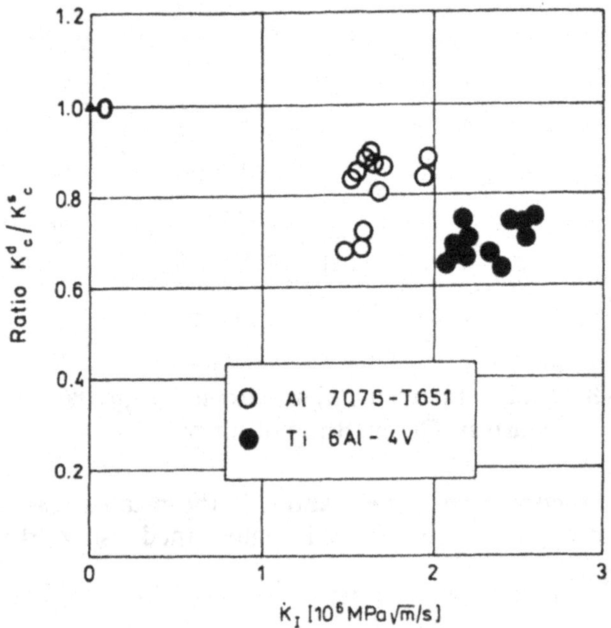

Fig. 6.9, Quasi-static and dynamic relative values of plane-stress fracture toughness for aluminum and titanium alloys ; quasi-static values : Al-alloy K_c = 42 MPa$\sqrt{m}$, Ti-alloy K_c = 83 MPa$\sqrt{m}$, [86], courtesy The Institute of Physics.

wave is heavily distorded. Thus, the principle of the force equilibrium in the bars does not hold, i.e. $P = 1/2\ (P_A + P_B)$ is not valid for a relatively long time.

It is clear from the review of the test methods for fast loading of stationary cracks that the Kolsky apparatus or split Hopkinson pressure bar is a very useful tool in determination of fracture resistance parameters like K_{Ic}, K_c or J_{Ic} in Mode I. Still new configurations are proposed. For example, Couque et al. [89] introduced an experimental method based on two coupled Hopkinson bars and two small CT specimens. The experimental technique involves the coupled Hopkinson bars preloaded in tension. A fast release of the elastic energy stored in the bars loads dynamically two CT specimens attached in between bars via a brittle notched fracture starter. Details of the attachment of CT specimens to the fracture starter and bars are shown in Fig. 6.10. Bars of diameter 38 mm are initially coupled with the fracture starter in the form of a hard steel ring with external circumferential notch, and two CT specimens with the effective width W varied from 44 mm to 63 mm are inserted into the starter. Following preloading of bars in tension the starter is fractured by introducing a sharp cut into the circumferential notch. Failure of the starter releases an unloading stress wave in the bars which develops a fast opening displacement of the CT specimens. The crack opening displacements of the CT specimens are measured by the strain gages cemented on the bars. In addition, an eddy current transducers can measure COD near the crack tips and the crack speed can be determined by the resistance crack gages. The instant of crack initiation is determined by a standard small SR gage cemented on the side of CT specimens.

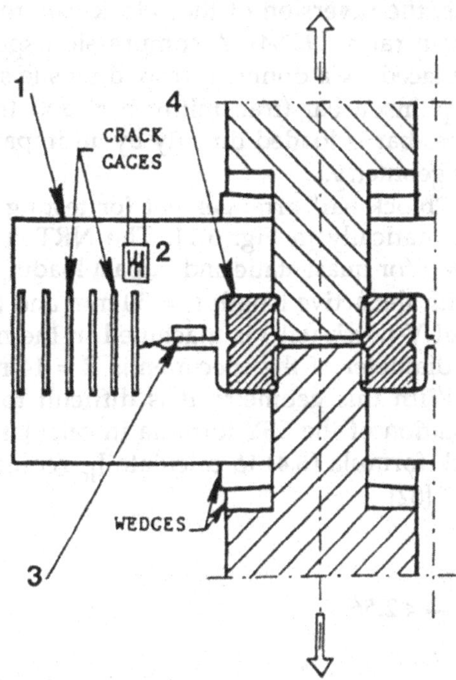

Fig. 6.10, Schematic of the specimen attachment in the coupled Hopkinson bar test, [89] ; 1-CT specimen ; 2-strain gage ; 3-eddy current COD transducer ; 4-notch round starter.

The fact that two specimens are tested at the same time does not assure symmetry of fracturing and elimination of bending waves in bars. The starter breaks at one point and next the crack propagates circumferentially. The situation is similar as it has been discussed by Lipinski and Klepaczko, [90], where the propagation transition temperature (PTT) was studied in 1045 steel using a long NRT specimen which behaves like the coupled Hopkinson bars. In such case both, longitudinal and bending unloading waves propagate along bar after fracture. On the other hand the test method with coupled Hopkinson

bars eliminates problems associated with the friction effects present in the WLCT technique.

Another experimental configuration with a short NRT specimen has recently been reported by Bensussan, [91]. A general idea behind this test method lies in an application of so-called "block-bar" arrangement due to Hauser and Dharan, [92]. In the "block-bar" technique the incident Hopkinson bar is discarded and impact is allowed to occur on a specimen directly. The specimen of any geometry is positioned at the face B of the transmitter bar. Although original application off the "block-bar" arrangement has been limited to compression tests, the inversion of the "block-bar" method permits for tension tests at high strain rates, [93,94]. A compression specimen, usually in the form of a disc, is replaced by a double threaded tensile specimen with one end attached to the Hopkinson bar (transmitter bar) and the second is fixed into a cross-bar. The cross-bar is loaded directly by an impact of a hammer or by a projectile of special geometry.

Application of the "block-bar" arrangement for testing of dynamic crack initiation is shown schematically in Fig. 6.11. The NRT specimen geometry applied in [91] can be used for quasi-static and impact loadings and it has been chosen as a short bar with the active length l_o = 30 mm and the total length l = 70 mm, a circumferential V-notch (60°) is machined in the middle of the active length. The external diameter of the specimen is R = 14 mm and the notch diameter d = 9.3 mm. With this geometry it is difficult to satisfy condition (6.1) which limits application of the SSY formula in calculation of K_{Ic}. On the other hand the dip crack formula (6.4) to calculate J_{Ic} requires two geometric conditions to be satisfied, [82]

$$r_o \geq 50 \frac{J_{Ic}}{\sigma_y} \quad \text{and} \quad \frac{\sigma_n}{\sigma_y} < 2.56 \tag{6.11}$$

where r_o is the radius of the remaining circular ligament after prefatiguing, σ_y is the yield limit and σ_n is the net section stress on the prefatigued ligament. It has been suggested in [7] that in order to assure an accuracy of J_{Ic} calculations

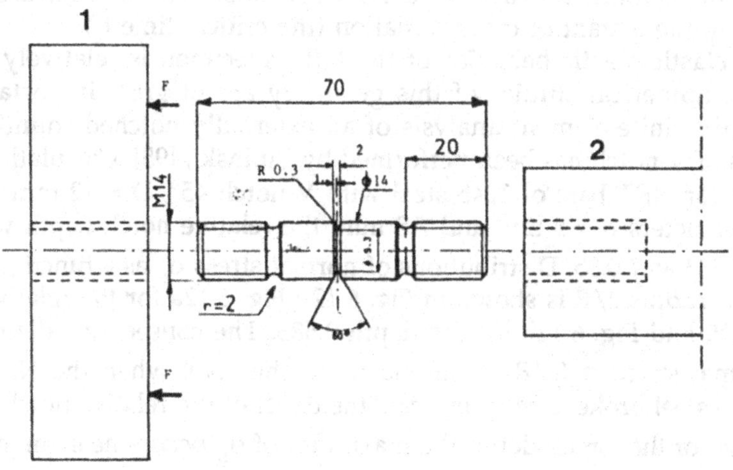

Fig. 6.11, NRT specimen and "block-bar" loading scheme for determination of fracture toughness at high rates, [91] ; 1-cross bar, end of Hopkinson bar.

by the deep crack formula the relative crack depth a_o/R, where $a_o = R - r_o$, should be

$$\frac{a_o}{R} > 0.70 \tag{6.12}$$

which replaces condition (6.11) for r_o. To satisfy condition (6.12) the NRT specimen with R = 7 mm should have the ligament radius after prefatiguing r_o = 4.2 mm.

Thus, optimization of the NRT specimen is equally difficult as other geometries, and diameter of Hopkinson bar must match geometric conditions in specimen designing.

Equally important is the way in which force P and displacement δ is measured. According to suggestions in [7] the deep crack formula based on the load and notch opening displacement $P(\delta)$ is more accurate than based on the load and load point displacement $P(\Delta)$. In the case of dynamic loading, the deep crack formula must be based on the load transmitted across the ligament (eq. 5.14) and the notch opening displacement δ. Of course, the load transmitted through the ligament can be determined from the transmitted wave $\varepsilon_T(t)$, but the notch opening displacement must be measured independently. In the scheme applied by Bensussan, [91], two-point optical displacement gage was used with the rise time ~ 1 μs. In addition, two sets of strain gages were cemented on the cylindrical parts of the NRT specimen. Those gages are helpful in measuring the instant of crack initiation (the critical time t_c).

Since elastic-plastic behavior of the NRT specimen is relatively complicated some numerical studies of this geometry are of great importance. An elastic-plastic, finite-element analysis of an externally notched round bar for two depths of V-notch has been performed by Lipinski, [95]. Calculations were carried out for NRT bars of 1045 steel with V-notch 45°, D = 12 mm and two ligament diameters d = 5 mm and 7.8 mm, the relative notch depth were respectively 0.583 and 0.35. Distributions of normal stress σ_n as a function of non-dimensional radius r/R is shown in Fig. 6.12 ; Fig. 6.12a for the relative depth of notch 0.35 and Fig. 6.12b for the depth 0.583. The consecutive distributions of the normal stress $\sigma_n(r/R)$, from elastic to the level when the NRT specimen of 1045 steel broke, clearly indicate the effect of the relative notch depth. In the case of the small depth the maximum of σ_n occurs near the notch tip through the process of yielding and up to fracture at $\sigma_n \approx 1$ GPa.

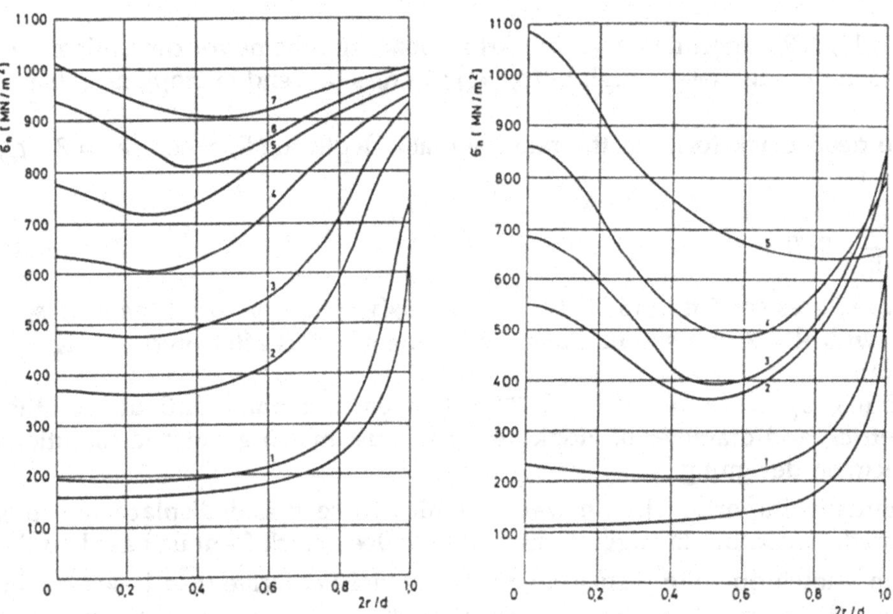

Fig. 6.12, Distribution of normal stress σ_n as a function of non-dimensional distance r/R ; a - results for 1045 steel for r = 7.8 mm and R = 12 mm ; b - results for r = 5.0 mm and R = 12 mm.

Whereas for the large depth, redistribution of the maximum of σ_n is observed and at the instant of fracture the maximum occurs at the axis of the NRT bar.

In the latter case the gross yielding of the ligaments is observed and fracture occurs at $\sigma_n \approx 1.08$ GPa. Such situation is analogous to fracture in a round bar with a neck as it has been discussed, for example, by Nadai, [96].

In conclusion, it is clear from the above considerations that a caution is recommended in interpretation of the fracture tests with the NRT specimens.

6.2 Application of Hopkinson bar to bending schemes in Mode I

Some testing procedures which utilize bending, such as Charpy test, have been used extensively well ahead of development of fracture mechanics. Because of simplicity of bending scheme and the existence of a large data base, many attempts have been made to modify specimen geometries to determine quantities which had been introduced by an advent of fracture mechanics. Equal attempts have been made to introduce new loading arrangements and to improve instrumentation.

Generally, bending is not well fitted to extract fracture mechanic informations from measurements of forces. Speeds of bending waves are relatively slow and vibrating masses of specimen halves develop a lot of ringing. Application of the Hopkinson bar concept can eliminate some of the problems inherent in bending scheme, especially in force measurements.

Some attempts have been reported to apply the Hopkinson bar concept in dynamic loading of the Charpy specimen. This idea is quite interesting since the wave loading eliminates the specimen "bouncing" in between tup and supports. More exactly, in the standard configuration of the instrumented impact test the gage which measures force is separated by some finite distance from the point of contact with specimen and the tup is part of a more rigid structure consisting of the pendulum and the mounting mechanism. Wave reflections and interactions from various free and fixed boundaries appear in the signals recorded with the strain gages. The superimposed signals caused by the elastic wave reflections are specially important as the total time of the test becomes shorter for a more brittle materials. In addition, the energy absorbed by the vibrations and elastic energy or the loading system must be considered in an exact evaluation of the test. As it has been shown by many studies, interpretation of the load-time or load-displacement records are rather difficult, specially when brittle materials are tested. The Hopkinson bar configuration applied as instrumented test shows some advantages, and also some limitations.

The apparatus used by Nicholas, [97], consisted of a long striker bar, an input bar, and a support fixture, is shown schematically in Fig. 6.13. In this configuration, the complete force-time and displacement-time histories at the end of the bar in contact with the three point bending specimen can be determined from records of elastic waves measured by a strain gage positioned at the center of the incident bar. Details of specimen geometry (Charpy specimen), tup and support geometries are shown in Fig. 6.14. The incident wave $\varepsilon_I(t)$ which loads the specimen is generated through the impact of the striker bar of length 1220 mm (the incident wave duration $\lambda \approx 500\ \mu s$). The input bar (the incident bar) of diameters from 12.3 mm to 19.0 mm (interchanged) are twice the length of the striker. The geometry of the loading fixture is that of the ASTM E23 Standard.

When the incident wave $\varepsilon_I(t)$ arrives at the specimen, part of it is reflected and part transmitted into and through the specimen into the rigid support fixture. The reflected wave $\varepsilon_R(t)$ is recorded sometime later of the same

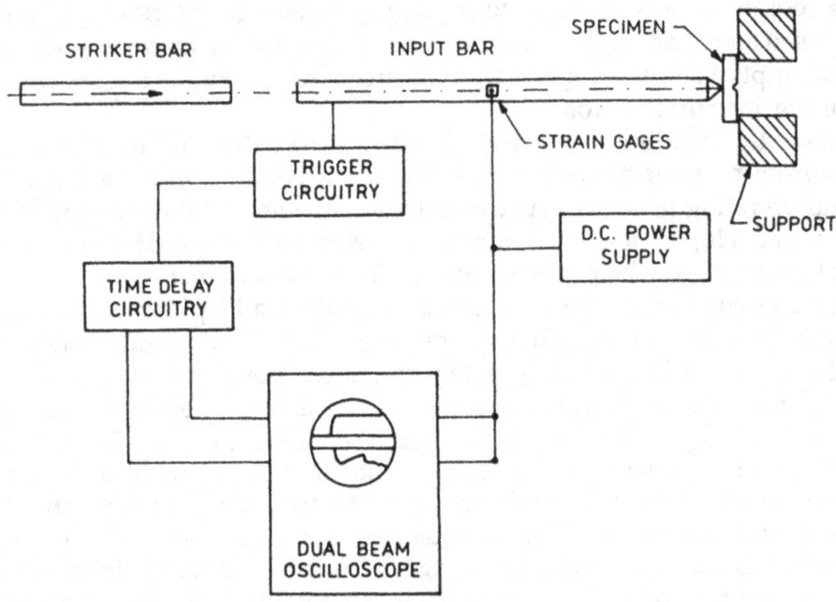

Fig. 6.13, Schematic diagram of apparatus with Hopkinson bar used for Charpy test ; [97].

strain gage at the center of the incident bar. The time delay t_s is defined by the distance between the specimen and the strain gage x_s

$$t_s = x_s/C_o \quad C_o \text{ - elastic wave speed in bar} \tag{6.13}$$

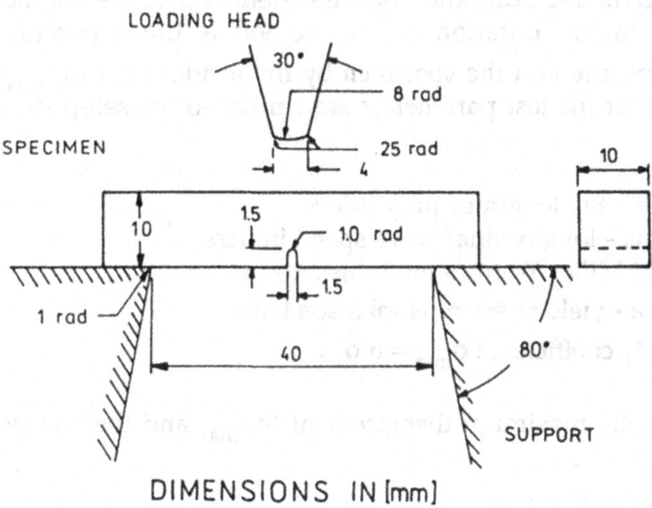

Fig. 6.14, Details of Charpy specimen with U-notch, tup and support, as used in [97] ; ASTM Standard E23.

The data reduction is quite simple and efficient. By shifting the incident and reflected waves by an amount of $2t_s$ the waves appear to be time coincident. The net force acting on the specimen is

$$P_s = EA\,[\varepsilon_I(t) - \varepsilon_R(t)] \tag{6.14}$$

where E is Young's modulus and A the cross-sectional area of the incident bar. The net particle velocity at active end of the incident bar (bar/specimen contact) is

$$v_s = C_o\,[\varepsilon_I(t) + \varepsilon_R(t)] \tag{6.15}$$

The specimen deflection at the center can be obtained by integrating $v_s(t)$, thus

$$\delta_s = C_o \int_0^t [\varepsilon_I(\theta) + \varepsilon_R(\theta)]\, d\theta \tag{6.16}$$

Since the striker bar is of the same material and cross-sectional area as the incident bar the dynamic calibration of the system is possible via the impact velocity v_o. The main limitation of the method is the maximum deflection which can be imposed on the specimen by the incident bar $(\delta_s)_{max}$. For example, assuming that the test parameters are similar to the setup used by Nicholas :

L = 1500 mm - the length of projectile ;
C_o = 5 mm/ms - longitudinal wave speed in bars ;
E = 2.06 x 10^5 MPa - Young's modulus ;
σ_y = 2500 MPa - yield stress of Hopkinson bar ;
α = 0.6 - safety coefficient ; $\sigma_{max} = \alpha\,\sigma_y$;

the relation for the maximum displacement $(\delta_s)_{max}$ and the wavelength λ_I is

$$(\delta_s)_{max} = C_o\,(\varepsilon_I)_{max}\,\lambda_I$$

$$\lambda_I = \frac{2L}{C_o}$$

$$(\delta_s)_{max} = 2\,\frac{\alpha}{E}\,\sigma_y\,L \qquad (6.17)$$

Introducing those numbers to the relation for $(\delta_s)_{max}$ one yields $(\delta_s)_{max}$ = 17.5 mm. In the case of shorter projectiles the maximum displacement is proportionately smaller.

Since force $P_s(t)$ and deflection $\delta_s(t)$ can be determined after each test the energy dissipated in the specimen is

$$W_s(t) = P_s(t)\,\delta_s(t) \qquad (6.18)$$

$$W_s(t) = \rho C_o^3 A\,[\varepsilon_I(t) - \varepsilon_R(t)] \int_0^t [\varepsilon_I(\theta) + \varepsilon_R(\theta)]\,d\theta \qquad (6.19)$$

In addition, elimination of time in $P_s(t)$ and $\delta_s(t)$ produces $P_s(\delta_s)$ diagram. Experimental data of Nicholas obtained for beryllium specimens also show some vibrations but the higher vibration frequencies are eliminated.

It is interesting to note that tests performed for different beryllium at different impact velocities show a minimum of energy as a function of velocity v_s imposed on Charpy specimens, [98].

A similar configuration of Hopkinson bar, projectile and test specimen has recently been reported by Mines and Ruiz, [99]. In addition, the photoelasticity methods were applied. It was found that Hopkinson bar arrangement permits for a clear evaluation of the test results, and is preferable to the pendulum/tup test.

Another configuration of the loading system for Charpy specimen has been developed by Tanaka and Kagatsume, [100]. The incident part of the apparatus is similar, i.e. the striker bar and the incident bar, however, the specimen is supported by the transmitter tube. The experimental set up is shown schematically in Fig. 6.15. Since cross sections of bars and a tube can differ the relation for specimen deflection is more complicated

$$\delta_s = \int_0^t \left[\frac{2\sigma_I(\vartheta)}{\rho_I C_{oI}} - \frac{1}{\rho_T C_{oT}}\left(1 + \frac{A_T\rho_T C_{oT}}{A_I\rho_I C_{oI}}\right)\sigma_T(\vartheta)\right] d\vartheta \tag{6.20}$$

where $\sigma_I = E\varepsilon_I$ and $\sigma_T = E\varepsilon_T$ are incident and transmitted stress elastic waves, the densities, the wave velocities and the sectional areas for the incident bar and transmitter tube are ρ_I, ρ_T, C_{oI}, C_{oT}, A_I and A_T.

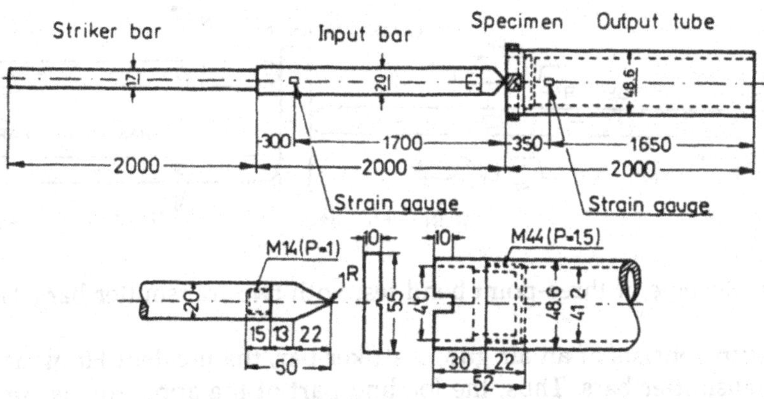

Fig. 6.15, Scheme of Charpy test with incident bar and transmitter tube, [100].

If the force equilibrium is applied,

$$A_I(\sigma_I - \sigma_R) = A_T\sigma_T \quad (6.21)$$

therefore, the load imposed on the specimen (the support tube) is

$$P_s = A_T\,\sigma_T(t) \quad (6.22)$$

Experimental results obtained for commercial 0.45% steel at different temperatures, 78 K ≤ T ≤ 298 K, demonstrated excellent load-deflection curves. Because of deflection limit in the incident bar the repeated tests have been carried out for the case of annealed specimens.

Temperature dependence of the absorbed energy obtained with Hopkinson bar arrangement and instrumented Charpy hammer reported in [100] demonstrate a high degree of precision attainable with the Hopkinson incident bar and transmitter tube arrangement.

Yet another apparatus, based on the Hopkinson bar concept, has been developed by Yokoyama and Kishida [101,102]. The arrangement is designed to test three point bend specimens which are larger than standard Charpy one. Fatigue precracked bent specimen has dimensions 100 x 20 x 10 mm (W = 20 mm ; B = 10 mm) and the support span is 80 mm. The scheme of specimen loading is shown in Fig. 6.16.

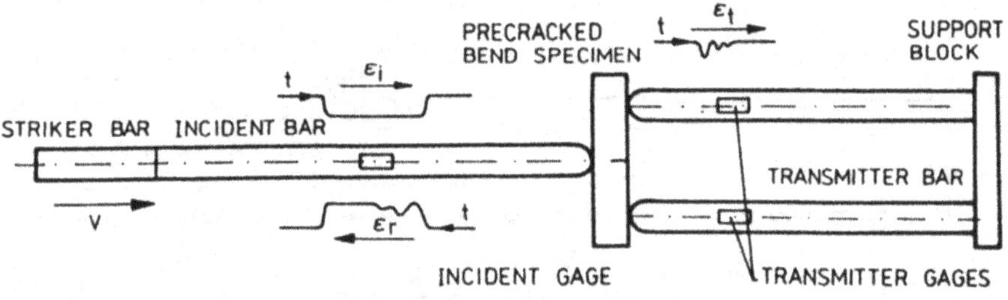

Fig. 6.16, Scheme of three-point bend test with two transmitter bars, [101].

The test setup consists of an air gun, a striker bar, the incident Hopkinson bar and two transmitter bars. Thus, the loading part of the apparatus is similar to original version by Nicholas. Since the points of support are apart 80 mm two symmetric transmitter bars have been introduced.

Interpretation of experiment in the form of oscillograms is practically the same as in precedent schemes. Since there are two symmetric transmitter bars the total transmitted force is

$$P_T(t) = EA\,[\varepsilon_{T1}(t) + \varepsilon_{T2}(t)] \qquad (6.23)$$

where the subscripts 1 and 2 are referred to the transmitter bars.

The authors of this method claim that fracture toughness K_{Ic} can be determined with the loading rate $\dot{K}_I \approx 6 \times 10^5$ MPa$\sqrt{m}$/s. However, oscillograms obtained for the aluminum alloy 7075-T651 MPa clearly indicate that quasi-static analysis is practically impossible. This is due to large specimen dimensions. The situation would be still worse for steel specimens. In order to avoid such difficulties a semiconductor strain gage was cemented on each specimen close to the crack tip, as shown in Fig. 6.17, and the finite element method was used in parallel to determine K_{Ic}. The incident and transmitted forces were used as input data for the FE computation to evaluate $K_I(t)$ and the critical time t_c was measured by the strain gage.

In conclusion, application of the three point bend specimens, even with the optimal measuring systems like Hopkinson bar techniques, does not warrant good results, mainly due to the fact that the three point bend specimen is not optimal in impact or stress wave loadings. Combined techniques with extra strain gages and the FE analyses may improve estimations of fracture toughness.

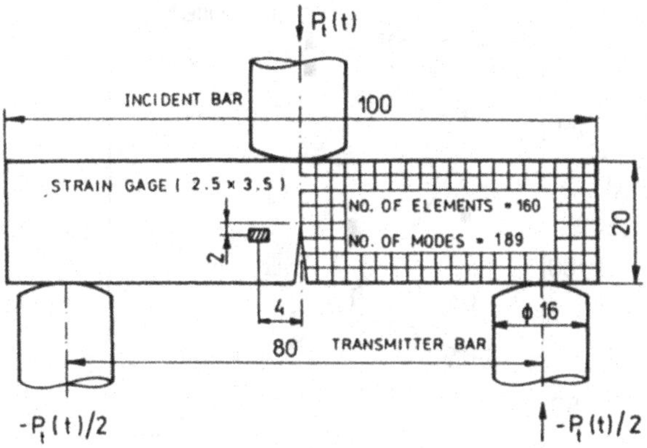

Fig. 6.17, Three point bend specimen instrumented for test in the setup shown in Fig. 6.15 ; the FE mesh is also shown, [101].

Perhaps an optimal solution would be application of the method of caustics, (only for flat sided specimens). The shadow-spot method (the method of caustics) is discussed by Kalthoff. While under quasi-static circumstances the use of caustics does not present any particular advantage over other more established experimental techniques, under dynamic conditions, however, the method of caustics is a valuable tool to determine critical values of K_{Ic} or J_{Ic}. In the case of metal the method of reflected caustics is of importance.

6.3 Application of torsional Hopkinson bar to Mode III

After discussions of experimental methods for determination of fracture resistance at different rates focussed on Mode I, there would be also an interest to review the other two modes, and especially the antiplane shear (Mode III). It has been found that in the case of Mode III the stress intensity factor K_{IIIQ} or K_{IIIc} INCREASES with increase of the material yield limit in shear τ_y. For example, such findings were reported by Hurd and Irving [103], and by Tsangarakis [104, 105]. Result of Hurd and Irving are shown in Fig. 6.18. Additional informations on quasi-static procedures for K_{III} determination can be found in [106].

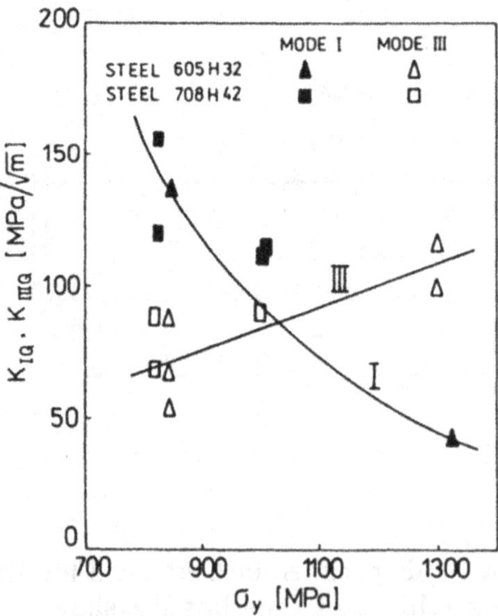

Fig. 6.18, Relation between K_{Ic} and K_{IIIc} and yield limit σ_y for a low alloy steel ; Δ - departure from T(φ) linearity ; □ - maximum force, after [103].

Hurd and Irving, [103], studied the fracture toughness of low alloy steels in Mode I and in Mode III. Different quenching and tempering provided a range of yield and strength limits. It has been found, as it is demonstrated in Fig. 6.18, that the fracture resistance to fracture initiation decreases with increasing yield stress in Mode I, whereas it INCREASES with increasing yield limit in Mode III. A similar result has been reported by Tsangarakis, [105], for hypereutectoid steel HF1 which is shown in Fig. 6.19. Those steels were tested in the quenched and tempered state and they are low alloy steels used for shafts and axles.

In the case of Mode III testing, values of the stress intensity K_{III} are calculated using the expression (assumption of SSY), [104]

$$K_{III} = \frac{2T}{\pi r^3}(\pi r)^{1/2} f\left(\frac{r}{R}\right) \quad ; \quad a = R - r \tag{6.24}$$

where T is the torque applied to a bar of diameter D = 2R, R is the specimen radius, r is the ligament radius and a = R - r is the crack length. The compliance function f(r/R) is given in the literature, for example Tada [107]. The expression for K_{III} is correct only if the plastic zone is small in comparison to

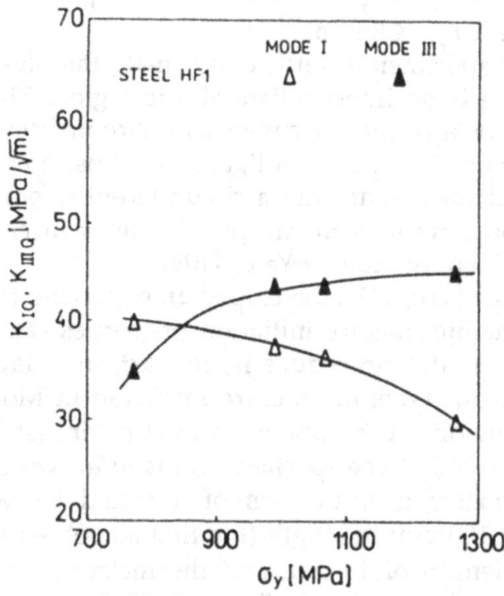

Fig. 6.19, Relation between K_{Ic} and K_{IIIc} vs. yield stress σ_y for HF1 steel, after [104].

the ligament radius r. As it was shown by Hult and McClintock [108], the locus of points limiting plastic zone in an infinite body with an isolated crack of length a is a circle of radius r_y around the edge of the crack,

$$r_y = \frac{K_{III}^2}{2\pi\tau_y^2} \quad \text{for} \quad \nu = \frac{1}{2} \; , \; \tau_y = \frac{\sigma_y}{\sqrt{3}} \tag{6.25}$$

and

$$r_y = \frac{3}{2\pi}\left(\frac{K_{III}}{\sigma_y}\right)^2 \quad ; \quad \frac{3}{2\pi} = 0.4775 \tag{6.26}$$

It appears that the plastic zone in Mode III calculated for SSY is much larger in comparison to Mode I zone for the same ratio of K/σ_y. Thus, it is difficult to obtain values K_{IIIc} using acceptable geometries of specimens, most of the results are invalid K_{IIIQ}, i.e. K_{IIIq}. For example, for σ_y = 1000 MPa and K_{IIIc} = 50 MPa$\sqrt{m}$ radius of plastic zone is $r_y \approx 1.2$ mm, but for K_{IIIc} = 100 MPa and the same σ_y, the plastic zone is $r_y \approx 4.8$ mm.

When a precracked specimen is sufficiently small, the plastic zone can be relatively large that there is no intermediate elastic region. The limit of non-existing elastic domain for a round specimen with circumferential notch is $r_y = r$, where r is the radius of the specimen ligament. Thus, $r_y/r < 1$ is the limit for the SSY in cylindrical specimens with a circumferential prefatigued notch. A more exact discussion how to estimate plastic zones in circumferentially cracked round bars are given recently by Yates [109].

Klepaczko and Nkule [110, 111], developed an experimental procedure by means of which the dynamic fracture initiation properties can be determined in Mode III. The experimental procedure is, in part, an adaptation of split Hopkinson torsional bar to dynamic fracture initiation in Mode III. A schematic diagram of the torsional bar applied to fracture initiation at high loading rates is shown in Fig. 6.20. The specimen is made in two consecutive stages and departure geometry is in the form of a round bar with a V-notch, 25 mm in diameter and 100 mm in length (the first stage). After fatiguing the specimen is cut to the length of 40 mm and the incisions are milled on its sides (the second stage), as it is shown in Fig. 6.21. Thus, the specimens of the identical geometry could be tested under quasi-static and stress wave loading rates. Before testing with the Hopkinson bar, specimen is fixed to the incident and transmitter Hopkinson bars by means of special clamps.

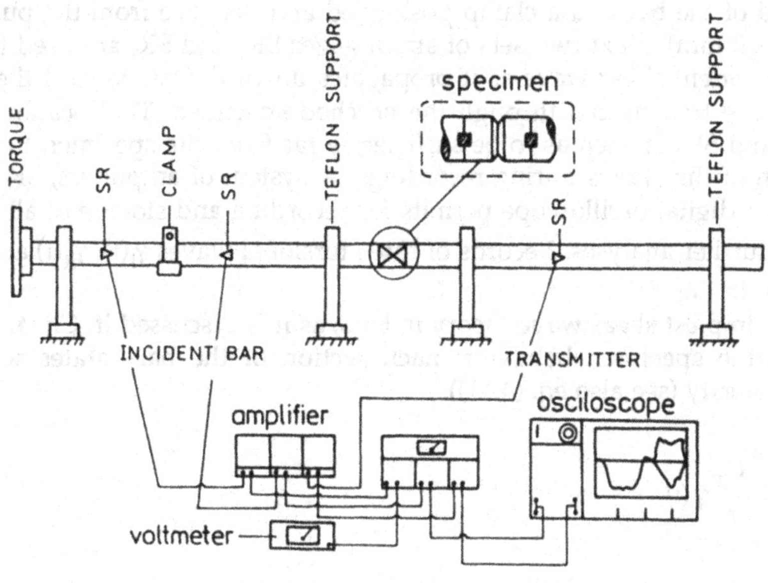

Fig. 6.20, Scheme of split Hopkinson torsional bar for testing K_{IIIc} at high loading rates.

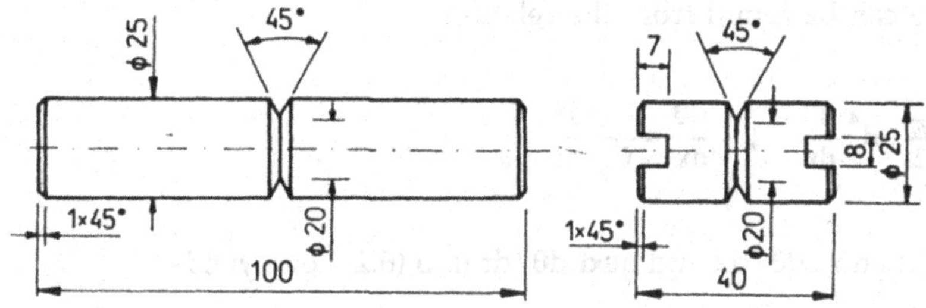

Fig. 6.21, Two statges of specimen preparation for K_{III} testing.

Three sets of strain gages are cemented to the bars, two to the incident bar and the third to the transmitter. The incident bar is divided into two parts, the first on the left serves as storage of elastic energy. The set of strain gage SR_1 cemented to the stored-torque bar measures the shear stress level to which this part of the incident bar is prestressed. The incident shear wave is produced by a sudden release of a stored torque. This requires a torque pulley at the end of the bar and a clamp positioned at a distance from the pulley (usually ~ 500 mm). Next two sets of strain gages SR_2 and SR_3 are used to measure the incident shear wave as it propagates down the bar toward the specimen and the transmitted through the notched specimen. The location of the sets SR_2 and SR_3 is such as to be sufficiently far from the specimen to ensure separation of the waves during recordings. A system of amplifiers, SR supply units and a digital oscilloscope permits for recording and storage of all signals used in further analyses. Records of three torsional waves $\gamma_I(t)$, $\gamma_R(t)$ and $\gamma_T(t)$ are shown in Fig. 6.23.

The simplest shear wave theory in bars, as it is discussed in Ch. 4, is used to determine specimen behavior. Each section of the bar rotates with the angular velocity (see also eq. (4.41))

$$\frac{d\theta}{dt} = \frac{C_2}{r}\gamma(t) \tag{6.27}$$

where γ is shear strain at radius r. Thus the angle of twist on the bar surface, r = R, can be found by integration

$$\theta(t) = \frac{C_2}{R}\int_0^t \gamma(\theta)\, d\theta \tag{6.28}$$

The torque can be found from the relation

$$T = \frac{\pi}{2}\mu R^4 \frac{d\theta}{dx} \quad ; \quad \frac{d\theta}{dx} = \frac{1}{C_2}\frac{d\theta}{dt} \tag{6.29}$$

after introduction $d\theta/dx$ and next $d\theta/dt$ into (6.29) one yields

$$T(t) = \frac{\pi}{2}\mu R^3 \gamma(t) \tag{6.30}$$

At r = R, i.e. on the bar surface, all quantities are defined as a function of shear strain $\gamma(t)$. If the surface shear strains $\gamma_I(t)$, $\gamma_R(t)$ and $\gamma_T(t)$ are recorded a further analysis is similar as for split Hopkinson pressure bar

$$\theta_I(t) = \frac{C_2}{R} \int_0^t \gamma_I(\theta)\, d\theta \tag{6.31}$$

$$\theta_R(t) = \frac{C_2}{R} \int_0^t \gamma_R(\theta)\, d\theta \tag{6.32}$$

$$\theta_T(t) = \frac{C_2}{R} \int_0^t \gamma_T(\theta)\, d\theta \tag{6.33}$$

Since the "thickness" of the crack in x direction is practically zero the condition of torque equilibrium is satisfied all the time during loading of the crack, i.e.

$$T_I + T_R - T_T = 0$$

the net increment of θ due to crack plastic zone and crack propagation inward the bar is $\Delta\theta$

$$\Delta\theta = -\frac{2C_2}{R} \int_0^{t_c} \gamma_R(\theta)\, d\theta \tag{6.34}$$

Value of $\Delta\theta$ is an analog to the crack opening displacement in Mode I. In the case of torque equilibrium the transmitted wave $\gamma_T(t)$ provides a direct measure of the torque transmitted through the circumferential ligament. Thus, the critical torque $T(t_c)$ acting on the ligament can be determined when the critical time t_c is known.

$$T(t_c) = \frac{\pi}{2} \mu R^3 \gamma_T(t_c) \tag{6.35}$$

For more brittle materials the critical time t_c coincides with maximum torque. Generally, for SSY the quasis-static formula can be used, eq. (6.24), and K_{IIIc} for K_{iiiQ} calculated using $\gamma_T(t)$ value

$$K_{IIIc} = \mu(\pi r)^{1/2} \left(\frac{r}{R}\right)^3 \gamma_T(t_c) f\left(\frac{r}{R}\right) \tag{6.36}$$

Preliminary experiments have been performed on XC48 steel (0.5% C, 0.67% Mn), according to AISI designation 1045 steel. Quasi-static as well as stress-wave loading tests were run at room temperature, 13 tests at the rate $\dot{K}_{III} \approx 1$ MPa$\sqrt{m}$/s and 6 tests at $\dot{K}_{III} = 1 \times 10^6$ MPa$\sqrt{m}$/s. Quasi-static records torque vs. time are shown in Fig. 6.21, and the records of three waves $\gamma_I(t)$, $\gamma_R(t)$ and $\gamma_T(t)$ are shown in Fig. 6.22. All transmitted pulses had well defined maxima and the criterion of maximum torque was applied in determination of K_{IIIQ}. The results for quasi-static and dynamic test are shown in Fig. 6.24.

The surprising result is that those preliminary experiments reveal a positive loading rate effect on critical stress intensity factor determined in Mode III. Moreover, a more exact inspection of the fractured ligaments revealed a network of adiabatic shear bands emanating as spirals from the circumferential crack tip.

In conclusion, the positive loading rate sensitivity of K_{IIIc} implies that the mechanism of material separation is controlled in Mode III by plasticity.

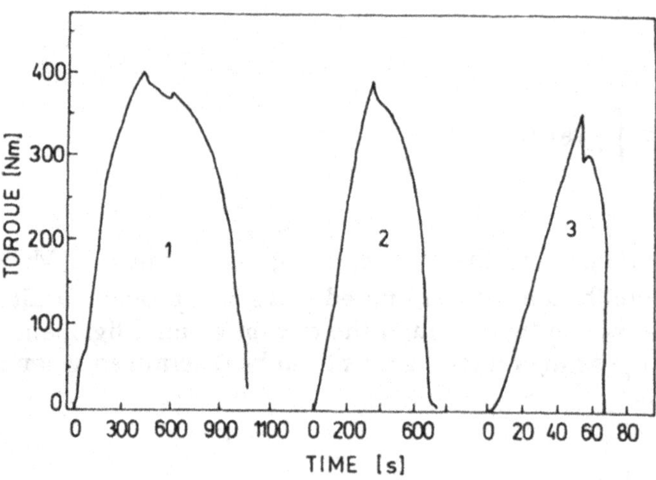

Fig. 6.22, Three records T(t) for specimen made of XC48 steel.

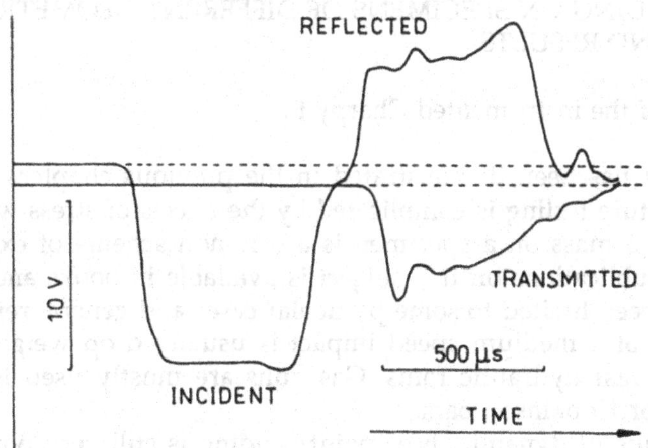

Fig. 6.23, Records of $\gamma_I(t)$, $\gamma_R(t)$ and $\gamma_T(t)$ for specimen of XC48 steel.

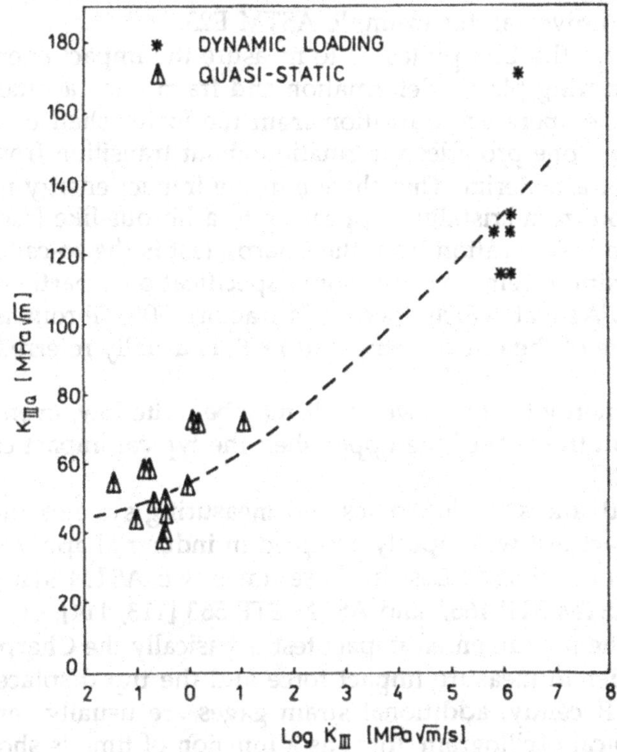

Fig. 6.24, Loading rate spectrum for Mode III, comparison of quasi-static and torsional bar tests for XC48 steel.

7. IMPACT LOADING ON SPECIMENS OF DIFFERENT GEOMETRIES, TEST METHODS AND RESULTS

7.1. Discussion of the instrumented Charpy test

Although it has been demonstrated in the previous chapters that high loading rate fracture testing is complicated by the effects of stress waves, still direct impact of a mass on a specimen is a common scheme of experiment. Because many publications on this subject is available in books and journals this review has been limited to some particular cases and general remarks.

The source of a medium speed impact is usually drop weights, weigh pendulums and fast hydraulic rams. Gas guns are mostly used for a high velocity impact or Hopkinson bars.

The technique of dynamic three-point bending is still very popular and this test is applicable to specimens with a wide range of dimensions. Specimens are usually loaded by a falling weight or by a large pendulum. The earliest proposition of such test is due to Charpy [112]. The Charpy impact test has been standaritized in many countries in mid thirties. Now the standard with V-notch is almost universal, for example ASTM E23.

Typical task for the Charpy test is to measure the impact energy, i.e. the energy absorbed during plastic deformation and fracture as a function of initial temperature. Temperature transition from the lower shelf of the impact energy to the upper one provides information about transition from brittle to ductile behavior of a material. This change in the impact energy is accompanied by a transition from cristalline-appearing to a fibrous-like fracture surface. Thus, the second observation from the Charpy test is the so-called Fracture Appearance Transition Temperature. Some specifications, particularly those of ASTM, define FATT at which specimens fracture 50% fibrous and 50% in cleavage, this value of the critical temperature T_c is usually referred to as 50% FATT.

The impact energy for the lower shelf may be quite low, even ~ 1 J for a very brittle steel, in the case of the upper shelf the typical impact energy is in between 50 J to 200 J.

Because of advances in electronics and measuring systems the so-called instrumented impact test was rapidly accepted in industrial applications. This trend is dated from mid seventies. In those times two ASTM symposia have been organized, ASTM STP 466, and ASTM STP 563 [113, 114], on instrumented Charpy test. The instrumented impact test is basically the Charpy test with extra gages in order to measure impact force and the tup displacement as a function of time. Recently, additional strain gages are usually cemented on the specimen. Typical oscillogram, force as a function of time is shown in Fig. 7.1, Ireland [115].

It appears that signal generated by the tup gage represents a complex combination of specimen response, inertia of the setup and characteristics of electronics. A complicated pattern of stress waves generated by the impact, in both specimen and pendulum, makes an exact interpretations of oscillograms difficult. The main frequency of oscillations is in between 30 to 50 kHz (period 20 to 30 μs). Those oscillations are superimposed on the main signal of force. Analyses suggest that the first oscillation is associated with specimen inertia and the next are the result of vibrations and elastic stress waves.

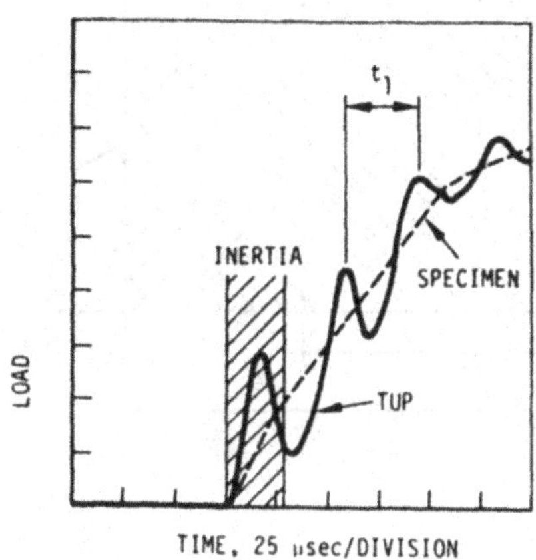

Fig. 7.1, Typical electric signal from the pendulum tup of instrumented Charpy tester, after [115] ; courtesy ASTM.

An idealized load-deflection record is shown schematically in Fig. 7.2. Since the ligament in the Charpy specimen is small, W-a = 8 mm, general yielding at F_{yg} is usually observed before fracture occurs. Fast fracture starts at the load F_1, and the crack is arrested at force F_a.

The energy of Charpy test

$$W = \int_0^{\delta} F(x)\, dx \tag{7.1}$$

can be subdivided into three components, W_y is the energy of elastic deformation and gross yielding, integrated within the range of displacement $0 \leq x \leq x_{max}$, where (F_{max}, x_{max}) is the point of maximum force, the energy W_i, found within the limits of integration $x_{max} \leq x \leq x_f$, where (F_f, x_f) and (F_a, x_f) are respectively the points of fast crack propagation and crack arrest, is the energy of unstable bending. Of course, when plasticity of specimen material is reduced, it consequently reduces the energies W_y and W_i. Finally, the third component W_h is the energy of plastic hinge. Some results of numerical analysis on plastic behavior of Charpy specimen can be found in [118].

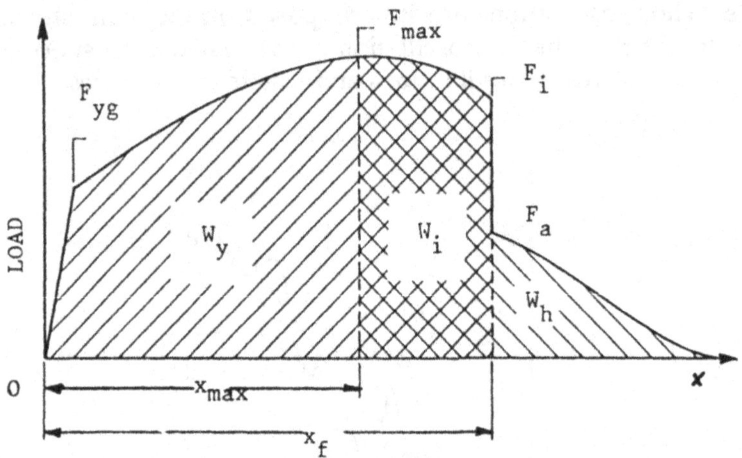

Fig. 7.2, Schematic record of force vs. deflection or time from impact on Charpy specimen, F_{yg} - beginning of gross yielding, F_{max} - instability point ; F_f - onset of fast fracture, F_a - crack arrest.

Facts given above are the main reasons why adaptation of Charpy specimen to the concepts of fracture mechanics is very limited. Those difficulties from one side and simplicity of the test from the other are causing a vast inflow of new studies and publications on instrumented Charpy test, for example reviews by Barsom and Rolfe, [116, 117].

It is out of the scope of this text to review all problems created by the instrumented impact test some of them are discussed by Kalthoff. It may be only mentioned that in the literature the following different approaches are represented how to obtain the most exact specimen response to impact loading :

i. force measurements is previewed not only in the tup but also in the supports, as exact as possible measurements of deflection as a function of time, application of additional strain gages on the specimen surface ;

ii. analytic modeling of the loading system and specimen response ;

iii. if fracture toughness determination is attempted the "K - calibration" is very useful, the so-called impact response curve, Kalthoff et al. [119].

A new attempt to reduce many disadvantages and limitations of the instrumented Charpy test was recently reported by Rintamaa et al. [120, 121]. A new pendulum tester has been designed which employs the so-called inverted geometry. In this case the specimen rests on the instrumented stationary tup and the ends of the specimen are loaded by pendulum impact. The purpose of the inverted configuration is to reduce inertia vibrations in the measuring system. The impact tester system has larger maximum impact energy, 500 J with impact velocity 5 m/s. In addition, the tester is equipped with optical COD measuring device based on assumption of a stationary rotation of bent specimen halves. The principle of the test is shown in Fig. 7.3. In spite of the inverted test geometry all oscillograms published in [120] and [121] do show a substantial vibrations. Those oscillations are due to the specimen geometry and their reduction is practically impossible.

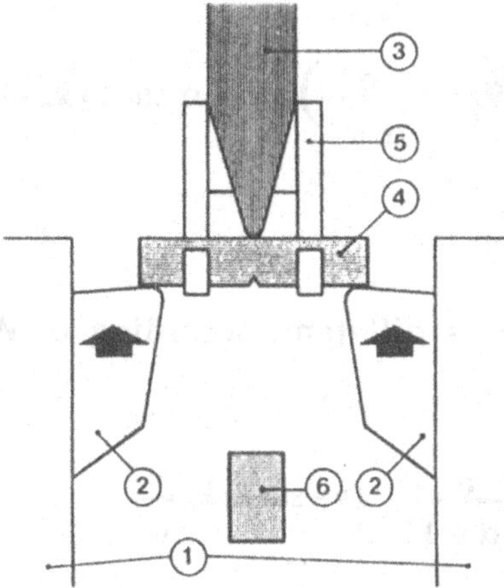

Fig. 7.3, Schematic diagram of the inverse impact geometry on Charpy specimen, 1 - moving hammer, 2 - anvil (span width 40 or 80 mm), 3 - instrumented tup, 4 - Charpy specimen, 5 - specimen support.

Another approach in analysis of the instrumented Charpy test is an analytic formulation of the problem. For example, Williams [122] published the analysis of instrumented impact tests using a mass-spring model. The test is modeled by a spring of stiffness k_2, the contact stiffness k_1 and an equivalent mass m. A damping factor can also be included in the model, [124]. The mechanical analogy considered is shown in the figure 7.4, where U is the mass displacement, while the contact point (tup/specimen) moves at constant velocity v. The equation of motion for the mass m

$$m\ddot{U} - k_1 v + (k_1 + k_2)\, U = 0 \tag{7.2}$$

with initial conditions U = 0 and v = 0 at t = 0

$$U = \left(\frac{k_1}{k_1 + k_2}\right)\left(t - \frac{\sin \omega t}{\omega}\right) v \tag{7.3}$$

where the characteristic frequency ω is

$$\omega = \left(\frac{k_1 + k_2}{m}\right)^{1/2} \tag{7.4}$$

The contact point force P_1 is given by

$$P_1 = k_1 (vt - U) \quad \text{and} \quad P_1 = \left(\frac{\alpha}{\alpha + 1}\right) (x + \alpha \sin \alpha)\, k_2 \frac{v}{\omega} \tag{7.5}$$

where $a = k_1/k_2$ and $x = \omega t$

The force in the specimen is different, according to Williams and Adams [123], and is

$$P_2 = k_2 U \quad ; \quad P_2 = \left(\frac{\alpha}{\alpha + 1}\right) (x - \sin x)\, k_2 \frac{v}{\omega} \tag{7.6}$$

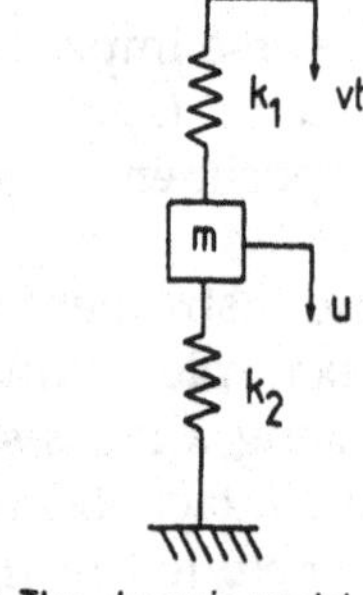

Fig. 7.4, Mass-spring model of Charpy test analysed in [122] and [123].

According to those authors P_2 is required for determination of the critical force and consequently K_I or K_{Ic}, whilst it is P_1 which is measured by load cell in the tup. Similarly, expressions can be written down for energies, one lost by the striker and the second gained by the specimen. Solutions for P_1 and P_2 give forces as a function of time, x = ωt. Fig. 7.5 shows dimensionless forces P_1 and P_2

$$\bar{P}_i = \bar{P}_i \left(\frac{\alpha + 1/\alpha}{k_2}\right) \frac{v}{\omega} \quad , \quad i = 1,2 \tag{7.7}$$

versus x = ωt for α = 10. Curves $P_1(x)$ and $P_2(x)$, although without damping, show certain similarity to real oscillograms recorded from the tup. However,

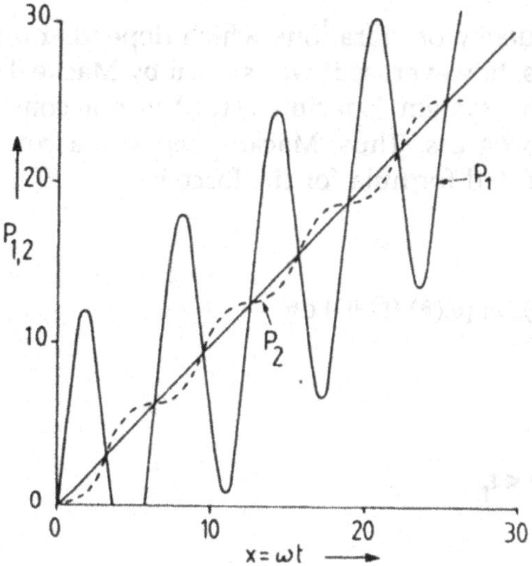

Fig. 7.5, Non-dimensional forces P_1 and P_2 calculated for α = 10, eq. (7.6), after [123].

real oscillograms do show some damping. The effect of damping can be artificially introduced by multiplication of each harmonic term by the factor, [123]

$$\alpha = \exp(-\eta x) \tag{7.8}$$

where η is the damping constant, for polymers η ≈ 0.2. The approach proposed by Williams et al. has been applied to analyse oscillograms for polymers which show rather brittle specimen response. For the case of metals the situation is more complicated because, due to plasticity and gross yielding, the spring characteristic k_2 will change during the test. A more rigorous treatment of damping has recently been published in [124].

A similar analysis has been proposed by Kishimoto et al. [125]. This analysis is based on the earlier work of Nash, [127]. Without specific mechanical analogy the integral harmonic equation is proposed which is supposed to describe specimen response during loading. More precisely, the force P(t) acting on the three point bending specimen is

$$P(t) = \omega_1 \int_0^t P(\theta) \sin [\omega_1 (t - \theta)] \theta \tag{7.9}$$

where ω_1 is the frequency of vibrations which depends on the compliance of the system. It appears, however, as it was shown by Macke, [128] and [129] that the compliance of the system (specimen/tup) is not constant but increases when the test is in progress. Thus, Macke proposed a continuous correction for compliance. Corrected formula for the force is

$$P(t) = \omega(t) \int_0^t P(\theta) \sin [\omega(\theta) (t - \theta)] \, d\theta \tag{7.10}$$

with

$$\omega (t, \theta) = \omega_1 \text{ if } \theta < t_E$$

and

$$\omega (t,\theta) = [m\, C (t,\theta)]^{-1/2} \quad \text{if } \theta \geq t_E$$

At the begining of the test the compliance function $\omega_1(t)$ is a constant, $\theta < t_E$, when fracturing starts or plasticity intervenes the compliance function $\omega(t,\theta)$ changes, where $C(t, \theta)$ is the actual compliance of specimen, it is simply the inverse slope of dP/dt, and m is the reduced mass of the system (specimen/tup). A numerical procedure has been developed for calculations of the current compliance and for correction of P(t) curve. An example is shown in Fig. 7.6, (PMMA specimen), [128]. It is worth to mention that during numerical analyses the first maximum is omitted and calculations start at the instant t_1.

The analysis proposed by Kishimoto and improved by Macke gives sometime good results, but its theoretical foundations are not well established.

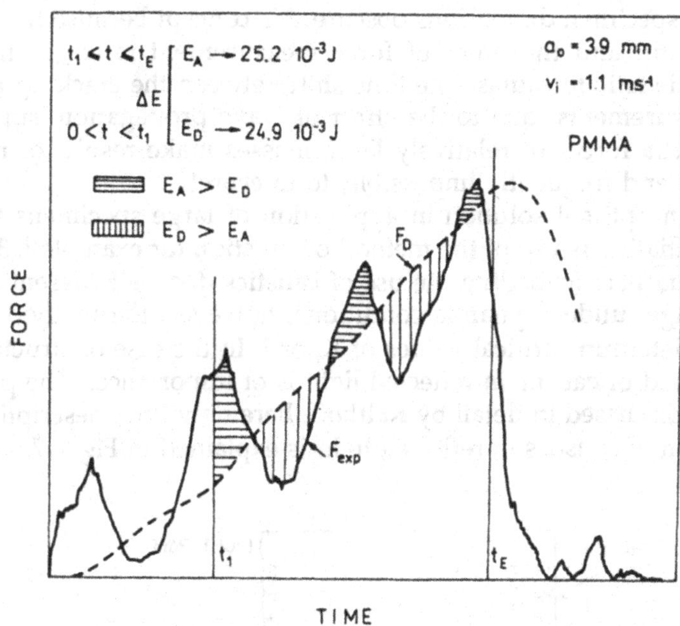

Fig. 7.6, Record of Charpy test, force as a function of time F(t) for PMMA specimen (solid line), and the response curve calculated with eq. (7.10), (broken line) ; after [128].

In conclusion, application of the Charpy specimen for determination of fracture resistance quantities introduced in fracture mechanics does not warrant good results. Even application of more modern loading schemes like different split Hopkinson bar/tube configurations and optimal measuring systems cannot eliminate inherent behavior of Charpy specimen. A low frequency of free vibrations of this specimen, slow bending waves always present in the 3-point bending schemes cannot be changed without a fundamental change of specimen geometry. Combination of numerical analyses with well designed experiment is one of possibilities to alleviate the situation. The fracture parameter which may be suitable for further studies, and combination with the instrumented Charpy test, is the critical value of J-integral, J_{Ic}. Analyses of Charpy geometry based on sound physically motivated models may also contribute to a better understanding of the test, [130].

7.2 Remarks on the use of large 3-point bend specimens for impact tests

Application of the quasi-static approach for interpretation of the impact tests is limited to relatively small specimens where it is assumed that the force for which a specimen was calibrated can be used in the case of impact loading. Large specimen dimensions obscure this concept because the distance between crack tip and the point of force measurement is large enough to invalidate quasi-static formulas. The time shift between the crack tip response and force measurements, due to the effect of wave propagation, superimposed on the inertia forces of relatively large masses make results of measurements difficult, and frequently impossible, to interpret.

Perhaps an optimal solution in application of large specimens to dynamic fracture initiation is use of the method of caustics, for example [131 - 134]. While under quasi-static loading the use of caustics does not present any particular advantage, under dynamic conditions, however, the method is a valuable tool to determine critical values of K_I or J. In the case of structural materials the method of caustic in reflected light is of importance. The principles of caustics are discussed in detail by Kalthoff, here only brief description is offered. Formation of caustics in reflected light is explained in Fig. 7.7.

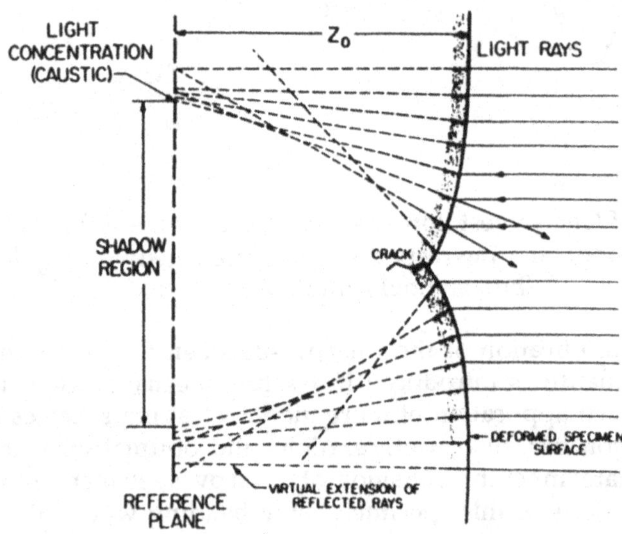

Fig. 7.7, Formation of caustic in reflected light from the polished, deformed surface near the crack tip, after [134].

A set of parallel light rays normally incident on planar, reflective specimen with a prefatigue crack, produces caustics if the crack is loaded by an external stress. Because of the non uniform contraction at the crack tip, the rerflected rays deviate from parallel traces and, under suitable conditions, the set of those reflected rays possesses an envelope in the form of a three dimensional surface in space. This surface, which is called the caustic surface, is the locus of points of maximum light density of reflected rays, as shown on the right of Fig. 7.7. If the 3D caustics is crossed by a screen placed parallel to the specimen surface the image will appear as a completely dark area (the shadow spot) sourrounded by a bright curve (the caustic curve). The locus of points on the specimen which map exactly into the caustic curve is called the initial curve. The size of the initial curve relative to the specimen thickness can be related to the actual stress intensity factor or to J-integral in the presence of more developed crack tip plasticity.

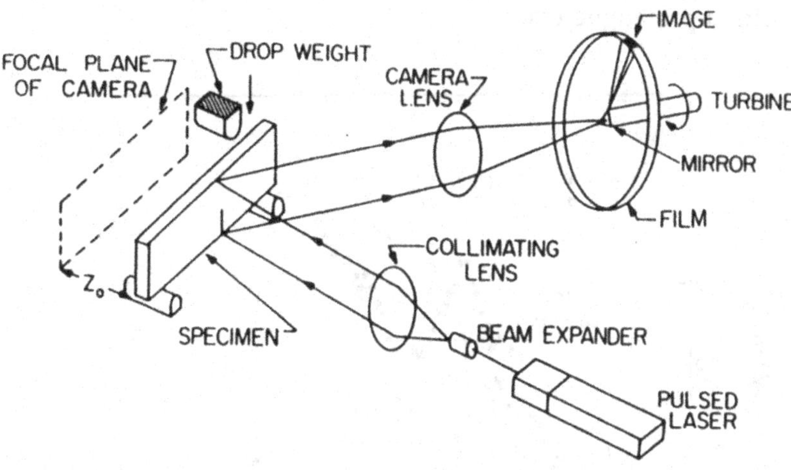

Fig. 7.8, Experimental setup used in Aeronautical Laboratories of Caltech for high speed photography of caustics, after [134].

The experimental setup, used by Rosakis et al. at the Caltech, [133, 134], consists of a drop weight tower, digital recording oscilloscope, pulsed laser, and high speed camera, is shown schematically in Fig. 7.8. The three point bend specimen is rather large and has dimensions 304 x 128 x 9.5 mm with initial crack length a_o = 37.3 mm. During preparation one side of the specimen is ground, lapped, polished to the mirror finish and vaccum coated with aluminum.

The tup is instrumented allowing the dynamic impact force to be recorded with a digital oscilloscope. All experiment is properly synchronized and the rotating mirror high speed camera can record from 2×10^2 to 2×10^5 frames per second. Although it operates as a streak camera, discreet pictures are obtained by pulsing the laser light source. The exposure time of each frame is ~ 15 μs, so the sharp pictures of caustics are obtained.

Analyses of caustics as a function of time and the tup signal, P(t), can provide a complete time history of the stress intensity factor, $K_I(t)$. Moreover, the critical time t_c can be determined, and after the crack tip starts to move $K_{ID}(t)$ can also be determined.

The records of two impact tests for 4340 steel are shown in Fig. 7.9. Two notch tip radiuses were applied r = 0.7 mm and r = 0.15 mm, respectively black and white circles. It is seen that the specimen with the blunted notch tip fractured at t_c = 640 μs and that the specimen with the sharper notch tip fractured after t_c = 240 μs. Up to time of fracture initiation for specimen with r = 0.15 mm the $K_I(t)$ records for both tests are nearly identical. Of course, value of $K_{Ic} \approx 120$ MPa√m is closer to that which would be determined for the same specimen with a prefatigue crack.

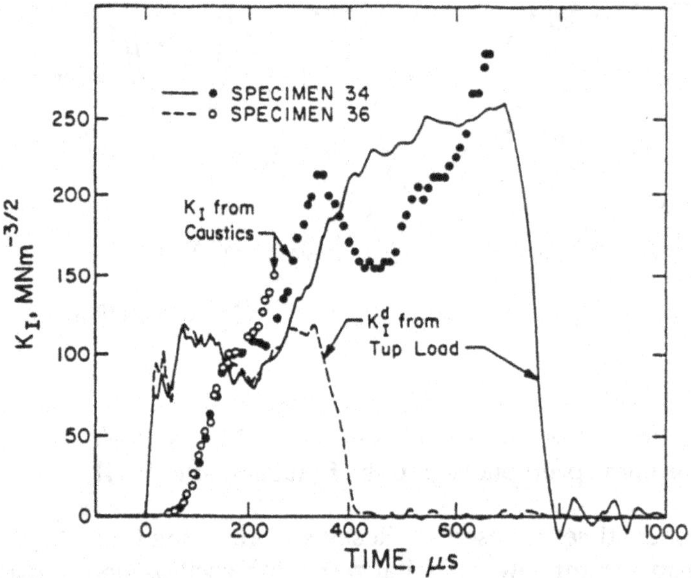

Fig. 7.9, Evolution of SIF in a large 3-point bend specimens of 4340 steel (two tests) ; specimen size 304 x 128 x 9.5 mm ; ● - crack tip radius r = 0.7 mm, O - crack tip radius 0.15 mm, after [134] ; loading rates : $\dot{K}_I = 4.4 \times 10^5$ MPa√m/s for r = 0.7 mm, $\dot{K}_I = 7.5 \times 10^5$ MPa√m/s for r = 0.15 mm.

Those experimental results demonstrate well known fact that for a large specimen fracture toughness K_{Ic} calculated by using the quasi-static formula and the maximum force P_{max} mesured by the tup is underestimated in comparison to the instantaneous real values. For the specimen which was used in those experiments the negative error is - 11% for r = 0.7 mm and - 20% for r = 0.15 mm. Those errors are much lower for smaller specimens. In conclusion they indicate the inefficiency of a large three point bend specimen in impact testing when the maximum of the tup force is used for K_{Ic} determination.

7.3 Impact on Notched Round Tensile specimen

Klepaczko and Solecki [135], redesigned a part of the Charpy hammer with pendulum to perform impact test on precracked NRT specimens. The hammer was equipped with a special device to load the NRT specimen axially and with minimum bending. Maximum impact velocity V = 5.4 m/s provided impact energy W = 270 J. A schematic diagram of this device is shown in Fig. 7.10. The cross bar CB is connected in series with the NRT specimen S and dynamometer D. One end of dynamometer is attached to the hammer base B. The impact velocity is provided by a pendulum set to a desired energy, usually ~ 1.5 x 10^2 J. The cross bar undergoes the impact symmetrically at two points A, and the total tensile force P is monitored by the dynamometer with two short base (3 mm) strain gages which eliminate bending component. To ensure axial impact the cross bar CB is guided at both sides into slots (not shown in the figure). Accurate measurement of the cross bar displacement δ as a function of time is achieved by using an optical device based on frequency modulation coding. The cross bar has a set of grids that was produced mechanically on its bottom surface, the grid having the pitch h/λ = 0.25 mm. The black screen S has one slot parallel to the rulings with the width $\lambda/2$. Since a beam of parallel light is supplied from the bottom to the grids and part of the light is reflected through the screen to the photodiode F through bifurcated fiber-optic tubes, the displacement δ is coded as a periodic voltage signal from the photodiode. Also, a moiré grids can be used to increase the accuracy of displacement measurements.

During impact the photodiode responds to changes in light intensity caused by fringes passing across the screen slot. The displacement resolution in this case is approximately $\lambda/8$, i.e. ~ 3.1 x 10^{-2} mm (31 µm). This measurement technique has many advantages which makes it ideally suited to impact fracture testing. For example, no dynamic calibration is needed when the distance between grids λ is once exactly measured, and the photodiode system reacts practically with no inertia.

Both electric signals, one from dynamometer and the second from photodiode, are recorded with a digital oscilloscope. It was found that displacement recording δ(t) are more important for less brittle materials to monitor COD. For a more brittle specimen, force as a function of time P(t) is adequate to determine critical force P_c and impact fracture toughness.

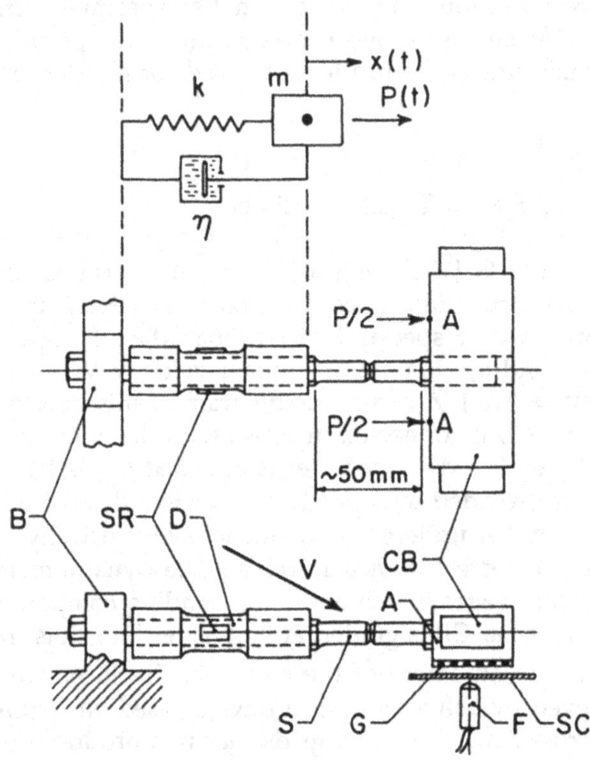

Fig. 7.10, Schematic diagram of device for impact fracture initiation ; S - NRT specimen, CB - cross bar with impact points A, D - dynamometer with strain gages SR, B - base, G - grid on the crossbar, SC - screen with a slot, F - photodiode, and V- impact velocity ; [135].

In the case of more brittle specimens the accuracy of force determination can be improved by a more detailed analysis of the force-time characteristics. Because some masses are involved during impact and they are connected in series (cross bar-specimen-dynamometer) the loading chain starts to vibrate. Those vibrations are superimposed on the main tensile pulse which breaks the specimen. The vibrating system can be reduced to the system shown in the upper part of Fig. 7.10, i.e. reduced mass m, usually the mass contained in between the notch plane and the plane located on the dynamometer at half length of strain gages. The spring constant is k, and the damping coefficient η.

If the system has one step of freedom, then its vibrations are described by the differential equation

$$m\ \ddot{x} + \eta\ \dot{x} + kx = P(t) \tag{7.11}$$

where P(t) is the real force acting as a function of time at the notch cross section. The recorded signal P(t) from the dynamometer will differ slitghtly as compared to P(t). Signal P(t) is the input signal, and P(t) is a recorded response of the loading system. The problem is that the response P(t) is known as a recorded oscillogram, and force P(t) acting on the notch cross section is unknown. It is very frequently assumed that P(t) = P(t). But in the discussed case the loading system has been analyzed by a computer solving the vibration equation with identified parameters : m, k and η. The solution with the identified parameters leads to finding P(t). The typical result of such an analysis is that P(t), concerning the first peak, is lower than measured P(t), this is shown in Fig. 7.11. Thus, for a brittle specimen the maximal real force P_{max} acting on the notch cross-section is lower than the maximal response force P_{max} recorded by the dynamometer. Assumed scheme of calculations refers to the case of the triangular input pulse P(t) when initially force increase in proportion to the loading time and further drops instantaneously (fracture takes place).

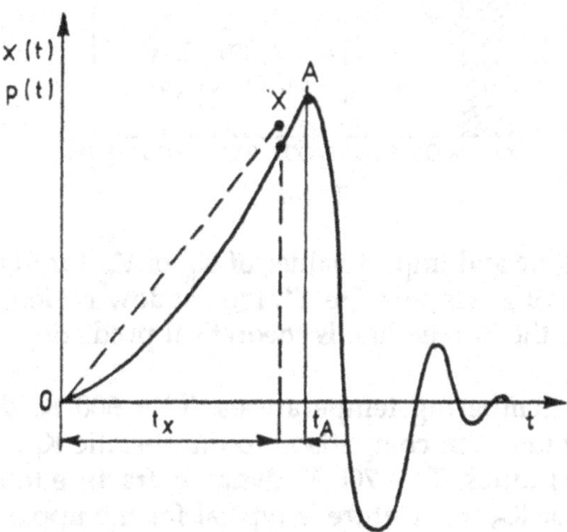

Fig. 7.11, Idealized force-time impact function which is unknown - broken line, and recorded dynamometer response - solid line ; t_x - fracture time ; t_A - apparent fracture time with delay $t_d = t_A - t_x$, ΔP - force overshoot $\Delta P = P_x - P_A$, P_x - true fracture force, P_A - apparent fracture force, [135].

The numerical analyses have shown that for the specimen geometry with diameter D = 12 mm and length L = 55 mm with notch diameter d = 9.6 mm, P_{max} do not exceed the maximum real force P_{max} by 10%.

For example, one of many results obtained with this experimental setup is shown Fig. 7.12. The effect of tempering temperature T* on fracture toughness K_{Ic} was investigated for 5140 H steel, Klepaczko and Solecki, [135].

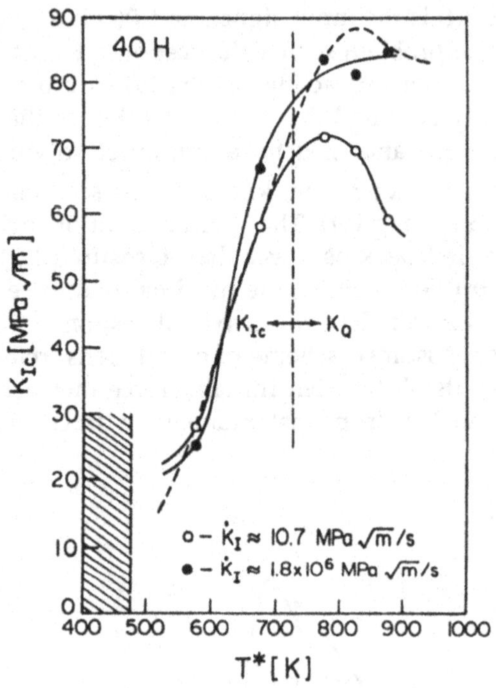

Fig. 7.12, Quasi-static and impact values of K_{Ic} or K_Q for 5140 H steel as a function of tempering temperature T*, the shadow region shows brittle behavior ; the broken line is theoretical prediction, [135].

For the case of low tempering temperatures, T* ≈ 600 K, dynamic fracture toughness is slightly lower in comparison to quasi-static K_{Ic}. Whereas for higher tempering temperatures, T* ≈ 700 K, dynamic fracture toughness is higher that quasi-static K_{Ic} or K_Q this feature is typical for the upper shelve behavior of fracture toughness. A noteworthy trend being noted that at higher temperatures of fracture testing T or after higher temperatures of tempering T*, where a fibrous fracture starts to dominate, fracture toughness is generally higher under high loading rate than under quasi-static loading.

Since the experimental technique with the NRT prefatigued specimens loaded by a pendulum hammer seems to be relatively inexpensive and effective it should be recommended for screening tests of large series of specimens. The techniques which apply NRT scheme, and in addition use the Hopkinson bar concept, for example [82, 83, 89, 91], may establish a sound experimental basis to determine fracture toughness spectra within a wide range of loading rates and at different temperatures. Those techniques, however, are not suitable for application of the method of caustics where flat specimens must be used.

8. SOME CORRELATIONS BETWEEN UNIAXIAL STRESS PROPERTIES AND PARAMETERS FOR FRACTURE INITIATION

8.1. The master plot

The main question arises how to correlate the unnotched specimen properties, i.e. the state of uniaxial stress, characterized by constitutive relation (2.7), with the fracture toughness properties, in the form of eq. (2.8).

One of many possibilities is to search for a master plot which woul relate, for example, yield stress σ_y at any strain rate $\dot{\varepsilon}$ and temperature T to fracture toughness K_{Ic} at any loading rate $\dot{K}_I$ and temperature T. Since it is generally accepted that the rate and temperature effects are closely related to the thermally activated micromechanical processes, the correlation of (2.7) and (2.8) can be attempted using this approach, [73, 74]. The concept that thermal activation is involved in material separation during fracture processes has been pursued by Zhurkov [136]. In the other formulation it means that in a stressed body the creation of fracture surfaces occurs with the assistance of thermal motion of atoms. At constant temperature the decisive parameter must be a critical time of plastic deformation and fracturing. Assuming that both the yield stress and flow stress from one side and fracture toughness from the other are dependent via the same thermal activation processes, the hypothesis will be used here that there exists a characteristic time t_c through which the correlation can be achieved. This concept is just a consequence of the process of fracture being plasticity induced.

The strain rate $\dot{\varepsilon}$ and the loading rate $\dot{K}_I$ can be correlated by the characteristic time concept as follows

$$t_c = \frac{\varepsilon_c}{\dot{\varepsilon}} \quad \text{and} \quad t_c = \frac{K_{Ic}}{\dot{K}_I} \tag{8.1}$$

where ε_c is critical strain. Elimination of t_c from eq. (8.1) yields

$$\dot{K}_I = \frac{K_{Ic}}{\varepsilon_c}\dot{\varepsilon} \tag{8.2}$$

Expression (8.2) can be modified introducing linear strain hardening

$$\sigma = \alpha E \varepsilon_c \tag{8.3}$$

where $E_t = \alpha E$ is the secant modulus with α less than one with the likely range $1 < \alpha < 0.1$ at not so large strains. When ε_c is eliminated from (8.2), the final expression relating ε and $\dot{K}_I$ is obtained

$$\dot{K}_I = \alpha \frac{EK_{Ic}}{\sigma_c} \dot{\varepsilon} \tag{8.4}$$

this is eq. (1.43) in the inverted form with $\alpha = 1/k$. For $\alpha = 1$ the case is reduced to the elastic one, eq. (1.40). Expressions like eq. (1.43) and (8.4) were also discussed by Hahn et al. [137], where value $\alpha = 0.1$ was recommended. It is worth to note that Eftis and Krafft, [63], related $\dot{K}_I$ and $\dot{\varepsilon}$ by introducing $\varepsilon_c = n$ in eq. (8.2), where n is the strain hardening index

$$n = \left(\frac{\partial \log \sigma}{\partial \log \varepsilon}\right)_\theta \tag{8.5}$$

thus in eq. (8.4)

$$\alpha = \frac{1}{n}\frac{\sigma_o}{E} \quad \text{or} \quad \alpha = \frac{1}{n} \tag{8.6}$$

where σ_o is stress at $\varepsilon = 1$ and $\sigma_o/E = 1$.

The concept of characteristic time enables the strain rate $\dot{\varepsilon}$ at the elastic-plastic boundary to be related to the loading rate $\dot{K}_I$ of the precracked specimen. Thus, at constant temperature a change in strain rate $\Delta\dot{\varepsilon}$ will develop an equivalent change in flow stress $\Delta\sigma$. On the other hand, the same change in strain rate $\Delta\varepsilon$ will develop an equivalent change in loading rate $\Delta\dot{K}_I$, which in turn will be the source of the equivalent change in K_{Ic}. A similar approach can be applied to a change in temperature ΔT at constant $\dot{\varepsilon}$ or $\dot{K}_I$. If equivalence between change of $\Delta\dot{\varepsilon}$ or ΔT from one side and $\Delta\sigma$ or ΔK_{Ic} from the other indeed exists, then the construction of a unique master plot is possible in the form of K_{Ic} (σ_y) in a variety of sets of coordinates. The existence of the master plot leads to the graphic representation shown in Fig. 3.13. But in order to prove the existence of such correlation, two sets of experimental data must be investigated, i.e. σ_y $(T,\dot{\varepsilon})$ and K_{Ic} $(T, \dot{K}_I)$.

Holtzmann et al., [33], tried to correlate tensile properties and fracture toughness for seven low alloy steels using the coordinates K_{Ic}/σ_y vs. σ_F/σ_y, where σ_F is the critical cleavage stress. It was shown that all data can be distributed along one band with a reasonable scatter. The study was limited to two loading rates. More recent evaluation of experimental results for Fe E 460 steel (Fig. 3.9) by Dahl et al. [137], has led to a good correlation via one master curve in the coordinates $\sigma_{0.03}$ vs. K_{Ic}, where $\sigma_{0.03}$ denotes the flow stress at ε = 0.03. Thus, different authors were relatively successful in finding a unique correlation in different coordinate systems. All those correlations prove that an increase of yield stress due to decrease of temperature, or increase of strain rate, results in an adequate decrease of fracture toughness. This statement is, of course, true for the lower shelf behavior, and is as well limited to the medium-high loading rates, say up to $\dot{K}_I \approx 10^4$ MPa$\sqrt{m}$/s. For the upper shelf a similar correlation may exist but with the positive loading rate sensitivity η, eq. (3.6).

Further studies on the existence of unique correlations based on thermal activation strain rate analyses were pursued by Klepaczko, [29, 34]. In order to demonstrate such correlations, the experimental results for the pressure vessel steel A508 Cl.3 were analysed (Fig. 3.7) in the paper [29]. The result of the analysis is shown in Fig. 8.1 for four loading rates. The coordinates used log K_{Ic} vs. log σ_y demonstrate that the experimental points can be cast in yet another correlation in the form of the negative slope. Every set of points represents one loading rate as indicated in the figure. This figure proves that it is basically possible to obtain an approximate master plot for this steel. The coordinates log K_{Ic} and log σ_y are convenient to characterize transition from lower to upper shelf of fracture toughness. The lowest value of K_{Ic} is characteristic for the absolute zero temperature, and it is estimated from Fig. 8.1 as $(K_{Ic})_{min}$ = 22.3 MPa$\sqrt{m}$. This is an absolute material constant which should be rate independent. The upper shelf is determined for specimen thickness B = 20 mm. It is also clear that the upper shelf is a function of specimen thickness B and cannot be understood as an absolute material constant. In the double logarithmic coordinates (log K_{Ic}, log σ_y) both shelves can be approximated by two horizontal lines. A crude approximation of the transition region is by the third straight line with the negative slope - κ, i.e.

$$-\kappa = \frac{d\,(\log K_{Ic})}{d\,(\log \sigma_y)} \tag{8.7}$$

The right side of the master plot is limited by the critical cleavage stress σ_y. When the level $\sigma_y = \sigma_F$ is reached the unnotched specimen will fracture in a brittle manner, for A508 steel the estimated value for σ_F is σ_F = 1740 MPa.

Experimental data of Dahl et al. [137] have also been analysed in [29] and the result in the form of the master plot is shown in Fig. 8.2. The loading rate range was within the limits : 3×10^{-1} MPa$\sqrt{m}$/s < $\dot{K}_I$ < 1×10^5 MPa$\sqrt{m}$/s, and

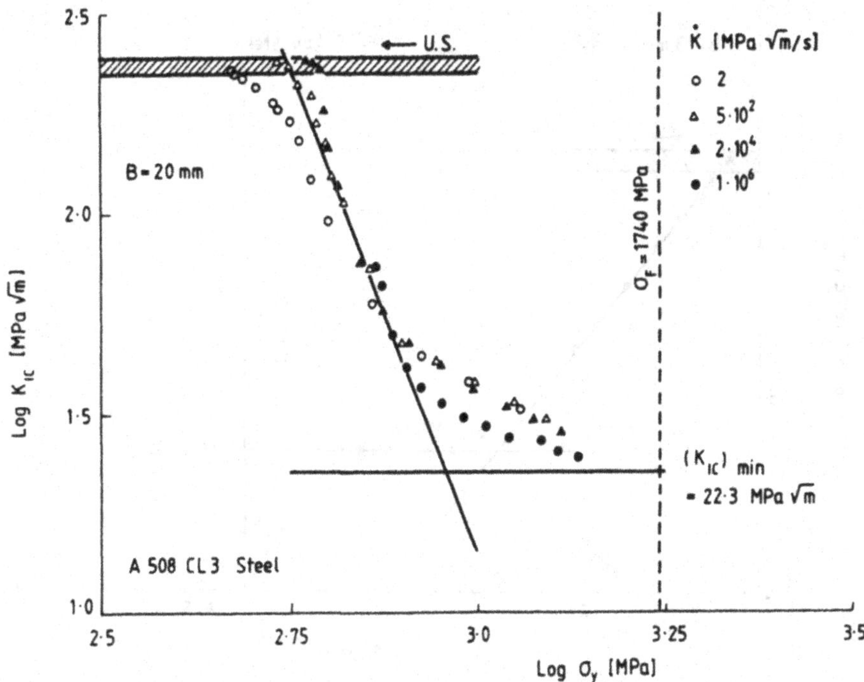

Fig. 8.1, Master plot for A508 Cl.3 steel, specimen thickness B = 20 mm ; US denotes the upper shelf, after [29].

the flow stress $\sigma_{0.03}$ (for ε = 0.03) instead of σ_y was taken into analysis. The points in Fig. 8.2 indicate the mean value of K_{Ic}. Generally, the scatter of the experimental points analysed was smaller in comparison to Fig. 8.1. The absolute minimum of fracture toughness $(K_{Ic})_{min}$ = 22.7 MPa$\sqrt{m}$ and the critical cleavage stress σ_F = 1550 MPa were taken as acceptable values, [73]. The whole picture for Fe E 460 steel is similar to that constructed for A508 steel and the linear approximation of the transition region by the straight line seems to be quite satisfactory.

Another example of the master plot is given in Fig. 8.3, where the points indicate the mean values of K_{Ic} determined for pressure vessel steel 15H2NMFA (Soviet Standards) [18]. The master plot is constructed for only two loading rates, the quasi-static $K_I = 1 \times 10^2$ MPa$\sqrt{m}$/s and the fast one K_I = 3 x 10^5 MPa√m/s. The whole picture is again consistent with the previous master plots shown in Figs 8.1 and 8.2. The absolute minimum of fracture toughness was estimated as $(K_{Ic})_{min}$ = 30.9 MPa√m and the critical cleavage stress was

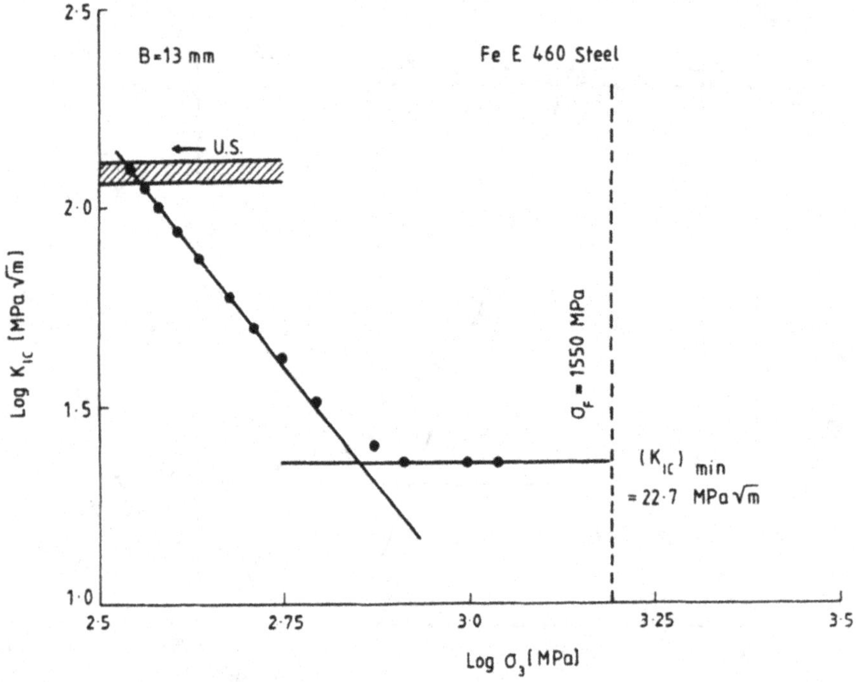

Fig. 8.2, Master plot for Fe E 460 steel (German Standards), specimen thickness B = 13 mm, U.S. denotes the upper shelf, after [29].

also determined in ref. [18] as σ_F = 1850 MPa. The linear approximation of the transition region is not so good for this case and to some extent is similar to A508 steel. The effect most important is the specimen thickness B on the upper shelf level which is indicated as a short horizontal line for each B and for both loading rates : slow - S and fast - D. The levels of the upper shelf for each B differ and the numbers above each line indicate thickness in millimeters. As expected, the upper shelf level of fracture toughness is very sensitive to specimen thickness B. This observation, for example, clearly indicates the limitation of the instrumented Charpy test method in determination of fracture toughness with B = 10 mm.

The main conclusion drawn after the analysis of figures from Fig. 8.1 to Fig. 8.3, i.e. the master plots, is that the construction of a master plot K_{Ic} (σ_y) is possible. This in turn indicates that the dislocation mechanisms which are involved in yielding and plastic deformation in the uniaxial state of stress also participate in the quasi-cleavage separation during fracture. More specifically, there exists the same reciprocity between strain rate and temperature for both plastic deformation and fracture, i.e. the main role of thermally activated dislocation mechanisms is expected [16, 136]. On the other hand, the double

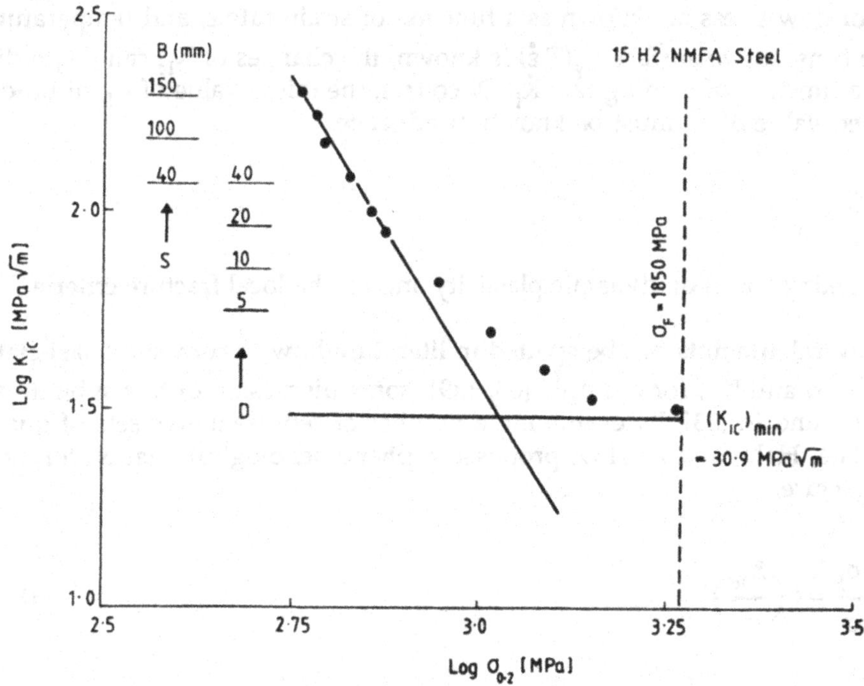

Fig. 8.3, Master plot for 15H2MMFA pressure vessel steel (Soviet Standards), the horizontal short lines denote levels of the upper shelf for different specimen thickness B [mm] ; S - quasi-static loading with $\dot{K}_I = 1\text{x}10^2$ MPa√m/s ; D - fast loading with $\dot{K}_I = 3\text{x}10^5$ MPa√m/s, after [29].

logarithmic coordinates log K_{Ic} vs log σ_y provide a simple approximation of the transition region. The concept of characteristic time enables one to rationalize the effects of loading rate and temperature in plastic flow and fracture.

Another way of constructing the master plot K_{Ic} (σ_y) was discussed in ref. [18]. Within the limits of phenomenology for thermal activation analysis it is possible to correlate flow stress σ_y at different temperatures and strain rates through the so-called temperature modified by strain rate T_ε* [138]. On the other hand, fracture toughness K_{Ic} at different temperatures and loading rates can be correlated for the case of lower shelf by loading rate parameter T_K^* with the dimension of temperature. The first step of the master plot construction is to obtain plots $\sigma_y(T_\varepsilon^*)$ and $K_{Ic}(T_K^*)$ which are rate-independent. The final step is to eliminate T^* for equivalent values T_ε^* and T_K^*. Elimination of T^* produces the master plot such as those shown in Figs 8.1 - 8.3.

The practical meaning of the master plot is that if changes of the yield stress or flow stress are known as a function of strain rate ε, and temperature, i.e. the constitutive surface $\sigma_y(T,\dot{\varepsilon})$ is known, the changes of K_{Ic} can be predicted as a function of loading rate $\dot{K}_I$. Of course, the initial value of K_{Ic} at predetermined value of $\dot{K}_I$ must be known in advance.

8.2 Modeling based on dynamic plasticity and on the local fracture criteria

Several attempts can be spotted in literature how to correlate quasi-static values of σ and K_{Ic}, for example [63, 139], some discussion of this subject can be also found in [33]. By examining a number of representative sets of quasi-static data, Hahn and al. [139] proposed a phenomenological relation for cleavage fracture

$$\frac{\sigma_F}{\sigma_y} = f\left(\frac{K_{Ic}}{\sigma_y}\right) \tag{8.8}$$

where σ_F is the critical cleavage stress. Many authors have shown that σ_F appears to be independent of temperature and to some extent of loading rate, [66]. A more specified form of eq. (8.8) has been also proposed in [139], i.e.

$$\frac{\sigma_F}{\sigma_y} = \alpha\left(\frac{K_{Ic}}{\sigma_y}\right)^\beta \tag{8.9}$$

where α and β are empirical constants. The representative values of α and β are found to be $\alpha = 1.274\ m^{-1/2}$ and $\beta = 1/3$, with K_{Ic} [MPa$\sqrt{m}$] and σ_y [MPa]. The rate dependence of K_{Ic} in an explicit form, can be predicted by introducing eq. (2.7) into (8.9). At the same time the relation $\dot{\varepsilon} = f_3\ (\dot{K}_I)$ must be known in explicit form. Before discussing of such procedure it may be mentioned that the constitutive surface as defined by eq. (2.8) is usually characterised by two regions, the low and a medium temperature domain with a negative loading rate sensitivity, and the higher temperature domain with a mild, positive sensitivity [73]. The established behavior has its source in a different mechanism of decohesion. Reduction in toughness at decreasing temperatures and increasing loading rates is associated with fracture by cleavage, the lower shelf behavior. Whereas so-called the upper shelf behavior, where the loading rate sensitivity is positive, is associated with ductile fracture by forming of dimples. Thus, to analyse both types of behavior two fracture criteria are needed. At present the most successful are the local criteria in predicting fracture toughness, they are based on critical fracture stress σ_F and critical fracture strain ε_F, [140]. Both criteria can be used in modeling of the lower and upper shelf toughness. The σ_F concept establishes local criterion for unstable cleavage fracture as proposed by Ritchie, Knott and Rice [141], i.e. RKR criterion. It has been assumed by those authors that for the cleavage failure the magnitude of K_{Ic} is governed by the size of the plastic zone when the stress component σ_{yy} normal to the crack plane exceedes σ_F over the characteristic distance l_F. The distribution of σ_{yy} component acting directly ahead of a sharp crack in plane strain can be found from singularity solution for hardening material due to Rice and Rosengreen [142] and Hutchinson [143], i.e. HRR solution. The RKR criterion together with the HRR solution leads to the following correlation

$$K_{Ic} = A_L \, \sigma_y \left(\frac{\sigma_F}{\sigma_y} \right)^{\lambda(n)} \tag{8.10}$$

where $\lambda(n)$ is a simple function of strain hardening index n, and A_L is the lower shelf constant. The constant A_L and other quantities in eq. (8.10) will be discussed further on. The relation (8.10) is similar to that of proposed earlier by Hahn and al. (eq. 8.9). This coincidence supports strongly an universality of eq. (8.10) in further studies of rate effects in steels.

Crack-tip blunting and local plastic instability is usual starting point to formulate a fracture criterion for the upper shelf of fracture toughness. Ductile fracture can be modeled as strain induced via linking of voids, [144]. On the basis of such concept the relationship for fracture toughness has been formulated by Ritchie and al. [140], where it was assumed that the critical strain ε_F must be exceeded over the characteristic distance l_F. Thus, the critical strain model can be expressed in the form

$$K_{Ic} = A_u (\varepsilon_F l_F E \sigma_y)^{1/2} \qquad (8.11)$$

where A_u is a constant which is characteristic for a particular microprocess of fracturing.
Those two formulations for upper and lower shelves, eq. (8.10) and eq. (8.11), were used together with the following constitutive relation for uniaxial yield properties

$$\sigma_y = \sigma_\mu \left[1 + \frac{\sigma_p^*}{\sigma_\mu}\left(1 - \frac{\beta(T)}{\sigma_p^*} \log \frac{\dot{\varepsilon}_o}{\dot{\varepsilon}}\right)\right] \qquad (8.12)$$

to compute the constitutive surface (2.8), [28]. In eq. (8.12) σ_μ and $\sigma_p{}^*$ are respectively the internal stress and the Peierls stress, β (T) is temperature dependent strain rate sensitivity of yield stress and ε_o is the pre-exponential factor [16], $\dot{\varepsilon}$ is the current value of strain rate. All details of these calculations are given elsewhere, [28]. In Fig. 8.4 the complete theoretical surface $K_{Ic} = f_2 (\dot{K}_I, T)$ is shown as computed for a carbon steel.
Although results of numerical calculations shown in Fig. 8.4 are in general agreement with experimental observations, there is one point which would be discussed further. Krafft and Coworkers, [63], have indicated, in their early experiments on rapid crack propagation for a mild steel, that fracture toughness for a runing crack K_{ID} increases substantially when velocity of propagation v increases. On the other hand K_{Ic} estimates for the same steel based on a simple model proposed by Krafft have lead to results that fracture toughness of stationnary crack would decreases when $\dot{K}_I$ increases. A comparison of K_{Ic} estimates with the rapid propagation results suggests a plateau or a minimum of K_{Ic} at relatively high values of loading rate $\dot{K}_I$. An advent of new experimental techniques make it possible to determine the loading rate spectra for a wide range of loading rates $\dot{K}_I$, and for some steels the minima of K_{Ic} have indeed been observed. Experimental results collected for two steels where the the minima were found are shown in Fig. 5.16. The upper shelf behavior is manifested over the low loading rate region where the positive loading rate sensitivity is observed, i.e. K_{Ic} increases with log $\dot{K}_I$. The lower shelf behavior is characteristic for decrease of K_{Ic} when log K_I increases. At the loading rate $\dot{K}_I \cong 1 \times 10^4$ MPa$\sqrt{m}$/s the minima of K_{Ic} are observed for both steels. When the loading rate exceedes value ~ 1×10^4 MPa$\sqrt{m}$/s fracture toughness rises again.

An independent confirmation that for some steels the minimum does occur has been provided by Kalthoff, [145]. The shadow optics has been applied along with one-side notch specimen loaded by the incident stress wave

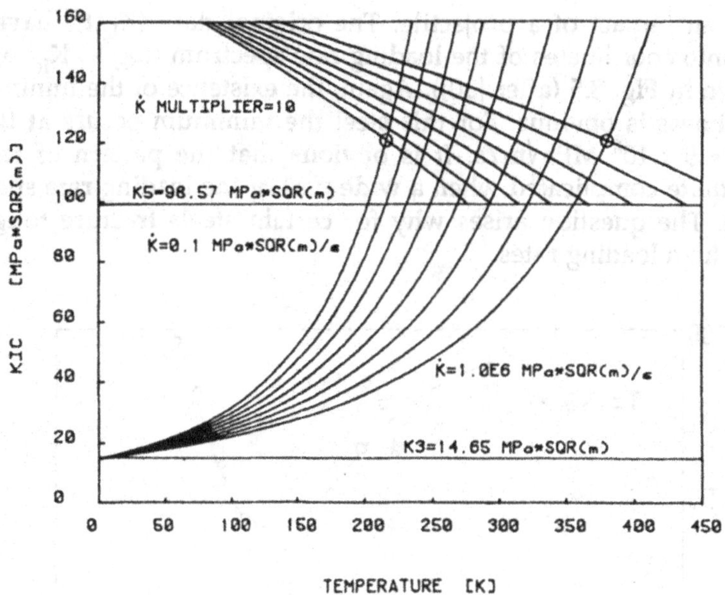

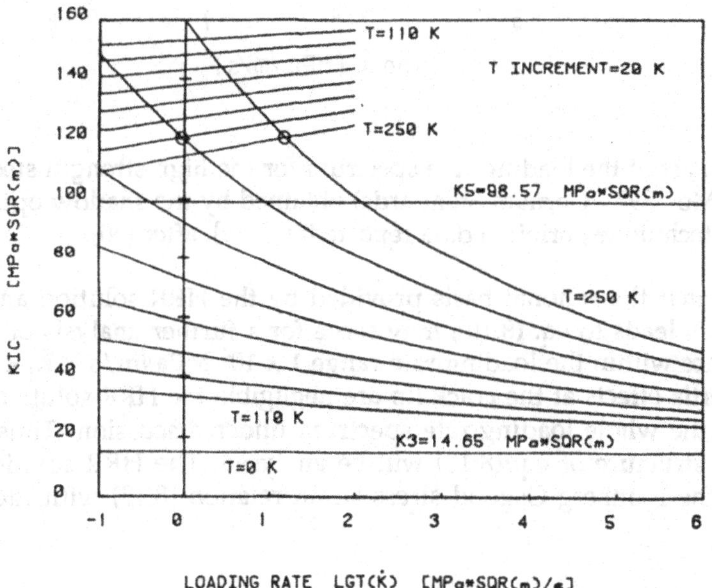

Fig. 8.4, Results of numerical calculations of fracture toughness K_{Ic} ; a - as a function of temperature for eight different loading rates $\dot{K}_I$; b - as a function of log $\dot{K}_I$ for eight temperatures, 110 K < T < 250 K, after [28].

produced by an impact of a projectile. The original data (K_{Ic},t_c) have been transformed into coordinates of the loading rate spectrum (log $\dot{K}_I$,K_{Ic}) and the result is shown in Fig. 8.5 (after [30]). Again, the existence of the minimum of fracture toughness is obvious. For this steel the minimum occurs at the loading rate $\dot{K}_I \approx 5 \times 10^6$ MPa$\sqrt{m}$/s. It is obvious that the pattern of fracture toughness is quite complicated when a wide part of the loading rate spectrum is considered. The question arises why for certain steels fracture toughness rises again at high loading rates.

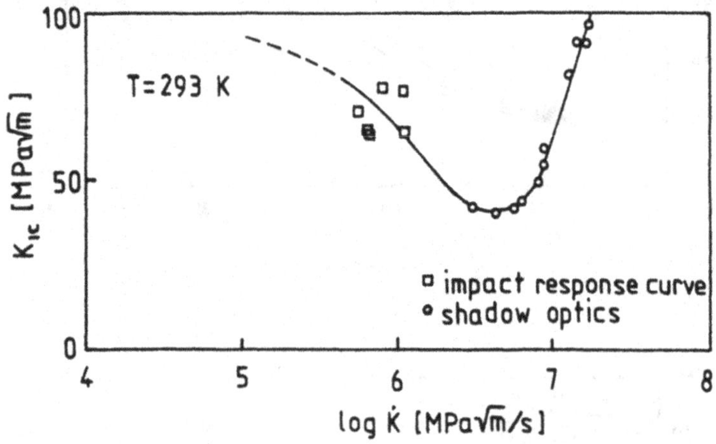

Fig. 8.5, Part of the loading rate spectrum for the high strength steel (X2NiCoMo 1895 - German Standards) obtained by the shadow optics technique ; original data reported in [145], after [30].

It is believed that the rational basis provided by the HRR solution and RKR criterion, which leads to eq. (8.10), may serve for a further analysis of the K_{Ic} minimum. Since within the loading rate range 1×10^4 MPa$\sqrt{m}$/s < $\dot{K}_I$ < 1×10^6 MPa$\sqrt{m}$/s inertia effects at the crack tip are negligible the HRR solution could be used over the whole loading rate spectrum under discussion. Thus, some details in the structure of eq. (8.10) will be analysed. The HRR solution was obtained for the Ramberg-Osgood stress-strain relation (8.13) with the exponent N,

$$\varepsilon = \frac{\sigma}{E} + \left(\frac{\sigma}{\sigma_o}\right)^N \tag{8.13}$$

and the stress intensification σ_{yy}/σ_y can be expressed as

$$\frac{\sigma_{yy}}{\sigma_y} = f\,(N,\theta)\left[\frac{(1-\nu^2)}{\varepsilon_y\, I(N)}\right]^{\frac{1}{N+1}}\left[\frac{x}{(K_I/\sigma_y)^2}\right]^{-\frac{1}{N+1}} \tag{8.14}$$

where f (N,θ) is the angular dependence term, ν is Poisson's ratio, I(N) is the numerical constant, and x is the distance from the crack tip. Application of the RKR criterion, i.e. $K_I = K_{Ic}$ when $\sigma_{yy} = \sigma_F$ at the distance $x = l_F$ leads to the following relation

$$K_{Ic} = \left[\frac{I(N)\,\sigma_y\, l_F}{E(1-\nu^2)\, f^{N+1}}\right]^{1/2} \sigma_y \left(\frac{\sigma_F}{\sigma_y}\right)^{\frac{N+1}{2}} \tag{8.15}$$

This is obviously relation (8.10) with

$$\lambda(N) = \frac{N+1}{2} \quad \text{or} \quad \lambda(n) = \frac{1+n}{2n} \quad \text{since} \quad N = \frac{1}{n} \tag{8.16}$$

where n is the strain hardening index equal to the uniform deformation ε_f in the tensile test. The first term in eq. (8.15) is the lower shelf constant A_L. Expressions for I(N) and f(N,θ) were analysed in [140], I(N) is decreasing function of N at small N, $0 < N < 20$, and f(N,θ) is increasing function of N also at the small N region, $0 < N < 20$. At higher values of N they are almost constant, $I(N) \cong 4.0$ and $f(N,\theta) \cong 2.8$. An analysis of experimental data for twenty two kinds of steel has shown that a linear relation exists between exponent N and yield stress σ_y ; $\sigma_y = C_T(N-N_o)$, this fact was also reported earlier [146]. If A_L in eq. (8.10) had to be a constant the critical distance l_F would be an increasing function of N according to the relation

$$l_F(N) = \frac{E\,(1-\nu^2)\, A_L^2\, f^{N+1}}{C_T\,(N-N_o)\, I(N)} \tag{8.17}$$

The term $\sigma_y\, l_F$ in eq. (8.15) should remains approximately constant with a tendency to decrease at large N. To analyse variations of K_{Ic} at different temperatures and loading rates, under assumption that A_L is a constant, variations of σ_y and l should be discussed further. As it was shown by numerical calculations for the lower shelf (Fig. 8.4) an increase of σ_y, which reflects an increase of strain rate $\dot{\varepsilon}$ or decrease of temperature T, will lead to a decrease of K_{Ic}. However, the loading rate dependency is more complex due to the fact that fracturing at high loading rates is the adiabatic process and the exponent λ is directly or indirectly rate and temperature dependent. The constancy of λ is an approximation. It was shown in [147] that the uniform deformation ε_f in tensile tests of mild steel is a decreasing function of strain rate. Since $\varepsilon_f \cong n$ the rate dependence per-se will develop an increase of $\lambda(n)$. An estimation of decrease of n due to adiabatic heating was attempted by Krafft et al. [148, 149]. Thus, both contributions-direct and indirect may develop an increase of $\lambda(n)$. It can be shown that a proper combination of σ_F/σ_y and λ can produce, due to specific rate dependence, a minimum of K_{Ic}. The favorable conditions to develop the minimum are associated with small values of strain hardening index n together with a small σ_F/σ_y. Thus, a more harder materials, for example after cold work, would show a minimum of K_{Ic} at specific ranges of K_I and T. The typical example are titanium alloys, [28].

The other factor which may contribute to the minimum of fracture toughness is the cleavage, or quasi-cleavage stress σ_F. The critical stress σ_F is characteristic for the transgranular cleavage fracture when σ_F is locally exceeded over the normal stress σ_{yy} in some part of the material, [66], as depicted in Fig. 8.6. There is a strong evidence that cleavage fracture in steels is controlled by microcrack propagation [14, 15, 147]. Thus, microcrack dynamics is the key factor for understanding cleavage separation. Many models for cleavage mechanism in steel have been proposed, and it is out of scope to review them, but one point will be rised concerning Griffith-Orowan formulation for σ_F. This formulation is discussed in Ch. 2, eqs (2.1) and (2.2).

It is known that σ_F remains fairly contant over the range of cleavage separation at different temperatures, typical values are 800 MPa < σ_p < 3800 MPa [66]. It is to be shown whether σ_F changes at high loading rates. Since in eq. (2.2) the plastic work contribution w_p dominates over the surface energy γ, typically $\gamma \cong 2\ J/m^2$ versus $w_p \cong 10\ J/m^2$, and within the region of strain rate in excess of $\dot{\varepsilon} \cong 5 \times 10^3\ s^{-1}$ the micromechanism of plastic deformation changes [13], it may cause an increase of σ_F at loading rates in excess of $\dot{K}_I \cong 5 \times 10^6$ MPa$\sqrt{m}$/s (the logarithmic conversion factor is equal 3), but not at $\dot{K}_I \cong 1 \times 10^4$ MPa$\sqrt{m}$/s. A small rate sensitivity of σ_F was reported in [140] for a mild steel, the quasi-static values of σ_F were σ_F = 1830 MPa for SA 533 B and σ_F = 1650 for SA 302 B steel ($\dot{K}_I \cong 5 \times 10^1$ MPa$\sqrt{m}$/s), respectively for impact loading σ_F = 2000 MPa and σ_F = 1900 MPa ($\dot{K}_I \cong 1 \times 10^5$ MPa$\sqrt{m}$/s), the increases are 9.3% and 15.2%. Typical values of exponent λ in eq. (8.10) are $2 < \lambda < 5$ and even small increase of σ_F may substantially increase K_{Ic}. In order to verify such possibility experimental results of tensile tests performed at different strain rates on mild steel reported in [150] were evaluated and exponent λ was determined as a function

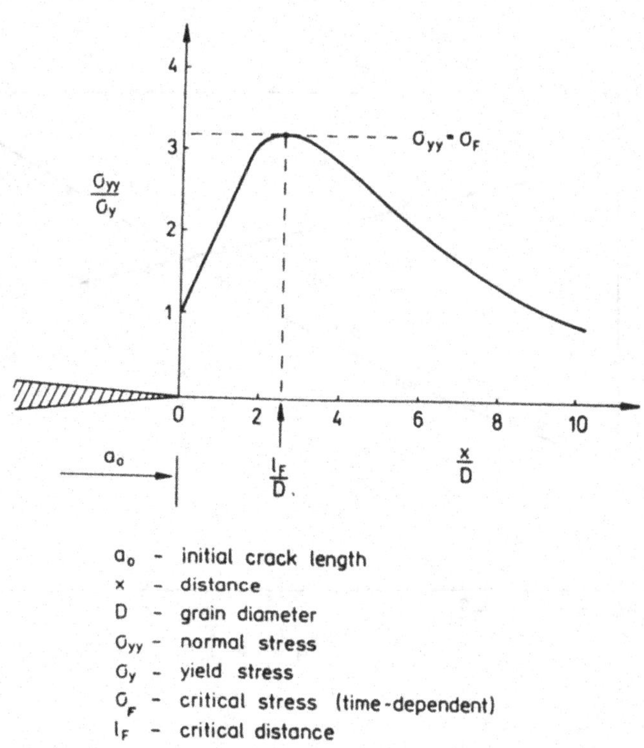

Fig. 8.6, Illustration of the local fracture criterion $\sigma_{yy} = \sigma_F$ at the critical distance from the crack tip l_F/D.

of strain rate. The results obtained in [150] have demonstrated a substantial reduction in the uniform deformation ε_f when strain rate was increased. Using relation $\varepsilon_f = n$ the strain hardening index n was determined at different

$\dot{\varepsilon}$, and exponent λ was calculated from eq. (8.16). The result is shown in Fig. 8.7. Next, values of K_{Ic} were calculated using material constants provided in [28] together with actual values of $\lambda(n)$ characteristic for each $\dot{\varepsilon}$ or $\dot{K}_I$. The result is shown in Fig. 8.8 by the broken line. Indeed, the minimum is present but it occurs at $\dot{K}_I \cong 1 \times 10^7$ MPa$\sqrt{m}$ and the shape of the curve is different as those observed experimentally. Thus, some further considerations are of importance why a sharp increase of K_{Ic} occurs beginning from $\dot{K}_I \approx 1 \times 10^4$ MPa$\sqrt{m}$/s. A short-pulse fracture mechanics can be helpful in explaining this feature.

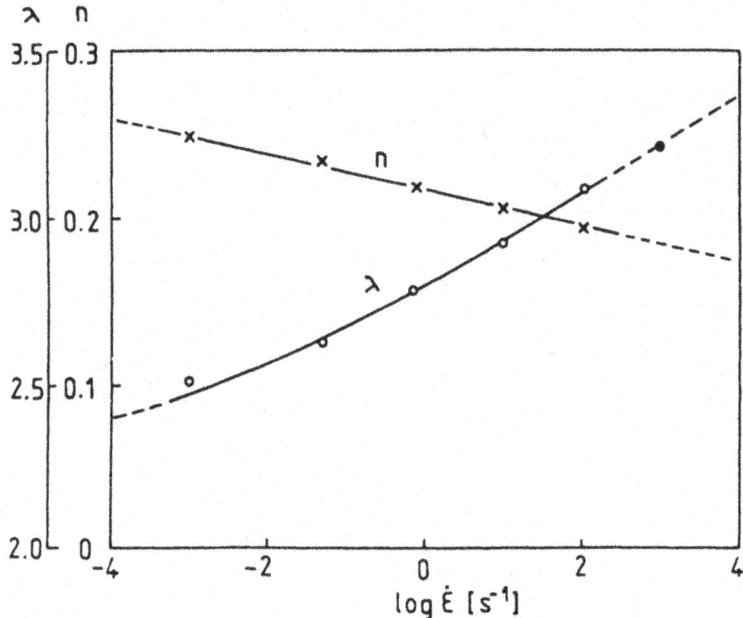

Fig. 8.7, Values of strain hardening index $n \approx \varepsilon_f$ as determined from tensile tests of different $\dot{\varepsilon}$; data after [150], values of λ calculated from eq. (8.16).

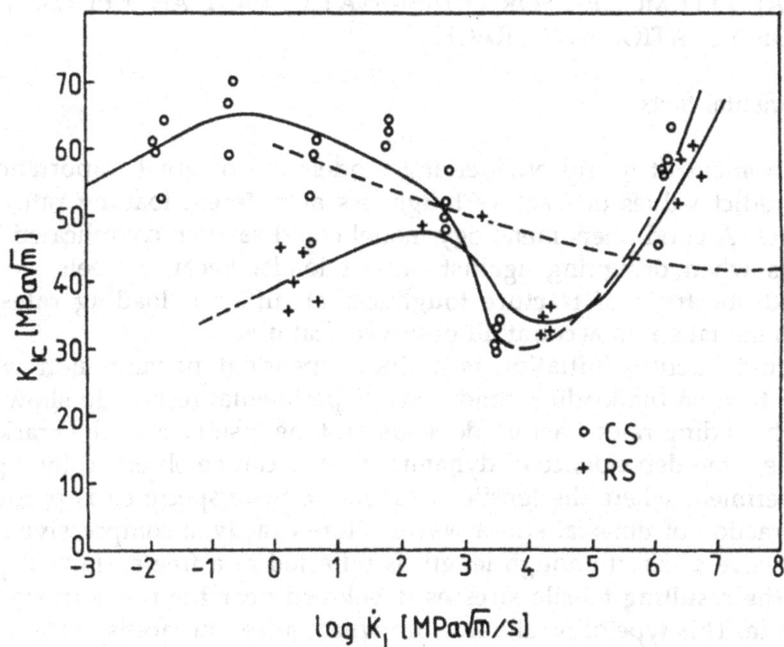

Fig. 8.8, Result of numerical calculations of K_{Ic} - broken line, based on eq. (8.10) with rate dependence of λ and constancy of σ_F ; the experimental data are of Fig. 5.16.

9. GENERALIZED MODEL FOR QUASI-STATIC, FAST AND PULSE LOADING OF A STATIONARY CRACK

9.1. Experimental facts

Development of useful engineering models are of great importance in order to predict values of fracture toughness at different loading rates and temperatures. A good phenomenology model could save on complicated FEM calculations when designing against impact loads. Because steels show a complicated spectrum of fracture toughness at different loading rates the modeling must take into account all observed features.

Dynamic fracture initiation is a time-dependent phenomenon which depends in turn on the loading conditions. Experimental results do show that at very high loading rates fracture depends on time history near the crack tip. For example, time dependence of dynamic fracture can be observed by a plate impact experiment where the tensile stress in the plate specimen is produced by the interaction of uniaxial strain waves. More exactly, a compressive uniaxial strain wave of short enough length is reflected at a free surface of plate specimen, the resulting tensile stresses developed near the free surface may cause fracture. This type of tensile fracture where a free surface is created inside the specimen material is called spalling or scabbing. Another example of time-dependent fracture can be shown for proportional loading with K_I = const. Finally, a more complicated test for time-dependent fracture is the pulse loading. The all three cases are shown in Fig. 9.1.

In the case of tensile wave loading of uniaxial strain the bulk material separates when the critical tensile stress (cleavage or fibrous) σ_F is exceeded. However, for the particular value of σ_F separation occurs after well defined short time interval t_c. This feature is similar to the phenomenon of delayed plasticity in yielding of BCC metals [152]. For a long duration pulse the stress to cause fracture is lower than for a short one. This is illustrated in Fig. 9.2 for three polycrystalline metals. In the case of longer pulses than several microseconds σ_F remains nearly constant for this type of loading. It is interesting to note that the aluminum alloy is "rate-insensitive" as observed in many experiments in dynamic plasticity. Since steel represents BCC structure it is more sensitive than copper which belongs to FCC structures. This figure demonstrates that fracture, i.e. spalling, is not instantaneous and a finite time is required for crack initiation. The representative parameters are the fracture stress σ_F and the critical time t_c.

For the case of an isolated crack loaded at a constant opening rate (kinematic loading), case 2 in Fig. 9.2, the delayed fracture is manifested by a positive loading rate sensitivity η at high values of $\dot{K}_I$. In this case the governing

1. Uniaxial-strain tensile wave

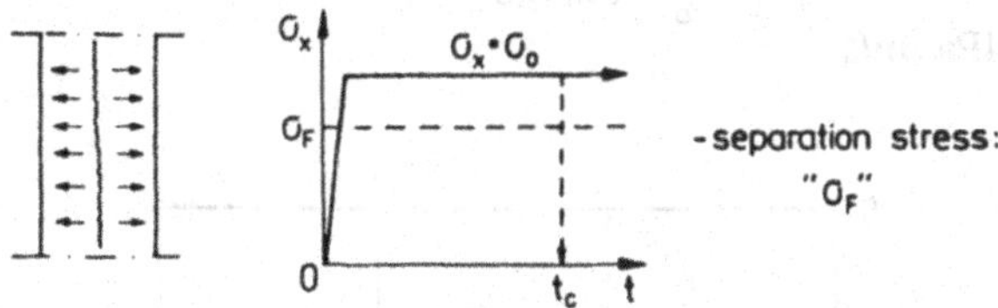

2. Proportional loading of an isolated crack

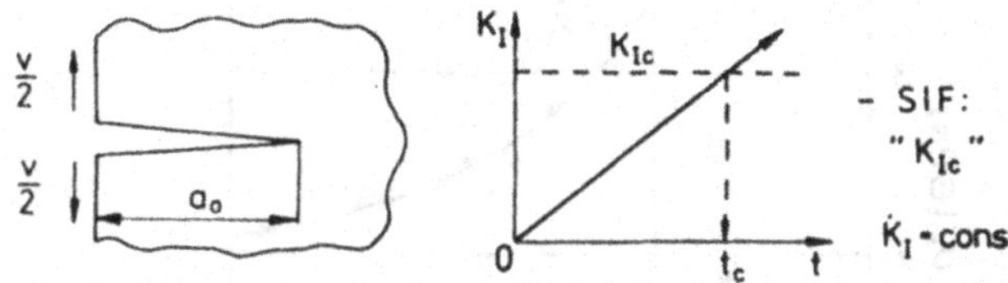

3. Pulse loading of an isolated crack

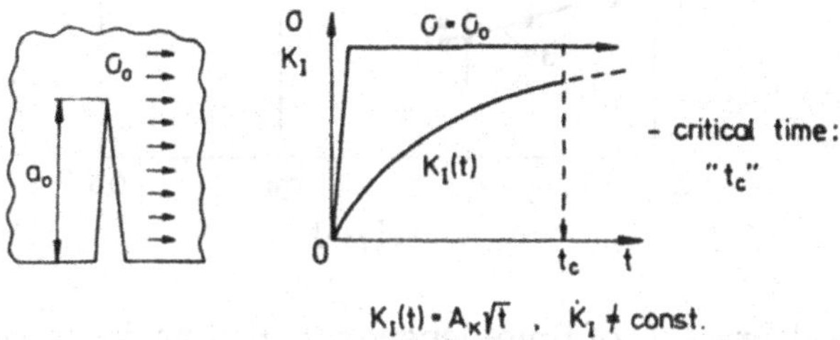

Fig. 9.1, Three types of dynamic loading which lead to time-dependent fracture ; 1 - tensile wave of uniaxial strain (spalling) ; 2 - proportional loading of an isolated crack (for example WLCT technique) ; 3 - short pulse loading (delayed fracture).

parameters are K_{Ic} and t_c. The delayed fracture occurs for critical times shorter than t_{co}, where

$$t_{co} = \frac{(K_{Ic})_{min}}{\dot{K}_{Imin}} \tag{9.1}$$

For example, t_{co} can be determined for 1045 steel from Fig. 8.8

$$t_{co} = \frac{32 \text{ MPa}\sqrt{m}}{10^5 \text{ MPa}\sqrt{m}/s} \qquad t_{co} = 320\ \mu s$$

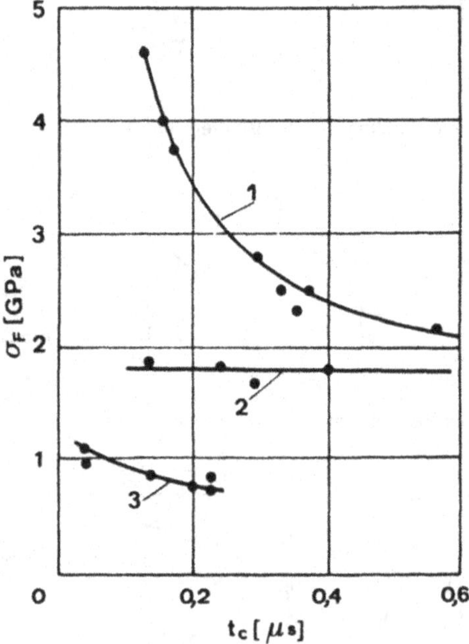

Fig. 9.2, Critical stress σ_F of time-dependent spalling for three polycrystalline metals ; 1 - low carbon steel ; 2 - aluminum alloy ; 3 - copper.

It must be remembered that this is proportional loading as shown in Fig. 9.1 and the critical times are much longer as compared to the instantaneous case (spalling).

The third case of loading of an isolated crack is possible by the short-time pulse, as it is shown in Fig. 9.3. This type of experiment was first pursued by Shockey et al. [153, 154] and later by Buchar [156]. Both schemes of loading are shown in Fig. 9.3 and Fig. 9.4.

A fatigue crack produced in the single edge notched (SEN) specimen is loaded by a tensile pulse $p(t)^+$. Due to a short projectile the pulse has a maximum and the crack of length a_o starts to move AFTER the maximum tensile stress is reached. The time of loading is t_o as it is shown in the phase diagram in Fig. 9.3. (Lagrange plot). Impact of the projectile on the cylindrical plate welded to one of the specimen ends generates at the instant of impact an compressive

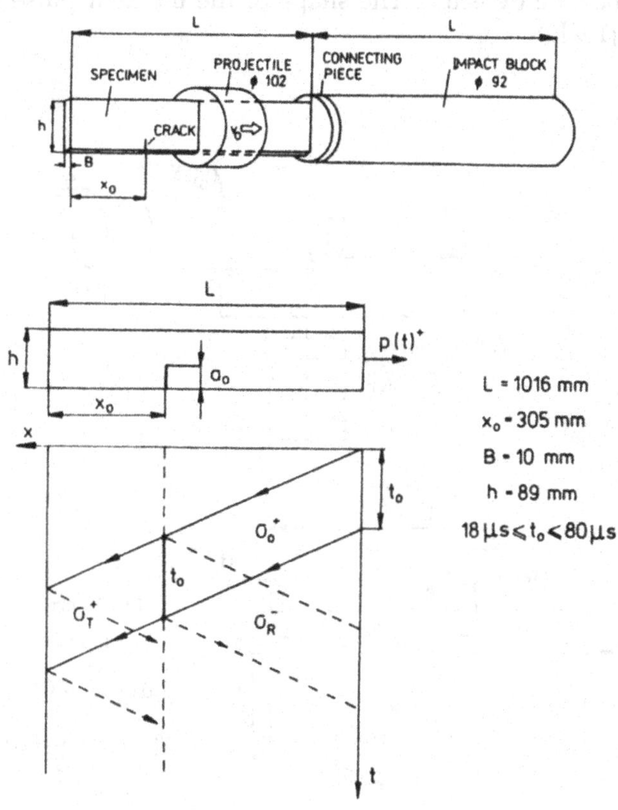

Fig. 9.3, Edge cracked specimen (SEN) loaded by projectile impact with incident tensile pulse, [157].

wave in the projectile and impact block and a tensile wave in the SEN specimen. Since the projectile is much shorter than the SEN specimen and the impact block, the compressive wave in the projectile is reflected from its free end and returns to the point of impact. The time of contact is t_o, $t_o = 2\,L_p/C_o$, where L_p is the projectile length. Consequently, the duration of the resulting tensile wave in the specimen is t_o. The amplitude of the incident pulse in the specimen, at the specific impact velocity v_o of projectile, is a combination of the cross sections of projectile, specimen and impact block. Application of three projectiles of different lengths and SEN specimens with different lengths a_o of prefatigued cracks, $0.2 \leq a_o/h \leq 0.8$, permitted for testing at different combinations of pulse lengths t_o, $80\ \mu s \leq t_o \leq 18\ \mu s$. Several impact velocities were applied to each specimen with increasing v_o, i.e. in creasing stress amplitudes. The procedure was repeated until incremental crack growth could be detected. The shape of the incident pulse is monitored by a strain gage, [157].

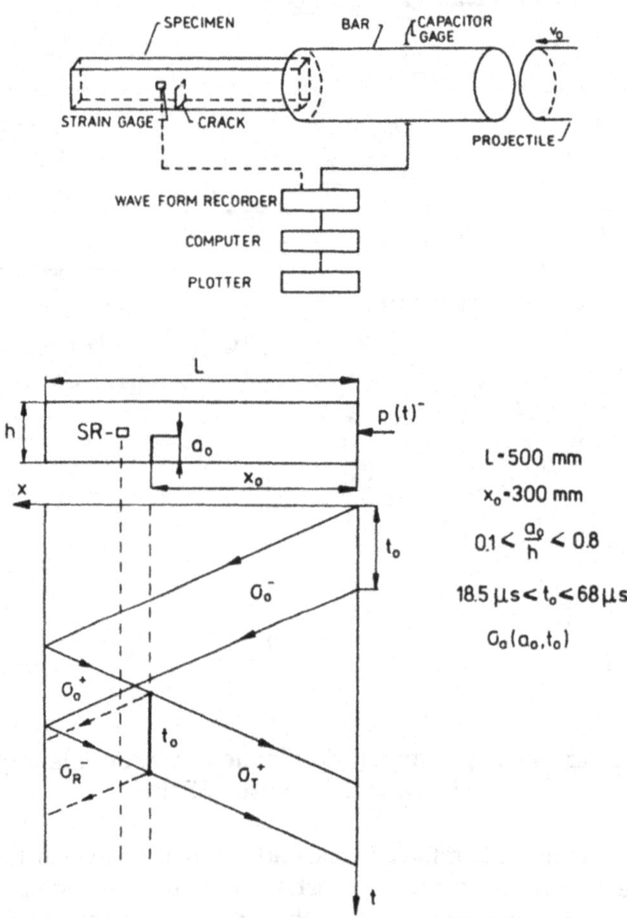

Fig. 9.4, Edge cracked specimen (SEN) loaded by projectile impact with incident compressive pulse, [155, 156].

A similar scheme of loading for SEN specimen but with initial compressive pulse $\sigma(t)^-$ was applied in [155] and [156]. Since one end of the SEN specimen is in contact with an incident bar and the incident bar is hit by a projectile of length L_p with velocity v_o, the crack is first loaded by compressive wave of duration $t_o = L_p/C_o$, and after reflection from the free end the crack is loaded in Mode I. It is assumed that the compressive pulse is harmless to the prefatigued crack and the short pulse test starts as it is shown in the lower part of Fig. 9.4 (Lagrange plot).

Those two schemes for the short pulse loading are practically the same. If the incident pulse has a rectangular form the governing parameters in this test are the stress amplitude σ_o and the critical time t_c after which crack starts to move. It is important to note that elastodynamic solution for a semi-infinite stationary crack loaded with a step normal tensile wave predicts variation of K_I as a function of time in the form, [158]

$$K_I(t) = \frac{2\sigma_o}{(1-\nu)}\left[\frac{C_1(1-2\nu)}{\pi}\right]^{1/2} t^{1/2} \tag{9.2}$$

where σ_o is the stress amplitude. Thus, the elastodynamic solution predicts that for such loading the SIF will increase in proportion to $\sqrt{t}$; i.e. $K_I(t) = A_K\sqrt{t}$, this is shown schematically in Fig. 9.1 (the third case of loading).

Results of short pulse experiments reported in [153-157] have lead to the conclusion that for short crack lengths the stress amplitude σ_o for crack instability decreases with increasing crack length, which is in agreement with quasi-static formulation, i.e.

$$K_I = A\,\sigma_o\sqrt{\pi a_o} \tag{9.3}$$

This occurs for relatively long pulses. But when the duration of the incident stress pulse is decreased the critical stress amplitude σ_o tends to remain at constant value for larger crack lengths. This situation is shown schematically in Fig. 9.5. The shorter the pulse duration the smaller the critical crack length a_o from which a constant critical amplitude σ_o is observed. At the same time for a shorter pulse a higher stress amplitude is needed to move the crack tip. Some implications to fracture mechanics which follow from experimental findings discussed above have been reviewed by Kalthoff, [157].

The short pulse experiments show that the short duration, low intensity stress pulses develop delayed fracture. In another words, if the pulse duration of amplitude σ_o is decreased below the limit $t_o = t_c$ the critical condition of fracturing does not depend on crack length. Such situation is not predicted by the rate-independent fracture mechanics. When the critical points like 1, 2, 3 in the schematic figure 9.5 are plotted as a function of the pulse duration t_o the

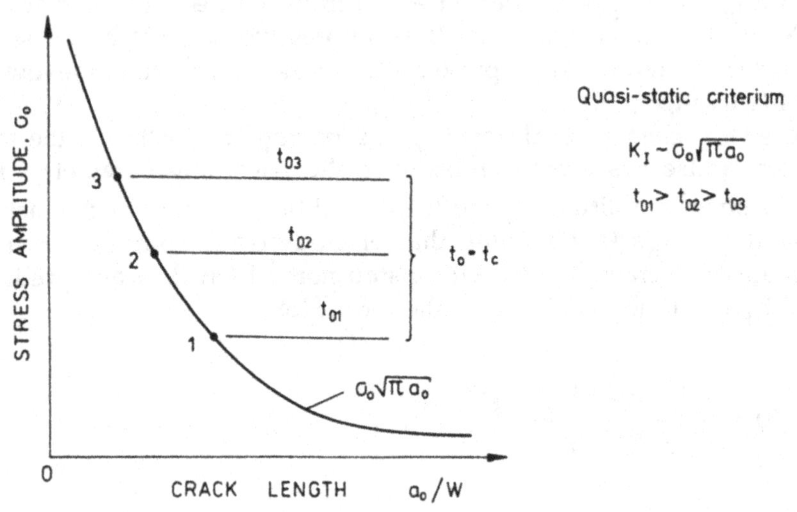

Fig. 9.5, Instability stress σ_o as a function of crack length a_o/W ; points 1, 2 and 3 show departure from quasi-static behavior, $t_o = t_c$ is the critical time of departure from $K_I \sim \sigma_o\sqrt{\pi a_o}$.

results are obtained as shown in Fig. 9.6. To construct $\sigma_o(t_c)$ plots experimental results for 4340 steel reported in [155] and for bainitic steel reported in [156] have been analysed.

It is interesting to note that those results can be approximated by relation $A_o = \sigma_o\sqrt{t_c}$ shown in Fig. 9.6 by the broken line with the mean value of constant A_o ; A_o = 556.5 MPa$\sqrt{\mu s}$.

Comparison of figures 9.2 and 9.6 indicates for similarity of $\sigma_F(t_c)$ from one side and $\sigma_o(t_c)$ from the other. Of course, the scales of time are different due to different type of loading. However, both figures clearly indicate for the existence of delayed material separation not necessarily developed by inertia forces. Another indirect confirmation of this conclusion is the positive rate sensitivity η for 1045 steel at loading rates $\dot{K}_I$ in excess of ~ 10^4 MPa$\sqrt{m}$/s, Fig. 8.8. When the experimental points obtained in excess of 10^4 MPa$\sqrt{m}$/s of Fig. 8.8 are plotted in the coordinates (t_c, K_{Ic}) a similar picture emerges to that shown
in figures 9.2 and 9.6.

The three types of experiments provide data in the form (σ_F, t_c) for spalling, (σ_o, t_c) for short pulse experiments and (K_{Ic}, t_c) for fast proportional loading.

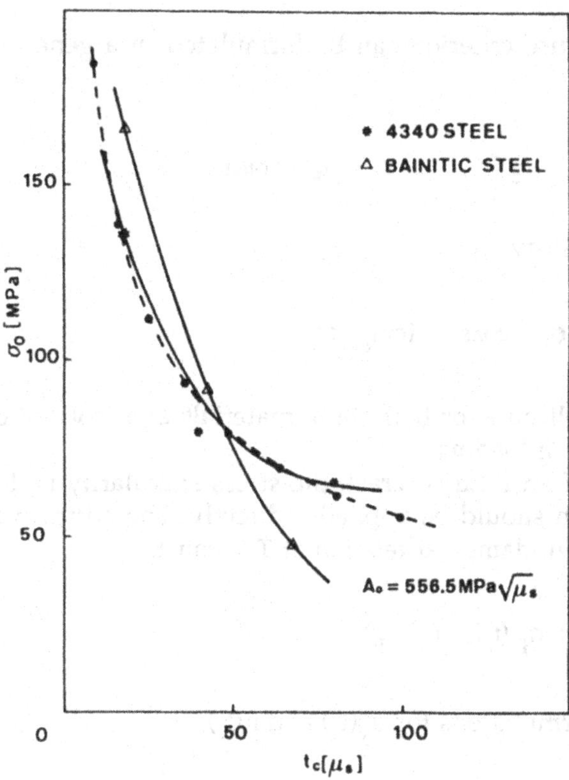

Fig. 9.6, Result of analysis of short-pulse experiments ; * - 4340 steel, data after [155] ; Δ - bainitic steel, data after [156].

A fundamental question arises, if it is possible to construct a simple model which would take into account all effects discussed above. Since for the lower rates of loading the incubation time for fracture can be neglected, i.e. $t_{co} > t_c$, where t_{co} is the characteristic time for particular type of loading, a constant value of the quasi-cleavage stress σ_F is a satisfactory assumption. It has been shown in Ch. 8 that the model derived using HRR singularity along with the local fracture criterion, and a constant value for the cleavage stress σ_F, is adequate for the range of medium loading rates, up to ~ 10^4 MPa$\sqrt{m}$/s. The spall and short stress pulse experiments demonstrate that the critical separation stress σ_F increases when the time of loading decreases. This observation can be implemented into the model based on the HRR singularity and the local fracture criterion.

9.2. A phenomenology model

The local fracture criterion can be formulated in a general form as follows

$$t_c = f_3\,[\sigma_F\,(t)] \qquad t_c \le t_{co} \qquad , \qquad l_F = \text{const.} \tag{9.4}$$

with the loading history

$$\sigma_F = f_4\,(t) \qquad \sigma_F = \text{const.} \quad \text{for } t_{co} < t_c \tag{9.5}$$

The criterion is applicable for both, bulk materials and isolated cracks with a complicated history of loading.

For the case of an isolated crack the stress singularity is developed and the fracture criterion should be applied indirectly. The critical value of SIF is obtained using the fundamental relation at T = const.

$$K_{Ic} = f_1\,[\sigma_y\,(t_c),\, \sigma_F\,(t_c),\, n\,(t_c),\, l_F]_T \tag{9.6}$$

with appropriate formulations for $\sigma_y(t_c)$ and $n(t_c)$,

$$\sigma_y = f_2\,(\dot{\varepsilon},T) \quad \text{or} \quad \sigma_y = f_2\,(t_c,T) \tag{9.7}$$

In addition, strain hardening index depends on strain rate or time

$$n = f_5\,(t_c) \tag{9.8}$$

The meaning of all symbols has been explained earlier.

The explicit form of (9.6) is similar to that based on HRR singularity

$$K_{Ic} = A_L \sigma_y\,(t_c,T) \left[\frac{\sigma_F\,(t_c)}{\sigma_y\,(t_c,T)}\right]^{\lambda(n)} \tag{9.9}$$

with σ_y and σ_F depending on time, which is modified eq. (8.12)

$$\sigma_y(t_c,T) = \sigma_\mu \left[1 + \frac{\sigma_p^*}{\sigma_\mu}\left(1 - \frac{\beta(T)}{\sigma_p^*} \log\left(\frac{\dot{\varepsilon}_o n(t_c)}{3t_c}\right)\right)\right] \tag{9.10}$$

In equation (9.10) linear temperature dependence of rate sensitivity of the flow stress is assumed

$$\beta(T) = \beta_o T \tag{9.11}$$

where β_o is a constant. The local strain rate in eq. (9.10) is eliminated by introduction $\dot{\varepsilon} = n/3t_c$. The critical strain at the crack tip or in uniaxial strain separation ε_{cr} is reduced due to stress triaxiality, [144], to $\varepsilon_{cr} = n/3$, where n is the strain hardening index.

The set of equations (9.4), (9.5), (9.8), (9.9), (9.10) and (9.11) constitute all relations needed to determine $\sigma_F(t_c)$, $\sigma_o(t_c)$ and $K_{Ic}(K_I)$. The formula (9.9) for K_{Ic} is highly non linear with the respect to all parameters : σ_y, σ_F, $\lambda(n)$. Thus, for each loading history of the crack tip those parameters must be specified on the individual basis. Some simplifications are also possible. Numerical calculations shown in Fig. 8.4 are one example of simplified approach with σ_F = const., λ = const. and l_F = const. Such simplification is inacceptable at short critical times t_c and explicit form of the local fracture criterion with delay time must be formulated.

The following cumulative fracture criterion in the integral form is proposed

$$t_{co} = \int_o^{t_c} \left(\frac{\sigma_F(t)}{\sigma_{Fo}}\right)^{\alpha(T)} dt \quad ; \quad \begin{array}{l} t_c \le t_{co} \\ \sigma_F \le \sigma_{Fo} \end{array} \tag{9.12}$$

where σ_{Fo}, t_{co} and $\alpha(T)$ are three material constants, t_{co} is the longest critical time when $\sigma_F(t_{co}) = \sigma_{Fo}$, for $t_c > t_{co}$ $\sigma_F = \sigma_{Fo}$ and σ_F = const. The exponent α is temperature dependent and is related to the activation energy of material separation, [136]. A similar fracture criterion has been proposed earlier by Tuler and Butcher, [159]. This criterion, however, is incorrect for a longer critical times since it is based on the overstress concept. In the present notation the criterion from [159] has the form

$$t_{co} = \int_{0}^{t} (\sigma_F - \sigma_{Fo})^{\delta}\, dt \qquad (9.13)$$

in the limit $\sigma_F \to \sigma_{Fo}$ the integral goes to zero and not to t_{co}.

On the other hand the proposed criterion in the form of eq. (9.12) has some physical motivation based on the thermally activated rate processes.

At the begining the instantaneous loading of a bulk material will be discussed. For this case the loading history is defined by

$$\sigma_F(t) = \sigma_F\, H(t)$$

where H(t) is Heaviside step function. The criterion takes the form

$$\sigma_F(t_c) = \sigma_{Fo} \left(\frac{t_{co}}{t_c}\right)^{1/\alpha} \quad ; \quad t_c \le t_{co} \qquad (9.14)$$

or

$$t_c = t_{co} \left(\frac{\sigma_F}{\sigma_{Fo}}\right)^{-\alpha} \qquad (9.15)$$

Analysis of the results for the low carbon steel of Fig. 9.2 yields the following values of α, σ_{Fo} and t_{co} : $\alpha = 2.31$, $\sigma_{Fo} = 1250$ MPa and $t_{co} \approx 2$ µs. Since in the reality the pulse which causes spalling is not ideal and the rise time always occurs, value of t_{co} is usualy greater than for the ideal case. When the loading of the bulk material is not instantaneous with respect to time the fracture criterion should be integrated accordingly. For example, in the linear case

$$\sigma_F(t) = A_F t \qquad (9.16)$$

where A_F is the rate of loading and $\sigma_F(t_c) = A_F t_c$

$$t_{co} = \int_{0}^{t} \left(\frac{A_F t}{\sigma_{Fo}}\right)^{\alpha} dt$$

$$t_c = (\alpha + 1)\, t_{co} \left(\frac{\sigma_F(t_c)}{\sigma_{Fo}}\right)^{-\alpha} \tag{9.17}$$

and

$$\sigma_F(t_c) = \sigma_{Fo} \left[(\alpha + 1) \frac{t_{co}}{t_c}\right]^{1/\alpha} \tag{9.18}$$

In this case the longest critical time is $t_{co}^* = (\alpha + 1)t_{co}$, where t_{co} is referred to the instantaneous loading.

When the proportional loading is considered for an isolated crack, the loading history $\sigma_F(t)$ is defined as time change of the normal stress σ_{yy} at CONSTANT distance l_F from the tip of the stationary crack. If the K_{Ic} is calculated the other parameters, i.e. σ_y and λ change accordingly. It is assumed in further analyses that the critical distance remains constant, l_F = const. Assuming in addition that λ = const., $\sigma_F(t_c)_T$ and $\sigma_y(t_c)_T$ can be introduced into the fundamental relation (9.9), the result is

$$K_{Ic} = A_L \sigma_y(t_c)_T \left(\frac{\sigma_{Fo}}{\sigma_y(t_c)}\right)^{\lambda} \left[(\alpha + 1) \frac{t_{co}}{t_c}\right]^{\lambda/\alpha} \tag{9.19}$$

for $t_c \leq t_{co}^*$, with $t_{co}^* = (\alpha + 1)\, t_{co}$. For $t_c > t_{co}^*$ relation for K_{Ic} with σ_F = const. must be applied

$$K_{Ic} = A_L \sigma_y(t_c)_T \left[\frac{\sigma_F}{\sigma_F(t_c)_T}\right]^{\lambda} \tag{9.20}$$

An inverse problem can also be analyzed when the experimental data in the form $K_{Ic}(\dot{K}_I)$ for proportional loading are known and $\sigma_F(t_c)$ has to be calculated. Such analysis has been carried out for the 1045 carbon steel using experimental data of the form $K_{Ic}(\dot{K}_I)$ shown in Fig. 8.8. Details of calculations are skipped here and the result in the form $\sigma_F(t_c^*)$; $t_c^* = (\alpha + 1)t_c$ is shown in Fig. 9.7. The overall similarity of the $\sigma_F(t_c^*)$ curve to the $\sigma_F(t_c)$ curves shown in Fig. 9.2 and Fig. 9.6 is encouraging.

The most interesting case is the pulse loading. An approximate analysis can be performed using elastodynamic solutions, eq. (9.2). The solution enables to find the time evolution of SIF due to loading the faces of a crack. If the initial conditions at the crack tip are $\sigma(t) = \sigma_o H(t)$, where σ_o is the amplitude of the tensile stress the elastodynamic solution (9.2) can be reformulated

$$K_I(t) = 2\sigma_o f(v)\sqrt{C_1 t} \quad , \quad f(v) = \left(\frac{1-2v}{\pi(1-v)^2}\right)^{1/2} \tag{9.21}$$

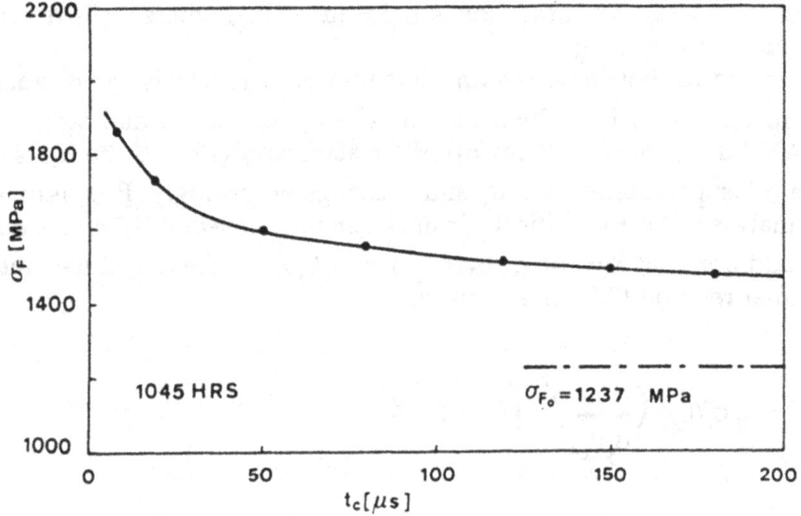

Fig. 9.7, Calculated values of $\sigma_F(t_c)$ for 1045 steel from eqs. (9.16), (9.17) and (9.19) using data of Fig. 8.8.

Application of the following equations permits for determination $\sigma_F(t_c)$ for the case of pulse loading (decreasing part of the $\dot{K}_I(t)$ is neglected),

a. elastodynamic solution, eq. (9.21)

$$K_I(t_c) = 2\sigma_o f(v)\,(C_1 t_c)^{1/2}$$

b. yield stress as a function of critical time (proportional loading approximation), eq. (9.10)

$$\sigma_y(t_c) = \sigma_\mu \left[1 + \frac{\sigma_p^*}{\sigma_\mu}\left(1 - \frac{\beta(T)}{\sigma_p^*}\log\left(\frac{\dot{\varepsilon}_o}{3t_c}\right)\right)\right]$$

c. fundamental relation (constitutive relation), eq. (9.9)

$$K_{Ic}(t_c) = A_L \sigma_y(t_c) \left(\frac{\sigma_F(t_c)}{\sigma_y(t_c)}\right)^\lambda$$

Introduction of the elastodynamic solution (a) into the constitutive relation (c) permits for determination $\sigma_F(t_c)$ in the form

$$\sigma_F(t_c) = \left[\frac{2\sigma_o f(v)\, C_1^{1/2}}{A_L}\right]^{\frac{1}{\lambda}} \left[\sigma_y(t_c)\right]^{1-\frac{1}{\lambda}} t_c^{\frac{1}{2\lambda}} \tag{9.22}$$

Analyses of experimental data of Homma et al. [155] and Kalthoff et al. [157] for 4340 steel and also data by Buchar, [156], for bainitic steel have led to the following phenomenology relation, shown by the broken line in Fig. 9.6,

$$\sigma_o(t_c)^{1/2} = A_o \tag{9.23}$$

with A_o = 556.5 MPa$\sqrt{\mu s}$ for 4340 steel and A_o = 562.0 MPa$\sqrt{\mu s}$ for bainitic steel. Taking into account $A_o = \sigma_o \sqrt{t_c}$ in the $\sigma_F(t_c)$ relation one obtains

$$\sigma_F(t_c) = \left(\frac{2A_o}{A_L} f(v)\, C_1^{1/2}\right)^{\frac{1}{\lambda}} \left[\sigma_y(t_c)\right]^{1-\frac{1}{\lambda}} \tag{9.24}$$

The result of calculations of $\sigma_F(t_c)$ for 4340 steel is shown in Fig. 9.8.
Again, a reliable dependence of $\sigma_F(t_c)$ has been obtained using the pulse loading data in the form eq. (9.23).

Assuming that $\sigma_F(t) = A_F K_I(t)$ defines the loading history of σ_F, at l_F = const.

$$\sigma_F(t) = 2A_F \sigma_o f(v)\,(C_1 t)^{1/2} \tag{9.25}$$

the fracture criterion can be integrated as follows

$$t_c = \left[\frac{1}{2}(\alpha + 2)t_{co}\left(\frac{2A_F\sigma_o f(v)\sqrt{C_1}}{\sigma_{Fo}}\right)^{\alpha}\right]^{\frac{2}{\alpha+2}} \tag{9.26}$$

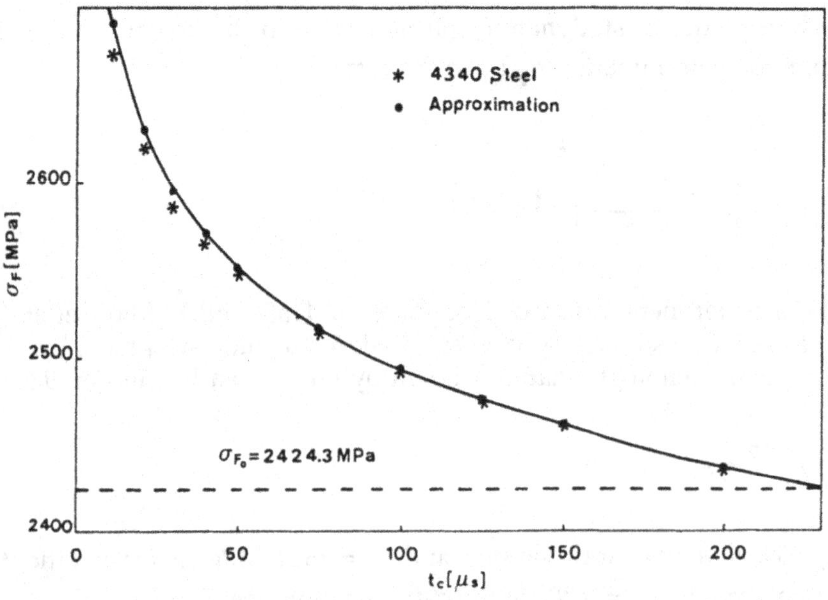

Fig. 9.8, Calculated values of $\sigma_F(t_c)$ for 4340 steel from eq. (9.24) and eq. (9.27) using the short-pulse data of Fig. 9.5, o - approximation of $\sigma_o(t)$ by eq. (9.23), * - exact formulation with eq. (9.25).

Finally, substituting for t_c in eq. (9.25)

$$\sigma_F(t_c) = 2A_F\sigma_o f(v)\sqrt{C_1}\left[\frac{1}{2}(\alpha+2)t_{co}\left(\frac{2A_F\sigma_o f(v)\sqrt{C_1}}{\sigma_{Fo}}\right)^{\alpha}\right]^{\frac{1}{\alpha+2}} \tag{9.27}$$

The modified time t* in eq. (9.27) is

$$t_c^* = \frac{1}{2}(\alpha+2)\, t_{co}$$

The latest relation for $\sigma_F(t_c)$ is shown by stars in Fig. 9. Relation (9.27) indicates that the local value of the quasi-cleavage stress σ_F is a power function of the stress amplitude σ_o. Thus, eq. (9.27) can be reduced to the form

$$\sigma_F(t_c) = A\,[\sigma_o(t_c)]^{\frac{2(1+\alpha)}{2+\alpha}} \quad ; \quad t_c \le t_{co} \tag{9.28}$$

where

$$A = 2A_F f(v)\sqrt{C_1}\left[\frac{1}{2}(\alpha+2)t_{co}\left(\frac{2A_F f(v)\sqrt{C_1}}{\sigma_{Fo}}\right)^{\alpha}\right]^{\frac{1}{\alpha+2}}$$

In conclusion, analyses performed for three different cases of loading in dynamic fracture initiation yield a clear picture that the process of material separation, more specifically the quasi-cleavage stress σ_F, increases at short loading times. This is shown by $\sigma_F(t_c)$ plots for three different kind of experiment, Figs (9.2), (9.7) and (9.8). Moreover, the quasi-cleavage stress σ_F is also time-history dependent which has lead to the cumulation fracture criterion in the form of eq. (9.12). The basis to explain time-dependece of σ_F needs to be sought from the point of science of materials.

ACKNOWLEDGEMENTS

Material presented in this review constitutes result of studies performed by the Author through many years. During those period the following colleagues participated in some parts of the projects : A. Andrzejewski, M.R. Bayoumi, P. Lipinski and L. Nkule. Thanks are also due to Ms. L. Lallemand for typing and correcting the text.

REFERENCES

1. Irwin, G.R. : Plastic zone near a crack and fracture toughness, in : Proc. 7th Sagamore Conf. 1960, IV-63.
2. Hult, J.A.H. and F.A. Mc Clintock : Elastic-Plastic Stress and Strain Distributions Around Sharp Notches Under Repeated Shear, in : Proc. of the Ninth Int. Congr. of Appl. Mech., University of Brussels, Brussels 1957, 8, 51.
3. Mc Clintock, F.A. and G.R. Irwin : Plasticity Aspects of Fracture Mechanics, in : Fracture Toughness Testing and Its Applications, ASTM STP 381, American Society for Testing and Materials, Philadelphia 1965, 84.
4. Begley, J.D. and J.A. Landes : The J integral as a fracture criterion, in : ASTM STP 514, American Society for Testing and Materials, Philadelphia 1972, 1.
5. Landes, J.D. and J.A. Begley : The effect of specimen geometry on J_{Ic}, in : ASTM STP 514, American Society for Testing and Materials, Philadelphia 1972, 24.
6. Bui, H.D. : Dual path independent integrals in the boundary-value problems of cracks, Engng. Fracture Mech., 6 (1974), 287.
7. Nakamura, T., C.F. Shih and L.B. Freund : Computational methods based on an energy integral in dynamic fracture, Int. J. Fract., 27 (1985), 229.
8. Atkinson, C. and J.D. Eshelby : The flow of energy into the tip of a moving crack : Int. J. Fract., 4 (1968), 3.
9. Freund, L.B. : Crack propagation in an elastic solid subjected to general loading, I. Constant rate of extension, J. Mech. Phys. Solids, 21 (1973), 47.
10. Wells, A.A. : Application of fracture mechanics at and beyond general yielding, British Welding J., 10 (1963), 563.
11. Dugdale, D.S. : Yielding of steel containing slits, J. Mech. Phys. Solids, 8 (1960), 8.
12. Klepaczko, J.R. : A general approach to rate sensitivity and constitutive modelling of FCC and BCC metals, in : Impact : Effects of Fast Transient Loadings (Eds. W.J. Amman et al.), A.A. Balkema, Rotterdam, 1988, 3.
13. Campbell, J.D. and W.G. Ferguson : The temperature and strain-rate dependence of the shear strength of mild steel, Phil. Mag., 81 (1970), 63.
14. Hahn, G.T., B.L. Averbach, W.S. Owen and M. Cohen : Initiation of cleavage microcracks in polycrystalline iron and steel, in : Fracture (Eds. B.L. Avervbach et al.), The Technology Press of MIT and J. Wiley, Cambridge/New York 1960, 91.
15. Louat, N. and H.L. Wain : Brittle fracture and the yield-point phenome non, in : Fracture (Eds. B.L. Averbach et al.), The Technology Press of MIT and J. Wiley, Cambridge/New York 1960, 161.
16. Kocks, U.F., A.S. Argon and M.F. Ashby : Thermodynamics and Kinetics of Slip, Pergamon Press, Oxford 1975.

17. Wilson, M.L., R.H. Hawley and J. Duffy : The effect of loading rate and temperature on fracture initiation in 1020 hot-rolled steel, Engng. Fract. Mech., 13 (1980), 371.
18. Krasovsky, A.J., Yu. A. Kashtalyan and V.N. Krasiko : Brittle-to-ductile transition in steels and the critical transition temperature, Int. J. Fract., 23 (1983), 297.
19. Marandet, B., G. Phellipeau and G. Sanz : Experimental Determination of Dynamic Fracture Toughness by J-Integral Method, in : Advances in Frac ture Research (Ed. D. François), Proc. ICF-5, Pergamon Press, Oxford 1981, 375.
20. Krabiell, A. and W. Dahl : Influence of Strain Rate and Temperature on the Tensile and Fracture Properties of Structural Steels, in : Advances in Fracture Research (Ed. D. François), Proc. ICF-5, Pergamon Press, Oxford 1981, 393.
21. Kussmaul, K., C. Zimmermann, T. Demler and D. Kraemer : On the Use of Opto-electronic Components for the Registration of Crack Tip Behaviour Under Dynamical Loading Conditions, in : Proc. Conf. DYMAT 85, Les éditions de physique, Les Ulis, France 1985, C5-219.
22. Klepaczko, J.R. : Displacement gauge with a photodiode, J. of Measurements, Automatics and Control, 12 (1966), 466, (in Polish).
23. Klepaczko, J.R. : An extensometric gauge, Patent of Poland No 54728, 1967.
24. Shoemaker, A.K. and S.T. Rolfe : Static and dynamic low-temperature K_{Ic} behavior of steels, J. Basic Engng, Trans. ASME Ser. D., (1969), 512.
25. Madison, R.B. and G.R. Irwin : Dynamic K_c testing of structural steel, J. of the Struct. Div., ASCE, 100, No ST 7, Proc. Paper 10653n, (1974), 1331.
26. Kalthoff, J.F., S. Winkler, W. Böhme and W. Klemm : Determination of the dynamic fracture toughness K_{Id} in impact test by means of impact response curves, in : Advances in Fracture Research (Ed. D. François), Proc. ICF-5, Pergamon Press, Oxford 1981, 368.
27. Kalthoff, J.F., S. Winkler and W. Böhme : A novel procedure for measuring the impact fracture toughness K_{Id} with precracked Charpy specimens, in : Proc. Conf. DYMAT 85, Les éditions de physique, Les Ulis, France 1985, C5-179.
28. Klepaczko, J.R. : Loading rate spectra for fracture initiation in metals, Theoretical and Applied Fracture Mechanics, 1 (1984), 181.
29. Klepaczko, J.R. : Fracture initiation under impact, Int. J. Impact Engng., 3 (1985), 191.
30. Klepaczko, J.R. : Fracture initiation of metals over a wide range of loading rates, loading rate spectra, presented on IUTAM Symposium on Macro and Micro Mechanics of High Velocity Deformation and Fracture, Aug. (1985), Tokyo, Japan.
31. Nunomura, S., T. Kashiwamura, K. Machida and S. Sakui : Fracture

toughness of ball bearing steel, in : Fracture Mechanics and Technologyf (eds. G.C. Sih and Y.A. Chow), Sijthoff and Noordhoff, Leyden, 1977, 553.
32. Kanninen, M.F. and C.H. Popelar, Advanced Fracture Mechanics, Oxford University Press, New York 1985.
33. Holtzman, M., B. Vlach and Z. Bilek : The effect of microstructure on the fracture toughness of structural steels, Int. J. Press. Ves. Piping, 9 (1981), 284.
34. Klepaczko, J.R. and G. Pluvinage : Fracture Toughness of Some Structural Steels at High Loading Rates and Different Temperatures, in : Proc. Conf. DYMAT 85, Les éditions de physique, Les Ulis, France 1985, C5-145.
35. Hopkinson, B. : A method of measuring the pressure produced in the detonation of high explosives and by the impact of bullets, Phil. Trans. Roy. Soc. (London), Ser. A, 213 (1914), 437.
36. Hopkinson, B. : Collected Scientific Papers, University Press, Cambridge 1921.
37. Davies, R.M. : A critical study of the Hopkinson pressure bar, Phil. Trans. Roy. Soc. (London), Ser. A, 240 (1948), 375.
38. Kolsky, H. : An investigation of the mechanical properties of materials at very high rates of strain, Proc. Phys. Soc. (London), Ser. B, 62 (1949), 676.
39. Klepaczko, J.R. : The modified Hopkinson bar, Theoretical and Applied Mechanics, 9 (1971), 479 ; (in Polish).
40. Hauser, F.E., J.A. Simmons and J.E. Dorn, Strain rate effects in plastic wave propagation, in : Response of Metals to High Velocity Deformation, Interscience, New York 1960.
41. Lindholm, U.S. : Some experiments with the split Hopkinson pressure bar, J. Mech. Phys. Solids, 12 (1964), 317.
42. Duffy, J., J.D. Campbell and R.H. Hawley : On the use of a torsional split Hopkinson bar to study rate effects in 1100-0 aluminum, Brown Univ. Rep. NSF-GK-4242/1, Providence 1970.
43. Campbell, J.D. and J.L. Lewis : The development and use of a torsional split Hopkinson bar for testing materials at shear strain rates up to 15000 sec^{-1}, Univ. of Oxford Rep. No 1080/69, Oxford 1969.
44. Klepaczko, J.R. : Application of the split Hopkinson pressure bar for impact testing of rocks, Engineering Transactions, 28 (1980), 381 (in Polish).
45. Kolsky, H. : Stress Waves in Solids, Dover Publications, Inc., New York 1963.
46. Achenbach, J.D. : Wave Propagation in Elastic Solids, North Holland, Amsterdam 1976.
47. Follansbee, P.S. and C. Frantz : Wave propagation in the split Hopkinson pressure bar, J. Engng. Materials and Technology, 105 (1983), 61.
48. Kolsky, H. : Experimental Studies in Stress Wave Propagation, in Proc. 5-th U.S. Natl. Congr. of Appl. Mech., ASME, New York, 1966, 21.

49. Klepaczko, J.R. and R.J. Clifton : The Propagation of Plastic Wave Fronts in a Plastically Deforming Aluminum Alloy, Techn. Report ARO-D-G182/9, Brown University, Providence, 1974.
50. Malinowski, J.Z. and J.R. Klepaczko : A unified analytic and numerical approach to specimen behaviour in the split Hopkinson pressure bar, Int. J. Mech. Sci., 28 (1986), 381.
51. Costin, L.S., E.E. Crisman, R.H. Hawley and J. Duffy : On the localisation of plastic flow in mild steel tubes under dynamic torsional loading, in : Mechanical Properties at High Strain Rates of Strain 1979 (Ed. J. Harding), The Institute of Physics, Bristol 1979, 90.
52. Hartley, K.A., J. Duffy and R.H. Hawley : Measurement of the temperature profile during shear band formation in steels deforming at high strain rates, Brown University Report No DAAG 29-85-K-0003/2, Providence 1986.
53. Klepaczko, J.R. : Application of the split Hopkinson pressure bar to fracture dynamics, in : Mechanical Properties at High Rates of Strain 1979 (Ed. J. Harding), The Institute of Physics, Bristol 1979, 201.
54. Gambin, W., P. Lipinski and G. Pluvinage : A singular element for a new experimental method of fracture toughness determination, Engng. Fract. Mech., 18 (1983), 567.
55. Klepaczko, J.R. : Discussion of a new experimental method in measuring fracture toughness initiation at high loading rates by stress waves, J. Engng. Materials and Technology, 104 (1982), 29.
56. Klepaczko, J.R. and A. Andrzejewski : Fracture toughness of some aluminum alloys at low and high loading rates, IFTR Rep. No 39/1979, Warsaco, 1979.
57. Klepaczko, J.R. : Determination of the critical value of the J-integral at high loading rates using the wedge-loaded specimen, J. of Testing and Evaluation, 13 (1985), 441.
58. Klepaczko, J.R. and J.Z. Malinowski : Dynamic frictional effects as measured from the split Hopkinson pressure bar, in : High Velocity Deformation of Solids (eds. K. Kawata and J. Shioiri), Springer-Verlag, Berlin 1979, 403.
59. Priest, A.H. : Influence of strain rate and temperature on the fracture and tensile properties of several metallic materials, in : Proc. Conf. on Dynamic Fracture Toughness, Welding Institute and ASM, London 1979, 95.
60. Dambrine, B., P. Lipinski, G. Pluvinage : Mesure de la ténacité en dynamique d'aciers pour rails, Mémoires et Etudes Scientifiques Revue de Métallurgie (1982), 329.
61. Shabbits, W.O. : Dynamic fracture toughness properties of heavy section A533 grade B Class 1 steel plate, Westinghouse Report, WCAP-7623, Dec. 1973.
62. Costin, L.S., J. Duffy and L.B. Freund : Fracture initiation in metals under

stress wave loading conditions, in : Fast Fracture and Crack Arrest, ASTM STP 627, American Society for Testing and Materials, Philadelphia, 1977, 301.

63. Eftis, J. and J.M. Krafft : A comparison of the initiation with the rapid propagation of a crack in a mild steel plate, J. of Basic Engng., 87 (1965), 257.
64. Rice, J.R. : A path independent integral and the approximate analysis of strain concentration by notches and cracks, J. Appl. Mech., 35 (1968), 379.
65. Rice, J.R. : Mathematical analysis in the mechanics of fracture, in : Fractu re (Ed. H. Liebowitz), Vol. II, Academic Press, New York, 1968, 191.
66. Knott, J.F. : Fundamentals of Fracture Mechanics, J. Wiley, New York 1979 ; (3-rd. ed.).
67. Broberg, K.B. : Crack growth criteria and non-linear fracture mechanics, J. Mech. Phys. Solids, 19 (1971), 407.
68. Rice, J.R., P.C. Paris and J.G. Merkle : Some further results of J-integral, analysis and estimates, ASTM STP 536, 1973, 231.
69. Merkle, J.G. and H.T. Corten : A J-integral analysis for compact specimen, considering axial force as well as bending effects, J. Press. Vessel Techn., Trans. ASME, J 96 (1974), 286.
70. Kanazawa, T., D. Machida, M. Onozuka and S. Kaneda : A preliminary study on the J-integral fracture criterion, Report of the University of Tokyo, IIW X-779-75, Tokyo 1975.
71. ASTM Standard : Standard Test for J_{Ic}, A Measure of Fracture Toughness, E813-81, ASTM, 1981.
72. Clarke, G.A. and J.D. Landes : Evaluation of J for the compact specimen, J. of Testing and Evaluation, Philadelphia, 7 (1979), 264.
73. Klepaczko, J.R. : Loading rate spectra for fracture initiation in metals, Theoretical and Applied Fracture Mechanics, 1 (1984), 181.
74. Klepaczko, J.R. : Fracture initiation under impact, Int. J. Impact. Engng., 3 (1985), 191.
75. Bui, H.D. : Mécanique de la Rupture Fragile, Masson, Paris 1978.
76. Ehrlacher, A. : Path independent integral for the calculation of the energy release rate in elastodynamics, in : Advances in Fracture Research, (Ed. D. François), Pergamon Press, Oxford 1981, 2187.
77. Mall, S. : A finite element analysis of transient crack problems with a path-independent integral, in : Advances in Fracture Research, (Ed. D. François), Pergamon Press, Oxford 1981, 2171.
78. Klepaczko, J.R. : An experimental procedure to determine J-integral under high loading rates, in : Mechanical Properties at High Rates of Strain, 1979 (Ed. J. Harding), The Inst. of Physics, Bristol 1979, 201.
79. Klepaczko, J.R., P. Lipinski and G. Pluvinage : A numerical analysis of yield stress and strain hardening effects on elastic-plastic zone growth around crack tip, unpublished report, Metz University, Metz 1984.

80. Bagoumi, M.R. and M.N. Bassim : Experimental correlation between ductility and J-integral in the transition region of 1045 steel, Engng. Fract. Mech. 18 (1983), 468.
81. Klepaczko, J.R. : Quasi-static and dynamic compression behavior of materials, Technical Report No 1, Dept. of Mech. Engng., The University of Manitoba, Winnipeg 1982.
82. Costin, L.S. : The effect of loading rate and temperature on the initiation of fracture in a mild, rate sensitive steel, Brown University Report No NSF ENG77-07798/2, Providence 1978.
83. Dormeval, R., J.M. Chevalier and M. Stelly : Fracture initiation of metals at high loading rates, in : Advances in Fracture Research, (Ed. D. François), Pergamon Press, Oxford 1981, 355.
84. Couque H., J. Duffy and R.J. Asaro : Effects of prior austenite and ferrite grain size on fracture properties of a plain carbon steel, Brown University Report No DAAG 29 81-K-0121/7, Providence 1984.
85. Ohlson, N.G. : Determination of crack initiation at high strain rates, in : Mechanical Properties at High Rates of Strain, (Ed. J. Harding) 1979, The Institute of Physics, Bristol 1979, 215.
86. Kishida, K., T. Yokoyama and M. Nakano : Measurement of dynamic fracture toughness based on the split Hopkinson bar technique, in : Mechanical Properties at High Rates of Strain, 1984 (Ed. J. Harding), The Institute of Physics, Bristol 1984, 221.
87. Lindholm, U.S. and L.M. Yakley : High strain-rate testing : tension and compression, Exp. Mech., 8 (1968), 1.
88. Corran, R.S.J., F.G. Benitez, J. Harding and C. Ruiz : A discussion of pro blems encountered in the dynamic fracture toughness, in : Mechanical Properties at High Rates of Strain, 1984 (Ed. J. Harding), The Institute of Physics, Bristol 1984, 253.
89. Couque, H., S.J. Hudak and U.S. Lindholm : On the use of coupled pressure bars to measure the dynamic fracture initiation and crack propagation toughness of pressure vessel steels, in : Proc. Int. Conf. on Mech. and Phys. Behaviour of Materials Under Dynamic Loading, Les éditions de physique 1988, C3-347.
90. Lipinski, P. and J.R. Klepaczko : A new experimental method in determining crack propagation transition temperature in steel, Int. J. Solids and Struct., 18 (1982), 1129.
91. Bensussan, Ph. : Fracture dynamics of 35 NCD 16 steel, in : Proc. Int. Conf. on Mech. and Phys. Behaviour of Materials Under Dynamic Loading, Les éditions de physique 1988, C3-199, (in French).
92. Dharan, C.K.H. and F.E. Hauser : Determination of stress-strain characteristics at very high strain rates, Exp. Mech., 10 (1970), 370.
93. Tobota, A., J.R. Klepaczko and J. Gronostajski : Application of the rotational hammer for dynamic tensile tests, Theoretical and Applied Mecha-

nics, 18 (1980), 258.
94. Kawata, K., S. Hashimoto, K. Kurokawa and N. Kanayama : A new testing method for the characterization of materials in high-velocity tension, in : Mechanical Properties at High Rates of Strain, 1979 (Ed. J. Harding), The Institute of Physics, Bristol 1979, 71.
95. Lipinski, P. : Propagation transition temperature and crack dynamics in structural steel, Report of Institute Fundamental Technological Research, Warsaw 1980, (in Polish).
96. Nadai, A. : Theory of Flow and Fracture of Solids, McGraw-Hill, New York 1950, 81.
97. Nicholas, T. : Instrumented impact testing using a Hopkinson bar apparatus, Technical Report AFML-TR-7554, Wright-Patterson AFB, Ohio 1975.
98. Nicholas, T. : Notched bend behavior of beryllium over a wide range of strain rates, Technical Report AFML-TR-75-177,Wright-Patterson AFB, Ohio 1975.
99. Mines, R.A.W. and C. Ruiz : The dynamic beheviour of the instrumented Charpy test, in : Proc. Int. Conf. on Mech. and Phys. Behaviour of Materials Under Dynamic Loading, Les éditions de physique 1985, C5-187.
100. Tanaka, K. and T. Kagatsume : Impact bending test on steel at low temperatures, Bull. JSME, 23 (1980), 1736.
101. Yokoyama, T. and K. Kishida : A novel impact three-point bend test method for determining dynamic fracture initiation toughness, Proc. Int. Conf. on Fracture and Fracture Mechanics, Shangai 1987, 553.
102. Yokoyama, T. and K. Kishida : Measurement of dynamic fracture initiation toughness by a novel impact three-point bend test technique using Hopkinson pressure bars, in : Impact Loading and Dynamic Behaviour of Materials, DGM Informationsgesellschaft Verlag, Oberursel 1988, 273.
103. Hurd, N.J. and P.E. Irwing : A comparison of Mode III and Mode I toughness in quenched and tempered steels, in : Fracture and Fatigue, Proc. 3rd Colloquium on Fracture (Ed. J.C. Radon), Pergamon Press, Oxford 1980, 239.
104. Tsangarakis, S. : Fracture behavior of 4340 steel under Mode III loading, Engng. Fract. Mech., 16 (1982), 569.
105. Tsangarakis, S. : The dependence of Mode III fracture initiation toughness on strength and microstructure, Engng. Fract. Mech., 19 (1984), 903.
106. Gupte, K.A. and S. Banerjee : Fracture of round bars loaded in Mode III and a procedure for K_{IIIc} determination, Engng. Fract. Mech., 19 (1984), 919.
107. Tada, H. : The Stress Analysis of Cracks Handbook, Del Res. Corp., Hellertown 1973.
108. Hult, J.A.H. and F.A. McClintock : Elastic-plastic stress and strain distributions around sharp notches under repeated shear, Proc. 9thf IUTAM Int. Congress, University of Brussels, 8 (1957), 51.

109. Yates, J.R. : Crack tip plastic zone sizes in cylindrical bars subjected to torsion, Fatigue Fract. Engng Mater. Struct., 10 (1987), 471.
110. Nkule, L. : Tests on Fracture Toughness in Mode III at High Rate for XC48 Steel, Ph. D. Thesis, ENSM Nantes 1985 ; (in French).
111. Klepaczko, J.R. and L. Nkule : A new experimental test method for dynamic fracture initiation in Mode III, in preparation.
112. Charpy, G. : On testing metals by the bending of notched bars (translation of the original paper from Mémoires de la Société des Ingénieurs Civils de France, 1904, p. 468), Int. J. Fract., 25 (1984), 287.
113. Impact Testing of Materials, ASTM STP 466, American Society for Testing and Materials, Philadelphia 1969.
114. Instrumented Impact Testing, ASTM STP 563, American Society for Testing and Materials, Philadelphia 1974.
115. Ireland, D.R. : Procedures and problems associated with reliable control of the instrumented impact test, ASTM STP 563, Philadelphia 1974, 3.
116. Barsom, J.M. and S.T. Rolfe : Fracture Fatigue Control in Structures, Prentice-Hall, Englewood Cliffs 1987.
117. Barsom, J.M. and S.T. Rolfe : Correlations between K_{Ic} and Charpy V-notch test results in the transition temperature range, ASTM STP 466, American Society for Testing and Materials, Philadelphia 1969, 281.
118. Norris, D.M. : Computer simulation of the Charpy V-notch toughness test, Engng. Fract. Mech., 11 (1979), 261.
119. Kalthoff, J.F. : Determination of the dynamic fracture toughness K_{Id} in impact tests by means of response curve, in : Advances in Fracture Research, (Ed. D. François), Pergamon Press, Oxford 1981, 363.
120. Rintamaa, R. and C. Zimmermann : Advanced instrumented impact testing facility for characterization of dynamic fracture behavior, Nucl. Engng. and Design, 96 (1986), 159.
121. Rintamaa, R., K. Rahka, K. Wallin, K. Ikonen, H. Talja, H. Kotilainen and E. Sirkkola : Instrumented impact testing machine with reduced specimen oscillation effects, Research Report of Technical Research Centre of Finland, Espoo 1984.
122. Williams, J.G. : The analysis of dynamic fracture using lumped mass-spring models, Int. J. Fract., 33 (1987), 47.
123. Williams, J.G. and G.C. Adams : The analysis of instrumented impact tests using a mass-spring model, Int. J. Fract., 33 (1987), 209.
124. Williams J.G. and M.N.M. Badi : The effect of damping on the spring-mass dynamic fracture model, Int. J. Fract., 39 (1989), 147.
125. Kishimoto, K., S. Akoi and M. Sakata : Simple formula for dynamic stress intensity factor of precracked Charpy specimen, Engng. Fract. Mech., 13 (1980), 501.
126. Kishimoto, K., S. Akoi and M. Sakata : Dynamic fracture mechanics parameter estimation for three-point bend specimen in large scale

yielding, in : Impact Loading and Dynamic Behaviour of Materials, DGM Informationsgesellschaft Verlag, Oberursel 1988, 129.

127. Nash, G. : An analysis of the force and bending moments generated during the notched beam impact test, Int. J. Fract. Mech., 5 (1969), 269.

128. Macke, T. : Development and analysis of a method for material characterization under impact ; Application for testing of fracture toughness of composites with ceramic or metallic matrix, Ph. D. Thesis, University of Bordeaux, Talence 1988 ; (in French).

129. Macke, T., J.J. Balette and J.M. Quenisset : A method for evaluation of dynamic toughness and impact loading resistance, in : Impact Loading and Dynamic Behaviour of Materials, DGM Informationsgesellschaft Verlag, Oberursel 1988, 289.

130. Tvergard, V. and A. Needleman : Effect of material rate sensitivity on failure modes in the Charpy V-notch test, J. Mech. Phys. Solids, 34 (1986), 213.

131. Theocaris, P.S. and G.A. Papadopoulos : Interrelation between static and dynamic stress intensity factors and their evaluation by caustics, J. of Strain Analysis, 19 (1984), 127.

132. Kalthoff, J.F. : The shadow optical method of caustics, in : Handbook on Experimental Mechanics, Ch. 9, (Ed. A.S. Kobayashi), Prentice Hall, Englewood Cliffs, 1985.

133. Rosakis, A.J. and A.T. Zehner : Caustics by reflection and their application to elastic-plastic and dynamic fracture mechanics, in : Proc. SPIE Conference on Photomechanics and Speckle Metrology, San Diego, 1987.

134. Zehner, A.T. and A.J. Rosakis : Dynamic fracture initiation and propagation in 4340 steel under impact loading, Calif. Inst. of Technology, GAL, SM Report 86-6, 1986.

135. Klepaczko, J.R. and A. Solecki : Effect of tempering on quasi-static and impact fracture toughness and mechanical properties for 5110 H steel, Met. Trans. A, 15A (1984), 901.

136. Zhurkov, S.N. : Kinetic concept of the strength of solids, Int. J. Fracture, 1 (1965), 311.

137. Dahl, W., W. Hesse, A. Krabiell and H.J. Rosezin : Influence of yielding behaviour and stress-strain law on the failure analysis, Nucl. Engng. Design, 76 (1983), 309.

138. McGregor, C.W. and J.C. Fisher : A velocity-modified temperature for the plastic flow of metals, J. Appl. Mech., 68 (1946), A11.

139. Hahn, G.T., R.G. Hoagland and A.R. Rosenfield : The variation of K_{Ic} with temperature and loading rate, Met. Trans., 2 (1971), 537.

140. Ritchie, R.O., W.L. Server and R.A. Wullaert : Critical fracture stress and fracture strain models for the prediction of lower and upper shelf toughness in nuclear pressure vessel steels, Met. Trans. A, 10A (1979), 1557.

141. Ritchie, R.O., J.F. Knott and J.R. Rice : On the relationship between critical

tensile stress and fracture toughness in mild steel, J. Mech. Phys. Solids, 21 (1973), 395.
142. Rice, J.R. and G.F. Rosengreen : Plane strain deformation near a crack tip in a power-law hardening material, J. Mech. Phys. Solids, 16 (1968), 1.
143. Hutchinson, J.W. : Singular behavior at the end of a tensile crack in a hardening material, J. Mech. Phys. Solids, 16 (1968), 13.
144. McClintock, F.A. : Plasticity aspects of fracture, in : Fracture, An Advanced Treatise (Ed. H. Liebowitz), Vol. III, Academic Press, New York 1971, 48.
145. Kalthoff, J.F. : Fracture behavior under high rate of loading, Rep. Z5/85 Fraunhofer-Institut für Werkstoffmechanik, Freiburg 1985.
146. Rosenfield, A.R. and G.T. Hahn : Numerical description of the ambient low-temperature, and high-strain rate flow and fracture behavior of plain carbon steel, Trans. ASM, 59 (1966), 962.
147. Hahn, G.T. : The influence of microstructure on brittle fracture toughness, Met. Trans. A, 15A (1984), 947.
148. Krafft, J.M. and G. Irwin : Crack velocity considerations, in : Fracture Toughness Testing and Its Applications, ASTM STP 381, American Society for Testing and Materials, Philadelphia 1965, 84.
149. Krafft, J.M. : Correlation of plane strain crack toughness with strain hardening characteristics of a low, a medium and a high strength steel, Appl. Mat. Res., 3 (1964), 88.
150. Campbell, J.D. : Dynamic Plasticity, CISM Udine, 197.
151. Broek, D. : Elementary Engineering Fracture Mechanics, Martinus Nijhoff Publ., Dordrecht 1987.
152. Campbell, J.D. : The dynamic yielding of mild steel, Acta Metall., 1 (1953), 706.
153. Shockey, D.A., J.F. Kalthoff and D.C. Erlich : Evaluation of dynamic crack instability criteria, Int. Journ. of Fracture, 22 (1983), 217.
154. Shockey, D.A., J.F. Kalthoff, H. Homma and D.C. Erlich : Short pulse fracture mechanics, in : Dynamic Fracture (Eds. W.G. Knauss, K. Ravi-Chandar and A.J. Rosakis) Caltech, Pasadena 1983, 57.
155. Homma, H., D.A. Shockey and Murayama : Response of cracks in structural materials to short pulse loads, J. Mech. Phys. Solids, 31 (1983), 261.
156. Buchar, J. : The effect of strain rate sensitivity on crack initiation under dynamic loading, in : IUTAM Symposium (Eds. K. Kawata and J. Shioiri) Springer Verlag, Berlin 1985, 428.
157. Kalthoff, J.F., D.A. Shockey and H. Homma : Short pulse fracture mechanics, in : Mechanical Properties at High Rates of Strain, 1984 (Ed. J. Harding), The Institute of Physics, Bristol 1984, 205.
158. Freund, L.B. : Crack propagation in an elastic solid subjected to general loading, III stress wave loading, J. Mech. Phys. Solids, 21 (1973), 47.
159. Tuler, F.R. and B.M. Butcher : A criterion for the time dependence of dynamic fracture, Int. J. Fract. Mech., 4 (1968), 431.

MICROSTATISTICAL FRACTURE MECHANICS IN DYNAMIC FRACTURE

D.R. Curran

SRI International, Menlo Park, CA, USA

ABSTRACT

This section discusses attempts to construct a microstatistical fracture mechanics; that is, a methodology that relates the kinetics of material failure on the microstructural level to continuum mechanics. The approach is to introduce microstructural descriptions of damage into the continuum constitutive relations as internal state variables. The microstructural damage descriptions are based on dynamic and quasi-static experiments with carefully controlled load amplitudes and durations. The resulting constitutive relations describe the nucleation, growth, and coalescence of the microscopic voids and cracks, and therefore in principle describe both quasi-static and dynamic fracture on the continuum scale. We discuss several such kinetics models, show examples of several engineering applications, and discuss the link between microstatistical fracture mechanics and continuum fracture mechanics.

I. MICROSCOPIC FAILURE MODES

I.1 Introduction

"Microstatistical Fracture Mechanics" (MSFM) or the "Local Approach" to fracture mechanics describe material failure on the microscopic or "local" level. This series of four presentations on MSFM is largely excerpted from a recent review paper [1]: "Dynamic Failure of Solids", by D. R. Curran, L. Seaman, and D. A. Shockey, Physics Reports, Nos. 5 & 6, 253-388 (1987).

Mathematical attempts to describe fracture have taken two different approaches. The first approach is that pioneered by Griffith and Irwin in which a macroscopic crack is treated as a stress-free boundary in a mathematical boundary value problem. The threshold condition for crack instability is assumed to be a critical energy density, stress intensity factor, or other measure of the stress and strain fields near the crack tip. The growth kinetics of this macroscopic crack are also handled as part of a dynamic boundary value problem.

This approach, termed "fracture mechanics", has been extremely successful in cases where the behavior of a single, large crack in fairly brittle material is of prime interest. "Linear elastic fracture mechanics" (LEFM) is the term used when the plastic zone at the macro-crack tip is very small compared with specimen dimensions and crack size, whereas the term "elastic plastic fracture mechanics" (EPFM) is used when the plastic zone is larger but still contained at the crack tip. However, there are many cases where neither LEFM or EPFM applies. A classic example is the ductile failure of a smooth bar in tension. In this case there is no macroscopic crack, and failure occurs by the nucleation, growth, and coalescence of millions of microscopic voids in the necking region of the rod.

Another example is the case of large-scale plastic yielding in a structure originally containing a single large crack. In this case, microvoid activity may also be observed in a large region around the macrocrack tip.

For such cases, treatment of each microcrack individually becomes forbidding, and a second mathematical approach has been developed: namely to treat some key measures of average microscopic void behavior as internal state variables in the constitutive relations for the material. This is, in addition to strain, entropy, and temperature, one also specifies the microvoid concentration as well as orientation and size distribution functions in the description of the current state of a material particle. In this approach, the fracture kinetics involves the microvoid behavior and introduces specific rate dependence into the constitutive relations. A scale size, the void spacing, is also introduced in the constitutive relations. In this paper we will refer to this second approach as "microstatistical fracture mechanics" (MSFM).

I.2 Nucleation, growth, and coalescence of microdamage

Nucleation sites

We define nucleation of microscopic fracture as the appearance in a material element of microscopic voids or cracks with sizes comparable to that of the "graininess" that defines the continuum limit of the material. Thus, in polycrystalline metals with grains a few micrometers in size, the appearance of voids with sizes of about a micrometer in the grain boundaries or within the grains would constitute nucleation.

According to the above viewpoint, the nucleation of fracture occurs at microscopic heterogeneities in the material. Table 1.1 lists some of the most commonly observed types of heterogeneity along with corresponding nucleation mechanisms and governing continuum load parameters. In the following paragraphs we expand on the content of table 1.1

Table 1.1

Experimentally observed microscopic fracture nucleation processes

Nucleation Site	Nucleation Mechanism	Governing Continuum Load Parameters
Preexisting flaws (voids or cracks).	Growth of flaw.	Tensile stress and plastic strain
Inclusions and second phase particles	Cracking of inclusion, debonding at interface, or fracture of matrix material near the inclusion.	Tensile stress and plastic strain
Grain boundaries	Vacancy clustering, grain boundary sliding, or mechanical separation.	Tensile stress and plastic strain
Subgrain structure	Dislocation pileup.	Shear strain

Preexisting flaws

Many materials contain a population of naturally occurring flaws that exist in the material before the loading history of interest. For example, many geologic materials consist of loosely cemented grains, so that weak or debonded grain boundaries can be considered inherent flaws. On a larger scale, faults ranging in size from meters to kilometers may be considered inherent flaws if the "specimen" of interest is considerably larger than these flaws (which then defines the continuum limit for the particular application), and if the applied stresses vary slowly over these distances. Preexisting flaws and voids are also found in plastics, composites and metals.

Inclusions and second phase particles

Nucleation of fracture at inclusions and second phase particles is commonly observed at all loading rates. It occurs by fracture of the inclusions, debonding at the interface, or fracture of the matrix adjacent to a stronger inclusion.

In general, cracking and debonding of inclusions are the governing nucleation mechanisms in may structural materials at all loading rates.

A less common mechanism for nucleation at inclusions is fracture of the matrix material near the inclusion. Such a mechanism can operate when the fracture strength of the matrix is less than that of both the inclusion and the interface. Crack nucleation by this process has been seen in brittle beryllium, which has weak basal cleavage planes and hard, strongly bonded BeO inclusions.

Grain boundaries

Grain boundaries are favorable sites for fracture nucleation by several mechanisms. Under high temperature creep conditions vacancy clustering is observed to form cavities on grain boundaries, a process that is enhanced by grain boundary sliding. Cavities can also form on grain boundaries under impact loading.

Such cavities tend to be roughly spherical after some growth has occurred in high tempertaure creep experiments, but in lower temperature, dynamic experiments it is possible to observe brittle cracks in grain boundaries.

Subgrain structure

In materials that are relatively free of inclusions, it is often difficult to identify the nucleation site for observed microscopic voids or cracks. For example, in impact-loaded specimens of 1100 series aluminum, OFHC copper, and high purity aluminum and copper, we were unable to consistently correlate observed microscopic voids with inclusions or grain boundary nucleation sites. In such cases, the nucleation must occur at submicroscopic heterogeneities. At such heterogeneities, dislocation motion can be impeded and resultant dislocation pileups can cause void nucleation. A feature of this mechanism is that it requires plastic flow. Thus, subgrain structure nucleation is almost always associated with plastic flow.

Growth of microvoids, cracks, and shear bands.

Following nucleation, the microscopic damage features--voids, cracks, and shear bands--grow or extend under continued loading.

Growth occurs by atomic diffusion processes, ductile processes (plastic or viscous flow), brittle crack extension, or by shear slipping and extension. As these processes are considered in more detail, several types of each process can be identified, and often the damage shows a mixture of processes.

Growth of microscopic voids or cracks appears as a continuation of the nucleation process at pre-existing flaws, inclusions, second-phase particles, grain boundary junctions, or at submicroscopic sites. These flaws will subsequently grow if the applied load exceeds the growth threshold. In many cases, the growth threshold will be identical with the nucleation threshold, for example, when vacancy diffusion has formed a cluster large enough so that the applied tractions can overcome the surface tension. But in other cases, such as debonding of inclusions, subsequent growth may occur by a different mechanism, and the growth threshold can be either higher or lower than the nucleation threshold. In polycrystalline solids the microscopic void or crack growth can be conveniently classified into two modes: ductile void growth and brittle crack propagation. Either mode can occur intergranularly (in the grain boundaries) or transgranularly (inside the grains). At low temperatures or high loading rates the brittle mode is encouraged because plastic flow is inhibited; at high temperature and low loading rates ductile void growth is encouraged because the dislocation processes responsible for plastic flow are encouraged. As an example, Armco iron fractures through growth of ductile voids under quasi-static conditions at room temperture. Under impact loading the damage appears as microcracks. But if the impact occurs on a specimen at 300°C, voids appear and grow just as under static, room temperature conditions.

At some temperatures and loading rates (the specific temperatures and loading rates depend on the material), "mixed mode" growth is observed, wherein a fracture surface consists of a mixture of coalesced ductile voids and brittle cracks. If the brittle cracks occur transgranularly, cleavage cracks are observed where weak crystallographic planes have separated. If the brittle cracks occur intergranularly the grain boundary separates as a weak surface. In studies of impact fractures in a Charpy machine, as the temperature increases, the cleavage fracture surface decreases. At the "lower shelf" of charpy energy, the fracture is all cleavage, whereas at the "upper shelf" the fracture is all ductile. Detailed observations of room temperature impact fractures in Armco iron revealed that voids were nucleated at inclusions. Then cleavage cracks grew out of these voids. Hence, we have reached the following conclusions:

- Fracture under high rate loading conditions may involve several different ductile and brittle processes, depending on the strain rate, temperature, and loading conditions (state of stress and strain).
- These dynamic fracture processes are not basically different from those that can occur under quasi-static conditions.

<u>Coalescence of microvoids, cracks, and shear bands to form fragments</u>

Microscopic observation reveals a large array of coalescence processes. At the risk of oversimplification, we classify them in two groups: direct impingement and localization between voids.

Direct impingement

In some cases the coalescence process is simply direct impingement of the voids or cracks. In pure, ductile metals nearly spherical voids may grow to near impingement before localized ligament stretching completes the coalescence process. This process also may occur under dynamic loads.

Impingement is also the coalescence mode for cleavage cracks and adiabatic shear bands.

Finally, when the damage is confined to grain boundaries, such as in the high temperature creep specimen of silver, impingement is usually the coalescence mode.

Localization between voids

In some cases, notably structural steels, coalescence is observed to occur via strain localization in sheets between the voids. For example, in A533B pressure vessel steel tested at 1000°C, nucleation occurs at large MnS and Al_2O_3 inclusions, and the voids subsequently grow to a relative void volume of about 0.01. With increased loading, localization occurs in sheets between the voids, and these sheets contain smaller voids formed at small iron carbide inclusions.

In summary, planar flaws (such as cleavage or intergranular cracks) as well as shear bands will coalesce via direct impingment, whereas ellipsoidal voids may coalesce either by impingment or localization. Therefore, successful constitutive modeling of the coalescence process must be able to account for both coalescence modes.

I.3 Mathematical description of microdamage evolution.

Microdamage concentration (no./cm^3) as a function of microflaw size and orientation.

We have chosen a detailed description of the microscopic damage that is based on actual counts and measurements of microcrack numbers, sizes, and orientations in specimens that have been exposed to known stress and strain histories.

We first characterize each microcrack or void with a size R and orientation n. For basically planar microcracks, R is the average radius and n is a unit vector normal to the crack plane. For ellipsoidal voids, R can be chosen to be the average of the major and minor axes, and n can be chosen to be a unit vector normal to the plane of maximum cross-sectional area.

The description of damage is then chosen to be the function $\rho_f(X,t,R,n)$, where ρ_f is the concentration of active flaws, t is the time, and X is the Lagrangian coordinate that specifies the material point, the center of the material element of interest. (The size of the material element will be labeled δX)

For given values of X and t, the total number of active flaws per unit (in the reference system defined by X) is the integral of ρ_f over all values of R and n:

$$N_t = \int_R \int_n \rho_f dn\, dR \tag{1.1}$$

The distribution function ρ_f represents the transition from microscopic, discrete processes to continuum mechanics.

In the above equation N_t refers to the number of voids or cracks per unit volume whereas in the experiments discussed above we commonly observe only the number of voids or cracks per unit area of a metallographic section. In appendix A we describe analytical procedures used to convert surface observations to volume concentrations.

Another useful function is the total number of microcracks or voids $N_g(R)$ with sizes greater than or equal to R, obtained by integrating ρ_f over all orientations n and from R to infinity:

$$N_g(X,t,R) = \int_R^\infty \int_{\text{all } n} \rho_f(X, t, R, n) dn\, dR. \tag{1.2}$$

Experimental observations, as well as theoretical predictions show that N_g often has the form

$$N_g(X,t,R) = N_t(X,t) \exp\left[-R/R_o(X,t)\right] \tag{1.3}$$

where N_t is the total number of microcracks or voids per unit volume (in the reference configuration) and R_o is a characteristic size for this exponential size distribution. Equation (1.3) requires that

$$\rho_f(X,t,R) = \left[N_t(X,t)/R_o(X,t)\right] \exp\left[- R/R_o(X,t)\right] \tag{1.4}$$

where

$$\rho_f(X,t,R) \equiv \int_{\text{all } n} \rho_f(X,t,R,n)\, dn. \tag{1.5}$$

Evolution equations for the concentration function

The nucleation and growth of damage must be based on the evolution of ρ_f in time in response to applied continuum (remote) stress, strain, strain rate, temperature, etc. That is,

$$\dot{\rho}_f = \dot{\rho}_f(\sigma, \epsilon, \dot{\epsilon}, T). \tag{1.6}$$

In a given material element δX, $\dot{\rho}_f$ will have one contribution from the nucleation of new microcracks or voids and another contribution from the growth of previously nucleated ones.

To see how one may attempt to separate the contribution of nucleation from that of growth, consider the example given by eqs. (1.3) and (1.4). For size distributions of this exponential type, nucleation will increase the value of N_t. If the nucleation process maintains the size distribution shape, i.e., equal numbers of new cracks are added to each crack size, then R_o would be unchanged by the nucleation process. That is, from eq. (1.4)

$$\dot{\rho}_f \text{ (nuc.)} = (\dot{N}_t/R_o) \exp(-R/R_o)$$

or

$$\dot{\rho}_f \text{ (nuc.)}/\rho_f = \dot{N}_t/N_t. \tag{1.7}$$

On the other hand, microcrack growth does not change N_t. If the growth also preserves the form given by eq. (1.3) or (1.4) then

$$\dot{\rho}_f\text{(growth)}/\rho_f = (R/R_o - 1)\, \dot{R}_o/R_o. \tag{1.8}$$

That is, the rate of change in ρ_f for a given X and R due to growth would depend on $\dot{R}_o$, and would be negative for $R<R_o$ and positive for $R>R_o$.

It is usually more convenient to deal with $N_g(X,t,R)$ than with $\rho_f(X,t,R)$ because the experimental data are often displayed in this form and, as we shall see, it is easier to deduce nucleation and growth information from the N_g function.

Similiar to the approach that led to eqs. (1.7) and (1.8) we write

$$\dot{N}_g \equiv (\partial N_g/\partial t)_{X,R} = \dot{N}_g^N + \dot{N}_g^G \tag{1.9}$$

where the superscripts N and G refer to nucleation and growth, respectively. If we assume a distribution given by eq. (1.3), we obtain

$$\dot{N}_g^N/N_g = \dot{N}_t/N_t \tag{1.10}$$

and

$$\dot{N}_g^G/N_g = (R/R_o)\, (\dot{R}_o/R_o). \tag{1.11}$$

Solving eqs. (1.9), (1.10) and (1.11) for $\dot{R}_o/R_o$ yields

$$\dot{R}_o/R_o = (R_o/R) \left[(\dot{N}_g/N_g) - (\dot{N}_t/N_t)\right]. \tag{1.12}$$

Thus, to obtain the nucleation and growth rates, consider the experimental data shown schematically in fig. 1.1. The intercept of the $N_g(R)$ curve with the R=0 axis is N_t. Thus the rate this intercept moves up in time gives us $\dot{N}_t$. Then, for any other value of R (say, $R = R_o$) we measure the rate of increase of $N_g(R)$ to get $\dot{N}_g(R)$. Substitution into eq. (1.12) give us $\dot{R}_o/R_o$, the growth rate.

Furthermore, it is easy to see that preserving the form of eq. (1.3) implies that cracks of all sizes grow according to a growth equation

$$\dot{R}/R = A.$$

That is, if no nucleation occurs, a particular crack of size R is specified by the value of N_g (the 10000th largest crack, for example). Thus, according to eq. (1.3), during the growth of a crack of size R the

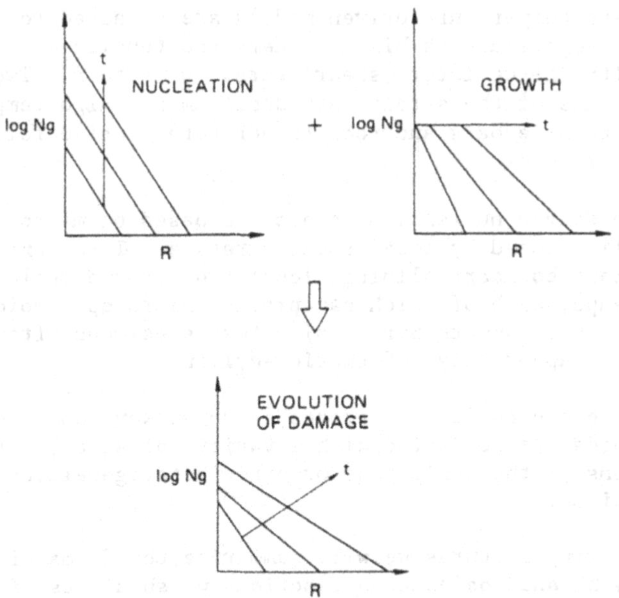

Figure 1.1. Evolution of microscopic damage.

corresponding value of N_g is constant, so

$$R/R_o = \text{constant} \qquad (1.13)$$

$$\dot{R}/R = \dot{R}_o/R_o.$$

Thus, every crack grows with the relative rate given by $\dot{R}_o/R_o$.

Let us return now to the nucleation process. The nucleation rate $\dot{N}_t$ may be written as

$$\dot{N}_t = N_o f/\tau, \qquad (1.4)$$

where N_o is the number of currently available potential nucleation sites, f is the fraction of those sites that are undergoing nucleation (i.e., f is the nucleation probability), and τ is the nucleation time (that is, τ^{-1} is the frequency of nucleation attempts). The above form for N_t will be useful later when we discuss the nucleation process in terms of Arrhenius rate theory.

Specific microstructural models of void nucleation at grain boundaries, inclusions, or other microscopic heterogeneities are basically of two types, (1) tensile stress/temperature-driven and (2) deformation (plastic strain)-driven.

Stress/temperature-driven nucleation models are based on the effect of stress on the free energy functions for thermally activated atomic bond breaking or, from a related but more practical viewpoint, for the diffusion of vacancies into a cluster to form and grow a microscopic void. Such stress/temperature-driven models are expected to apply to cases where the temperature is high or where the tensile mean stress is high compared with the deviator (shear) stress or strain. Two such cases at the opposite ends of the strain rate spectrum are high temperature creep-ruture of tensile bars and room temperature tensile failure in plate-impact experiments.

Deformation-driven nucleation models are based on microscopic, heterogeneous slip caused by local shear stresses. Three examples of such processes are grain boundary sliding, local flow around inclusions, and dislocation pileups, each of which can produce microsopic voids. Ductile failure in the neck of smooth bar tensile test specimens often exhibits nucleation that is apparently deformation-driven.

Finally, once the nucleation proces is under way, one usually observes that voids are nucleated with a variety of sizes, reflecting the size distributions of the inclusions or other heterogeneities that form the nucleation sites.

In the following lectures we will summarize the forms of the nucleation, growth, and coalescence functions we should use for computational modeling.

II. EXPERIMENTS TO MEASURE MICRODAMAGE NUCLEATION, GROWTH AND COALESCENCE

II.1 Plate impact experiments

The most direct way to experimentally measure microfracture nucleation rates is by metallographic examination of sectioned samples that have been exposed to known stress and strain histories.

Figure 2.1 shows a method commonly used to produce tensile pulses in laboratory specimens. A thin plate is accelerated in a light gas gun and strikes a flat plate specimen of the same material. The target plate is about twice the thickness of the impactor. Figure 2.2 schematically shows the stress wave reverberations plotted in distance-time space. In the figure, F refers to the impactor or "flyer," and S refers to the specimen. Elastic and slower plastic compressive waves run forward in the target and backward in the impactor (lines 1 and 2 in fig. 2.2). They reflect as unloading or rarefaction waves at the impactor and specimen free surfaces (lines 3 and 4 in fig. 2.2). Each rarefaction fan reduces the compressive stress to zero behind it. The rarefaction fans intersect in the specimen (point A in fig. 2.2), and the region labeled T in fig. 2.2 finds itself in tension. Figure 2.2 shows that the duration of the tensile stresses will vary through the thickness of the specimen, being a maximum in the center region and decreasing toward the edges.

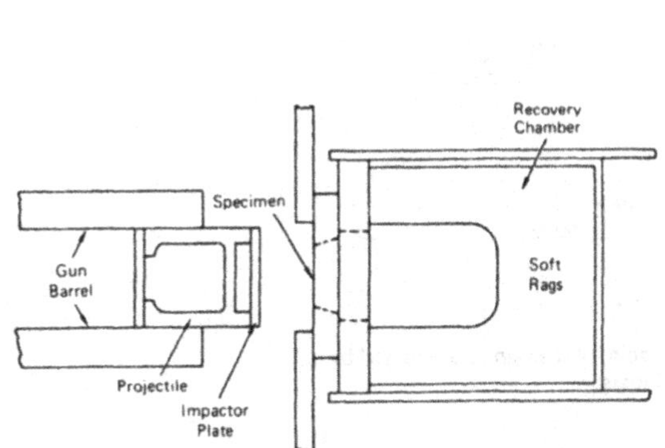

Figure 2.1 Plate impact experimentsfor studies of dynamic fracture.

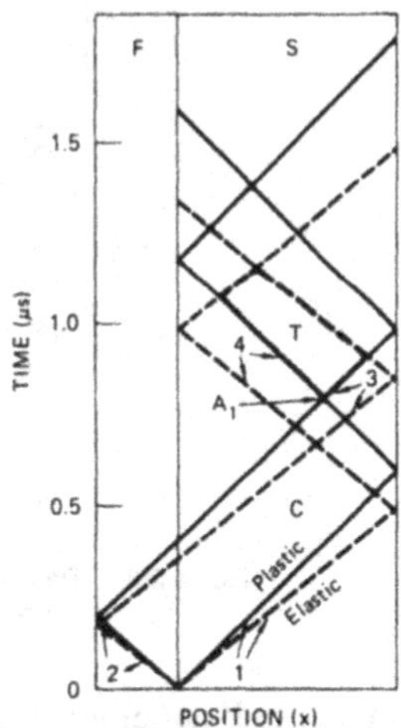

Figure 2.2. Distance-time plot showing wave paths and compressive (C) and tensile (T) regions in a one-dimensional impact.

By varying relative specimen and impactor thicknesses and the impactor velocity, one can vary the amplitudes and duration of the tensile pulse in laboratory-sized specimens over about one order of magnitude. Typical values in metals would be tensile stresses of a few GPa for a few tenths of a microsecond.

The plates must be wide enough so that rarefaction waves from the plate edges do not arrive at the central region until after the through-the-thickness stress waves have caused the microscopic damage. In practice, this requires aspect ratios of 6 to 10.

Using the above impact experiments and the methods of analysis described in the previous lecture, we were able to deduce for a variety of metals a nucleation rate of the form

$$N_t = N_0 \exp\left[(\sigma - \sigma_{n0})/\sigma_1\right] , \tag{2.1}$$

where N_t is the total number of nucleated voids per unit volume, σ is the mean tensile stress, and N_0, σ_{n0}, and σ_1 are material properties.

At lower loading rates, many experiments can be used to produce tensile or shear loads of variable amplitude and duration. Some of the most common experiments are shown (albeit somewhat schematically) in stress-strain rate space in fig. 2.3.

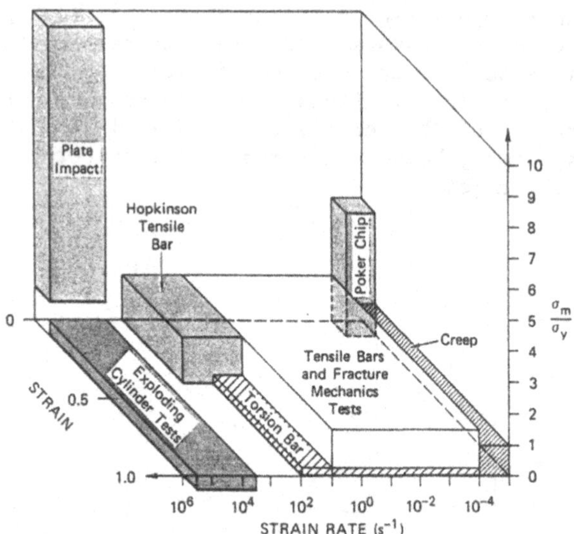

Figure 2.3. Regions of stress, strain, and strain rate attained by various mechanical tests.

II.2 Round and notched bar tensile experiments

Significant information has been obtained from metallographic examination of sectioned specimens from tensile bar experiments. Many workers have correlated observed inclusion cracking or debonding in sectioned tensile bars with estimated levels of stress and strain.

Figure 2.4 shows data of the type discussed in the previous lecture for reactor grade steel.

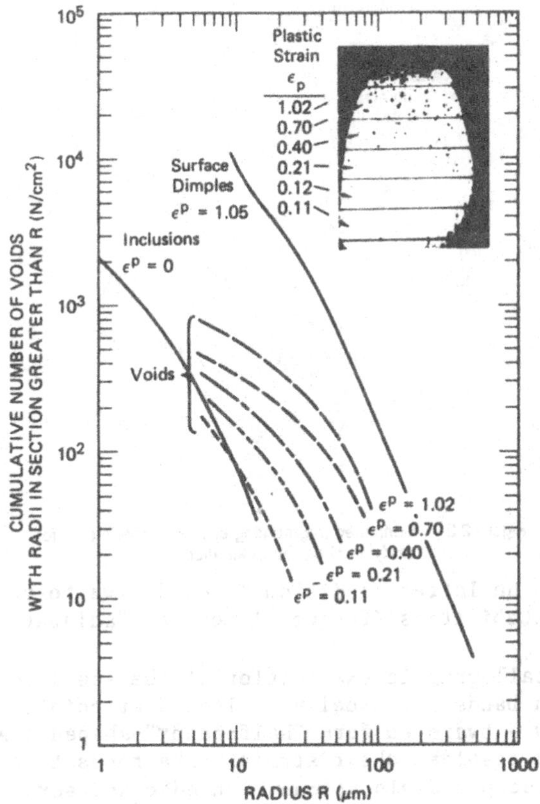

Figure 2.4. Size distributions of inclusions, voids, and surface dimples in a smooth tensile bar of A533B steel, heat CDB, at various plastic strains.

II.3 Controlled fragmenting cylinders

Another experiment from the group of experiments shown in fig. 2.3 that has proved useful in our laboratory for measuring microflaw kinetics involves dynamically expanding a thick-walled cylinder of specimen material by internal detonation of an explosive. The expansion is stopped at a desired strain level with a confining cylinder arrangement, as shown in fig. 2.5. The Plexiglas layer between the expanding specimen cylinder and the massive containment cylinder softens the impact and reduces the effects of reverberating stress waves in the radial direction.

Computer simulations of this experiment are performed to determine the histories of the stress and strain components as a function of location in the specimen cylinder.

The stress and strain histories vary along the cylinder axis because there is less confinement of the explosive gases at the ends. The outer surface of the specimen cylinder, after a short reverberation time, experiences mostly hoop tension and large plastic strain, whereas the inner portion of the cylinder experiences compression and large

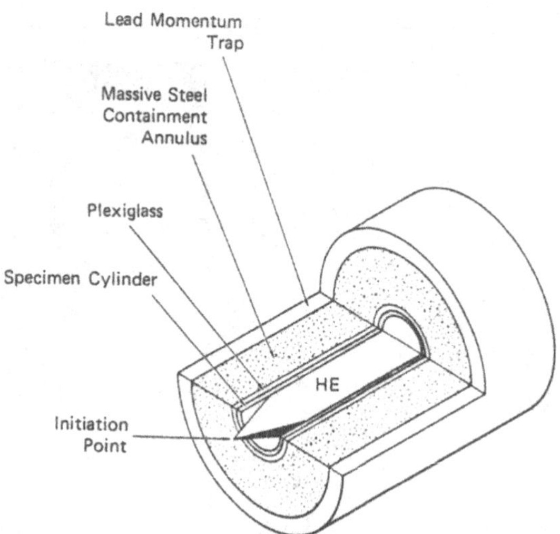

Figure 2.5. Contained fragmenting cylinder experiments for studying shear band kinetics.

plastic strain. The latter condition is conducive to the formation of plastic shear instabilities (frequently called "adiabatic shear bands").

Posttest metallographic examination of the specimen cylinders reveals that such bands are usually nucleated at points on the inner surface* and grow outward to form "half-penny"-shaped shear bands in planes of maximum resolved shear strain. The bands thus have the geometry of macroscopic dislocations with edge and screw components.

These bands can then be counted and measured to obtain nucleation and growth kinetics information as discussed earlier for microcracks and voids. This is shown schematically in fig. 2.6.

II.4 Examples of damage evolution functions for several materials

i. Nucleation

Nucleation can occur by diffusion processes and by mechanical debonding of inclusions. A combined nucleation rate function based on both experimental data and theory is [1]:

$$\dot{N}_i = \dot{N}_0 \exp \frac{\sigma_m \Omega - 4\gamma^3 F_v/\sigma_m^2}{kT} + \frac{N_0 \varepsilon_0 \dot{\varepsilon}_i^p}{(\varepsilon_i^p)^2} \exp(-\varepsilon_0/\varepsilon_i^p) + \frac{N_0 \sigma_0 \dot{\sigma}_m}{\sigma_m^2} \exp(-\sigma_0/\sigma_m) \tag{2.2}$$

I II III

*Occasional nucleation at internal sites is observed.

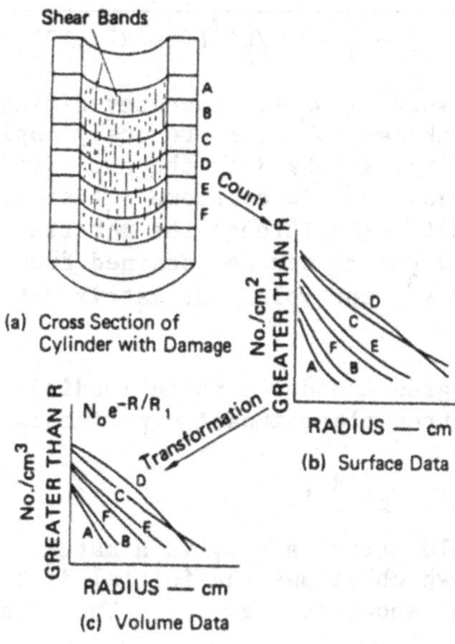

Figure 2.6. Steps in obtaining cumulative shear band distributions from contained fragmenting cylinder data.

In this equation, term I describes the thermal rate theory diffusion-driven nucleation, term II describes mechanical debonding of inclusions due to applied continuum plastic strain, and term III describes debonding of inclusions caused by applied continuum mean tensile stress. The parameters in the above equation are defined in Ref. [1]. Note that term I is similar to eq. (2.1). In fact, eq. (2.1) is a special case of term I above. Of course, each term can be modified to better describe specific processes. The main point of eq. (2.2) is that it divides the nucleation process into Eyring rate theory processes and mechanical debonding processes.

ii. Growth

Under conditions of moderate triaxial (mean) tensile stress and high temperatures, void growth tends to occur on grain boundaries by vacancy diffusion. But at the extreme of high mean tensile stress and lower temperatures, void growth tends to occur by localized plastic flow around the voids. Therefore, to form continuum constitutive relations, we must adequately model the change in growth mode with changes in the stress and strain environment.

As reviewed in [1], for vacancy diffusion in grain boundaries, Raj and Ashby and others considered the case of a periodic array of voids with spacing 2ℓ in the grain boundary, and obtained the following expression for the rate of volume change for a single void, due to grain boundary vacancy diffusion:

$$\frac{dV}{dt} = \frac{2\pi\Omega D_B \delta(1 - R^2/\ell^2)^2(\sigma - 2\gamma/R)}{kT[\ln(\ell/R) - \frac{3}{4} + (R/\ell)^2(1 - R^2/4\ell^2)]} ; \tag{2.3}$$

where Ω is the atomic volume, D_B is a vacancy diffusion parameter, δ is the grain boundary thickness, σ is the remotely applied tensile stress across the grain boundary, and $2\gamma/R$ is the nucleation threshold stress at which the surface tension γ is overcome. A similar expression is obtained for vacancy diffusion through the material volume. The rate of change of relative void volume can be obtained from eq. (2.3) by dividing both sides by ℓ^3, the volume of matrix material associated with each void.

For ductile void growth under dynamic loading conditions, a viscous growth law is deduced from plate impact experiments,

$$dR/dt = T_1(\sigma - \sigma_{g0})R , \tag{2.4}$$

where σ_{g0} is a threshold stress and T_1 is a material constant with units of inverse viscosity, which allows the initial distribution to retain its exponential form as shown in fig. 2.9. The final radius R_1 is

$$R_1 = R_0 \exp[T_1(\sigma - \sigma_{g0})\Delta t] . \tag{2.5}$$

On the other hand, for quasistatic void growth under conditions of significant plastic deformation, the void growth observed is given to a good approximation by

$$v/v_0 = \exp[1.67\ \Delta\varepsilon_p\ \sinh(3\sigma_m/2\sigma_y)] , \tag{2.6}$$

where v and v_0 are the relative void volumes at the end and beginning of the strain increment $\Delta\varepsilon_p$, respectively. σ_m and σ_y are the mean stress and yield stress. The initial value of v_0 is that at nucleation (see fig. 2.4).

For very brittle materials, one expects the microcracks to obey classical linear elastic fracture mechanics.

Reference 1 reviews some growth laws proposed for linear elastic fracture mechanics. Using an energy balance approach for crack growth in elastic material, Mott derived the expression

$$\dot{R} = V_T(1 - R^*/R)^{1/2} , \tag{2.7}$$

where R^* = $\pi K_{Ic}^2/4\sigma^2$, the critical crack size for growth of a penny-shaped crack,

V_T = the terminal velocity, the limiting speed approached as the crack becomes long,

K_{Ic} = the plane strain fracture toughness.

Equation (2.7) can be integrated for any initial radius R_0 to obtain the radius R_1 after the passage of a tensile stress wave. The resulting distribution is shown in fig. 2.7. It is seen that with the Mott growth law, there is no growth for R less than R^*. As R increases beyond R^*, the total growth rapidly approaches $V_T\Delta t$, the value given by the constant velocity law.

Dulaney and Brace and Berry derived a more complex form for the growth law:

$$(\dot{R})^2 = V_T^2(1 - R^*/R)[1 - (2n^2 - 1)R^*/R] \ , \tag{2.8}$$

where $n = \sigma_c/\sigma$, the ratio of critical stress to applied stress. For an applied stress which is just critical (n = 1), the growth law is

$$\dot{R} = V_T(1 - R^*/R) \ . \tag{2.9}$$

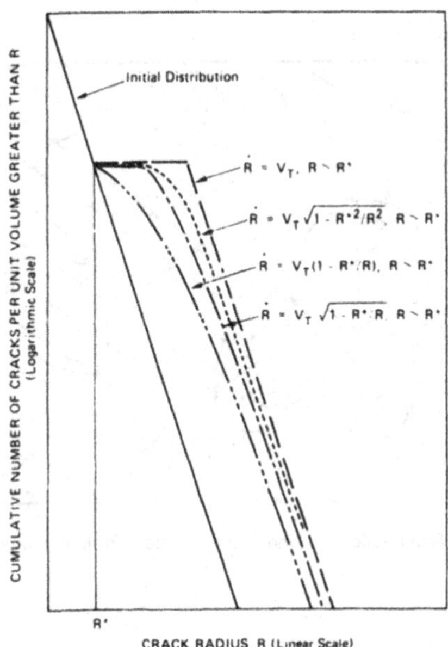

Figure 2.7. **Crack size distributions resulting from several growth laws.**

If the applied stress is far above critical (n << 1), then the law is

$$\dot{R} = V_T(1 - R^{*2}/R^2)^{1/2} \ . \tag{2.10}$$

For problems with a range of crack sizes, n will vary from 0 to 1 so that eqs. (2.9) and (2.10) are bounds on the elastic growth law. With all the final distributions shown in fig. 2.7, there is a sudden break at $R = R^*$ and for larger R the curve is concave downward.

In our dynamic plate impact work with Armco iron, beryllium and polycarbonate, we found that the evolution of $N_g(X,t,R)$ did not follow any of the forms shown in fig. 2.7, but more nearly pivoted out as shown in fig. 2.8, which indicates the same viscous growth relation as found for ductile voids, namely eq. (2.4).

iii. Coalescence

As reviewed in [1], the situation for modeling ductile void coalescence is somewhat confusing. The area requiring most work both for determining the onset of localization and for the resulting softening is that of describing material behavior inside bands of localized deformation. If unambiguous constitutive relations for such material could be constructed, analytical procedures exist for incorporating them into the constitutive relations for homogeneously softening material.

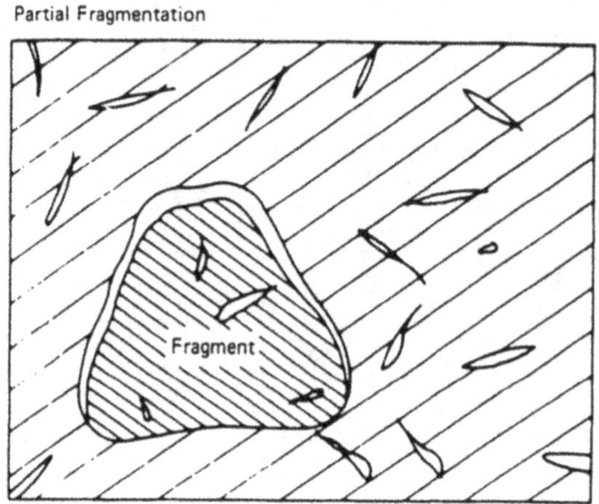

Figure 2.8. Schematic depiction of the fragment formation process.

The coalescence of brittle microcracks or shear bands results in the formation of fragments. Our computational fragmentation model has been reviewed in [1].

A fragment is produced by the intersection of cracks, each crack forming surfaces on two adjacent fragments. The crack radius is R_c and the fragment size is defined by a radius R_f. The ratio of these radii is γ:

$$\gamma = R_f/R_c , \tag{2.11}$$

where γ is a number near 1.0. For chunky fragments with a small number of faces, one can relate the number of fragments produced to the number of cracks. The parameter β is defined as

$$\beta = N_f/N_c \ , \tag{2.12}$$

where, for example, β is 1/4 for eight-sided fragments. The fragment size distribution has the same form as the crack size distribution. Thus,

$$N_f^{ij} = N_{f0}^{ij}\exp(-\ R_f/R_{f1}^{ij}) \ , \tag{2.13}$$

where N_f^{ij} and R_{f1}^{ij} are related to N^{ij} and R_1^{ij} through eqs. (2.11) and (2.12). The ij superscript refers to the crack orientation, and designates a point in laboratory coordinates on a unit sphere surrounding a material point. In short, it gives the direction of the unit vector n, discussed in lecture 1. Thus, the fragment size distribution is completely determined from the crack size distribution and the factors β and γ.

The volume V_f of a fragment of radius R_f is

$$V_f = T_f(R_f)^3 \ , \tag{2.14}$$

where T_f is a coefficient that reflects the shape and number of sides of the fragment. The total volume of fragments can be determined by multiplying the number of fragments of radius R, (dN/dR)dR from eq. (2.13) by the relative volume per fragment, and integrating the result:

$$V_f = T_f\Sigma_{ij} \int \frac{N_{f0}^{ij}}{R_1^{ij}} \exp\left(-\frac{R_f}{R_{f1}^{ij}}\right) R_f^3 dR = 6\beta\gamma^3 T_f \Sigma_{ij} N_0^{ij}(R_1^{ij})^3 \ . \tag{2.15}$$

The transition from slightly cracked to fully fragmented must occur gradually and with no discontinuities. In our computational model, the early stage of crack interaction is characterized by the concept of a "crack range," defined as a volume surrounding each crack. When the crack ranges of two cracks overlap, the cracks interact strongly, that is, the subsequent growth velocity and direction of one crack are altered by the presence of the other crack. The crack range is visualized as having a pancake or doughnut shape, as indicated in fig. 2.9. Crack range is a function of the crack size only, and therefore, its volume is

$$V_{cr} = T_c R^3 \ , \tag{2.16}$$

where T_c is a material constant. Fragmentation begins when the crack ranges of all the cracks completely fill the volume of the material. The total crack range in a material element is

$$V_{cr} = T_c \Sigma_{ij} \int_{R=0}^{\infty} R^3 dN \ . \tag{2.17}$$

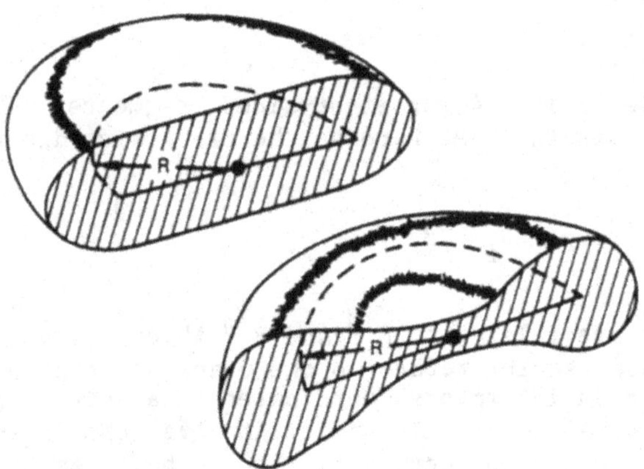

Figure 2.9. Possible cross sections and shapes of crack range surrounding penny-shaped cracks.

This is equivalent to

$$V_{cr} = 6T_c \Sigma_{ij} N_0^{ij} (R_1^{ij})^3 \ . \tag{2.18}$$

The criterion for beginning fragmentation is then

$$V_{cr} = 1 \ . \tag{2.19}$$

At the completion of fragmentation, the volume V_f from eq. (2.15) is unity. At the beginning of fragmentation, it has a value V^* obtained by solving eq. (2.15) with the conditions of eqs. (2.18) and (2.19):

$$V^* = \beta\gamma^3 T_f / T_c \ . \tag{2.20}$$

Now we define V_f as the potential fragment volume, because there are no fragments until V_f exceeds V^*. The actual fragment volume is computed according to the relation in eq. (2.16)

$$V_a^f = (V_f - V^*)/(1 - V^*) = 1 - f_u \ , \tag{2.21}$$

where f_u is the unfragmented fraction of material. We presume that this same fraction $(1 - f_u)$ pertains to all fragment sizes and orientations, so the actual number of fragments is

$$(N_{f0}^{ij})_a = N_{f0}^{ij}(1 - f_u) \ . \tag{2.22}$$

The stress-strain relations for material undergoing fracture and fragmentation must account for both the opening of cracks and the voids produced by internal fragments. In addition, as discussed above, the softening is anisotropic. As usual, the stress is separated into pressure and deviatoric components.

The pressure is related to the specific volume and internal energy in the same way as for ductile fracture. For brittle fracture, however, the porosity is the sum of the crack opening volume V_v and the volume of the fragmented material $(1 - f_u)V_s$, where V_s is the specific volume of the solid material. When f_u is greater than zero, the material in a fragment is treated as having the same specific volume as the remaining solid, which is under stress. When $f_u = 0$, the fragment density is set to the original density. The volume associated with the fragment, including the fragment volume at original density and the void around the fragment, is assumed to equal the volume that the fragment would have at the current specific volume of the stressed solid.

The deviator stress is computed from an expression which accounts for the anisotropy of damage:

$$\Delta\sigma_h' = 2G(\Delta\varepsilon_h' - \Delta\varepsilon_h'^c) \ , \tag{2.23}$$

where $\Delta\varepsilon_h'$ is the deviator strain (total strain less mean strain) in the h direction, and $\Delta\varepsilon_h'^c$ is the deviator strain taken up by changes in the porosity in the h direction. The contribution to $\varepsilon_h'^c$ from any element of the crack orientation array is the total crack opening and fragmented volume of the element multiplied by the square of the cosine of the angle between the element and the h direction. As in ductile fracture, the deviator stress from eq. (2.23) is modified if yielding occurs.

As for ductile fracture, an iteration procedure is required to solve simultaneously for damage and stresses. Here, however, the initial stress estimate is based simply on a purely elastic-plastic calculation. Damage is then computed based on these stresses. The computed damage leads to new stresses. These new stresses serve as estimated for the next cycle. The changes in crack strains $\Delta\varepsilon_h^c$ are heavily damped from cycle to cycle to force convergence. As discussed later, in lecture IV, the above coalescence model appears to match experimental data in many cases.

III. CONSTITUTIVE RELATIONS FOR THE EVOLUTION OF MICRODAMAGE

III.1 The "local" or "microstatistical" approach to fracture mechanics.

i. Introduction

Continuum fracture mechanics treats the instability and growth of a single macrocrack as a function of the near-tip stress or strain fields. However, from experiments we know that in most cases the macrocrack actually grows by coalescence of microcracks in the material element at the macrocrack tip. Therefore, it should be possible to link microcrack kinetics to continuum fracture mechanics. Here, we outline initial steps made toward this linkage.

A somewhat oversimplified synopsis of the present state of EPFM is that the conditions for macrocrack instability and extension in a structure in plane strain is well described by $J_1 > J_{1c}$, where J_{1c} is a material- and geometry-dependent* value of a path-independent integral J_1 given by

$$J_1 = \int_\Gamma (Wn_1 - \sigma_{mn} U_{m,1} n_n) ds \tag{3.1}$$

where W is the strain energy density, σ_{mn} the Cauchy stress tensor, and $U_{m,1}$ stands for components of the displacement gradient. The element ds is taken on a path Γ that begins on one macrocrack face, completely surrounds a tip, and ends on the other macrocrack face. The vector n is the unit exterior normal to the path element ds, and n_n is its component in the x_n direction. The coordinate normal to the plane is x_2, and x_1 lies in the direction of potential crack extension. Furthermore, conditions of stable crack growth are given by the slope of the $J_1(a)$ curve, where a is the macrocrack half-length. The J_1 versus a curve is called the J-resistance or J-R curve, and is often considered to a material-and geometry-dependent property on the same level as J_{1c} itself. A related concept is the "tearing modulus" dJ/da, which can be considered to govern the resistance

*Plane strain conditions must apply.

of the structure to stable macrocrack growth. Analysis of the stresses and displacements of a given structure can then be used to determine when the load in the structure begins to drop, which thereby triggers a structural instability and signals the transition from stable macrocrack growth to unstable extension and structural failure.

ii. Relationship of MSFM to LEFM and EPFM

As discussed in Ref. 1, in previous work on A533B pressure vessel steel we showed that correct values of J_{1c} could be calculated from a microvoid kinetics model by performing a full computational simulation of a center-cracked panel test. However, in such simulations it is difficult to see directly how changes in the microvoid kinetics parameters would change J_{1c}. In the following paragraphs, we derive an approximate relationship between the microvoid kinetics parameters and the J-R curve, which allows us to estimate the benefits in macrocrack initiation and stability gained by improving the microstructural properties. Furthermore, variations in the J-R curve to be expected from variations in inclusion concentrations can be determined.

In the center-cracked panel, the primary contributor to the J integral is the strain energy density term. That is,

$$J_1 \approx \int_{PZ} W\, dy \qquad (3.2)$$

where W is the strain energy density, y is normal to the crack plane, and the line integral is taken just outside the process zone (PZ).* If we use Green's theorem on the J_1 line integral, retaining only the dominant term, then the line integral becomes a volume integral:

$$J_1 \approx \frac{-1}{\Delta z} \int_{PZ} \frac{\partial W}{\partial x}\, dV \qquad (3.3)$$

where the volume integral is taken over the process zone, z is an increment of distance in the z direction along the crack front, and x is in the direction of the ligament.

At the leading edge of the process zone

$$W = \int_0^{\bar{\epsilon}_0^P} \tau\, d\bar{\epsilon}^P$$

where τ and $\bar{\epsilon}^P$ are the effective shear stress and equivalent plastic strain, respectively, and $\bar{\epsilon}_0^P$ is the threshold strain for void nucleation. At the trailing edge of the process zone,

---

*In this paper, we define the "process zone" as the volume of material where microcracks or voids have been nucleated. Other authors have used different definitions.

$$W = \int_0^{\bar{\epsilon}^p} \tau d\bar{\epsilon}^p$$

Therefore, a rough estimate of $\partial W/\partial x$ in the process zone is

$$-\frac{\partial W}{\partial x} \approx \frac{1}{\delta_{PZ}} \int_{\epsilon_0^p}^{\bar{\epsilon}_0^p} \tau d\bar{\epsilon}^p \tag{3.4}$$

where δ_{PZ} is the size of the process zone. Combining eqs. (3.3) and (3.4.) gives

$$J_1 \approx \frac{\Delta V}{\Delta z \delta_{PZ}} \int_{\epsilon_0^p}^{\bar{\epsilon}^p} \tau d\bar{\epsilon}^p = (\delta_{PZ}) \int_{\epsilon_0^p}^{\bar{\epsilon}^p} \tau d\bar{\epsilon}^p \tag{3.5}$$

where we assume $\Delta V = \Delta z(\delta_{PZ})^2$. Equation (3.5) thus expresses that J is basically a measure of the plastic work done in the process zone.

Next consider the most simplified version of the void kinetics model described in lecture 21 (see eq. 2.6):

$$\frac{\Delta v}{v_0} \approx 2.5 \int_{\epsilon_0^p}^{\bar{\epsilon}^p} \frac{\sigma_m}{\sigma_y} d\bar{\epsilon}^p \tag{3.6}$$

where Δv is the increase in relative void volume over its value at nucleation, v_0, σ_m is the mean stress, and σ_y is the yield stress.

If we next assume proportional loading in the process zone, eq. (3.6) can be written

$$\frac{\Delta v}{v_o} = 2.5 \frac{\sigma_m}{\sigma_y \tau} \int_{\epsilon_0^p}^{\bar{\epsilon}^p} \tau d\bar{\epsilon}^p \tag{3.7}$$

Comparisions of eqs. (3.7) and (3.5) yields

$$\frac{J_1}{\sigma_y \delta_{PZ}} = \frac{\tau}{2.5\sigma_m} \frac{\Delta v}{v_0} \tag{3.8}$$

At the crack tip, $\tau/2.5\sigma_m$ is about unity (1.2 for the uniaxial stress case), and in the small-scale yielding approximation should be fairly independent of crack length. In general, $\tau/2.5\sigma_m$ will be a functions of the geometrical constraints at the crack tip.

Since J_1/σ_y is proportional to the crack opening displacement, the left-hand side of eq. (3.8) is a measure of the crack tip opening angle. However, to keep the load characterizing parameter J_1 distinguished from the material properties, we write eq. (3.8) as

$$J_1 \approx \sigma_y \frac{\tau}{2.5\sigma_m} \left[\delta_{PZ}\Delta v/v_0\right] \tag{3.9}$$

Crack initiation occurs when $\Delta V \geq (\Delta v)_c$ over a volume N_0^{-1}, where N_0 is the number per unit volume of nucleated voids (debonded inclusions). That is, N_0^{-1} is the "material element". Thus, the expression for J_{1c} is

$$J_{1c} = \sigma_y \, (\tau/2.5\sigma_m) \left[\delta_{PZc} (\Delta v)_c / v_0\right] \tag{3.10}$$

where δ_{PZc} is the size of the process zone at initiation, and $(\Delta v)_c$ is the value of $v - v_0$ at onset of microvoid coalescence in the material element N_0^{-1}.

Furthermore, the size of the process zone is expected to be at least several inclusion spacings, i.e.,

$$\delta_{PZc} = \beta N_0^{-1/3} \tag{3.11}$$

where β is the number of inclusion spacings in the process zone. Since

$$v_0 \approx 8\pi N_0 R_0^3 \tag{3.12}$$

where R_0 is the center of the activated inclusion size distribution:

$$N_g = N_0 \exp(-R/R_0) \tag{3.13}$$

we can combine eqs. (3.10, (3.11) and (3.13) to obtain

$$J_{1c} \approx \left(\frac{\beta}{8\pi}\right) \left(\frac{\pi}{2.5\sigma_m}\right) \frac{\sigma_y (\Delta v)_c}{N_0^{4/3}/R_0^3}, \quad \beta \geq 1 \tag{3.14}$$

In this relation, both β and $\tau/2.5\sigma_m$ are functions of the geometrical constraints along the macrocrack front. For example, blunter cracks will have larger values of β because the enhanced strain field ahead of the macrocrack will extend farther and activate more inclusions. Both β and τ/σ_m vary during the preinitiation loading history; therefore, eq. (3.14) is only a crude approximation. However, we believe that eq. (3.14) may serve as a good rule of thumb to estimate the interdependence of J_{Ic} on the microstructural quantities $(\Delta v)_c$ and R_0, the continuum quantity σ_y and the geometrical quantities β and τ/σ. That is, eq. (3.14) demonstrates clearly that J_{1c} depends on specimen geometry as well as on micro and macro material properties, and provides quidance for interpreting fracture mechanics tests.

During stable macrocrack growth, eq. (3.14) tells us that changes in J_{1c} must come from changes in β and τ/σ since N_0 and R_0 are microstructural properties and $(\Delta v)_c$ is a local property operative over a distance about equal to the void spacing.

We now assume that β and τ/σ_m are functions of macrocrack length, a, and specimen geometry (SG). Thus, from eq. (3.14)

$$\left(\frac{\partial J_{1c}}{\partial a}\right)_{SG} = \frac{\sigma_y (\Delta v)_c}{20\pi N_0^{4/3} R_0^3} \left\{ \beta \left[\frac{\partial(\tau/\sigma_m)}{\partial a}\right]_{SG} + (\tau/\sigma_m)\left(\frac{\partial\beta}{\partial a}\right)_{SG} \right\} \tag{3.15}$$

Thus for a given specimen geometry, a strongly rising R curve suggests that either τ/σ_m, β, or both are increasing rapidly with crack length.

Computations indicate that τ/σ_m remains fairly constant in front of a slowly growing crack. That is, all the stress components increase proportionally with increased crack length. If this is the case, then the shape of the J-R curve is determined principally by the $(\partial\beta/\partial a)_{SG}$ term.

Increases in β with increasing macrocrack length are due to the increased extent of the plastic zone and the associated growth of the process zone through occurence of the threshold strain $\epsilon_0^{\bar{p}}$ at further distances from the macrocrack tip. In short, the work required to extend the macrocrack increases because the volume of the process zone, and the associated energy to bring the total number of voids therein to coalescence, increases. The vast majority of that work is not spent in creating new void surface, but in doing plastic work on the matrix material around the voids.

In summary, J_{1c} and $\partial J/\partial a$ are governed by a combination of microstructural properties and continuum boundary conditions, as shown by eqs. (3.14) and (3.15). Both J_{1c} and $\partial J/\partial a$ are inversely proportional to the increase in relative void volume from the inclusion volume to the volume at onset of void coalescence. The above properties are microstructural. However, J_{1c} is also directly proportional to the process zone size at macrocrack initiation (coalescence of the microvoids at the crack tip), and, during slow stable crack growth, $\partial J/\partial a$ depends directly on the change in process zone size with crack length. The process zone size depends, in turn, in both microstructural and continuum variables, That is, the threshold plastic strain $\epsilon_0^{\bar{p}}$ is a microstructural property, but the distance from the crack tip at which $\epsilon_0^{\bar{p}}$ is first attained, thus defining the process zone size, depends on the continuum strain field. The increase in process zone size with crack length is therefore purely a continuum property. In short, the height of the J-R curve depends on the microstructure, but the shape depends only on crack tip and specimen geometries.

An exciting area of future research would be to vary the microstructural and geometrical properties of test specimens to attempt to confirm the above picture.

iii. Use of internal state variables in continuum constitutive relatives to predict the evolution of damage.

Although a MSFM model aims to describe the evolution of microscopic damage, it is nonetheless a continuum mechanics model. Therefore, it must average the microdamage with a function that can be treated as an internal state variable in the material's constitutive relations, and when used in numerical computations, must be cell independent. As reviewed in Reference

1, many such damage functions have been proposed. As described in detail in Reference 1, we have chosen a damage functions, $N_g(R, n)$, the number of microcracks or voids per unit volume, associated with material point, with crack orientation in the material cordinates given by the unit vector n, and with crack size greater than R. Since N_g is defined at each material point, it is independent of computational cell size. It is commonly found experimentally that for a given orientation (or summed over all orientations), N_g is given by:

$$N_g = N \exp(-R/R_1) \tag{3.16}$$

Where N is the total crack or void concenration (number per unit volume), and R_1 is the characteristic size of the distribution.

Although N_g is cell size-independent, it only has physical meaning in a "material element" large enough to contain a statistical number of microcracks; otherwise the averaging implicit in the model will have no meaning. The nucleation and growth rates of the damage are described by $\dot{N}$ and $\dot{R}_1$, respectively. As discussed in Reference 1, the nucleation and growth rates are found experimentally to be a strong function of loading state. In the remainder of this paper we confine the discussion to the uniaxial strain conditions often enforced in shock physics experiments.

In uniaxial strain, the conservation of mass and momentum can be written (in Lagrangian coordinates) as:

$$\frac{\partial u}{\partial h} - \frac{\partial \epsilon}{\partial t} = 0 \tag{3.17}$$

$$\rho_0 \frac{\partial u}{\partial t} - \frac{\partial \sigma}{\partial h} = 0 \tag{3.18}$$

where σ and ϵ are the stress and strain (positive in tension), t is the time, ρ_o is the initial density, and h is the Lagrangian position coordinate. To simplify the following analysis, we omit the energy conservation equation, and discuss only mechanical behavior.

As mentioned above, the constitutive relations must contain a description of the damage evolution. A convient way to do this is to express the total strain rate as the sum of elastic, plastic, and damage components:

$$\dot{\epsilon} = \frac{\dot{\sigma}}{M(\sigma)} + \dot{\epsilon}_p + \dot{\epsilon}_f \tag{3.19}$$

where the elastic strain rate is given by the first term on the right hand side of the equation, $\dot{\epsilon}_p$ and $\dot{\epsilon}_f$ are the plastic and microfracture strain rates, respectively, and M is the uniaxial strain elastic modulus. The functional form of the microfracture strain rate, $\dot{\epsilon}_f$, depends on the specific microfracture mechanism. For size distributions given by Eq. (1), the void-induced volumetric strain is

$$\epsilon_f(\text{voids}) = 8\pi N R_1^3 \tag{3.20}$$

For cleavage cracks (see Ref. 1), the microfracture volumetric strain due to elastic crack opening is proportional to the product of the stress with the value of ϵ_f given in Eq. (5), or:

$$\epsilon_f(\text{cracks}) = 32NR_1^3(1-v^2)\sigma/E \tag{3.21}$$

where v is Poisson's ratio and E is Young's modulus. For adiabatic shear bands (see Ref. 1), the damage-induced strain is slip, and is proportional to the total shear band area in a slip plane, $2\pi NR_1^2$.

Thus, for microscopic voids, cracks, and adiabatic shear bands, $\dot{\epsilon}_f$ is a function of $\dot{N}$ and $\dot{R}_1$.

The plastic strain rate, $\dot{\epsilon}_p$, from dislocation dynamics considerations, is often written as a relaxation process:

$$\dot{\epsilon}_p = (\sigma - \sigma_o)/\tau \tag{3.22}$$

where σ_o is a threshold stress and τ is a viscosity. We now define Ψ as the total non-elastic strain rate:

$$\Psi = \dot{\epsilon}_p + \dot{\epsilon}_f \tag{3.23}$$

III.2 Requirements for damage evolution relations

i. Uniqueness

A valid MSFM model must not adversely affect the uniqueness or stability of the solution to Eqs. (3.17 - 3.19). We first discuss uniqueness. A significant simplification occurs if Ψ is not a function of derivatives of ϵ, u, or σ, but only of the variables themselves. In that case Eqs. (3.17-3.19) form a hyperbolic set with characteristic family wave speeds of $+\sqrt{M/\rho_o}$, 0, and $-\sqrt{M/\rho_o}$. Uniqueness is assured if M is positive, and characteristics of the same family do not intersect. Since M in Eq. (3.19) is the uniaxial strain elastic modulus, it is positive, and characterics of the same family do not intersect. Thus, although uniqueness can be lost through normal shocking-up processes due to the nonlinear behavior of M, no uniqueness problems will be introduced by the MSFM model if ψ does not depend on derivatives of the flow variables.

Each MSFM model must be examined individually to test its uniqueness. Eqs. (3.20) and (3.21) show that $\dot{\epsilon}_f$ for both voids and cracks will depend on $\dot{N}$ and $\dot{R}_1$. As discussed in Reference 1, under dynamic uniaxial strain conditions the nucleation and growth rates for both voids and cracks are often found experimentally to be well-described by:

$$N = N_0 \exp \frac{\sigma - \sigma_{no}}{\cdot\ \sigma_1} \tag{3.24}$$

$$R_1 = R_1 \frac{\sigma - \sigma_{go}}{4\eta} \tag{3.25}$$

where σ_{no} and σ_{go} are threshold stresses, and σ_1 and η are also material constants.

Inspection of Eqs. (3.20), (3.22), and (3.23-3.25) shows that ψ does not depend on derivatives of the flow variables for microscopic voids. Also, as discussed in Reference 1, the nucleation and growth rates for adiabatic shear bands are found experimentally to depend primarily on $\dot{\epsilon}_p$, which in Eq. 3.22 does not depend on derivatives of the flow variables.

For microscopic cracks, however, Eq. 3.21 shows that $\dot{\epsilon}_f$ will have a term in $\dot{\sigma}$. This term can be combined with $\dot{\sigma}/M$ term in Eq. (3.19) to soften the modulus. Since this softening will not cause intersection of characteristics of the same family, there will also be no MSFM-induced uniqueness problems in this case.

ii. Stability

We next discuss the stability conditions. Eqs. (3.17), (3.18), (3.19) and (3.23) can be combined to form a wave equation for σ:

$$\frac{\partial^2 \sigma}{\partial t^2} - \frac{M}{\rho_o} \frac{\partial^2 \sigma}{\partial h^2} + M - \frac{\partial \psi}{\partial t} = 0 \qquad (3.26)$$

where for simplicity we have taken M to be constant, that is, we have assumed the elastic response to be linear.

Inspection of Eqs. (3.20) - (3.23) shows that $M\dot{\psi}$ becomes:

$$M\dot{\psi} = \lambda(\sigma)\dot{\sigma} + B(\sigma) \qquad (3.27)$$

where ψ and B are positive, and B approaches zero as σ approaches the growth threshold σ_{go}. If we linearize the above equations around the state given by σ_b, N_b, and R_{1b}, and let $\sigma = \sigma_b + s$, where s is a small perturbation, then the solution to Eqs (3.26) and (3.27) is

$$s = \phi - B(\sigma_b)/\lambda(\sigma_b) \qquad (3.28)$$

where ϕ is the solution for the case when $B = 0$, and the B/λ term is a damping function.

G. B. Whitham* has discussed such equation systems in detail, and shows that the above system is stable if $\lambda > 0$. In the above examples, this stability condition is fulfilled, and expresses the requirement that an increase in $\dot{\epsilon}_f$ requires an increase in stress, that is, the fracture process must be dissipative.

We conclude that the experimentally-based MSFM models discussed above appear to pose no obvious uniqueness or stability problems.

---

*Comm. on Pure and Applied Math., Vol. XII, 113 (1959).

iii. "Active Softening"

As the damage grows, the stress may decrease as the partition of total strain shifts from elastic strain to fracture strain (see Eq. 3.19). Thus, "strain softening" can occur without causing stability problems.

III.3 Implementation of MSFM models in computer codes.

i. Modular; part of constitutive relations that calculate increments in the stress tense arising from an imposed deformation increment.

A Lagrangian "hydrocode" solves Eqs. (3.17) and (3.18) for a given material element (given Δh), together with a constitutive relation subroutine for the stresses. The solution procedure is, for a given time step, to calculate a trial stress increment based on allowing the imposed total strain increment to be elastic, i.e. $\dot{\epsilon}_p$ and $\dot{\epsilon}_f$ are set to zero in Eq. (3.19). Then Eqs. (3.20) - (3.25) are used to get a first estimate of $\Delta\epsilon_p$ and $\Delta\epsilon_f$ in Eq. (3.19). An iteration procedure is used to converge to final values of σ, ϵ_p, and ϵ_f. Then Eqs. (3.17 and (3.18) are used to obtain new values of υ and ϵ.

ii. Cumulative damage; best suited for Lagrangian codes with explicit solution procedures.

Since the damage is associated with particular elements of material, the above cumulative damage models are best suited for Lagrangian codes that base the computational mesh on the positions of specific material particles. When Eulerian meshes are used (for which material particles flow through the computational mesh), complicated mixture algorithms must be used that introduce unrealistic diffusion of damage.

IV. EXAMPLE APPLICATIONS OF MSFM

IV.1 Spall of copper under shock wave loading

Observations of plate impact data in soft aluminum and copper revealed that the size distributions had the form discussed earlier, namely

$$N_g(R) = N_0 \exp(-R/R_1) , \tag{4.1}$$

where N_g is the cumulative number/cm^3 of voids with radii larger than R, N_0 is the total number/cm^3 of voids, and R_1 is a parameter of the distribution.

The total relative void volume is obtained by integrating over the entire distribution

$$\begin{aligned} V_v &= \frac{4\pi}{3} \int_0^\infty R^3 \frac{dN}{dR} dR \\ &= \frac{4\pi}{3} \int_0^\infty R^3 \left(\frac{N_0}{R_1}\right) \exp\left(-\frac{R}{R_1}\right) dR = 8\pi N_0 R_1^3 . \end{aligned} \tag{4.2}$$

The void size distribution at any time and at any point can be represented by N_0 and either R_1 or N_v.

Nucleation in the model occurs as the addition of new voids to the existing set. These new voids are presumed to occur in a range of sizes with a size distribution given by eq. (4.1). At nucleation, the parameter R_1 equals R_n, the nucleation size parameter (a material constant). The number of voids nucleated is governed by the nucleation rate function:

$$\begin{aligned} \dot{N} &= \dot{N}_0 \exp[(P_s - P_{n0})/P_1] , & P_s > P_{n0} \\ &= 0 , & P \le P_{n0} \end{aligned} \tag{4.3}$$

where $\dot{N}_0$, P_{n0} and P_1 are material constants, P_s is the tensile pressure in the solid material (not the average pressure in the composite of solid and voids). The constant P_{n0} is the threshold for nucleation.

The void volume nucleated in a time interval Δt is found from eqs. (4.2 and 4.3)

$$\Delta V_n = 8\pi \dot{N} \Delta t R_n^3 . \tag{4.4}$$

Damage increases by nucleation of new voids and by growth of the existing voids. A viscous growth rate is predicted theoretically and observed experimentally, namely:

$$\dot{R} = \left[(P_s - P_{g0})/4\eta\right]R \tag{4.5}$$

where P_{g0} is the threshold pressure for growth, and η is the material viscosity.

The growth of a void during a time interval Δt is obtained by integrating eq. (4.5) to obtain the new radius

$$R = R_0 \exp\left(\frac{P_s - P_{g0}}{4\eta}\Delta t\right) , \tag{4.6}$$

where R_0 is the radius at the beginning of the time interval. Since every void in the distribution grows by the same exponential factor, even the size parameter R_1 grows according to eq. (4.6)

$$R_1 = R_{10} \exp\left(\frac{P_s - P_{g0}}{4\eta}\Delta t\right) , \tag{4.7}$$

where R_{10} is the size parameter at the beginning of the time interval. Then the new void volume can also be found from eqs. (4.2) and (4.7):

$$V_v = 8\pi N_0 R_1^3 = V_{v0} \exp\left(3\,\frac{P_s - P_{g0}}{4\eta}\Delta t\right) , \tag{4.8}$$

where $V_{v0} = 8\pi N_0 R_{10}^3$, the void volume at the beginning of the time interval.

The total change in void volume is the sum of the contributions associated with nucleation and growth. Thus, the total void volume at the end of the interval is

$$V_v = V_{v0} \exp\left(3\,\frac{P_s - P_{g0}}{4\eta}\Delta t\right) + \Delta V_n . \tag{4.9}$$

The stress-strain relations for material undergoing fracture must account for the presence of voids. As usual, the stress is separated into pressure and deviatoric components.

We assume that an average pressure in the solid material can be computed from the specific volume of the solid and the internal energy through use of a small-strain form of the Mie-Gruneisen equation:

$$P_s = C(\rho_s/\rho_0 - 1) + \Gamma\rho_s E , \tag{4.10}$$

where C is the bulk modulus, Γ is the Gruneisen ratio, E is internal energy, ρ_s is the solid density, and ρ_0 is the initial density of the solid. The pressure computed from eq. (4.10) is necessarily an average, or continuous stress, because the actual micro stress states will vary greatly through partially fractured material.

The average pressure on the cross section of the fractured material can now be related to the pressure in the solid components according to a relation derived by Carroll and Holt for porous material:

$$P = P_s \rho / \rho_s , \tag{4.11}$$

where P is the average pressure on a section and ρ is the average density of the porous material. A combination of eqs. (4.10) and (4.11) relates the average pressure P to the energy E and density ρ.

The deviator stresses are computed by the usual elastic and plastic relations. However, the damage that occurs is presumed to affect both the yield strength and the effective shear modulus of the material. The modulus is reduced as a function of the developing porosity according to:

$$G = G_0(1 - V_v \rho F) , \tag{4.12}$$

where G is the effective shear modulus, G_0 is shear modulus of the solid, V_v is the specific volume of voids, and

$$F = 5 \frac{3C + 4G_0}{9C + 8G_0} = 15 \frac{1 - \nu}{7 - 5\nu} , \tag{4.13}$$

where ν is the Poisson's ratio and C is the bulk modulus of the solid.

The yield strength reduces somewhat more rapidly than the modulus as the porosity increases. Dynamic calculations of void growth [1] indicated that the yield strength should reduce in the following way:

$$Y = Y_0(1 - 4V_v \rho) . \tag{4.14}$$

where Y_0 is the yield strength for nonporous material.

As the voids grow, the stress-free surfaces generated in the material diminish the amount of stress generated by the overall deformation. This stress relaxation occurs by two basically different processes. First, under tension the existing voids open to accommodate some of the imposed volumetric strain. The volumetric strain of the matrix material can therefore relax elastically and the associated mean tensile stress also relaxes. Thus, we first partition each imposed volumetric strain increment ΔV into the sum of a matrix strain increment ΔV_s and a void strain increment ΔV_v:

$$\Delta V = \Delta V_s + \Delta V_v . \tag{4.15}$$

A solution for the mean stress P_s is obtained by an iteration procedure based on eq. (4.15). A first estimate of P_s is made by replacing ΔV_v with linearized forms of the nucleation and growth volume change relation [eq. (4.9)]. With the estimated P_s, the void and matrix volume increments are computed exactly and their sum (ΔV_a) is compared to the imposed volume change ΔV. The difference $\Delta V - \Delta V_a$ (the convergence criterion) is used to construct the next estimae of P_s, until convergence is reached. The mean stress P_c, computed from this sharing of volume change between matrix strain and void opening, undergoes a natural relaxation or reduction as the voids grow. At high loading rates, the stress reduction is small, allowing very high stresses to occur. At low rates, the mean stress remains near the threshold stress for nucleation and growth.

In the second process of stress relaxation, the decrease of load-bearing area introduces the correction factors expressed by eqs. (4.11) through (4.14) above.

In general, the first stress relaxation effect is dominant until the last stages of coalescence.

The present model does not treat void coalescence explicitly. Therefore, the default coalescence behavior is one of direct impingement since at the correspondingly high relative void volumes (about 0.5) the stresses have relaxed to near-zero values.

The above models has been used to describe the time-dependent evolution of spall damage in specimens of aluminum and copper loaded by plate impacts as well as by electron beam and x-ray irradiation. We reproduce next some OFHC copper results [1].

Place impact experiments were performed, where a low impedance PMMA buffer containing a manganin stress gage was placed behind the target to record the fracture or "pull back" signal.

The nucleation and growth material properties of eqs. (4.1)-(4.9) are given below in table 4.1.

Table 4.1

Ductile fracture parameters for 1145 aluminum and OFHC copper*

Parameter	Units	Description	Cu (OFHC)
$T1(3/4\eta)$	cm^2/dyn-s	Growth parameter	-0.01
η	dyn-s/cm^2	Material viscosity	75
$T2(\sigma_{g0})$	dyn/cm^2	Growth threshold	-5×10^9
$T3(R_0)$	cm	Nucleation void radius	1×10^{-4}
$T4(N_0)$	no./cm^3-s	Nucleation coefficient	2.8×10^{12}
$T5(P_{n0})$	dyn/cm^2	Nucleation threshold	-5×10^9
$T6(P_1)$	dyn/cm^2	Nucleation parameter	-2×10^9

*Stresses are taken positive in compression.

The experimental configuration parameters for two experiments are given in table 4.2. Thus, table 4.2 shows that in Experiment S24 a 0.062-cm-thick copper flyer plate impacted a 0.159-cm-thick copper target plate at 1.59×10^5 cm/s, causing a 29.3-kbar compressive stress at the impact interface. Since the flyer plate thickness in Experiment S24 was about half the thickness of that for Experiment S27, the compressive and subsequent tensile pulse durations for S24 were also about half those for S27.

Table 4.2

Impact conditions for two experiments in OFHC copper

Shot number	Copper flyer plate Thickness (cm)	Copper flyer plate Velocity (cm/s)	Target thickness (cm)	Backing plate	Impact stress (kbar)
S24	0.062	1.59×10^4	0.159	PMMA	29.3
S27	0.114	1.29×10^4	0.318	PMMA	23.7

Comparisons of computed and observed posttest damage distributions through the thicknesses of the targets for the two experiments are shown in figs. 4.1 and 4.2. The experimental values were obtained by metallographic examination of polished sections, where the surface void distributions were converted to volume distributions.

Comparisons of the computed and measured transmitted stress histories in the PMMA backing plate are given in figs. 4.3 and 4.4.

The compression wave termed "fracture signal" in these records is caused by recompression of the material adjacent to the damage region as the tensile stresses there are relaxed. The amplitude of the signal is related to the amount of stress relaxation and thus to the level of damage. The arrival time and rate of rise of the signal are related to the damage growth rate: a sudden growth of damage produces an early, rapidly rising signal; a gradual fracturing causes a ramped signal.

IV.2 Spall of composites

i. Propellants

In modeling fracture of composites, one must account for nucleation of microcracks at heterogeneities in the composite, and development of microcracking in several anisotropic directions.

Our model allows several distinct orientations of brittle microcracks in the material. In a given material element, only one of these orientations is active, based on the state of tensile stress at the onset of nucleation.

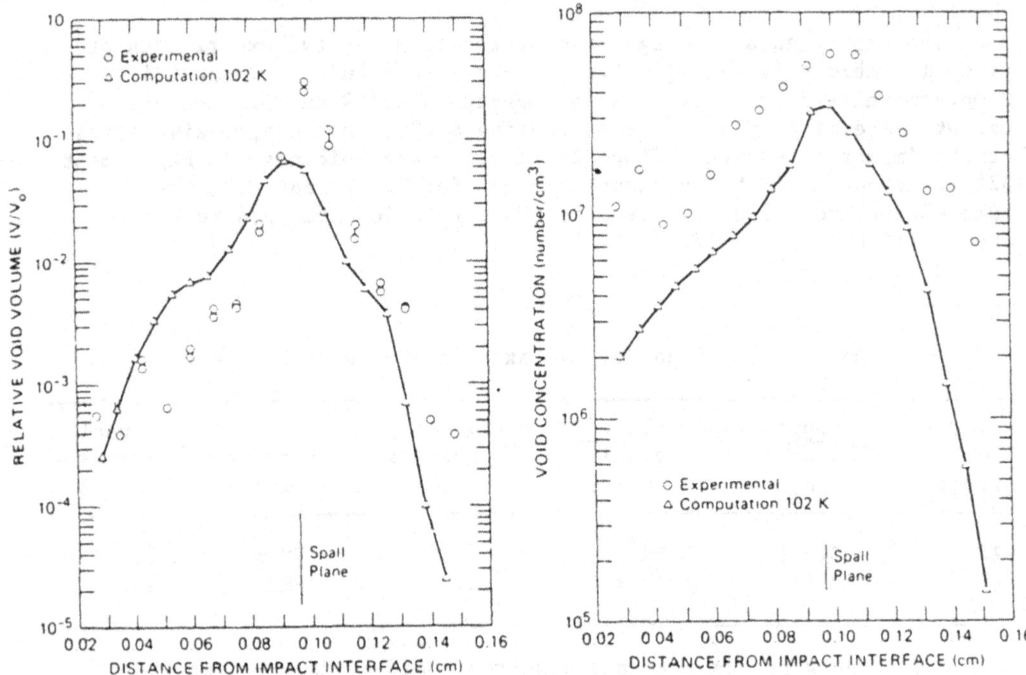

Figure 4.1. Computed and observed damage in OFHC copper (test S24).

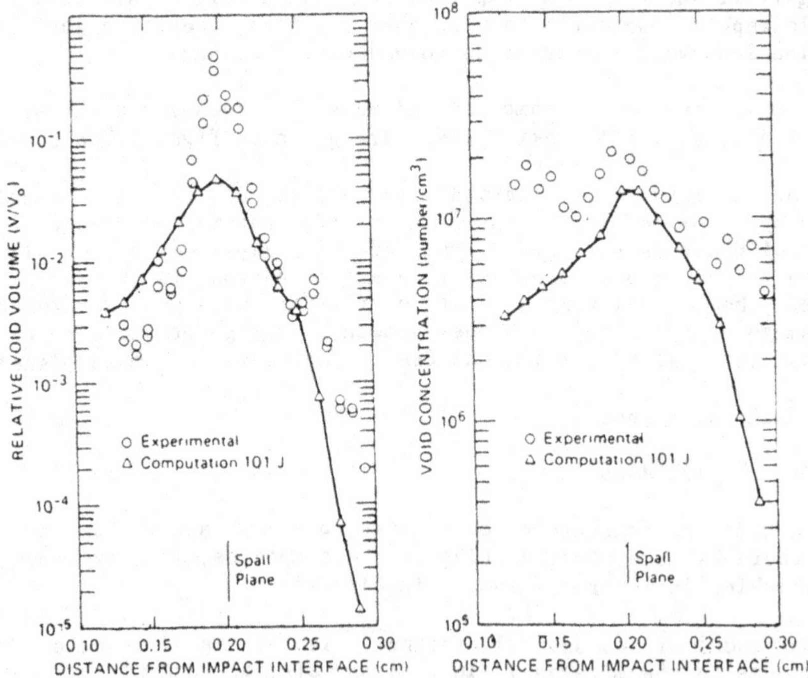

Figure 4.2. Computed and observed damage in OFHC copper (test S27).

Cracks and crack size distribution

Cracks are assumed to be penny shaped: circular and flat. To represent the observed crack sizes, the size distribution is given by the usual analytical form:

$$N_g = N_0 \exp(- R/R_1) \tag{4.16}$$

where N_g = the number/cm^3 of cracks greater than R
N_0 = total number/cm^3
R = the crack radius
R_1 = the shape parameter for the distribution.

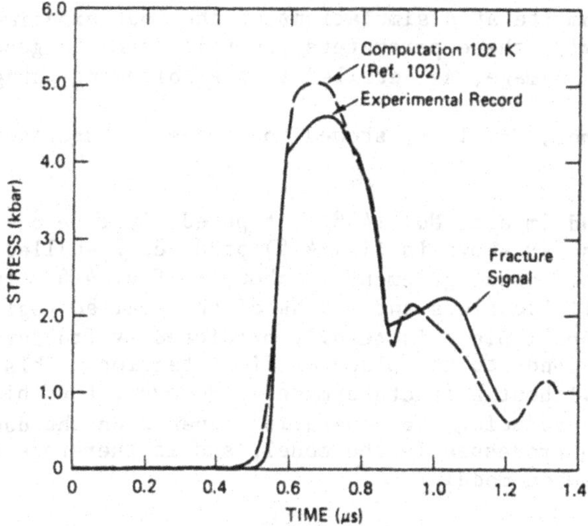

Figure 4.3. Comparison of stess record from manganin gage in PMMA behind OFHC copper with computed stress.

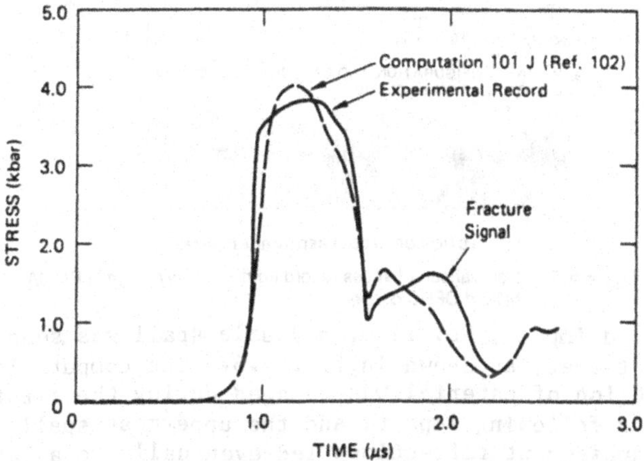

Figure 4.4. Comparison of stress record from manganin gage in PMMA behind OFHC copper with computed stress.

If the crack size distributions at nucleation or later do not fit this analytical form (as in brittle ceramics), the distribution may be provided numerically by size groups or line segments.

The microfracture kinetics model is of the type discussed in Lecture I for brittle cracks. It was applied to describing spall damage in a heterogeneous rocket motor propellant. The details are given in Ref. [1].

Four planar gas-gun (uniaxial strain) impacts were conducted on a prototype propellant, with projectile velocities from 44 to 191 m/s (computed impact stresses from 0.08 to 0.37 GPa). No fracture occurred in the lowest velocity impact, but various levels of spall and fragmentation took place in the other three. The fracture parameters were obtained from iterated simulations of the four experiments. The damage computed with these parameters generally shows a good agreement with the observed damage, as detailed in the following paragraphs.

The first shot, No. 1420, showed no damage in experiment or simulation.

In the second impact, No. 1458, a tapered flyer (projectile of varying thickness, as shown in fig. 4.5) produced a spalled portion of varying thickness. Good agreement is shown in fig. 4.5 between computed and observed spall locations and widths of the spalled region. The location of the spall plane is readily predicted by fracture models because it corresponds to the plane of first tension. This plane can be identified even without a fracture model. However, the thickness of the fractured region producing the separation depends on the damage-induced stress-relaxation processes in the model, and is therefore a better discriminator between models.

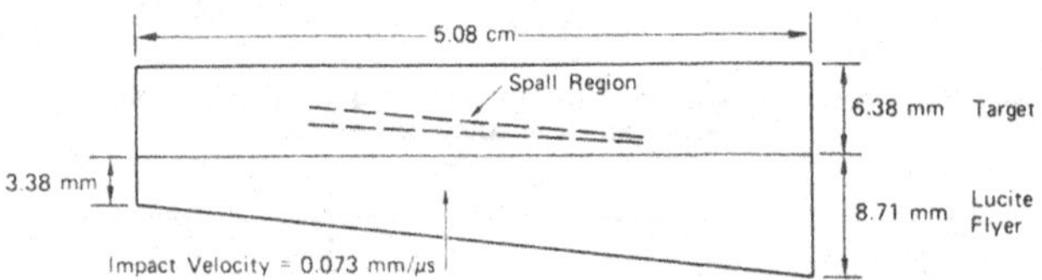

(a) INITIAL CONFIGURATION AND CALCULATED WIDTH AND LOCATION OF SPALL PLANE

(b) CROSS SECTION OF RECOVERED TARGET SAMPLE

Figure 4.5. Comparison of stress record from manganin gage in PMMA behind OFHC copper.

In the third impact, No. 1460, a double spall was seen in the cross section of the target, as shown in fig. 4.6. The computations showed that a large region of material was damaged during the first period of tensile stresses following impact, and the uppermost spall separation formed. Then subsequent reflections led eventually to a later spall plane closer to the impact plane. The correspondence between computed and observed regions of damage is good enough to suggest that the computed sequence may truly represent the actual sequence.

The fourth experiment, No. 1462, exhibited complete fragmentation. Figure 4.7 shows cumulative fragment size distributions; the radius shown is the radius of a spherical fragment of the same mass. Both the computed and observed distributions show a change of a slope around 0.1 to 0.2 cm; this effect is probably associated with the fracture mechanics concept that allows only cracks larger than a critical size to grow.

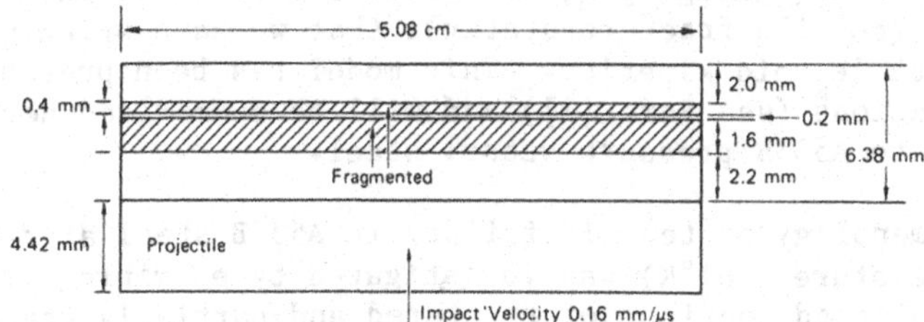

(a) CONFIGURATION AND COMPUTED RESULTS SHOWING TWO FRAGMENTED REGIONS

(b) CROSS SECTION OF SPECIMEN RECOVERED FROM EXPERIMENT 1460 SHOWING "DOUBLE" SPALL

Figure 4.6. Comparison of computed and observed fragmented regions in a propellant.

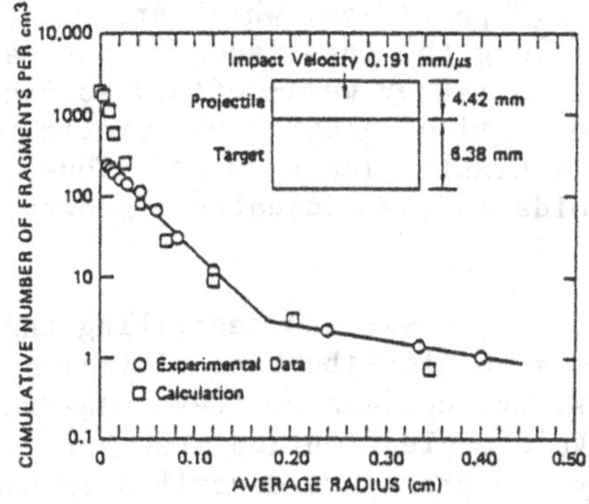

Figure 4.7. Comparison of computed and experimental fragment size distributions for propellant impact experiment.

ii. Structural composites

The above approach has also been used for several heatshield and structural composites with reasonable success. In each case, the microdamage modes were observed experimentally, and the evolution equations governing the nucleation and growth kinetics were found by the techniques described above.

IV.3 Quasistatic failure of reactor grade steel

In a room temperature round-bar tension test, the stress and strain histories will be significantly different from those in the impact cases discussed just previously. Inertial and viscous effects will be negligible, and there will be lower stresses with higher plastic strains.

In the following paragraphs, we describe a computational subroutine called DFRACTS (ductile fracture-static), that we have written to handle quasistatic ductile void kinetics. This model has been presented in several publications (see Ref. [1]) and will be summarized here. The application is to A533B pressure vessel steel.

The phenomenology of tensile failure in A533B steel at reactor operating temperature (561°K) was investigated by examining fracture surfaces and polished sections of fractured and partially fractured Charpy and round-bar tensile specimens tested at upper half shelf temperatures (355°K). Fracture surfaces showed ductile hemispherical dimples in a bimodal size distribution, and polished cross sections showed spherical voids. We concluded that tensile fracture occurred by nucleation, growth, and coalescence of voids.

Nucleation takes place predominantly at MnS and Al_2O_3 inclusions, which are typically 5 to 10 μm in diameter, and the voids grow plastically and independently of one another as the tensile strain is increased, maintaining a roughly spherical shape. As adjacent growing voids approach one another, a population of much smaller voids nucleates at the 0.2-μm-diameter Fe_3C particles, which are homogeneously dispersed throughout the material along the interfaces of the bainite laths. The submicrometer voids form on rather well-defined surfaces connecting the larger voids and grow to radii of about a micrometer before coalescing with neighboring voids by plastic impingement. Thus, the mechanism of coalescence of larger voids is the nucleation, growth, and coalescence of smaller voids.

To obtain quantitative expressions describing the microfailure process, we measured the size distributions of inclusions, voids, and fracture-surface dimples, and correlated these measurements with the imposed strain field. Void nucleation and growth rates were deduced from these data by using the procedure described in Lecture I.

Simulation of static ductile fracture experiments

The subroutine DFRACTS was incorporated into a Lagrangian, two-dimensional stress wave-propagation computer code and used to compute the failure behavior in a smooth tensile bar and in a severely notched tensile bar. Comparisons of measured and computed force-displacement curves and relative void volume as a function of location in the specimens were made to test the reliability of the model.

Next, to test the ability of the model to predict failure for cases when standard fracture toughness parameters apply, we performed a computational simulation of the center-cracked panel shown in fig. 4.8(a). The panel dimensions and crack length were chosen to give plane-strain conditions and J-controlled initiation and growth; that is, the half-crack length, ligament, and panel thickness exceeded 200 J_{Ic}/σ_0 where J_{Ic} is a critical fracture toughness parameter and σ_0 is the yield strength. For economy, a complete simulation using the coarse computational grid of fig. 4.8(b) was performed first to establish the boundary motions of a small butterfly-shaped region at the crack tip. This smaller region was then subdivided into much smaller cells, fig. 4.8(c), and the failure behavior of this region was computed by applying the boundary motions indicated in the first simulation. Figure 4.9 shows the calculated plastic-strain field and the process zone at the macrocrack tip within which the void activity occurs.

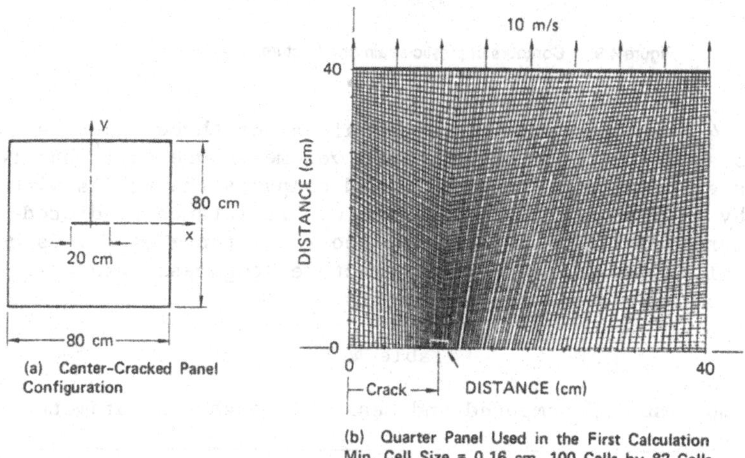

(a) Center-Cracked Panel Configuration

(b) Quarter Panel Used in the First Calculation
Min. Cell Size = 0.16 cm, 100 Cells by 82 Cells

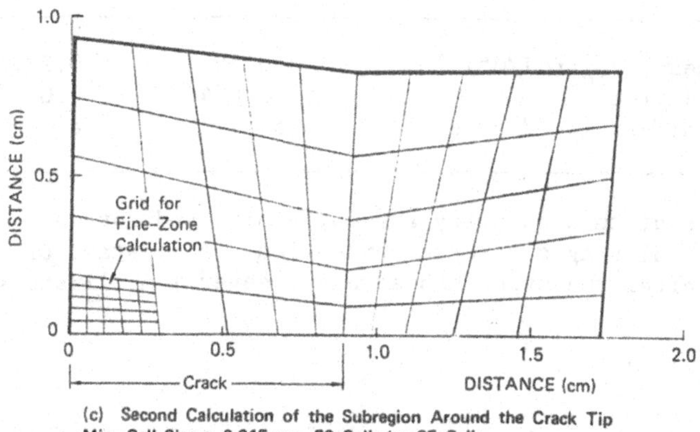

(c) Second Calculation of the Subregion Around the Crack Tip
Min. Cell Size = 0.015 cm, 50 Cells by 25 Cells

Figure 4.8. Specimen and cell geometries for the center-cracked panel simulation.

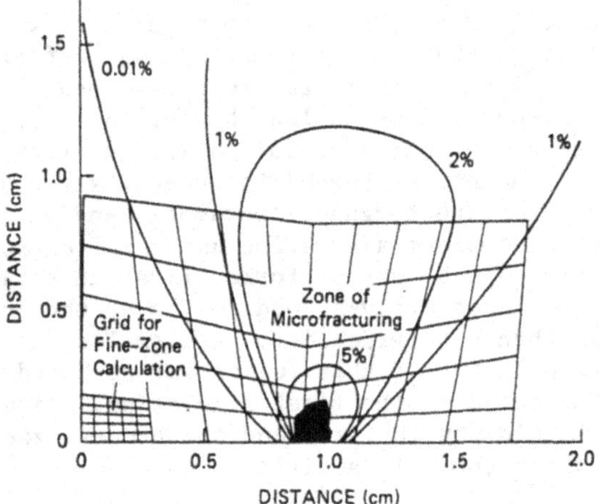

Figure 4.9. Contours of plastic strain and fracture in center-cracked plate simulation.

Table 4.3 presents the computed values for three fracture-toughness parameters at the point where the void volume in one cell* has reached 1% (the onset of macrocrack growth) and compares the values with values reported by Shih et al. [1]. The fracture resistance predicted by the microfracture model is only slightly above the range of values measured experimentally, demonstrating that fracture-toughness parameters can be computed from microvoid-kinetics models.

Table 4.3

Comparison of computed and measured toughness parameters

	Present computations	Experimental measurements
Fracture toughness, J_{Ic} (MJ/m^2)	0.45	0.23-0.43
Crack opening, δ_c (cm)	0.084	0.035-0.072
Crack opening half-angle (degrees)	54	a

[a]Measured values of this quantity are quite high and not well defined during the initial stages of crack extension. A constant of 11 to 17° is approached after the crack has advanced several millimeters.

*The cell size was chosen to be a few inclusion spacings; thus, the constitutive model contains a material-specific scale size, the "material element" discussed in chapter 1.

IV.4 Potential future applications of MSFM

Obvious future applications of MSFM are to problems of creep and fatigue.

In addition, two possible future applications of MSFM are to the shock physics problems of dynamic pore compaction and formation of heterogeneities in the shock front.

The subject of shock compaction of pores has become currently interesting because of the potential of combining the process with powder metallurgy to fabricate materials with unusual properties. The unusual properties are often postulated to be the result of rate processes occurring in the hot porewalls during and after compaction.

The MSFM modeling approach discussed above seems ideally suited to form the basis of computational descriptions of such processes, since the MSFM model keeps track of the evolving pore size distribution and associated pore volumes and surface areas.

The subject of heterogeneities in the shock front has long been of interest. Shear bands in the shock front form hot spots that influence metallurgical or chemical processes that occur there. In particular, the collapse of pores and the formation of shear bands have been postulated to form "hot spots" that initiate the detonation of high explosives. The MSFM modeling approach thus seems promising for application to detonation physics.

Printed in the United States
By Bookmasters